Plant Pathologist's Pocketbook
3rd Edition

Plant Pathologist's Pocketbook
3rd Edition

Edited by

J.M. Waller, J.M. Lenné and S.J. Waller

CABI *Publishing*

CABI *Publishing* is a division of CAB *International*

CABI Publishing
CAB International
Wallingford
Oxon OX10 8DE
UK

CABI Publishing
10 E 40th Street
Suite 3203
New York, NY 10016
USA

Tel: +44 (0)1491 832111
Fax: +44 (0)1491 833508
Email: cabi@cabi.org
Web site: www.cabi-publishing.org

Tel: +1 212 481 7018
Fax: +1 212 686 7993
Email: cabi-nao@cabi.org

A catalogue record for this book is available from the British Library, London, UK.

Library of Congress Cataloging-in-Publication Data
Plant pathologist's pocketbook / edited by J.M. Waller, J.M. Lenné and S.J.
Waller.   --3rd ed.
       p.   cm.
   Includes bibliographical references (p.).
   ISBN 0-85199-458-X (hardcover) -- ISBN 0-85199-459-8 (pbk.)
       1. Plant diseases--Handbooks, manuals, etc. 2. Plant diseases--Bibliography. I. Waller,
J. M. II. Lenné, Jillian M. III. Waller, S. J. (Sarah J.)

SB601 .P57 2001
632'.3--dc21

2001020356

Pbk ISBN 0 85199 459 8
Hbk ISBN 0 85199 458 X

Typeset by AMA DataSet Ltd, UK.
Printed and bound in the UK by Biddles Ltd, Guildford and King's Lynn.

Contents

Contributors

R. Bandyopadhyay, *International Crops Research Institute for the Semi-Arid Tropics (ICRISAT), Patancheru, Andhra Pradesh 502 324, India*

E. Boa, *CABI Bioscience UK Centre, Bakeham Lane, Egham, Surrey TW20 9TY, UK*

J. Bridge, *CABI Bioscience UK Centre, Bakeham Lane, Egham, Surrey TW20 9TY, UK*

P. Bridge, *Mycology Section, Royal Botanic Gardens, Kew, Richmond, Surrey TW9 3AE, UK and School of Biological and Chemical Sciences, Birkbeck, University of London, Malet St, London WC1E 7HX, UK*

P.F. Cannon, CABI Bioscience UK Centre, Bakeham Lane, Egham, Surrey TW20 9TY, UK

S. Chakraborty, *CSIRO Plant Industry, CRC Tropical Plant Protection, University of Queensland, Queensland 4072, Australia*

H.L. Crowson, *CAB International, Wallingford, Oxfordshire OX10 8DE, UK*

F.M. Dewey, *Plant Sciences Department, University of Oxford, South Parks Road, Oxford OX1 3RB, UK*

H.C. Evans, *CABI Bioscience UK Centre, Silwood Park, Ascot, Berkshire SL5 7PY, UK*

M. Holderness, *CABI Bioscience UK Centre, Bakeham Lane, Egham, Surrey TW20 9TY, UK*

D. Hollomon, *Long Ashton Research Station, University of Bristol, Long Ashton, Bristol BS18 9AF, UK*

W.C. James, *ISAAA AmeriCenter, c/o Cornell University, Ithaca, NY 14853, USA*

P. Jones, *Plant Pathogen Interactions (PPI), IACR-Rothamsted, Harpenden, Hertfordshire AL5 2JQ, UK*

J.M. Lenné, *International Crops Research Institute for the Semi-Arid Tropics (ICRISAT), Patancheru, Andhra Pradesh 502 324, India*

G.A. Matthews, *International Pesticide Application Research Centre, Imperial College at Silwood Park, Ascot, Berkshire SL5 7PY, UK*

L.A. McGillivray, *CAB International, Wallingford, Oxfordshire OX10 8DE, UK*

S.S. Navi, *International Crops Research Institute for the Semi-Arid Tropics (ICRISAT), Patancheru, Andhra Pradesh 502 324, India*

C. Parker, *5 Royal York Crescent, Bristol BS8 4JZ, UK*

J. Riley, *IACR-Rothamsted, Harpenden, Hertfordshire AL5 2JQ, UK*

B.J. Ritchie, *CABI Bioscience UK Centre, Bakeham Lane, Egham, Surrey TW20 9TY, UK*

G. Saddler, *CABI Bioscience UK Centre, Bakeham Lane, Egham, Surrey TW20 9TY, UK*

M.W. Shaw, *Department of Agricultural Botany, School of Plant Sciences, The University of Reading, Reading RG6 6AS, UK*

D. Smith, *CABI Bioscience UK Centre, Bakeham Lane, Egham, Surrey TW20 9TY, UK*

P.S. Teng, *Monsanto Company, Kings Court II Building, 2129 Chino Roces Avenue, Makati City, Philippines*

A.J. Termorshuizen, *Biological Farming Systems, Wageningen University, Marijkeweg 22, 6709 PG Wageningen, The Netherlands*

P.J. Terry, *Long Ashton Research Station, University of Bristol, Long Ashton, Bristol BS18 9AF, UK*

J.M. Waller, *CABI Bioscience UK Centre, Bakeham Lane, Egham, Surrey TW20 9TY, UK*

S.J. Waller, *45 St Michaels, Longstanton, Cambridge, UK*

G.W. Watson, *Entomology Department, The Natural History Museum, Cromwell Road, London SW7 5BD, UK*

A.R. Wellburn, *formerly of Division of Biological Sciences, Lancaster University, Lancaster LA1 4YQ, UK*

T.D. Williams, *4 Warren Park, West Hill, Ottery St Mary, Devon EX11 1TN, UK*

Preface to the 3rd Edition

The second edition of the *Plant Pathologist's Pocketbook* published in 1983 is now well out of date and this edition has been completely reorganized to create a format more suitable to the current needs of plant pathologists. The traditional and well-known title has been retained although the book is now rather larger than 'pocketbook' size. The contents have been arranged into five sections covering the main activity groups in plant pathology. Several new topics have been added including chapters covering epidemiology and disease forecasting, disease resistance, biochemical and molecular techniques, and electronic databases and information technology. Other topics covered in the second edition have all been revised or completely rewritten and the associated bibliographies updated. In order to keep the pocket book to size it has been necessary to restrict accounts of most subjects to succinct overviews, but hopefully the attached bibliographies will provide sufficient information to assist pathologists with their work. Inevitably decisions have had to be made on how much to include on particular aspects; the pocketbook is not intended as a general text on plant pathology but as a reference for information and relevant literature. As such the book will be used more in the laboratory than carried in the pocket and the larger size and rearranged format will hopefully enable better access to the contents.

Many chapters have been written or revised by authors both from CABI *Bioscience* and from outside; others have been revised or written by the editors. We acknowledge particular assistance with the chapter on viruses from Professor Mike Thresh of NRI, University of Greenwich.

Jim Waller
Jill Lenné
Sarah Waller

Landmarks in Plant Pathology

1

Early Recognition of Plant Diseases

c. 300 BC	Theophrastus of Lesbos – references to plant diseases in *Historia plantarum* and *De causis plantarum*.
c. AD 50	Caius Plinius Secundus – references to common diseases such as mildew and rusts of cereals in *Historia naturalis*.
c. AD 700	Cassianus Bassus – *Geoponica* – a compilation of Byzantine agriculture with many references to plant diseases.
c. AD 1200	Ibn-al Awam – *Kitab al-Felahah* – Arabic treatise from Seville with a chapter on problems of regional fruit crops.

Beginnings of Plant Pathology

1665	Robert Hooke – *Micrographia* contains first illustration of a microscopic plant pathogen (rose rust).
1755	Mathieu Tillet demonstrates seed-borne nature of wheat bunt (*Tilletia caries*).
1794	J.J. Plenk publishes *Physiologia et Pathologia Plantarum* containing a classification of plant diseases based on symptoms.
1802	William Forsyth introduces lime sulphur for control of mildew on fruit trees – first example of generally used fungicide.
1807	I.-B. Prevost publishes first experimental proof of fungal pathogenicity on plants.
1845–1849	Potato blight (*Phytophthora infestans*) epidemics in Ireland.

©CAB *International* 2002. *Plant Pathologist's Pocketbook*
(eds J.M. Waller, J.M. Lenné and S.J. Waller)

1853	Anton de Bary publishes *Unterschungen über die brandpilze* and established the role of fungi as plant pathogens.
1858	Julius Kuhn publishes *Die Krankheiten der Kulturgewachse* – the first plant pathology text.
1865/66	De Bary demonstrates heteroecism in *Puccinia graminis*.
1868–1882	Coffee rust (*Hemileia vastatrix*) epidemics in Sri Lanka and classic studies on its aetiology by Marshal Ward.
1874	Paul Soraur publishes first edition of *Handbuch für Pflanzenkrankheiten* of which there have been many subsequent editions.
1878–1885	Vine downy mildew (*Plasmopara viticola*) epidemics in France precipitate introduction of Bordeaux mixture by Millardet.
1885	J.C. Arthur provides first proof of bacterial pathogenicity to plants with fire blight (*Erwinia amylovora*).
1886–1898	Tobacco mosaic recognized as a virus disease transmitted by 'contagious living fluid' by Mayer, Ivanovski and Beijerinck.
1889	Hot water treatment of seed introduced for control of loose smut of wheat (*Ustilago tritici*).

Development of Basis for Modern Plant Pathology

1891/92	First plant pathology journals (*Zietschrift für Pflanzenkrankheiten* and *Revista di Patologia Vegetale*) published.
1894	Jacob Eriksson demonstrates host-specific special forms of cereal rusts.
1904	Rowland Biffen demonstrates Mendelian inheritance of resistance to rust in cereals.
1907	First university department of plant pathology founded at Cornell University, New York.
1908	American Phytopathology Society founded.
1913	Recognition of physiologic races of *Puccinia graminis* by Stakman. First use of organo-mercurial fungicides for seed application in Germany.
1914	First 'International Phytopathological Convention of Rome' held but resolutions never ratified.
1920	Imperial Bureau of Mycology founded in UK (a forebear of CABI Bioscience) – production of *Review of Plant Pathology* (as *Review of Applied Mycology*) commences 2 years later. E.F. Smith publishes *Introduction to Bacterial Diseases of Plants*.
1929	'International Convention for the Protection of Plants' in Rome produces first limited internationally agreed protocols for restricting the spread of plant diseases.
1934	Dithiocarbamate fungicides developed and patented by Du Pont.
1937	F.C. Bawden and N.W. Pirie establish nucleoprotein nature of viruses.

1946	Flor establishes 'gene-for-gene' hypothesis for genetic inter-action of resistance and virulence for flax rust *Melampsora lini*.
1951	European and Mediterranean Plant Protection Organization founded as first of several regional plant protection organizations.
1952	International Plant Protection Convention submitted to the Food and Agriculture Organization (FAO) and agreed in principle.
1963	*Annual Review of Phytopathology* begins.
	J.E. Vanderplank publishes *Plant Diseases: Epidemics and Control*.
	First example of biological control of a disease – *Heterobasidion annosum* by *Peniophora gigantea*.
1965	Discovery of first systemic fungicide, carboxin.
1967	First recognition of phytoplasmas (mycoplasma-like organisms) as plant pathogens.
1968	First International Congress of Plant Pathology held in London.
1969	First computer simulation program for plant disease epidemics.
1970	Southern corn leaf blight epidemic on hybrid (Tms) maize in USA.
1971	Deiner establishes potato spindle tuber as first recognized viroid disease.
1977/78	*Plant Disease: An Advanced Treatise* (3 Vols) by Horsfall and Cowling published.
1986	Plants genetically transformed to express virus coat proteins shown to be resistant to same virus.
1992	First examples of plant resistance genes isolated.

General References

Ainsworth, G.C. (1981) *An Introduction to the History of Plant Pathology*. Cambridge University Press, Cambridge, UK, 315 pp.

Carefoot, G.L. and Sprot, E.R. (1967) *Famine in the Wind*. Rand McNally, Chicago, USA.

Large, E.C. (1940) *Advance of the Fungi*. Jonathan Cape, London, UK, 488 pp.

Parris, G.K. (1968) *A Chronology of Plant Pathology*. Johnson & Sons, Mississippi, USA.

Some Major Plant Diseases

J.M. Lenné

ICRISAT, Patancheru, Andhra Pradesh 502 324, India

Diseases are listed under hosts, which are arranged alphabetically by common name.

In column two current scientific names are given in accordance with the CABI *Crop Protection Compendium* (*Global Module* 2nd edition, 2000); these may differ from those given in literature references given in column four where earlier synonyms may have been used.

In column three a broad indication of geographical distribution is given.

In most cases detailed information on distribution can be found in the relevant map in the series *CMI/IMI/CABI Distribution Maps of Plant Diseases*, indicated by 'Map 38' etc. and also in the CABI *Crop Protection Compendium* (*Global Module* 2nd edition, 2000).

In column four 'Desc. 147' etc. refers to the series of *CMI/IMI/CABI Descriptions of Pathogenic Fungi and Bacteria*; 'Virus Desc. 68' etc. refers to the series of *AAB Descriptions of Plant Viruses*. References in the form 58, 1234 are to abstracts in the *Review of Plant Pathology*. Further references can be found in the CABI *Crop Protection Compendium* (*Global Module* 2nd edition, 2000).

Common name	Pathogen	Distribution	Literature
Almond (*Prunus dulcis*)			
Brown rot	*Monilinia laxa*	Fairly widespread: Map 44	Desc. 619
Brown rot	*Monilinia fructicola*	Mainly New World: Map 50	Desc. 616
Apple (*Malus domestica*)			
Nectria canker	*Nectria galligena*	Widespread: Map 38	Desc. 147: 55, 1853
Mildew	*Podosphaera leucotricha*	Widespread: Map 118	Desc. 158
Brown rot, spur canker	*Monilinia fructigena*	Widespread: Map 22	Desc. 617: 56, 4099
Scab	*Venturia inaequalis*	Widespread: Map 120	Desc. 401
Fire blight	*Erwinia amylovora*	Scattered worldwide: Map 264	Desc. 44: 54, 1807
Chestnut mosaic	*Apple mosaic virus*	Widespread: Map 354	Virus Desc. 83
Apricot (*Prunus armeniaca*)			
Dieback, gummosis	*Eutypa lata*	Widespread: Map 385	Desc. 436
Aubergine (*Solanum melongena*)			
Tip over, blight, canker	*Phomopsis vexans*	Fairly widespread but few records for Africa and absent from Europe: Map 329	Desc. 338
Avocado (*Persea americana*)			
Anthracnose	*Glomerella cingulata*	Worldwide	Desc. 315
Root rot	*Phytophthora cinnamomi*	Worldwide: Map 302	Desc. 113
Banana (*Musa* spp.)			
Anthracnose/crown rot	*Colletotrichum musae*	Widespread	Desc. 222
Fusarium wilt	*Fusarium oxysporum* f.sp. *cubense*	Widespread: Map 31	Desc. 1115: 41, 610
Black Sigatoka	*Mycosphaerella fijiensis*	Widespread: Map 500	Desc. 413
Yellow Sigatoka	*Mycosphaerella musicola*	Widespread: Map 7	Desc. 414: 49, 2562
Moko disease	*Ralstonia solanacearum*	Widespread: Map 138	Desc. 1220
Bunchy top	*Banana bunchy top virus*	South Asia, Australia, Oceania, Africa: Map 19	
Streak	*Banana streak virus*	Widespread	
Barley (*Hordeum vulgare*)			
Spot blotch	*Cochliobolus sativus*	Widespread: Map 322	Desc. 701

Continued

Common name	Pathogen	Distribution	Literature
Powdery mildew	*Erysiphe graminis*	Widespread	Desc. 153:58, 41
Leaf stripe	*Pyrenophora graminea*	Widespread	Desc. 388
Net blotch	*Pyrenophora teres*	Widespread: Map 364	Desc. 390: 52, 269
Scald	*Rhynchosporium secalis*	Widespread: Map 383	Desc. 387: 54, 1237
Covered smut	*Ustilago hordei*	Widespread: Map 460	Desc. 749
Loose smut	*Ustilago nuda*	Widespread: Map 368	Desc. 280
Stripe mosaic	*Barley stripe mosaic virus*	Widespread	Virus Desc. 344
Yellow dwarf	*Barley yellow dwarf virus*	Widespread: Map 332	Virus Desc. 339
Bean, Broad (*Vicia faba*)			
Ascochyta blight	*Ascochyta fabae*	Widespread	Desc. 1164
Chocolate spot	*Botrytis fabae*	Widespread but in North America reported only from Canada: Map 162	Desc. 432
Rust	*Uromyces viciae-fabae*	Widespread: Map 200	Desc. 60
Bean, French (*Phaseolus vulgaris*)			
Angular leaf spot	*Phaeoisariopsis griseola*	Widespread: Map 328	Desc. 847
Web blight	*Thanatephorus cucumeris*	Widespread	Desc. 406
Anthracnose	*Colletotrichum lindemuthianum*	Widespread: Map 177	Desc. 316
Rust	*Uromyces appendiculatus*	Widespread: Map 290	Desc. 57
Halo blight	*Pseudomonas savastanoi* pv. *phaseolicola*	Widespread: Map 85	Desc. 45
Common blight	*Xanthomonas axonopodis* pv. *phaseoli*	Widespread: Map 401	Desc. 48, 49
Common mosaic	*Bean common mosaic virus*	Widespread: Map 213	Virus Desc. 337
Golden mosaic	*Bean golden mosaic virus*	America	Virus Desc. 192
Beet (sugar beet) (*Beta vulgaris* var. *saccharifera*)			
Leaf spot	*Cercospora beticola*	Widespread: Map 226	Desc. 721
Black leg	*Pleospora betae*	Widespread: Map 427	Desc. 149
Rust	*Uromyces beticola*	Widespread: Map 265	Desc. 177
Western yellows	*Beet western yellows virus*	Widespread	Virus Desc. 89
Yellows	*Beet yellows virus*	Widespread: Map 261	Virus Desc. 13

Disease	Pathogen	Distribution	Reference
Brassicas (*Brassica* spp.)			
Dark spot	*Alternaria brassicae*	Widespread: Map 353	Desc. 162
Dark leaf spot	*Alternaria brassicicola*	Widespread: Map 457	Desc. 163
Cabbage fusarium wilt	*Fusarium oxysporum* f.sp. *conglutinans*	Widespread: Map 54	Desc. 213
Black leg	*Leptosphaeria maculans*	Widespread: Map 73	Desc. 331
Ring spot	*Mycosphaerella brassicicola*	Widespread: Map 189	Desc. 468
Club root	*Plasmodiophora brassicae*	Widespread: Map 101	37, 746; 55, 3008
Black rot	*Xanthomonas campestris*	Widespread: Map 136	Desc. 47
Mosaic	*Cauliflower mosaic virus*	Widespread: Map 373	Virus Desc. 243
Cacao (*Theabroma cacao*)			
Witches' broom	*Crinipellis perniciosa*	Northern South America, Central America, Caribbean: Map 37	Desc. 223: 37, 151
Pod rot	*Moniliophthora roreri*	Colombia, Ecuador, Panama, Peru, Venezuela, Central America: Map 13	Desc. 226: 61, 150
Black pod, canker	*Phytophthora palmivora*	Widespread: Map 725	Desc. 831
	Phytophthora megakarya	West Africa	Desc. 832
Swollen shoot	*Cacao swollen shoot virus*	West Africa, Sri Lanka, Nothern South America and Sabah	Virus Desc. 10: 50, 591
Cassava (*Manihot esculenta*)			
Bacterial blight	*Xanthomonas axonopodis* pv. *manihotis*	Widespread: Map 521	Desc. 559: 58, 4604
Cassava mosaic	*Cassava common mosaic virus*	South America, Africa	Virus Desc. 90
Super-elongation disease	*Elsinoë brasiliensis*	Northern South America, Central America, Caribbean	56, 5308; 60, 615
African cassava mosaic	*African cassava mosaic virus*	Africa, South Asia	Virus Desc. 297
Cassava brown streak	*Cassava brown streak-associated virus*	Africa	
Cherry (*Prunus cerasus*)			
Bacterial canker	*Pseudomonas syringae* pv. *morsprunorum*	Widespread: Map 132	Desc. 125
Cherry leaf spot	*Blumeriella jaapii*	Europe, Asia, North America	

Continued

Common name	Pathogen	Distribution	Literature
Chestnut (*Castanea satva*)			
Blight	*Cryphonectria parasitica*	Eastern and Western North America, Europe, China, India, Japan, Korea, Turkey, USSR: Map 66	Desc. 704: 58, 2406 CABI/EPPO No.194
Chickpea (*Cicer arietinum*)			
Ascochyta blight	*Didymella rabiei*	Mediterranean, South and West Asia, East Australia: Map 151	Desc. 337
Rust	*Uromyces ciceris-arietini*	Mediterranean, Southern Europe, South Asia, Kenya, Malawi, Mexico: Map 235	Desc. 178
Wilt	*Fusarium oxysporum* f.sp. *ciceris*	Widespread	Desc. 1113
Grey mould	*Botryotinia fuckeliana*	Widespread	Desc. 431
Stunt	Bean leafroll virus (*Chickpea stunt virus*)	Europe, Asia, Africa	Virus Desc. 286
Citrus (*Citrus* spp.)			
Mal secco	*Phoma tracheiphila*	Europe, Asia, Africa	Desc. 399
Melanose	*Diaporthe citri*	Widespread: Map 126	Desc. 396
Citrus (sweet orange) scab	*Elsinoë australis*	Central and South America, Sicily, Oceania: Map 55	Desc. 440
Common (sour orange) scab	*Elsinoë fawcettii*	Widespread: Map 125	Desc. 438
Black spot	*Guignardia citricarpa*	Africa, China, Taiwan, Hong Kong, Indonesia, Australia, South America, New Hebrides: Map 53	Desc. 85
Brown rot, gummosis	*Phytophthora citrophthora*	Widespread: Map 35	Desc. 33
Black shank	*Phytophthora nicotianae* var. *parasitica*	Widespread: Map 613	
Canker	*Xanthomonas axonopodis* pv. *citri*	Widespread: Map 11	Desc. 11
Tristeza	Citrus tristeza virus	Widespread: Map 289	Virus Desc. 353 53, 550
Greening	Citrus huanglongbing (greening) bacterium	Fairly widespread in southern Africa and Asia	
Citrus scaly butt	Citrus exocortis viroid	South America, Australia, Mediterranean, North Africa, limited occurrence in USA, Oceania, New Zealand	Virus Desc. 226

Host / Disease	Pathogen	Distribution	Reference
Clovers (*Trifolium* spp.)			
Scorch	*Kabatiella caulivora*	Europe, North America, Morocco, Japan, Australia, New Zealand: Map 351	Desc. 1084
Rot	*Sclerotinia trifoliorum*	Fairly widespread with host: Map 274	
Yellow mosaic	*Clover yellow mosaic virus*	Western USA, Canada	Virus Desc. 111
Leaf spot	*Pseudopeziza trifolii*	Europe, Australia	Desc. 636
Phyllody	*Aster yellows phytoplasma group*	Europe	
Coconut (*Cocos nucifera*)			
Lethal yellowing	*Palm lethal yellowing phytoplasma*	West Africa, West Indies, Central America, USA	57, 4091; CABI/EPPO (1998) No.264
Cadang-cadang	*Coconut cadang-cadang viroid*	Philippines	52, 819
Bud rot	*Phytopthora palmivora*	Widespread: Map 725	Desc. 831
Coffee (*Coffea* spp.)			
Brown eye spot	*Mycosphaerella coffeicola*	Widespread with host: Map 59	Desc. 415
Wilt	*Gibberella xylarioides*	Africa: Map 464	Desc. 24
Coffee berry disease	*Colletotrichum kahawae*	Africa: Map 716	56, 4538
Rust	*Hemileia vastatrix*	Widespread: Map 5	Desc. 1: 52, 411
Cotton (*Gossypium* spp.)			
Blight	*Ascochyta gossypii*	Widespread: Map 259	Desc. 271
Wilt	*Fusarium oxysporum* f.sp. *vasinfectum*	Widespread: Map 362	Desc. 1120: 55, 2248
Wilt	*Verticillium dahliae*	Widespread, temperate and subtropical: Map 366	Desc. 256
Angular leaf spot	*Xanthomonas axonopodis* pv. *malvacearum*	Widespread: Map 57	Desc. 12
Cowpea (*Vigna unguiculata*)			
Anthracnose	*Colletotrichum lindemuthianum*	Widespread	Desc. 316
Brown blotch	*Colletotrichum destructivum*	Africa, Asia, South America	
Scab	*Elsinoë phaseoli*	Widespread	Desc. 317
Bacterial blight	*Xanthomonas axonopodis* pv. *vignicola*	Africa	Desc. 314
Stem rot	*Phytophthora vignae*	Asia, Africa, Australia	

Continued

Common name	Pathogen	Distribution	Literature
Black eye cowpea mosaic (Cowpea aphid borne)	Bean common mosaic virus strn blackeye cowpea	Widespread	Desc. 134
Cucurbits (*Cucumis, Cucurbita* spp., etc.)			
Wilt (on melon)	Fusarium oxysporum f.sp. cucumerinum	America, Europe, Taiwan, Iraq, Japan, Philippines, Australia: Map 259	Desc. 215
Downy mildew	Pseudoperonospora cubensis	Widespread: Map 285	Desc. 457: 60, 536
Powdery mildew	Sphaerotheca fuliginea	Widespread	Desc. 159: 58, 41
Bacterial wilt	Erwinia tracheiphila	Widespread in Canada and USA; Africa, Asia, Europe, Australia: Map 456	Desc. 233
Angular leaf spot	Pseudomonas syringae pv. lachrymans	Fairly widespread: Map 355	Desc. 124
Mosaic	Cucumber mosaic virus	Widespread, especially in temperate regions	Virus Desc. 213
Currant, gooseberry (*Ribes* spp.)			
Leaf spot, anthracnose	Drepanopeziza ribis	Europe, Australia, New Zealand: Map 187	
Powdery mildew	Sphaerotheca mors-uvae	Europe: Map 16	Desc. 254
Date palm (*Phoenix dactylifera*)			
Bayoud disease	Fusarium oxysporum f.sp. albedinis	Algeria, Morocco: Map 240	Desc. 1111; 52, 4184
Elm (*Ulmus* spp.)			
Dutch elm disease	Ophiostoma ulmi	Europe, West Asia, India, USA, Canada: Map 36	Desc. 361
Flax (*Linum usitatissimum*)			
Wilt	Fusarium oxysporum f.sp. lini	Widespread: Map 32	Desc. 51
Rust	Melampsora lini	Widespread: Map 68	
Pasmo disease	Mycosphaerella linicola	Widespread: Map 18	Desc. 709
Grape (*Vitis vinifera*)			
Grey mould	Botrytinia fuckeliana	Worldwide	Desc. 431
Black spot, anthracnose	Elsinoë ampelina	Widespread: Map 234	Desc. 439
Black rot	Guignardia bidwellii	Fairly widespread, absent from Australasia: Map 81	Desc. 710

Rust	*Physopella ampelopsidis*	South-East Asia, India, Sri Lanka, Central America and West Indies, USA, Colombia, Venezuela: Map 87	Desc. 173
Downy mildew	*Plasmopara viticola*	Widespread: Map 221	Desc. 980
Powdery mildew	*Uncinula necator*	Widespread	Desc. 160: 58, 41
Fanleaf	*Grapevine fanleaf virus*	Almost all temperate regions with host	Virus Desc. 28
Groundnut (peanut) (*Arachis hypogaea*)			
Early leaf spot	*Didymosphaeria arachidis*	Widespread: Map 166	Desc. 411
Late leaf spot	*Mycosphaerella berkeleyi*	Widespread: Map 152	Desc. 412
Rust	*Puccinia arachidis*	Widespread: Map 160	Desc. 53
Rosette (chlorotic, green)	*Groundnut rosette virus complex*	Africa	Virus Desc. 345 and 355
Peanut clump	*Peanut clump virus*	South Asia, West Africa	Virus Desc. 235
Peanut stripe	*Peanut stripe virus*	Asia, North America, West Africa	
Guava (*Psidium guajava*)			
Rust	*Puccinia psidii*	Central and South America, West Indies: Map 181	Desc. 56
Hop (*Humulus lupulus*)			
Downy mildew	*Pseudoperonospora humuli*	Europe; Asia, Canada, USA, Argentina: Map 14	Desc. 769
Lentil (*Lens culinaris*)			
Rust	*Uromyces viciae-fabae*	Widespread	Desc. 60
Ascochyta blight	*Ascochyta fabae* f.sp. *lentis*	Widespread	Desc. 461
Lettuce (*Lactuca sativa*)			
Downy mildew	*Bremia lactucae*	Widespread: Map 86	Desc. 682
Leaf spot	*Septoria lactucae*	Widespread: Map 485	Desc. 335
Mosaic	*Lettuce mosaic virus*	Widespread: Map 376	Virus Desc. 9
Necrotic yellows	*Lettuce necrotic yellows virus*	Australia, New Zealand: Map 480	Virus Desc. 43
Lucerne (alfalfa) (*Medicago sativa*)			
Rust	*Uromyces striatus*	Widespread: Map 342	Desc. 59
Wilt	*Clavibacter michiganensis* subsp. *insidiosus*	Widespread in USA, Europe, South Africa, Australia, New Zealand: Map 67	Desc. 13
Anthracnose	*Colletotrichum trifolii*	North America, Europe, Australia	
Fusarium wilt	*Fusarium oxysporum* f.sp. *medicaginis*	Widespread	

Continued

Common name	Pathogen	Distribution	Literature
Common leaf spot	*Pseudopeziza medicaginis*	Widespread	Desc. 637
Maize (corn) (*Zea mays*)			
Leaf spot	*Cochliobolus carbonum*	Fairly widespread: Map 380	Desc. 349
Southern leaf blight	*Cochliobolus heterostrophus*	Widespread: Map 346	Desc. 301
Stalk and ear rot, seedling blight	*Stenocarpella macrospora, Stenocarpella maydis*	Fairly widespread	Desc. 83, 84: 47, 1528
Rots and seedling blights	*Gibberella fujikuroi*	Widespread: Maps 102, 191	Desc. 22
Stalk and ear rot	*Gibberella zeae*	Worldwide	Desc. 384
Rust	*Physopella zeae*	South and Central America, West Indies: Map 469	Desc. 5
Rust	*Puccinia polysora*	Fairly widespread, absent from Europe, West and Central Asia: Map 237	Desc. 4
Downy mildews	*Peronosclerospora* spp.	Widespread: Maps 21, 179, 497	Desc. 453, 454, 761
Northern leaf blight	*Setosphaeria turcica*	Widespread: Map 257	Desc. 304: 55, 3002c
Blister smut	*Ustilago maydis*	Widespread: Map 93	Desc. 79: 47, 506
Bacterial wilt	*Pantoea stewartii*	Parts of North, South and Central America, Southern Europe, China, Thailand, Vietnam: Map 41	Desc. 123: 47, 1529
Mosaic	*Maize mosaic virus*	Americas, West Indies, East Africa, Hawaii, India	Virus Desc. 94
Rough dwarf	*Maize rough dwarf virus*	Europe, Israel (recorded only where vector present, see CIE Map 201)	Virus Desc. 72
Streak	*Maize streak virus*	Africa, Mauritius, Malagasy Republic, India, probably South-East Asia; absent from America	Virus Desc. 133
Millet (*Panicum miliaceum*)			
Head smut	*Sphacelotheca destruens*	Widespread in North America and Europe; scattered in Africa, Asia, South America; Australia: Map 219	Desc. 72
Millet, finger (*Eleusine coracana*)			
Downy mildew	*Sclerophthora macrospora*	North America, Australia, New Zealand; scattered in Africa, Asia, Europe: Map 287	60, 3641
Blast	*Magnaporthe grisea*	Widespread: Map 51	Desc. 169

Host / Disease	Pathogen	Distribution	Reference
Millet, pearl (*Pennisetum typhoides*)			
Ergot	*Claviceps microcephala*	Africa, India, Latin America, Australia	56, 693
	Claviceps africana		
Downy mildew	*Sclerospora graminicola*	Europe, Asia, Africa, Fiji: Map 431	Desc. 452: 60, 3641
Oak (*Quercus* spp.)			
Wilt	*Ceratocystis fagacearum*	East and Central USA: Map 254	
Oats (*Avena sativa*)			
Blight	*Cochliobolus victoriae*	North America, Argentina, Australia, Brazil, The Netherlands, Switzerland, UK: Map 267	Desc. 703
Speckle blotch	*Phaeosphaeria avenaria* f.sp. *avenaria*	Fairly widespread: Map 323	Desc. 312
Crown rust	*Puccinia coronata*	Widespread	
Leaf stripe, blotch	*Pyrenophora chaetomioides*	Widespread: Map 105	Desc. 389
Loose smut	*Ustilago avenae*	Widespread: Map 238	Desc. 279: 51, 2405
Oil palm (*Elaeis guineensis*)			
Freckle	*Cercospora elaeidis*	Africa, Australia: Map 487	Desc. 464
Wilt	*Fusarium oxysporum* f.sp. *elaeidis*	Central and West Africa, Colombia: Map 471	Desc. 1116
Onion etc. (*Allium* spp.)			
Grey mould neck rot	*Botrytis allii*	Widespread: Map 169	Desc. 433
Downy mildew	*Peronospora destructor*	Widespread: Map 76	Desc. 456
White rot	*Sclerotium cepivorum*	Widespread but few records for Africa and Asia: Map 331	Desc. 512
Smut	*Urocystis cepulae*	Widespread but not reported from Africa except Egypt and Morocco, eradicated from India: Map 12	Desc. 298
Passion fruit (*Passiflora edulis*)			
Brown spot	*Alternaria passiflorae*	East Africa, Australia, New Zealand, Fiji, Hawaii, Papua New Guinea, Canada: Map 479	Desc. 247
Woodiness	*Passion fruit woodiness virus*	Brunei, Kenya, Australia, Surinam: Virus Map 518	Desc. 122
Pea (*Pisum sativum*)			
Root rot	*Aphanomyces euteiches*	Australia, Denmark, France, Norway, Sweden, UK, USSR (European), Japan, USA: Map 78	Desc. 600: 54, 1915

Continued

Common name	Pathogen	Distribution	Literature
Leaf, stem and pod spot	*Ascochyta pisi*	Widespread: Map 273	Desc. 334
Blight, foot rot	*Didymella pinodes*	Widespread: Map 316	Desc. 340
Enation mosaic	*Pea enation mosaic virus*	Widespread in North temperate zone	Virus Desc. 372
Peach (*Prunus persica*)			
Brown rot	*Monilinia fructicola*	America, Australia, New Zealand, Japan: Map 50	Desc. 616
Leaf curl	*Taphrina deformans*	Widespread: Map 192	Desc. 711
Pear (*Pyrus communis*)			
Scab	*Venturia pirina*	Widespread: Map 367	Desc. 404
Brown rot	*Monilinia fructigena*	Widespread	Desc. 617
Fire blight	*Erwinia amylovora*	Widespread in North America, parts of Asia and Europe, New Zealand, Guatemala, Colombia: Map 1264	Desc. 44
Pigeonpea (*Cajanus cajan*)			
Wilt	*Gibberella indica*	India, Indonesia, Thailand, parts of Africa	Desc. 575
Phytophthora blight	*Phytophthora drechsleri* f.sp. *cajani*	India, Kenya, Central America	Desc. 840
Sterility mosaic	*Pidgeonpea sterility mosaic virus*	India, Bangladesh, Sri Lanka, Nepal	
Pine (*Pinus* spp.)			
Rust	*Cronartium fusiforme*	South-East Asia, USA: Map 47	58, 402
Dieback	*Gremmeniella abietina*	Europe, North America: Map 423	Desc. 369
Butt rot	*Heterobasidion annosum*	Worldwide in temperate areas: Map 271	Desc. 192
Needle rust	*Lophodermium pinastri*	Widespread, particularly in northern hemisphere: Map 371	Desc. 567
Blight, red band	*Mycosphaerella pini*	Fairly widespread: Map 419	Desc. 368
Plum (*Prunus domestica*)			
Pox, Sharka disease	*Plum pox virus*	Europe, Turkey: Map 392	Virus Desc. 70
Poplar (*Populus* spp.)			
Canker	*Cryptodiaporthe populea* (*Chondroplea populea*)	Europe, Turkey, North America: Map 344	Desc. 364

Disease	Pathogen	Distribution	Reference
Potato (*Solanum tuberosum*)			
Bacterial wilt	*Ralstonia solanacearum*	Widespread: Map 138	Desc. 1220
Gangrene	*Phoma foveata*	Australia, Europe: Map 210	Desc. 838
Late blight	*Phytophthora infestans*	Widespread: Map 109	Desc. 406
Black scurf	*Rhizoctonia solani*	Worldwide	Desc. 755: 56, 4655
Wart	*Synchytrium endobioticum*	Europe, Asia, South Africa, North and South America: Map 1	
Black leg	*Erwinia carotovora* var. *atroseptica*	Widespread: Map 131	Desc. 551
Leaf roll	*Potato leafroll virus*	Worldwide with crop	Virus Desc. 291
Mild mosaic	*Potato virus X*	Worldwide with crop	Virus Desc. 354
Leaf-drop streak, rugose mosaic	*Potato virus Y*	Worldwide with crop	Virus Desc. 240
Rice (*Oryza sativa*)			
Brown spot	*Cochliobolus miyabeanus*	Widespread: Map 92	Desc. 302
Bakanae	*Gibberella fujikuroi*	Widespread: Map 102	Desc. 22
Stem rot	*Magnaporthe salvinii*	Widespread: Map 448	Desc. 344
Blast	*Magnaporthe grisea*	Widespread: Map 51	Desc. 169: 44, 2131
Bacterial leaf blight	*Xanthomonas oryzae* pv. *oryzae*	South and East Asia, Togo, Australia, Brazil: Map 304	Desc. 239
Bacterial leaf streak	*Xanthomonas oryzae* pv. *oryzicola*	Asia, Australia: Map 463	Desc. 240
Hoja blanca	*Rice hoja blanca virus*	USA, Mexico, Central and South America: Map 359	Virsu Desc. 299: 54, 4922
Tungro	*Rice tungro virus*	Bangladesh, India, Indonesia, Malaysia, Philippines, Thailand: Map 516	Virus Desc. 67
Rose (*Rosa* spp.)			
Black spot	*Diplocarpon rosae*	Widespread: Map 266	Desc. 485
Rubber (*Hevea brasiliensis*)			
Red root rot	*Ganoderma philippii*	South-East Asia, Central and West Africa, New Caledonia, Papua New Guinea: Map 98	Desc. 446
South American leaf blight	*Microcyclus ulei*	Tropical South and Central America	Desc. 225: 49, 2617
Powdery mildew	*Oidium heveae*	South East Asia, Central Africa, Brazil, Papua New Guinea: Map 4	Desc. 508

Continued

Common name	Pathogen	Distribution	Literature
Brown root rot	*Phellinus noxius*	South-East Asia, Australia, Fiji, Vanuatu, Papua New Guinea, Africa: Map 104	Desc. 195
White root rot	*Rigidoporus microporus*	Fairly widespread in tropics but identity doubtful in America: Map 176	Desc. 198
Rye (*Secale cereale*)			
Ergot	*Claviceps purpurea*	Widespread: Map 10	
Safflower (*Carthamus tinctorius*)			
Rust	*Puccinia carthami*	Widespread: Map 424	Desc. 174
Sorghum (*Sorghum bicolor*)			
Zonate leaf spot	*Gloeocercospora sorghi*	Fairly widespread: Map 339	Desc. 300
Downy mildew	*Peronosclerospora sorghi*	Africa, South-East Asia, USA, Mexico, South America: Map 179	Desc. 761: 60, 3641
Loose smut	*Sphacelotheca cruenta*	Widespread: Map 408	Desc. 71
Head smut	*Sphacelotheca reiliana*	Widespread: Map 69	Desc. 73
Covered smut	*Sphacelotheca sorghi*	Widespread: Map 220	Desc. 74
Long smut	*Tolyposporium ehrenbergii*	Africa, South and West Asia, China: Map 377	Desc. 76
Stripe	*Burkholderia andropogonis*	Fairly widespread: Map 495	Desc. 372
Anthracnose	*Colletotrichum sublineolum*	Widespread: Map 586	Desc. 132
Soybean (*Glycine max*)			
Stem blight and canker, pod blight	*Diaporthe phaseolorum* f.sp. *caulivora*	North America: Map 360	Desc. 336
Downy mildew	*Peronospora manshurica*	Fairly widespread: Map 268	Desc. 689
Rust	*Phakopsora pachyrhizi*	South-East Asia, Africa, Australia, New Caledonia, Papua New Guinea, West Indies, Venezuala: Map 504	Desc. 589
Brown spot	*Septoria glycines*	USA, Canada, East Asia, Romania, Colombia: Map 361	Desc. 339
Mosaic	Soybean mosaic virus	Widespread: Map 390	Virus Desc. 93
Anthracnose	*Colletotrichum truncatum*	Widespread	
Charcoal rot	*Macrophomina phaseolina*	Widespread	Desc. 275

Disease	Pathogen	Distribution	Reference
Phytophthora root and stem rot	*Phytophthora sojae*	North America, Europe, Australia	Desc. 115
Bacterial blight	*Pseudomonas savastanoi* pv. *glycinea*	Widespread	
Spruce (*Picea* spp.)			
Butt rot	*Heterobasidion annosum*	Worldwide in temperate areas: Map 271	Desc. 192
Strawberry (*Fragaria* spp.)			
Grey mould	*Botryotinia fuckeliana*	Worldwide	Desc. 431
Leaf scorch	*Diplocarpon earliana*	Widespread: Map 452	Desc. 486
Red core	*Phytophthora fragariae*	Europe, Japan, Australia, New Zealand, North America: Map 62	57, 700
Sugarcane (*Saccharum officinarum*)			
Eye spot	*Bipolaris sacchari*	Widespread: Map 349	Desc. 305
Red rot	*Glomerella tucumanensis*	Widespread: Map 186	Desc. 133
Smut	*Ustilago scitaminea*	Fairly widespread, absent from North and Central America: Map 79	Desc. 80: 57, 5106
Leaf scald	*Xanthomonas albilineans*	Fairly widespread: Map 33	Desc. 18
Fiji disease	*Sugarcane Fiji disease virus*	Australia, Oceania, Malagasy, Malaysia, Philippines, Thailand	Virus Desc. 119
Mosaic	*Sugarcane mosaic virus*	Widespread: Map 299	Virus Desc. 342
Sunflower (*Helianthus annuus*)			
Downy mildew	*Plasmopara halstedii*	Widespread: Map 286	Desc. 979
Rust	*Puccinia helianthi*	Widespread: Map 195	Desc. 55
Tea (*Camellia sinensis*)			
Blister blight	*Exobasidium vexans*	South-East Asia, Japan: Map 45	Desc. 779
Tobacco (*Nicotiana tabacum*)			
Brown spot	*Alternaria longipes*	Widespread: Map 63	Desc. 245
Frog eye	*Cercospora nicotianae*	Widespread: Map 172	Desc. 416
Blue mould	*Peronospora tabacina*	America, Europe, North Africa, East Asia, Australia: Map 23	60, 2185
Black streak	*Phytophthora nicotianae* var. *nicotianae*	Widespread	Desc. 34
Black root rot	*Thielaviopsis basicola*	Widespread: Map 218	Desc. 170

Continued

Common name	Pathogen	Distribution	Literature
Bacterial wilt	*Ralstonia solanacearum*	Widespread: Map 138	Desc. 1220
Wildfire	*Pseudomonas syringae* pv. *tabaci*	Widespread: Map 293	Desc. 129
Mosaic	*Tobacco mosaic virus*	Worldwide	Virus Desc. 151
Ringspot	*Tobacco ringspot virus*	Widespread: Map 144	Virus Desc. 17
Tomato (*Lycopersicon esculentum*)			
Leaf spot	*Cercospora fuligena*	South and East Asia, Oceania, Africa, USA: Map 382	Desc. 465
Stem rot	*Didymella lycopersici*	Widespread: Map 324	Desc. 272
Leaf mould	*Mycovellosiella fulva*	Widespread: Map 77	Desc. 487
Fusarium wilt	*Fusarium oxysporum* f.sp. *lycopersici*	Widespread	Desc. 1117: 50, 3174
Bacterial wilt	*Ralstonia solanacearum*	Widespread: Map 138	Desc. 1220
Wilt	*Verticillium albo-atrum, Verticillium dahliae*	Widespread, temperate and subtropical: Maps 365, 366	Desc. 255, 256
Bacterial canker	*Clavibacter michiganensis* subsp. *michiganenesis*	Widespread: Map 26	Desc. 19
Spotted wilt	*Tomato spotted wilt virus*	Widespread: Map 8	Virus Desc. 363
Trees			
Armillaria root rot	*Armillaria mellea*	Worldwide: Map 143	Desc. 321: 56, 5810
Wheat (*Triticum* spp.)			
Root and foot rot	*Cochliobolus sativus*	Widespread	Desc. 701
Powdery mildew	*Erysiphe graminis*	Widespread	Desc. 153: 58, 41
Take-all	*Gaeumannomyces graminis* vars. *avenae* and *tritici*	Widespread: Map 334	Desc. 382, 383: 54, 3243
Scab	*Gibberella zeae*	Worldwide	Desc. 384
Glume blotch	*Phaeosphaeria nodorum*	Widespread: Map 283	Desc. 86
Eyespot	*Pseudocercosporella herpotrichoides*	Europe, North Africa, South Africa, South Australia, New Zealand, Canada, USA: Map 74	Desc. 386
Black or stem rust	*Puccinia graminis*	Widespread	Desc. 291: 55, 1185
Yellow or stripe rust	*Puccinia striiformis*	Widespread: Map 97	
Brown rust	*Puccinia recondita*	Widespread: Map 226	
Leaf blotch	*Mycosphaerella graminicola*	Widespread: Map 397	Desc. 986
Bunt (stinking smut)	*Tilletia tritici*	Widespread: Map 294	Desc. 719
Bunt	*Tilletia laevis*	Widespread: Map 295	Desc. 720
Loose smut	*Ustilago nuda*	Widespread: Map 368	Desc. 280

Surveys and Sampling

M. Holderness

*CABI Bioscience UK Centre, Bakeham Lane, Egham, Surrey
TW20 9TY, UK*

General Principles

Surveys of diseases in crops are required for many purposes: to determine general levels of crop health; or the presence of particular diseases of quarantine significance; prioritization of problems to enable proper allocation of crop protection resources; and to assess the losses caused by crop disease as discussed in Chapter 4. However, they are all concerned with determining the type and severity of diseases affecting crops. The structure of disease surveys and protocols for sampling of crops varies widely between different scientific needs, e.g. quarantine officers may require subsamples of seed lots for detection of specific pathogens, whereas extension workers may require rapid assessment of diseases and of the farmers' priorities in regard to the disease complex in his/her fields. The diagnostic survey thus combines identification of diseases with a measure of their intensity. Qualitative aspects are concerned with determining what is there. This may be either the type of damage, symptoms, etc., or which particular species of pest or pathogen is present. Samples will often need to be collected in order to confirm identities of pests and pathogens by laboratory examination. Quantitative aspects determine how much of a particular pest or pathogen is present, particularly in relation to how much damage is being caused to the crop. Incidence and severity need to be measured according to a predetermined sampling pattern. Surveys may need to be undertaken several times during the season as particular crop stages may differ in their susceptibility or diseases may develop in characteristic ways. The data may also need to relate to particular crop growth stages in order to determine the effect of disease on yield. Patterns of disease spread may be critical in diagnosis or in devising control methods. Aerial surveys are sometimes used to assess crop health or patterns of disease development over large areas. These may use true light or infra-red images, but usually need to be followed up on the ground for sampling and diagnostic procedures. Diseases do not usually occur in isolation and measures of the relative importance of different components of the disease complex are required to determine which would justify control measures and whether such measures are likely to be economic. Pests and abiotic factors may also be present and causing damage to the crop and these need to be taken into account during the survey.

Planning the Survey

Before any survey starts, decisions are required on:

1. The population or region to be covered;
2. The characteristics to be estimated;
3. The subclasses (e.g. varieties, soil types) to be covered;
4. The degree of accuracy required.

These factors determine the basic structure of the survey. Of these, point 2 involves the techniques to be used and the skills required, whereas 3 and 4 determine the scale of the survey.

The survey must cover sites that are truly representative of the region and must be of sufficient size to obtain the required accuracy. Planning the survey in relation to the resources (people, time, money, transport) available is crucial and the collection of extraneous information should be avoided at all costs. The planning should decide the minimum level of information required and match that against the maximum level of information which could be gathered given the resources available.

To ensure survey objectives can be met, the planning process should take account of all four stages of the survey:

1. Planning: decide what data are required.
2. Sampling and collection: the gathering of the data.
3. Analysis: conversion of the data into accessible information.
4. Report: use of the information to achieve the original objective.

If necessary, survey objectives can be modified during stages 1, 3 or 4, but such flexibility is difficult, if not impossible, during the sampling and collection stage. Before data collection begins (and before the training of data collectors begins), it is essential to have determined the survey goals including detail on the variables for which data will be collected, how often they will be collected and the level of accuracy which is required. One very useful way to do this is to undertake a small 'pilot' survey in advance of the actual survey, so that a realistic assessment can be made of what will be feasible and achievable, before extensive resources are committed.

A survey is not an experiment; it is feasible to use a survey to determine the likelihood of different pest and environmental factor combinations, but one cannot expect to make very accurate estimates of the effects of different factor combinations on the yields of crops from surveys alone. The objectives of a diagnostic survey may thus perhaps best be limited to the discovery of what combinations of pests and hosts exist and a relatively crude assessment of the damage caused in the cropping systems under study.

Sampling

Appropriate sampling procedures comprise a combination of a suitable sampling method and a representative sample size. For a sample to be representative it must relate to the spatial distribution of disease and for it to have the desired level of accuracy, it must be of an adequate size in relation to the survey objectives and logistics.

Sampling procedures should be: simple; representative; reliable; and ideally applicable in subsequent standard disease surveillance procedures.

Sample size

The sample is usually made up of several small sampling units (e.g. leaf, plant or plot) which are often predetermined by the purpose of the study, as are sampling dates and intervals. Stratified sampling can be appropriate to the detection of very low disease incidences, but care is needed in the bulking of samples and subsequent analysis to avoid losing true replication, or conversely to avoid creating spurious levels of accuracy. If the standard deviation of estimated pest incidence is: +/– SD and the mean is to be estimated with standard error: +/– SE, the number (n) of independent random samples required is:

$$n = (\text{SD})^2/(\text{SE})^2$$

However, the standard deviation is often unknown and must be guessed, using information from a pilot study or comparable pathosystems. A rough estimate of the sample sizes required can be obtained from a pilot study in which disease intensities are estimated for a number of sampling units and the running means and their corresponding standard deviations plotted against the number of sampling units. Both such curves tend to flatten after a certain number of entries and this point can be taken as the minimum (and thus in many cases optimum) sample size to obtain a reliable estimate.

Sampling patterns

For a sample to be truly representative, the spatial distribution of the disease must be taken into account. Depending on the mode of transmission, diseased plants or lesions (and thus appropriate sample points) may have a random, regular or clustered distribution. In the case of diagnostic surveys, the optimal choice of sampling techniques thus requires some pre-knowledge of the diseases present and their behaviour, such as can be acquired through a pilot study. Random sampling generally yields good results, but as many plant diseases have a clustered distribution, systematic methods give more precise coverage in most situations. Random sampling also requires true randomization

Table 3.1.   Disease symptoms associated with likely pathogens.

Disease symptom	Likely pathogen
Sown seeds	
Failure to emerge	*Fusarium, Rhizoctonia*
Damping-off	*Pythium, Phytophthora*, some seed-borne diseases
Roots (and tubers)	
Necrosis of young root tips and fine feeder roots	*Pythium, Phytophthora*, nematodes
Necrosis of main roots with:	
dark wet rot	*Phytophthora*
dark dry rot	*Ceratocystis, Gaeumannomyces*
dry rot with red or purple tinge	*Fusarium*
ashy grey rot	*Macrophomina*
white to brown mycelial sheets or fans	*Rosellinia, Rhizoctonia*
rhizomorphs	*Armillaria, Rigidoporus*
patchy cortical necrosis	Nematodes
Malformation with:	
stubby much-branched fine roots	Nematodes
galls, knots, proliferation	Nematodes, *Agrobacterium*
Stems	
Necrosis at soil level:	
dark with no obvious mycelium or sclerotia	*Fusarium, Phytophthora*, bacterium
brown with mycelium and/or sclerotia evident	*Rhizoctonia, Sclerotium* or other basidiomycete
Canker, often with gummosis	*Phytophthora*
Lesions pale and scabby	*Elsinoë, Sphaceloma*
Lesions often cankerous or anthracnose-like with minute dark-fruiting bodies	*Colletotrichum, Phomopsis*, etc.
Pustules of yellow, brown or orange spores	Rust
Sclerotia or mycelium running along stem or enveloping it	'Web blights', *Corticium, Marasmius* spp.
White powdery surface sporulation	Powdery mildew
White downy surface sporulation	Downy mildew
Malformation:	
swelling	Virus, systemic fungal infection
shortening of internodes (stunting)	Virus, etc.
elongation	*Fusarium*, some viruses, smuts
shoot proliferation (witches' broom)	Systemic fungal infection, some bacteria and viruses
Galls at soil level	*Agrobacterium*
Internal discoloration (of vascular tissue)	*Fusarium, Verticillium dahliae*
Die back (without primary necrosis at base of die back)	Secondary symptom of root rot, vascular disease or virus
Leaves	
White powdery sporulation on surface	Powdery mildew
Downy sporulation on surface	Downy mildew
Pustules with yellow, orange or brown spores	Rust
Pale scabs	*Elsinoë, Sphaceloma*
Discrete lesions:	
angular, often with chlorotic halo	Bacteria
more or less circular, grey centre, may be zonate (elongated in monocots)	*Cercospora, Drechslera, Corynespora*, etc.

Table 3.1. *Continued*

Disease symptom	Likely pathogen
irregular, containing dark-fruiting bodies	*Phomopsis, Colletotrichum, Septoria*
irregular large, starting at leaf edge with water-soaked margins	*Phytophthora, Peronospora*, bacteria
marginal scorch with chlorosis and wilting	Secondary symptom of vascular disease, systemic virus infection or root rot
Malformation	
stunted, often chlorotic	Virus, vascular infection, some downy mildews
chlorotic mottling, mosaic, vein banding, etc.	Virus
puckering, curling	Virus, insect damage
twisting, shredding	Downy mildew (Graminaceous), *Fusarium moniliforme*, insect damage
Wilting with milky exudate from cut stem below	Vascular bacterial disease (e.g. *Ralstonia*)
Fruits	
Anthracnose spots which eventually crack open	*Colletotrichum, Phomopsis*, some bacteria, feeding punctures of Hemipteran insects
Distorted, often with irregular ripening	Virus, vascular infection
Scabs on surface (other symptoms as for leaves)	*Elsinoë, Sphaceloma*
Seed pods, ears, etc.	
Sclerotia in ear (ergot)	*Claviceps*
Seeds on flower parts converted into dark spore mass	Smuts
Discoloration of glumes, etc.	*Fusarium, Septoria, Drechslera*
Virescence (conversion of floral parts to leafy structure)	Virus, downy mildew

and this is not always easy to achieve in the field. For systematic sampling, a simple diagonal is acceptable for very low or very high disease incidences, but a W-shaped sampling path gives better coverage of the field in other cases. Simple systematic approaches such as these can have an inherent bias, but are easier to implement than randomized or complex systematic approaches and the extra attention required for more complex approaches may significantly reduce the number of samples which can feasibly be taken. Again, a pilot study can be used to determine biases of the sampling system, allowing these to be taken into account in subsequent analyses. Instructions for sampling should be made as precise as possible to eliminate personal selection of the units to be sampled, e.g. specify the number of paces between samples.

Data collection

The two main parameters usually recorded on disease surveys relate to what is there and how frequently it occurs. Diseases are recognized in the field primarily by the symptoms they produce and these can give a guide to the type of pathogen involved. There are some excellent field guides to diseases of the main crop species (See Chapter 7). In order to achieve an accurate diagnosis it may be necessary to collect samples and return them to a plant clinic for examination (see Chapters 8 and 20), but a record should always be made of the field symptoms seen. The scheme in Table 3.1 goes some way to assist in recognizing the main signs and symptoms of disease and relating them to likely pathogens. Further details on collecting specimens for examination in the plant clinic are given by Waller *et al.* (1998).

Assessing the amount of disease or crop damage present depends on counting the frequency with which diseased plants occur and on measuring how severely they are damaged. Assessment of crop damage, which is covered more fully in Chapter 4, is usually based on a visual scale indicating the proportion of the plant affected and classified into categories. In rapid diagnostic surveys, the assessments need to be made as simple as possible, with the minimum number of categories necessary (but balanced against the need to ensure data remain of value). Thus for wilts, a percentage rating scale of 0, 0.1–5, 6–25 and 25% or more is often appropriate, whereas for other diseases, ratings of 0, < 1, 1–25, 26–75 and > 75% can often be used for incidence, and 0, < 1, 1–5, 6–25 and > 25% is more appropriate for severity.

Further Reading

Barnett, V. (1991) *Sample Survey Principles and Methods*. Edward Arnold, London, UK, 173 pp.

Binns, M.R., Nyrop, J.P. and van der Werf, W. (2000) *Sampling and Monitoring in Crop Protection: the Theoretical Basis for Designing Practical Decision Guides*. CABI Publishing, Wallingford, UK.

Chiarappa, L. (1971) *Crop Loss Assessment Methods: FAO Manual on the Evaluation and Prevention of Losses by Pests, Disease and Weeds*. CAB International, Wallingford, UK.

Cochran, W.G. (1977) *Sampling Techniques*. John Wiley & Sons, New York, USA, 428 pp.

Disthaporn, S., Hau, B. and Kranz, J. (1993) Comparison of sampling procedures for two rice diseases: leaf blast and tungro. *Plant Pathology* 42, 313–323.

James, W.C. (1974) Assessment of plant diseases and losses. *Annual Review of Phytopathology* 12, 27–48.

Kranz, J. (1988) Measuring plant disease. In: Kranz, J. and Rotem, J. (eds) *Experimental Techniques in Plant Disease Epidemiology*. Springer-Verlag, Berlin, Germany, pp. 35–50.

Nilsson, H.E. (1995) Remote sensing and image analysis in plant pathology. *Annual Review of Phytopathology* 33, 489–527.

Perry, J.N. (1994) Sampling and applied statistics for pests and diseases. In: Brain, P., Hockland, S.H., Lancashire, P.D. and Sim, L.C. (eds) *Sampling to Make Decisions, Aspects of Applied Biology 37*. Association of Applied Biologists, Wellesbourne, UK, pp. 1–14.

Prinsley, R.T. (1989) *Sampling and Analysis of Relationships Between Insects, Diseases, Nematodes, Weeds and Intercropping Systems: Method Development*. Technical Paper 270, Commonwealth Science Council, London, UK, 67 pp.

Teng, P.S. (1987) *Crop Loss Assessment and Pest Management*. APS Press, St Paul, Minnesota, USA, 270 pp.

Waller, J.M., Ritchie, B.J. and Holderness, M. (1998) *Plant Clinic Handbook*. IMI Technical Handbook No. 3. CAB International, Wallingford, UK, 94 pp.

Disease and Yield Loss Assessment

P.S. Teng[1] and W.C. James[2]

[1]Monsanto Company, Kings Court II Building, 2129 Chino Roces Avenue, Makati City, Philippines; [2]ISAAA AmeriCenter, c/o Cornell University, Ithaca, NY 14853, USA

Introduction

Plant diseases were first studied because of the yield losses they cause, yet the development of techniques and programmes for estimating these losses have been relatively neglected compared to other specializations in plant pathology. If we are not in a position to estimate the yield losses from diseases, how can we rationally decide on how much to spend on control? How can decision makers allocate resources and assign plant pathogen priorities in the most meaningful and efficient way, if there are no reliable data to indicate the relative importance of diseases and the corresponding benefits resulting from control? Reliable yield loss estimates are a prerequisite to the development of any rational and economical plant protection programme, at the field, farm, regional or national level, and are irrespective of whether the control measures involve fungicides, resistant varieties, cultural practices or integrated control. The cost of the loss must be known so that it can be compared with the cost of control.

On a global basis, the need for effective loss assessment programmes to quantify the constraints to production associated with plant diseases is extremely important within the context of global food security, acknowledging that 800 million people suffer from malnutrition today. Furthermore, the losses due to plant diseases in the developing countries of the world are substantially higher than for the developed countries, and the most severe losses usually occur in the poorer countries, which can least afford the loss. These substantial losses due to plant diseases occur despite the fact that a considerable and an increasing quantity of fungicides is used annually. The current total cost of fungicides annually on a global basis is estimated to be of the order of US$6 billion. In a world that on the one hand is short of food and on the other is conscious of protecting the environment and utilizing finite resources responsibly, loss estimates can play a vital role by providing the objective database for critically evaluating the use of resources in plant pathology and the corresponding gains resulting from control measures.

Several international meetings on crop losses have been convened and have resulted in the publication of proceedings updating the literature, e.g. the E.C. Stakman Commemorative Symposium in Minnesota, USA, in 1980 (Teng and Krupa, 1980), and the Crop Loss Assessment workshop at the International Rice Research Institute, Philippines, in 1988 (IRRI, 1990). Reviews on disease and loss assessment have been published by Chester (1950), Large (1966), James (1974), James and Teng (1979), and Teng (1986, 1987) and the reader is referred to these for more detailed discussion. Most of the published material on crop loss assessment is on foliage diseases of annual crops, particularly the cereal crops. Furthermore, most studies have reported the effect of a single disease on crop production. More recent work has concentrated on the effect of a complex of more than one disease or factor as constraints within the total production system, and on incorporating the influence of socio-economic factors on yield losses (Savary et al., 1994, 1997).

It is important to distinguish between the terms 'yield loss' and 'crop loss' (Nutter et al., 1993). 'Crop loss' refers to the damage caused by one or more plant diseases, quantified in social and economic terms to include the external effects on prices and overall production of a reduction in yield of the crop being studied. Crop loss assessment, therefore, requires input from economists and other social scientists. 'Yield loss' refers to the biological damage caused by one or more diseases, quantified in biological terms, and often expressed in economic terms, but without recognizing the external effects of pricing and the marketplace. 'Yield loss assessment' is a practical first step towards crop loss assessment and in itself provides valuable information for decision making. This chapter will focus on yield loss assessment.

This chapter uses a 'tools'-based approach to discuss the topic. We will first describe techniques for measuring disease and yield loss, since these measurements provide the basis for estimating the yield losses. Next, we will describe the techniques for quantifying disease–yield loss relations in two parts: data generation through experiments and on-farm surveys, and statistical analysis of the disease–yield data. Finally, the chapter is completed with some suggestions of how to deal with multiple disease or pest situations.

Techniques for Measuring Disease and Yield Loss

Disease assessment

Disease assessment is the process that generates all the data that quantify the progress of disease. It is therefore critical that the disease assessment methods are well defined and standardized at the earliest possible stage (Nutter et al., 1991). The requirements of an acceptable disease assessment method are demanding, but there are two principal criteria that must be satisfied during the testing stage, prior to using the method for experimental or survey work. Firstly, different observers using the method to assess a sample of diseased plants must be able to record similar assessments consistently, which are also well correlated with the actual or measured diseased area. Secondly, the assessments must be achieved simply and quickly, the latter being particularly important for large-scale surveys.

Disease can be measured by direct methods, i.e. measuring disease on the plant, or by indirect methods, e.g. monitoring the spore population. Direct methods have been more widely used because they are better correlated with losses in production than the indirect methods, which are rather laborious and time-consuming (James and Teng, 1979). Direct methods measure disease as incidence or severity, as defined below. The term disease intensity is often used to denote either incidence or severity.

$$\begin{array}{c} \text{Disease incidence } (I) \\ \text{(Frequency)} \end{array} = \frac{\text{Number of infected plant units}}{\begin{array}{c}\text{Total number (healthy and infected)}\\ \text{of units assessed}\end{array}} \times 100$$

$$\begin{array}{c} \text{Disease severity } (S) \\ \text{(Area)} \end{array} = \frac{\text{Area of plant tissue affected by disease}}{\text{Total area}} \times 100$$

Disease incidence is usually used for assessing systemic type infections, e.g. wilts or viruses, or when the diseased plants or parts thereof result in total loss, e.g. cereal bunts or smuts.

Incidence can also be used to measure the growth of disease during the very early part of some epidemics when both incidence (number of tillers infected) and severity (leaf area affected) are increasing simultaneously. However, all the tillers may become infected (100% I) at a very early stage and thereafter the growth of the epidemic can be measured only in terms of increased severity. Although disease incidence has been assessed with uniformity, disease severity has certainly not. A related term, 'disease prevalence', refers to disease incidence within the context of a geographical area. For example, ten fields in an area are inspected for disease and six are found to be infected; the disease prevalence for that area is 60%.

Lack of standardization is the major problem with most of the keys used by plant pathologists and plant breeders for assessing disease. For example, diseases are often graded as slight, moderate or severe with no indication to quantify the actual level of disease. Therefore 'slight' to one assessor may mean 'severe' to another, or the same assessor may assign the same specimen to slight in one season, and to moderate or severe in the following season. Similarly, other disease assessors will arbitrarily assign specimens to one of six ill-defined categories where 0 = healthy, 1 = slight and 6 = severe. In the absence of standardized assessment methods it is not possible to collate data from different observers or compare corresponding data from other workers.

Assessment keys

Methods for assessing severity can be conveniently separated into two types. The first is a descriptive type of key, which describes plants with different levels of disease and assigns a category, number, index, grade or percentage infection to each description. The most well known is the British Mycological Society (Anon., 1947) key for assessing late blight of potato (Table 4.1).

Standard area diagrams

The second type of assessment key utilizes standard area diagrams, which typify the development of disease on a whole plant or part of a plant. The simplest assessment method is the one least prone to error and whenever possible assessments should be made on single leaves rather than whole plants or plots. Since the objective is to relate disease to yield loss, the individual leaves or plant parts chosen for assessment should be the most important contributors to yield. Figure 4.1 (James *et al.*, 1968) shows a sample standard area diagram for assessing *Rhynchosporium secalis* on the top two leaves of barley; instructions for using the key are given in the caption. Many standard area diagrams have been published (Chiarappa, 1971–1981; James, 1971a, b) and can be prepared quite simply by using graph paper to estimate the infected area as a percentage of the total leaf area. Modern computer tools are also available to produce standard area diagrams and may be found in the catalogues of professional society newsletters such as that of the American Phytopathological Society.

In the interest of standardization it has been suggested that, irrespective of whether descriptive keys or standard area diagrams are used, the percentage scale should be used in preference to grades, categories, indices, etc. The percentage scale has been chosen because of the following advantages (James, 1974):

• universally known;
• the upper and lower limits are uniquely defined;
• can be conveniently divided and subdivided into, for example, 50%, 25%, 12%;

Table 4.1. Assessment key for late blight of potato (*Phytophthora infestans*). (After British Mycological Society (Anon., 1947).)

Blight (%)	Nature of infection
0.0	No disease observed.
0.1	A few scattered plants blighted; no more than one or two spots in 12-yard radius.
1.0	Up to 10 spots per plant; or general slight infection.
5.0	About 50 spots per plant; up to one in ten leaflets infected.
25	Nearly every leaflet infected, but plants retain normal form; plants may smell of blight; field looks green although every plant is affected.
50	Every plant affected and about 50% of leaf area destroyed; field appears green, flecked with brown.
75	About 75% of leaf area destroyed; field appears neither predominantly brown nor green.
95	Only a few leaves on plants, but stems green.
100	All leaves dead, stems dead or dying.

- can be used to assess incidence and severity although only a few percentage infections are given in most keys, e.g. 1, 10, 20 and 50. Interpolation can easily be practised and recorded; the extent of interpolation is dictated by the ability of the observer to detect particular differences;
- can easily be transformed for any subsequent epidemiological analysis, e.g. transformation to logits for calculation of r, the apparent infection rate.

In tropical Asia, disease-assessment techniques have been in use for screening crop cultivars against specific pathogens, e.g. The Standard Evaluation System for rice developed by IRRI (IRRI, 1996), which uses a 0–9 scale for determining plant reaction against diseases and incorporates notions of symptom type and severity or incidence. We recommend that this type of technique is not used for yield loss studies as it does not accurately reflect disease progress. Instead, percentage keys or standard area diagrams should be used. However, if lesion type is a desired additional piece of information in yield loss experiments, then techniques such as those of Pinnschmidt *et al.* (1993) for rice leaf blast may be developed.

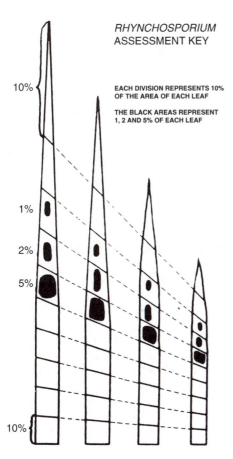

RHYNCHOSPORIUM
ASSESSMENT KEY

EACH DIVISION REPRESENTS 10%
OF THE AREA OF EACH LEAF

THE BLACK AREAS REPRESENT
1, 2 AND 5% OF EACH LEAF

Fig. 4.1. Assessment key for *Rhynchosporium* leaf blotch or scald of barley. Match the leaf to one of the diagrams and use the black areas (representing 1, 2, and 5% of each leaf) as a guide in assessing the percentage leaf (lamina) area covered by small isolated lesions, and the 10% sections for the larger lesions that have coalesced.

Whenever disease assessments are recorded, the growth stage of the crop should be noted so that disease progress can be related to host development at several locations where the time and length of the growing seasons may be different. Growth stage keys have been published for most of the major crops (Chiarappa, 1971–1981; Teng, 1987).

Yield and loss measurement

Yield measurement is as important as disease measurement. Since the objective is to measure yield losses under farming conditions, it is important to determine yield using the same harvesting techniques as farmers. Failure to do this may underestimate the yield loss. At each location in the network of experiments, the actual yield of the healthy plot will be different due to climatic, edaphic and other factors. Therefore, there is a need to adopt a technique that will allow all the data to be standardized and collated (Teng, 1987). This is usually achieved by designating the yield of the healthy plot at each location as the reference yield. It is also important that statistical and agronomic considerations on plot size be duly considered. It is important that the required precision for the yield estimates be determined at the outset; the results of uniformity trials can be used to establish the required size and shape of plots as well as the number of replicates (James and Shih, 1973).

Yield loss is calculated as the difference in yield between a diseased and a healthy treatment expressed as a percentage of the yield of the healthy plot at each location. Yield loss is defined as the measurable reduction in yield and/or quality, calculated as the difference between attainable yield (i.e. yield of healthy plot or field obtained under the best set of farming practices to maintain zero or minimal constraints on yield) and actual yield (i.e. yield of plot or field with the known disease incidence or severity). Potential yield loss is sometimes calculated from the difference between potential/maximum attainable yield (i.e. highest yield obtainable in small plots or estimated using computer models) and actual yield (Nutter et al., 1993).

Techniques for Quantifying Disease–Yield Loss Relations

Data generation through experiments and on-farm surveys

Data generation using plot experiments

The main objective is to generate data that will allow the relationship between disease and yield loss to be characterized in the form of a disease–yield loss relationship, commonly a statistical model. Field experiments should be conducted in all areas where the crop is important for a minimum of three seasons using the current major cultivars grown under normal farming practice. In the tropics, environmental differences due to wet and dry seasons exert strong influences on crop yield and disease development, and will have to be taken into account before deciding how many seasons of trials are needed.

Design of experiments

In each experiment it is essential to have one treatment free from disease to measure the yield of the healthy crop in the absence of disease. One or more additional treatments with varying levels of disease are used to determine the decrease in yield associated with different levels of disease. Although the paired plot design, e.g. untreated and treated with fungicides, has been widely used, it is not as efficient as multiple treatment experiments, which allow the effects of more than one epidemic to be studied at the same time (Teng, 1987). Interplot interference due to movement of airborne spores from one plot to another may preclude achieving complete control in the healthy treatments or the desired level of disease in other treatments (James *et al.*, 1973b). Interplot interference can be reduced by including the necessary buffer areas of 3–5 m between plots preferably cropped with a different, taller and denser canopy plant species than the one under study to act as a filter for spores moving between plots. Some workers (Richardson *et al.*, 1975; Shane *et al.*, 1987) have suggested that single tillers, as opposed to traditional plots, can be used for measuring the yield reductions associated with foliage diseases of cereals. The technique has advantages, which include economy of labour and land, but there are unresolved problems, which include inability to take into account compensation, and reduction in yield due to number of tillers/unit. A more serious disadvantage is that the yield loss models developed with the technique explain a very low percentage of the yield variability, which results in an inherently weak predictive model.

Methods for generating different epidemics

Although it is preferable to conduct experiments with natural infection, the success rate can be low because of the unpredictability of epidemic development. To increase the chances of success, many workers have resorted to artificial inoculation. However, if possible, only perimeter areas around the experiment should be inoculated so that infection can spread 'naturally' into the plots. Inoculum should not be applied directly to the plots because this may not simulate the natural epidemic. It is important to note that, irrespective of the method used to vary the epidemic, ideally the only effect should be to promote or delay the epidemic (Teng and Johnson, 1987). Secondary effects leading to a decrease or increase in yield not related to disease are a source of error. Epidemics with different characteristics can be generated by the following means.

Both the number and timing of fungicide applications can be varied to achieve different disease progress curves. Fungicides have been more widely used than any other technique and provide maximum flexibility for varying disease level. However, there are several disadvantages, which include secondary effects resulting from phytotoxicity, or beneficial effects, which are often related to the presence of micro-elements such as copper or manganese. The modern eradicant and systemic fungicides are better than the wide-spectrum protectant fungicides because they are more target-specific and efficient. With the simultaneous use of more than one target-specific fungicide, the effect of more than one disease can be studied at the same time in the same crop.

Isogenic lines or cultivars of varying susceptibility can also be used to generate different epidemics of aerial or soil-borne pathogens. Isogenic lines are similar in every respect except for susceptibility to disease. The use of a resistant and susceptible isogenic line in conjunction with fungicides probably affords the most accurate means of determining disease–loss relationships. Although not as accurate, a similar procedure can be followed with a group of cultivars of varying susceptibility but whose yields under disease-free conditions are reasonably similar. If the group includes a cultivar which expresses tolerance, defined as 'the same epidemic resulting in different percentage yield loss for different cultivars', the loss model for that particular variety will, of necessity, be different.

Soil-borne diseases present more problems than airborne pathogens because it is more difficult to create different levels of disease on different plots within one experiment. Inoculum added to soil does not always result in an increase in infection. An alternative is to use different rotations to build up different levels of inoculum or to select individual plants with different levels of disease within a large population. Micro-plots are the most common technique for experimentation with nematodes and soil-borne fungal diseases (Teng, 1987). The location of experiments in different geographical areas, expected to have different levels of disease, can also be used in conjunction with any of the above techniques to increase the chances of achieving variability in disease level. Different proportions of plants can be removed at different points in the growing season to simulate damage at early, mid or late season, e.g. removal of potato plants at emergence and later stages to simulate diseases such as those caused by *Verticillium* and *Rhizoctonia* showed that losses were not proportional to the percentage of missing plants, indicating that compensation had taken place (James *et al.*, 1973a).

Data generation using on-farm surveys

This aspect is also discussed in Chapter 3. In comparison to field experiments, there has been relatively little effort in the development of survey methodology, although it is equally, or perhaps even more important. It is unfortunate that many of the *ad hoc* badly planned surveys have led to the conclusion that surveys are not rewarding whereas, in fact, they are the only means of determining the economic importance of diseases within a regional or national context. Not only can surveys be used to identify and quantify the constraints to production associated with plant diseases, they can also be used to evaluate the effectiveness and the acceptability of recommended control practices. Information on diseases and control practices should be collected simultaneously in the same survey and can lead to a further understanding of how far, and in what way, farmers have accepted control practices. It may also indicate which key factors motivate farmers to introduce plant protection programmes (Savary *et al.*, 1994, 1997).

A series of surveys on barley foliage diseases, initiated by James (1969) in England and Wales in 1967 and continued by King (1977), is probably the most convincing demonstration of the usefulness of surveys. The first barley survey conducted in 1967 demonstrated the significance of powdery mildew and the need for chemical control measures, hitherto considered unnecessary. The annual barley surveys have confirmed that powdery mildew is the principal disease despite the fact that a substantial acreage had been treated for mildew with fungicides by 1974 and 1975. In the tropics, a comprehensive survey of rice constraints was initiated in 1987, from which a large database has enabled estimation of yield losses, identification of yield gaps, determination of intervention points to improve farmer decision making, and the setting of priorities for research and action on plant protection (IRRI, 1990; Savary *et al.*, 1994).

Some of the important factors to consider in survey methodology are variability and distribution of the disease and the expected range of severity, and the precision and cost of estimates derived from different sampling methods. Also it is very important that the identical technology developed in the experimental phase is strictly adhered to in the survey phase. Surveys generate a large database, which is expensive to collect and therefore should be fully interpreted. Much microcomputer-based software is available to assist field workers store, manipulate and analyse large, multivariate databases (Teng and Rouse, 1984); their statistical power and speed increase each year as prices decline. Lastly, the continued success of any survey is dependent on the feedback of information to the farming community, which can profit greatly from the exercise.

Statistical analysis of disease–yield data

Modelling the disease-loss relationship

The database generated on yield loss experiments or surveys can be simply classified into two groups: the independent variables (disease assessments) and the corresponding dependent variables (percentage yield loss estimates). Regression analysis has been the most widely used technique for describing the relationship between the two or more variables, although its use is subject to certain conditions of the original data (Teng, 1987). Three types of loss model have been developed to relate disease and yield loss. The critical-point (CP) models require only one disease assessment to be made whereas the area under the disease progress curve (AUDPC) models and multiple point (MP) models require more than one assessment.

CP models commonly predict yield loss from an assessment done at a crop growth stage that is particularly sensitive to the effects of disease. Yield loss due to rice neck blast was estimated by Katsube and Koshimizu (1970) with the equation $Y = 0.57X$, where X is percentage blasted nodes 30 days after heading, and Y is percentage loss. For barley leaf rust, several CP models were developed based on assessments made at the decimal crop growth stages designed by Zadoks *et al.* (1974) of 39/40 (flag leaf or ligule just visible), 49/50 (first awns or spikelets visible), 58/59 (emergence of inflorescence complete), 64/65 (anthesis completed), 73/74 (early milk) and 83/84 (early dough); the growth stage of 83/84 was found to provide the best estimate of yield loss caused by leaf rust when rust severity was assessed on the flag leaf (Teng *et al.*, 1979). CP models have been developed for potato late blight based on the number of blight-free days (Olofsson, 1968) and the time when 70% of potato foliage was blighted (Large, 1952).

AUDPC models are more advanced than CP models in that they can be used to distinguish two epidemics with different disease progress curves but which have the same percentage severity at a critical date. Schneider *et al.* (1976) working with *Cercospora* leaf spot of cowpea in Nigeria showed that percentage yield loss was equivalent to 0.43 (area under curve) + 14.95.

CP and AUDPC models have been successfully applied to situations where the disease is a late epidemic and where the infection rate is reasonably stable. Also they are more applicable to crops where the yield is accumulated during a short period at the end of the growing season, e.g. cereals. These restrictions do not apply to the multiple point models, which can be used for diseases with high variability in infection rates and where the disease progress curves can be markedly different. They can be used for epidemics that develop over a long time relative to the crop life and where yield accumulation is a prolonged process, e.g. potatoes. Burleigh *et al.* (1972) developed a multiple point model for wheat leaf rust where percentage yield loss, *Y*, was related to the percentage leaf rust at the boot (*X*2), early berry (*X*5) and early dough (*X*7) growth stages. The equation is: $Y = 5.3788 + 5.5260X2 - 0.3308X5 + 0.5019X7$.

James *et al.* (1972) used increments of disease during weekly intervals as the independent variable as opposed to estimates of disease at particular growth stages to estimate losses due to potato late blight. Their equation had the general form of

$$Y\ (\%\ \text{loss}) = b1X1 + 62 \times 2 \ldots\ldots\ldots bnXn$$

where b1 and b*n* are partial regression coefficients for the 1st and *n*th week, respectively, and where *X*1 and *Xn* are the corresponding weekly disease increments for the 1st and *n*th week, respectively. Using the equation the estimated loss is within 5% of the actual in nine cases out of ten.

The required precision for loss estimates will be one of the major factors governing choice of model. MP and AUDPC models require more inputs of disease assessments than the CP, and consequently they are more precise. The MP model provides the maximum flexibility and accuracy to deal with situations where the onset, rate of infection and level of infection may vary.

Explaining disease-loss relationships

Disease-loss models should not just be statistical 'best-fit' models but should also be meaningful in a biological sense. The physiology of yield in cereals for example is sufficiently understood that it is possible to explain why specific crop growth stages may be expected to show greater sensitivity to disease infection (Gaunt, 1987). Path coefficient analysis is a technique used to determine the correlation between disease variables assessed during field experiments or on-farm surveys, and yield components (e.g. 1000-grain weight, mean spike weight, number of fertile spikes per unit harvested area), and allows explanation of how a disease such as blast can influence the final outcome of rice yield through its effects on yield components (Torres and Teng, 1993). Similarly, correlation techniques have been used to determine how rust on different leaf positions at several growth stages affects final wheat yield

(Seck *et al.*, 1991). It is important that statistical models do not violate known biological phenomena, such as disease having a positive effect on yield when the partial regression coefficients in a model show a negative sign in a disease–yield loss equation.

Concluding Remarks

Biologists who have attempted to quantify the losses due to more than one disease or pest in the same crop have adopted a technique which is equivalent to integrating the experimental and survey phases. This is done by monitoring the development of the causal agents in a survey where yield is also measured, but where no treatments are used to control disease (Pinstrup-Andersen *et al.*, 1976). The approach has many merits and in an overall programme it could represent a useful first step to assess the relative importance of different diseases, and their overall importance versus other factors in the total production system. However, it should be noted that it is not a substitute for the experimental phase, which must be implemented to establish a cause and effect relationship. The synoptic approach used by Stynes (1975) to relate wheat yield in Australia to a large set of influencing variables, among which disease is only a small component, is largely based on a statistical technique called principal components analysis. The use of principal components analysis has now been tested on several other cropping systems and found useful in partitioning the relative importance of sets of variables on yield, and allowing more realistic regression models to be developed from the same data set for individual disease-loss relationships.

Recently, sophisticated simulation models have been developed for rice and other major food crops that allow the estimation of yield loss caused by individual or simultaneous multiple pest (insect, disease, weed) infestations (Pinnschmidt *et al.*, 1995). The incorporation of social-economic variables (e.g. farmer's social status, farmer attitudes towards risk) into the set of influencing variables to explain crop yield is also now possible through a set of statistical techniques called correspondence analysis (Savary *et al.*, 1994). A manual detailing survey protocol for data collection to analyse complex situations using ordinal and cardinal variables has been developed for use on rice in tropical Asia, and illustrates an increased role for yield loss assessment studies in integrated pest management (Savary *et al.*, 1996).

There is a growing literature on yield loss studies but still relatively few studies on crop loss assessment. Among published literature on yield losses, much variability exists in data quality and their potential use for interpretation. With rice, Savary *et al.* (1998) have found that much of the data published on yield losses are very location-specific with limited extrapolation potential, or they reflect 'worst-case scenarios' with little corresponding information of prevailing disease status in farmers' fields. There is still an unmet need to develop large area databases on crop yield and disease losses, so that rational decisions may be made on resource allocation for crop protection.

References and Further Reading

Anon. (1947) The measurement of potato blight. *Transactions of the British Mycological Society* 31, 140–141.

Burleigh, J.R., Roelfs, A.P. and Eversmeyer, M.G. (1972) Estimating damage to wheat caused by *Puccinia recondita tritici*. *Phytopathology* 62, 944–946.

Calpouzos, L., Roelfs, A.P., Madson, M.E., Martin, F.B., Welsh, J.R. and Wilcoxson, D.R. (1976) A new model to measure yield losses caused by stem rust in spring wheat. *Technical Bulletin*, University of Minnesota Agricultural Experimental Station, No. 307, 23 pp.

Chester, K.S. (1950) Plant disease losses: their appraisal and interpretation. *Plant Disease Reporter Supplement* No. 193, 189–362.

Chiarappa, L. (1971–1981) *Crop Loss Assessment Methods: FAO Manual on the Evaluation and Prevention of Losses by Pests, Diseases and Weeds.* Farnham Royal, Commonwealth Agricultural Bureaux, Supplements 1 and 2 loose-leaf. Supplement 3. 123 pp.

Church, B.M. (1971) The place of sample survey in crop loss estimation. In: Chiarappa, L. (ed.) *Crop Loss Assessment Methods 2.2/1–8.* Farnham Royal, Commonwealth Agricultural Bureaux, UK.

FAO (1967) Papers presented at symposium on crop losses. FAO, Rome, Italy.

Gaunt, R.E. (1987) A mechanistic approach to yield loss assessment based on crop physiology. In: Teng, P.S. (ed.) *Crop Loss Assessment and Pest Management.* American Phytopathological Society, USA, 270 pp.

IRRI (1990) *Crop Loss Assessment in Rice.* IRRI, Los Banos, Philippines, 334 pp.

IRRI (1996) *Standard Evaluation System for Rice.* IRRI, Los Banos, Philippines, 52 pp.

James, W.C. (1969) A survey of foliar diseases of spring barley in England and Wales in 1967. *Annals of Applied Biology* 63, 253–263.

James, W.C. (1971a) An illustrated series of assessment keys for plant diseases, their preparation and usage. *Canadian Plant Disease Survey* 51, 39–65.

James, W.C. (1971b) A manual of disease assessment keys for plant diseases. *Publication, Canadian Department of Agriculture* No. 1458, 80 pp.

James, W.C. (1974) Assessment of plant diseases and losses. *Annual Review of Phytopathology* 12, 27–48.

James, W.C. and Shih, C.S. (1973) Size and shape of plots for estimating yield losses from cereal foliage diseases. *Experimental Agriculture* 9, 63–71.

James, W.C. and Teng, P.S. (1979) The quantification of production constraints associated with plant diseases. *Applied Biology* 4, 201–267.

James, W.C., Jenkins, J.E.E. and Jemmett, J.L. (1968) The relationship between leaf blotch caused by *Rhynchosporium secalis* and losses in grain yield of spring barley. *Annals of Applied Biology* 62, 273–288.

James, W.C., Shih, C.S., Hodgson, W.A. and Callbeck, L.C. (1972) The quantitative relationship between late blight of potato and loss in tuber yield. *Phytopathology* 62, 92–96.

James, W.C., Lawrence, C.H. and Shih, C.S. (1973a) Yield losses due to missing plants in potato crops. *American Potato Journal* 50, 345–352.

James, W.C., Shih, C.S., Callbeck, L.C. and Hodgson, W.A. (1973b) Interplot interference in field experiments with late blight of potato (*Phytophthora infestans*). *Phytopathology* 63, 1269–1275.

James, W.C., Teng, P.S. and Nutter, F.W. (1991) Estimated losses of crops from plant pathogens. *CRC Handbook of Pest Management 1*, 2nd edn. CRC Press, Boca Raton, pp. 15–51.

Katsube, T. and Koshimizu, Y. (1970) Influence of blast disease on harvests in rice plant I: effect of panicle infection on yield components and quality. *Bulletin of the Tohoku National Agricultural Experiment Station, Japan* 39, 55–96

King, J.E. (1977) Surveys of foliar diseases of spring barley in England and Wales, 1972–75. *Plant Pathology* 26, 21–29.

Large, E.C. (1952) The interpretation of progress curves for potato blight and other plant diseases. *Plant Pathology* 1, 109–117.

Large, E.C. (1966) Measuring plant disease. *Annual Review of Phytopathology* 4, 9–28.

Nutter, F.W. Jr, Teng, P.S. and Shokes, F.M. (1991) Disease assessment terms and concepts. *Plant Disease* 75, 1187–1188.

Nutter, F.W. Jr, Teng, P.S. and Royer, M.J. (1993) Terms and concepts for yield, crop loss and disease thresholds. *Plant Disease* 77, 211–215.

Olofsson, B. (1968) Determination of the critical injury threshold for potato blight (*Phytophthora infestans*). *Meddelanden Vaxtskyddsanstalt, Stockholm* 14, 81–93.

Pinnschmidt, H.O., Teng, P.S. and Bonman, J.M. (1993) A new assessment key for leaf blast. *International Rice Research Notes* 18(1), 45–46.

Pinnschmidt, H.O., Batchelor, W.D. and Teng, P.S. (1995) Simulation of multiple species pest damage in rice using CERES-RICE. *Agricultural Systems* 48, 193–222.

Pinnschmidt, H.O., Chamarerk, V., Cabulisan, N., de la Pena, F., Long, N.D., Savary, S., KleinCiebbinck, H.W. and Teng, P.S. (1997) Yield gap analysis of rainfed lowland systems to guide rice crop and pest management. In: Kropf, M.J., Teng, P.S., Aggarwal, P.K., Buoma, J., Boumari, B.A.M., Jones, J.W. and van Laar, H.H. (eds) *Applications of Systems Approaches at the Field Level*. Kluwer Academic, Dordrecht, The Netherlands, pp. 321–328.

Pinstrup-Andersen, P., de Londcho, N. and Infante, M. (1976) A suggested procedure for estimating yield and production losses in crops. *PANS* 22, 359–365.

Richardson, M.J., Jacks, M. and Smith, S. (1975) Assessment of losses caused by barley mildew using single tillers. *Plant Pathology* 24, 21–26.

Savary, S., Elazegui, F.A., Moody, K. and Litsinger, J.A. (1994) Characterization of rice cropping practices and multiple pest systems in the Philippines. *Agricultural Systems* 46, 385–408.

Savary, S., Elazegui, F.A. and Teng, P.S. (1996) A survey portfolio for the characterization of rice pest constraints. *IRRI Discussion Paper Series* No. 18.

Savary, S., Elazegui, F.A., Pinnschmidt, H.O. and Teng, P.S. (1997) On characterization of rice pest constraints in Asia: an empirical approach. In: Teng, P.S., Kropff, M.J., ten Berge, H.F.M., Dent, J.B., Lansigan, F.P. and van Laar, H.H. (eds) *Applications of Systems Approaches at the Farm and Regional Levels*. Kluwer Academic, Dordrecht, The Netherlands, 468 pp.

Savary, S., Elazegui, F.A. and Teng, P.S. (1998) Assessing the representativeness of data on yield losses due to rice diseases in tropical Asia. *Plant Disease* 82, 705–709.

Schneider, R.W., Williams, R.J. and Sinclair, J.B. (1976) *Cercospora* leaf spot of cowpea models for estimating yield loss. *Phytopathology* 66, 384–388.

Seck, M., Roelfs, A.P. and Teng, P.S. (1991) Influence of leaf position on yield loss caused by wheat leaf rust in single tillers. *Crop Protection* 10, 222–228.

Shane, W.W., Baumer, J.S. and Teng, P.S. (1987) Crop losses caused by Xanthomonas streak on spring wheat and barley. *Plant Disease* 71, 927–930.

Stynes, B.A. (1975) A synoptic study of wheat. PhD Thesis, University of Adelaide, South Australia.

Teng, P.S. (1986) Crop loss appraisal in the tropics. *Journal of Plant Protection in the Tropics* 3(1), 3950.

Teng, P.S. (1987) *Crop Loss Assessment and Pest Management*. American Phytopathological Society Press, USA.

Teng, P.S. and Johnson, K.B. (1987) Analysis of epidemiological components in yield loss assessment. In: Rotem, J. and Kranz, J. (eds) *Techniques in Plant Disease Epidemiology*. Springer-Verlag, Berlin, Germany, pp. 179–190.

Teng, P.S. and Krupa, S.V. (1980) Crop loss assessment. In: *Miscellaneous Publication No. 7*. Agricultural Experiment Station, University of Minnesota, USA, 396 pp.

Teng, P.S. and Revilla, I.M. (1996) Technical issues in using crop loss data for research privatisation. In: Evenson, R.E., Herdt, R.W. and Hossain, M. (eds) *Rice Research in Asia: Progress and Priorities*. CAB International, Wallingford, UK, pp. 261–276.

Teng, P.S. and Rouse, D.I. (1984) Understanding computers – applications in plant pathology. *Plant Disease* 68, 539–543.

Teng, P.S., Close, R.C. and Blackie, M.J. (1979) A comparison of models for estimating yield loss caused by leaf rust (*Puccinia hordei* Otth) on Zephyr barley in New Zealand. *Phytopathology* 69, 1239–1244.

Teng, P.S., Batchelor, W.D., Pinnschmidt, H.O. and Wilkerson, G.J. (1998) In: Tsuji, G., Hoogenboom, G.P. and Thornton, P.K. (eds) *Understanding Options for Agricultural Production*. Kluwer Academic, Dordrecht, The Netherlands, pp. 225–270.

Torres, C.Q. and Teng, P.S. (1993) Path coefficient and regression analysis of the effects of blast on rice yield. *Crop Protection* 12, 296–302.

Zadoks, J.C., Chang, T.T. and Konzak, C.F. (1974) A decimal code for the growth stages of cereals. *Weed Research* 14, 415–421.

Postharvest Diseases

Revised by J.M. Waller

CABI Bioscience UK Centre, Bakeham Lane, Egham, Surrey TW20 9TY, UK

Substantial crop losses occur between harvesting and final utilization of many agricultural/horticultural products and these may be greater than the gains achieved by improvements in primary production. Most attention to post-harvest crop losses has been paid to protecting food stocks, particularly emergency grain supplies, so that more information is available concerning the handling and storage of grain and other 'durable' crops such as oilseeds and dry legumes than about non-grain staples such as starchy root crops, fruit and vegetables. Postharvest losses of these 'perishable' crops are greater, and although techniques such as controlled temperature storage, controlled atmosphere storage, chemical treatments, and improved packing and handling methods have considerably reduced these losses they still remain surprisingly high, even in well-developed temperate countries. In the tropics and subtropics postharvest losses of both 'durable' and 'perishable' crops are even greater and less well appreciated, even though in these regions demand for food is greatest.

Magnitude and Causes of Losses

Overall postharvest losses of durable crops including cereals, oilseeds and pulses have been established at 20% of the harvested crop in Africa, Asia and Latin America. An FAO estimate puts losses of these commodities at 10% on a worldwide basis. In individual cases losses may be much greater and it is suggested that losses at the farm of 35–50%, followed by 10–12% in traders' stores and a further 5% in centralized stores, may not be uncommon. A conservative estimate suggests that in the tropics around 25% of all perishable food crops harvested are lost between harvest and consumption. Both quantitative and qualitative losses occur after harvest and often result from inefficient methods of handling, storage and transportation that prevent effective control of physical and biological factors. Durable produce is generally stored in the dry state at a moisture content commonly of below 12% and losses are largely the result of external pests such as insects, moulds and rodents. Losses of perishable produce, which frequently have moisture contents of 50% or more, are more extensive as the name suggests and result largely from physical, physiological and pathological damage.

Insect and rodent damage

A wide range of storage insect pests can damage durable produce. Besides direct losses, insect infestations can lead to secondary damage through mould growth. Rats and mice are also serious pests of stored durable foods and can also cause damage and loss of perishable produce. They are not only destructive because of the amount of produce they consume and soil but also because they may transmit disease. A consideration of these agents is outside the scope of this review.

Mould damage

All durable and dried agricultural commodities, if not properly dried after harvest, are subject to attack by fungal moulds. Because dry produce can reabsorb moisture from the atmosphere, unless stored in sealed containers, loss from fungal attack is generally more common in the humid tropics where fungal growth is also stimulated by high ambient temperatures and relative humidities. Insect infestations also raise moisture levels through respiration and thus enable mould growth to occur. Moulds most commonly encountered include species of *Aspergillus*, *Penicillium*, *Mucor* and *Rhizopus*. Mould growth causes discoloration, production of bad odours and off-flavours, reduction in quality, and loss of viability of seeds. In some cases highly toxic substances (mycotoxins) are produced. Amongst the best known are aflatoxins produced by *Aspergillus flavus* and zearalenone produced by species of *Fusarium* which poison and kill livestock and are potential health hazards to humans. Mould damage can also occur during the processing of some commodities such as coffee where off-flavours are of particular significance.

Physical damage

Loss due to mechanical damage occurs in both durable and perishable produce but because of their higher moisture content and lower mechanical strength, perishable products are generally more susceptible. Mechanical injury occurs at all stages, from preharvest operations, through harvesting, grading, packing and transport to exposure in the market and domestic activities. Losses caused by such injury are frequently overlooked as they predispose produce to enhanced secondary physiological and pathological loss and are thus difficult to estimate. The susceptibility of crops such as soft fruit and leafy vegetables is conspicuous but the magnitude of losses in more robust produce such as root crops can be high. For example, as much as 33% of the potato crop can be so seriously damaged through harvesting and grading as to be fit only for stock feed, and a further 12% loss can occur during transit to market. In bananas, which are commercially harvested unripe, all mechanical injuries are not apparent on the green fruit and are revealed only after ripening when desiccation, discoloration and infections by pathogens lead to downgrading or total loss of the ripe fruit. Physical damage can result from exposure to either high or low extremes of temperature. Disruption of tissues occurs at temperatures below the freezing point of the produce but low temperatures above freezing may damage tropical fruit. Exposure to high temperature radiation also disrupts tissue and can lead to physiological changes that reduce quality and storage life.

Physiological losses

Physiological losses are diverse in nature. Because produce is alive, endogenous respiratory losses of dry matter and loss of water through transpiration always occur in perishable produce but the magnitude of these varies with the crop and is greatly influenced by the storage environment. Losses are great and storage life short when produce is stored in the tropics at high ambient temperature without refrigeration. The faster a product respires and ripens, the greater the quantity of heat generated, and the storage life of different commodities varies inversely as the rate of evolution of heat. Generally respiration is greatest during the first 24 h after harvest in vegetables. Fruits generally have a gradually increasing rate of respiration as they ripen; in some, such as apples and bananas, the rate rises to a peak, the climacteric, and then rapidly declines. Moisture loss represents a reduction in saleable weight and losses of 3–6% are generally enough to cause a marked decline in quality, although certain commodities are still marketable following a 10% loss of moisture. The drier the storage atmosphere, the more rapid is the loss of water from stored products.

Many tropical and subtropical fruit and vegetables, when stored at temperatures well above their freezing point, are subject to chilling injury, which is distinct from freezing injury. Chilling injury, like high temperature damage, appears to be a function of the temperature and period of exposure and is expressed in several ways, commonly involving surface pitting, internal discoloration, tissue breakdown, increased susceptibility to decay and impaired quality. These are accompanied by many biochemical changes that can reduce quality or increase susceptibility to pathogens.

Optimum harvest maturity for storage and handling purposes also affects physiological losses. Most root crops are more susceptible to losses when harvested immature, whereas many fruit and leafy vegetables become more susceptible with increasing maturity and ripening. Sprouting is a physiological problem characteristic of root, tuber and bulb storage. Certain physiological disorders may also arise during storage that reduce the value of particular crops. Examples are vascular streaking of cassava, which commonly prevents the storage of fresh roots for more than a few days, and the development in store of bitter pit, water core and scald in apples, all of which are greatly influenced by preharvest growing conditions.

Pathological attack

Attack by microorganisms (fungi, bacteria and to a lesser extent viruses) is probably the most serious cause of postharvest loss in perishable produce, although physical and physiological damage frequently predispose material to pathogenic attack. Losses from pathogens cause both quantitative losses of sound produce and reduction in quality.

Quantitative pathogenic losses result from the frequently rapid and extensive breakdown of host tissues by microorganisms. The pattern of attack is usually initial infection by one, or a few, specific pathogens followed by massive infection by a broader spectrum of secondary invaders such as species of *Fusarium*, *Botrytis*, *Rhizopus*, *Botryodiplodia* and *Erwinia*. These are generally weakly pathogenic or are saprophytic on the dead or moribund tissue remaining from the primary infection. However, they may be very aggressive once established in susceptible tissue and can play an important role in postharvest pathology, greatly exacerbating the damage commenced by primary pathogens. Qualitative pathogenic losses are typically the result of blemish or other surface diseases, such as apple scab, which render the produce less attractive even though relatively little destruction of tissue has occurred. Such diseases are particularly important in the fruit industry, where emphasis is placed on visual quality and even a small blemish may render the produce unmarketable.

Many microorganisms responsible for postharvest pathology originate in the field and either infect the produce before or during harvest or are present as surface contaminants. Contamination with pathogens may also occur during postharvest operations. Where infection occurs prior to harvesting, rotting may develop immediately after the infection is established, as in blight and pink rot of potatoes. However, some preharvest infections produce latent infections, which only become active following ripening or damage, as in the development of anthracnose during the ripening of bananas and other tropical fruit. Where infection occurs at or after harvesting, it is frequently at the sites of mechanical injury inflicted during harvesting and handling operations, often at the point where the produce is detached from the parent plant. Some postharvest pathogens are, however, capable of infecting produce through natural skin openings such as stomata or lenticels, and yet others can penetrate undamaged epidermal tissues. Spread of disease in storage is often slow but can be rapid between adjacent ripe fruit so that over time substantial losses can ensue.

Bacteria are often the most important causal agents in the spoilage of vegetables, particularly the bacterial soft-rot group such as species of *Erwinia*. Fungi are most frequently involved in fruit and root crop spoilage, but here the spectrum of pathogens is more complicated. Between 20 and 25 different fungi have been associated with postharvest decay in tomatoes and in yams, and 25 fungi have been isolated from crown rot of bananas in different countries.

Virus diseases, although not normally of postharvest significance, may detract from the market value of produce as, for example, spraing in potatoes, internal cork in sweet potatoes, internal brown spot of yams, apple ringspot and stony pit virus of pears. A list of some of the major market diseases and their causative agents are given for a range of common perishable produce in Table 5.1.

Methods of Loss Reduction and Control

Postharvest diseases may be controlled or avoided by various physical and chemical means. Biological control is also being developed but is currently not commercially exploited and more attention is now given to host-plant resistance to postharvest diseases in the selection and breeding of crop cultivars. Although not discussed here, the importance of socio-economic and farming system considerations must be taken into account when making recommendations and applying established technical control measures.

Loss avoidance

By using the produce immediately or processing it into some easily protected inert form, the need for fresh storage and risk of losses due to postharvest problems can be greatly reduced. In several regions of the world, favourable environmental conditions enable crops to be produced continually throughout the year, thus avoiding the need for storage of many commodities such as food. In many regions, however, this is not possible and where a continuous demand for a particular crop exists, a pattern of production, storage and transportation has to be adopted. By processing produce into a readily storable form less susceptible to attack by pathogens or more easily protected, storage requirements for fresh produce and consequential postharvest losses may often be reduced, although rarely eliminated. Any processing plant, no matter how simple, will generally be more viable the longer it can be kept in operation through the year. This necessitates a continuous supply of produce which, as mentioned above, can be achieved either by continuous production or by seasonal production combined with effective storage.

Physical control methods

The first measure to take to reduce risk of losses caused by postharvest diseases is to ensure that crops are harvested correctly and that only good quality produce is taken into storage. Storage properties can be influenced by cultivar, climate, soil, cultural conditions and maturity, as well as by pest and disease control on the growing crop. Maximum storage life can be obtained only by the efficient storing of high quality produce shortly after harvest. The physical and physiological condition and health of produce when it is placed in store are major factors governing losses during storage.

Table 5.1.   Some market diseases/disorders of selected perishable produce.

Commodity	Disease/disorder	Causal organism or condition
Apple	Bitter rot	*Glomerella cingulata*
	Brown rot	*Monilinia fructigena*
	Bull's eye rot	*Pezicula alba*
	Blue mould	*Penicillium expansum*
	Bitter pit	Calcium imbalance in field
	Brown heart	Carbon dioxide injury in store
	Superficial scald	Hot dry growing conditions and restricted ventilation in store
	Water core	Moisture stress in field
Aroids	Corm rot	*Botryodiplodia theobromae*
		Botrytis spp.
		Fusarium spp.
		Pythium spp.
		Rhizoctonia spp.
	Powdery grey rot	*Fusarium solani*
	Sclerotium rot	*Corticium rolfsii*
	Soft rot	*Erwinia carotovora* subsp. *caratovora*
		Phytophthora colocasiae
		Pythium splendens
Aubergine	Alternaria rot	*Alternaria* spp.
	Grey mould	*Botryotinia fuckeliana*
	Phomopsis rot	*Phomopsis vexans*
	Phytophthora rot	*Phytophthora* spp.
	Rhizopus rot	*Rhizopus stolonifer*
Avocado	Anthracnose	*Glomerella cingulata*
	Rhizopus rot	*Rhizopus stolonifer*
	Stem-end rot	*Botryodiplodia theobromae*
		Phomopsis spp.
		Dothiorella spp.
Banana	Anthracnose	*Colletotrichum musae*
	Crown rot	*Botryodiplodia theobromae*
		Colletotrichum musae
		Fusarium palidoroseum
		Verticillium theobromae
	Pitting disease	*Magnaporthe grisea*
Cassava	Primary deterioration	*Cassava brown streak virus*
		Also physiological
	Secondary deterioration	Many secondary invaders or weak pathogens; particularly species of *Fusarium, Rhizoctonia, Aspergillus, Penicillium, Botryodiplodia, Erwinia*, etc.
Citrus	Alternaria black rot	*Alternaria citri*
	Black pit	*Pseudomonas syringae* pv. *syringae*
	Blue mould	*Penicillium italicum*
	Green mould	*Penicillium digitatum*
	Brown rot	*Phytophthora citrophthora* and other *Phytophthora* spp.
	Melanose	*Diaporthe citri*
	Scab	*Elsinoë fawcettii, Elsinoë australis*
	Septoria spot	*Septoria depressa*
	Sooty mould	*Capnodium* spp.

Table 5.1. *Continued*

Commodity	Disease/disorder	Causal organism or condition
Citrus *(continued)*	Stem-end root	*Diaporthe citri*
		Botryodiplodia theobromae
	Internal gumming	Boron deficiency
	Storage spot	Physiological chilling injury
Crucifers	Alternaria leaf spot	*Alternaria* spp.
	Bacterial soft rot	*Erwinia carotovora* subsp. *carotovora*
	Black rot	*Xanthomonas campestris*
	Downy mildew	*Peronospora parasitica*
	Ring spot	*Mycosphaerella brassicicola*
	Soft rot	*Rhizopus* spp.
	Watery soft rot	*Sclerotinia sclerotiorum*
	White rust	*Albugo candida*
Cucurbits	Anthracnose	*Colletotrichum orbiculare*
	Bacterial soft rot	*Erwinia* spp.
	Black rot	*Didymella bryoniae*
	Charcoal rot	*Macrophomina phaseolina*
	Fusarium rot	*Fusarium* spp.
	Leak	*Pythium* spp.
	Sclerotium rot	*Corticium rolfsii*
	Soft rot	*Rhizopus* spp.
	Soil rot	*Thanatephorus cucumeris*
	Stem-end rot	*Botryodiplodia theobromae*
Grape	Alternaria rot	*Alternaria* spp.
	Anthracnose	*Elsinoë ampelina*
	Bitter rot	*Melanconium fuligineum*
	Black rot	*Guignardia bidwellii*
	Downy mildew	*Plasmopara viticola*
	Grey mould	*Botryotinia fuckeliana*
	Rhizopus rot	*Rhizopus stolonifer*
Mango	Anthracnose	*Glomerella cingulata*
	Black rot	*Aspergillus niger*
	Stem-end rot	*Botryodiplodia theobromae*
Onion	Bacterial soft rot	*Erwinia carotovora* subsp. *carotovora*
	Black rot	*Aspergillus niger*
	Fusarium/basal rot	*Fusarium oxysporum* and other species
	Grey mould/neck rot	*Botrytis allii*
	Smudge	*Colletotrichum circinans*
	Smut	*Urocystis cepulae*
	White rot	*Sclerotium cepivorum*
Pawpaw (Papaya)	Anthracnose	*Glomerella cingulata*
	Black rot	*Phoma caricae-papayae*
	Ripe fruit rot, stem-end rot	*Botryodiplodia theobromae*
Pepper (Capsicum)	Alternaria rot	*Alternaria* spp.
	Anthracnose	*Colletotrichum* spp.
	Bacterial soft rot	*Erwinia carotovora* subsp. *carotovora*
	Grey mould	*Botrytis* spp.
	Rhizopus rot	*Rhizopus* spp.

Continued

Table 5.1. Continued

Commodity	Disease/disorder	Causal organism or condition
Pear	Bitter rot	*Glomerella cingulata*
	Blue mould	*Penicillium* spp.
	Late storage rot	*Nectria galligena*
	Bitter pit	Calcium imbalance
	Stony pit	Stony pit virus
	Brown heart	Carbon dioxide injury
	Scald	Extended cold storage
Pineapple	Black rot/water blister	*Ceratocystis paradoxa*
	Marbling/brown rot/ fruitlet core rot	*Gibberella fujikuroi* *Penicillium funiculosum*
	Endogenous brown spot	Physiological
Potato	Bacterial wilt/brown rot	*Ralstonia solanacearum*
	Bacterial ring rot	*Clavibacter michiganensis* subsp. *sepedonicus*
	Black scurf	*Thanatephorus cucumeris*
	Blight	*Phytophthora infestans*
	Common scab	*Streptomyces scabies*
	Dry rot	*Fusarium* spp.
	Gangrene	*Phoma* spp.
	Pink rot	*Phytophthora erythroseptica*
	Skin spot	*Polyscytalum pustulans*
	Soft rots	*Erwinia* spp.
	Tuber rot	*Corticium rolfsii*
	Wart	*Synchytrium endobioticum*
	Mop top	*Potato mop-top virus*
	Spraing	*Tobacco rattle virus*
Pulses	Anthracnose	*Colletotrichum* spp.
	Bacterial soft rot	*Erwinia carotovora*
	Leak	*Pythium* spp.
	Pod spot	*Ascochyta* spp.
	Powdery mildew	*Erysiphe polygoni*
	Soil/pod rot	*Thanatephorus cucumeris*
	Soft rot	*Rhizopus* spp.
	Watery soft rot	*Sclerotinia* spp.
Stone fruit	Brown rot	*Sclerotinia* spp.
	Rhizopus rot	*Rhizopus stolonifer, Rhizopus nigricans*
Strawberry	Grey mould	*Botryotinia fuckeliana*
	Rhizopus rot	*Rhizopus* spp.
Sweet Potato	Black rot	*Ceratocystis fimbriata*
	Blue-mould rot	*Penicillium* spp.
	Charcoal rot	*Macrophomina phaseolina*
	Dry rot	*Diaporthe batatatis*
	Java black rot	*Botryodiplodia theobromae*
	Soft rot	*Rhizopus* spp.
	Surface/end rot	*Fusarium* spp.
	Internal cork	Virus?

Table 5.1. *Continued*

Commodity	Disease/disorder	Causal organism or condition
Tomato	Alternaria rot	*Alternaria* spp.
	Anthracnose	*Colletotrichum* spp.
	Bacterial canker	*Clavibacter michiganensis* subsp. *michiganenesis*
	Bacterial soft rot	*Erwinia carotovora* subsp. *carotovora*
	Blight	*Phytophthora infestans*
	Cladosporium rot	*Cladosporium herbarum*
	Early blight	*Alternaria solani*
	Fruit rot	*Didymella lycopersici*
	Grey mould	*Botryotinia fuckeliana*
	Phoma rot	*Phoma destructiva*
	Rhizopus rot	*Rhizopus stolonifer*
	Sclerotium rot	*Corticium rolfsii*
	Soil rot	*Thanatephorus cucumeris*
	Watery rot	*Geotrichum candidum*
Yams	Watery rot	*Erwinia* spp.
	Soft rot	*Fusarium* spp.
		Penicillium spp.
		Macrophomina phaseolina
		Rhizopus spp.
		Botrytis spp.
	Internal brown spot	Virus?

Mechanical

Very considerable reductions in losses can be achieved by careful handling and improved packaging techniques in order to minimize injury. Even small injuries can have large effects; for example, the rate of water loss from an apple may be increased by as much as 400% due to a single bad bruise. Produce should be transported and stored in containers with good ventilation, adequate cushioning materials, stacking strength, and durability to withstand crushing pressure and high humidity conditions. In the design of containers, due regard should be paid to the mechanical strength of the produce itself and the effect of load forces. For example, a reduction in losses from 16% to 3% was achieved in top-quality tomatoes in Ghana simply by reversing traditional inverted cone-shaped baskets so that they had broad bases and narrow necks. Moisture retentive plastic packaging materials can often be used to reduce desiccation and injury.

In the majority of starchy root crops, the adverse effects of injury may be reduced by a simple process of curing through exposure to a warm, well-aerated but humid environment. This promotes cell suberization and the formation of a cork layer over wounds, which not only reduces moisture loss but also acts as a physical barrier against wound pathogens. For example, the average storage loss of 33% (16% water loss and 17% rot) in sweet potatoes in the USA was reduced by curing to 17% (11% water loss and 6% rot). The physical removal and destruction of any damaged or decaying produce from the vicinity of grading and packaging areas is an important measure to reduce the source of pathogenic inoculum and prevent spread to healthy produce.

Careful attention must always be paid to general sanitation and cleanliness of implements, handling machinery, containers and stores.

Temperature

Refrigerated storage is probably the most common means of reducing postharvest losses in perishable produce. In a properly designed and managed cold store it retards:

1.   Respiration and other metabolic activities;
2.   Ageing due to ripening, softening, textural and colour changes;
3.   Moisture loss and resultant wilting;
4.   Undesirable growth, such as sprouting of root crops;
5.   Spoilage due to microorganisms.

Storage temperature has various effects on produce quality. The rate of respiration in fresh produce increases by a factor of 2–3 with every rise in temperature of 10°C; thus apples, held at 10°C, respire and ripen three times as fast as when held at 0°C, with a corresponding rise in heat production. High respiration rates are associated with high sugar loss; sweetcorn can lose as much as 14% of its sugar in 3 h when held at 20°C, but this loss can be reduced to 7–8% in 72 h at 0°C. In addition to sugar, the loss of other important components, such as vitamins, is influenced by storage temperature. Since all metabolic activities are slowed down by reduced temperatures, the sprouting of root crops and bulbs is generally retarded under refrigeration although other factors can influence this. Low storage temperatures can have detrimental effects; in potatoes stored below 5°C, an undesirable build-up of sugars occurs as starch is converted into sugar at low temperature. Yet other produce, as discussed above, including many tropical and subtropical fruit and vegetables are subject to chilling injury at temperatures lower than 10–15°C, and all produce is damaged by freezing. All of these factors can increase susceptibility to storage pathogens.

   With few exceptions the metabolism of microorganisms is reduced at refrigerated storage temperatures so rotting is normally retarded. Low storage temperatures, however, rarely kill the pathogens and when produce is returned to ambient conditions rotting may recommence and develop rapidly. Certain postharvest diseases such as anthracnose of mango and pawpaw, *Cladosporium* rot of apples, *Phytophthora* rot of citrus and tomatoes, and black rot of sweet potatoes may be controlled by heat treatment, but this may affect the subsequent ripening of fruit and is difficult to operate on a commercial scale.

   To maximize the benefits of low temperature storage, produce should be cooled as soon as possible after harvest but refrigerated storage rooms usually do not have either the refrigeration capacity or sufficient air circulation for rapid removal of field heat. To achieve this, various methods of field or precooling have been developed. The rate of cooling of any commodity is dependent on four factors:

1.   The accessibility of the produce to the cooling medium.
2.   The difference in temperature between the produce and the refrigerating medium.

3. The velocity of the refrigerating medium.
4. The nature of the cooling medium.

Rapidly moving cold air is a widely used method applicable to refrigerated rooms, trucks, conveyor tunnels, etc. Crushed ice, either mixed with or placed on top of produce in containers can be used, but this technique is now being replaced by hydrocooling, in which the produce is flooded with or immersed in iced water. Provided that the water is cold enough, the flow fast enough and the produce exposed to the cold water for long enough, this is the most effective means of removing field heat. Simply standing vegetables in a tank of cold water does not bring about fast cooling. Asparagus, sweetcorn, carrots, radishes and peaches are amongst the produce commonly hydrocooled. Vacuum cooling has recently become more popular with commodities such as lettuce and other leafy crops. The method involves cooling by the rapid evaporation under vacuum of water from the surface of the prewetted produce. It is costly, requires skilled operation and is economically feasible only where there is a large output of fresh produce. Regardless of the method of precooling, much of the benefit will be lost if produce is not refrigerated promptly after cooling.

Atmosphere

Modifying the oxygen and carbon dioxide content of the storage atmosphere can also reduce deterioration during storage. This controlled atmosphere (CA) storage involves use of lower oxygen and higher carbon dioxide levels than normal and their careful regulation. Its principal commercial use is in the storage of certain varieties of apples, but it is also of considerable benefit in the storage of other deciduous fruits. Carbon dioxide is introduced into the closed, refrigerated system either by sublimation from solid carbon dioxide or as a gas from a cylinder; alternatively, respiratory carbon dioxide is allowed to accumulate in the closed system giving at the same time a reduction of oxygen. CA storage retards respiration, quality changes, tissue breakdown, changes in colour, and the development of decay and scald. The amount of carbon dioxide required to reduce spoilage by pathogens is usually quite high and will rarely control decay after infection has taken place. It is not always clear whether it is the oxygen or carbon dioxide levels that play the major role. Excess carbon dioxide can cause skin injury and internal disorders, whereas very low oxygen may produce off-flavours and alcohol injury due to fermentation. 'Modified atmosphere storage' refers to storage in a beneficial atmosphere other than air, but not under closely regulated conditions, such as that which develops in polythene-lined boxes of pears, or by the addition of dry ice to transit containers. Hypobaric (subatmospheric pressure) storage of horticultural produce is sometimes used commercially. Lowering of pressure gives a proportionate reduction in oxygen level, which can be accurately controlled and gives many of the benefits of CA storage. Water vapour must be supplied to the system to avoid loss of water from the produce. The high cost of the equipment required for hypobaric storage limits its use to high-value crops.

Moisture loss from fresh produce is largely determined by the difference between the vapour pressure of the product and that of the surrounding air (vapour pressure deficit, VPD). The drier the air the more rapid is the water loss, and as VPDs are lower at lower temperatures, moisture loss is less at lower temperatures. Onions, for example, when kept in open air stores in Israel, lost 42% of their original weight every four weeks, but only 2.7% in sheds protected from high ambient temperature and 1.4% in cold stores at 0°C. The optimum relative humidity for the storage of most perishable produce is 85–95%.

Durable or dried perishable produce should be stored in a dry and cool environment to prevent mould growth. The simplest way to achieve this is by sealed storage in vapour-proof containers but this is not always practical or appropriate and other means of dehumidification, such as refrigeration or chemical absorption, must be used.

Radiation

Radiation is used for food preservation, particularly in wheat disinfestation, the inhibition of sprouting in potatoes and onions, and delayed ripening in mangoes and bananas. Under certain conditions radiation can kill insects, reduce populations of, or eliminate, microorganisms and retard physiological processes such as ripening and sprouting. It may, however, adversely affect quality. Radiation sources that have been used include gamma rays (cobalt-60 or caesium-137) and fast electrons (linear accelerators); each has its merits and limitations. Radiation systems are costly but can be easily integrated with other storage and handling methods and are now used mostly in the food packaging industries.

Chemical control methods

Before chemicals can be used successfully to reduce postharvest disease losses, a thorough knowledge of the aetiology and epidemiology of the diseases involved is required. For diseases where infection occurs in the field prior to harvest, chemical and other control measures should normally be directed at preventing field infection before harvest and are not discussed in this review. Nevertheless, application of pesticides in the field has important postharvest implications as permitted residue levels for most pesticides on marketed produce are now extremely small and many countries only tolerate zero detectable levels. Furthermore, field treatments may not be economically effective when used to control postharvest disease initiated from latent infections that occur before harvest, such as anthracnose of bananas or mangoes, so postharvest chemical control measures become necessary. The so-called wound parasites are more easily controlled by postharvest chemical treatments that are used to protect the produce from infection or eradicate new early infections.

Quite different properties are required for postharvest pesticides than for field-applied compounds and although active ingredients may be similar, formulations differ. Regulations governing the use of postharvest pesticides are very stringent and they must only be used in strict accordance with the manufacturer's recommendations and food additive regulations of the country or countries concerned. Besides lack of mammalian toxicity, postharvest fungicides and bactericides need to have good surface distribution and adherence. They should also be able to penetrate epidermal tissues to eradicate latent infections. Nevertheless, a great many chemicals have been used in reducing postharvest disease losses. These can be classified, according to their method of application, into three groups: fumigants; treated wraps and waxes; and dips, sprays or occasionally dusts.

Fumigants

Fumigants are particularly useful in treating very delicate produce, and for produce transported or stored in closed containers. They have the added advantage of greater powers of penetration. The best-known fumigant is sulphur dioxide (SO_2) used primarily to control *Botrytis* and other rots of grapes. This gas may be applied directly from cylinders, by burning sulphur, or by release from sodium acid sulphite. SO_2 kills fungal spores present on the fruit surface but does not destroy infections present in the tissues prior to fumigation. Overtreatment results in off-flavours and bleached skin spots; SO_2 is phytotoxic to most fruit and vegetables, and the gas is highly corrosive. Other fumigants include ozone and nitrogen trichloride (NCl_3). NCl_3 is dangerously unstable and must be generated on the spot and used with forced air circulation as it is highly corrosive to bare metal and poisonous in high concentrations. It has been successfully used to control stem-end rot, blue and green moulds of citrus, and common postharvest diseases of cantaloupe melons, tomatoes and onions. Overtreatment frequently causes surface damage.

Wraps

Chemically treated wraps have been used predominantly in the citrus and apple industries for localizing diseases and preventing blemish diseases from developing or spreading, by inhibiting sporulation of the causal organisms on the fruit surface. Such treatments are most effective when the chemical also acts as a vapour-phase fumigant, as in the case of biphenyl-impregnated wrappers. These have been used extensively, but are reported to affect flavour adversely and to produce undesirable odours. Other chemicals used to impregnate wraps are pine oil, sodium-*o*-phenyl phenate and various *o*-phenylphenol esters, copper sulphate and some active halogen compounds. Some success has also been obtained in treating container liners or shredded paper packaging material, instead of individual wrappers. The major effect of waxes is to improve the appearance of certain produce and to reduce moisture loss, but they commonly have little effect in reducing decay, and in some cases may stimulate it. Waxes are used commercially on citrus, cucumbers and to a lesser extent on other crops such as sweet peppers, tomatoes, melons, apples and root crops such as turnips and sweet potatoes.

Dips and sprays

Many chemicals have been used as dips, sprays or occasionally dusts to control postharvest diseases. These include borax, sulphur compounds, phenolic compounds, positive halogen compounds, dithiocarbamates, organic acids, antibiotics, various systemic fungicides such as, benzimidazole derivatives, dicarboximide fungicides, and more recently developed organic compounds.

The development of resistance to fungicides in pathogen populations has reduced the efficacy of some of the earlier systemic fungicides, especially the benzimidazole derivatives (see also Chapter 32). Comprehensive reviews of such chemicals and the diseases they control have been published, but the range of pesticides available for postharvest application is being steadily reduced in most countries.

Postharvest fungicides are most frequently applied as aqueous suspensions or solutions, which have the advantage of ease of preparation and application, with penetrating power approaching that of fumigants. In many cases the operation can also be readily incorporated into mechanical handling systems. Solutions or suspensions are conveniently applied to commodities that are wetted for other purposes such as cleaning of apples and citrus, washing and removal of latex from bananas, and hydrocooling of fruit and vegetables. It is often necessary to add fungicides and/or bactericides to the water to prevent the spread of disease-causing organisms. However, when a commodity is not normally treated with water, the application of a solution or suspension should be approached with caution since some types of produce decay more rapidly when wetted. True solutions have a definite advantage over suspensions or dispersions as they do not require continuous agitation to maintain uniformity and are more readily applied as sprays, which are more economical in both chemical and water. Dip application has the advantage of totally submerging the commodity so that maximum opportunity for penetration of the infection sites is always afforded. Chemical dipping may be combined with heat treatment and can be very effective in controlling latent infections such as anthracnose in avocados. The main disadvantage of tank dip treatment is that it requires a relatively large volume of dip which must be used over a period to reduce the unit cost of treatment; this method is best suited to chemicals that are both inexpensive and stable.

Growth regulators

The chemical manipulation of the physiological condition of produce in an attempt to reduce postharvest losses is being developed. Growth regulators can be used to adjust physiological maturity, postharvest ripening and senescence. Chemicals, including the methyl ester of α-naphthaleneacetic acid, tetra-chloronitrobenzene, isopropyl-*N*-phenyl carbamate, isopropyl-*N*-chlorophenyl carbamate, 3,5,5-trimethylhexan-1-ol, and maleic hydrazide, are used on a commercial scale for the control of sprouting in potatoes but all except tetrachloronitrobenzene have the disadvantage that they inhibit the development of wound cork. Hence water loss and infection via wounds may not be prevented. Bananas and other tropical fruit for commerce are harvested while still unripe, in which condition they may be more safely handled and transported over long distances. On arrival at its destination the green fruit is ripened under controlled conditions in special ripening rooms, using ethylene gas to precipitate simultaneous and uniform ripening. Techniques for delaying banana ripening by the chemical removal of endogenous ethylene, combined with modified atmosphere conditions in polythene bags, have been developed in Australia for the long-distance transport of the fruit at ambient temperatures.

Biological control

The manipulation of antagonistic fungi and bacteria for control of plant diseases generally is being widely investigated (see Chapter 34) and this is likely to find a place in postharvest control. Replacement of chemical pesticides by biological antagonists has perceived safety advantages for consumers and the environment. Direct application of antagonists to harvested produce as dips or sprays is a relatively straightforward technique, but methods to establish populations of antagonists on fruit surfaces prior to harvest is also feasible.

Conclusion

Postharvest losses are important from both an economic and nutritional standpoint, and the problem is technologically and scientifically complex. Nevertheless, greater success in increasing the net production of finally utilizable produce may frequently be achieved by reducing such losses than by increasing field or gross production. The methods selected to reduce losses will often have to be a compromise between a number of conflicting and regularly changing requirements. Thus the most appropriate method of storage should not be regarded as absolute, and different techniques will be more or less appropriate in different circumstances. The storage and handling method selected should yield maximum returns on the investment made. The difference between the price of a commodity at the time of harvest and that paid by the consumer may be considerable, so the introduction of the farmer and produce handler to efficient methods of handling and storage can result in more money being available as a result of their ability to sell good quality produce when prices are favourable.

The knowledge and application of methods used to avoid and reduce postharvest spoilage are generally much greater in developed countries, and in many developing nations the acquisition of knowledge has outstripped its understanding and application. The solution of many existing handling and storage problems thus lies in an educational and sociological approach, utilizing and adapting existing knowledge within many disciplines in a multidisciplinary approach.

Select Bibliography

Adams, J.M. (1977) A bibliography on post-harvest losses in cereals and pulses with particular reference to tropical and subtropical countries. *Report of the Tropical Products Institute* No. G110.

Anon. (1974) Methods of precooling fruits and vegetables. In: *Handbook and Product Directory, Applications*, Chapter 26. American Society of Heating, Refrigeration and Air-conditioning Engineering, New York, USA.

Bent, K.J. (1979) Fungicides in perspective: 1979. *Endeavour*, N.S. 3, 7–14.

Booth, R.H. (1974) Post-harvest deterioration of tropical root crops: losses and their control. *Tropical Science* 16, 49–63.

Burton, W.G. (1974) Some biophysical principles underlying the controlled atmosphere storage of plant material. *Annals of Applied Biology* 78, 149–168.

Dennis, C. (1983) *Post-harvest Pathology of Fruits and Vegetables.* Academic Press, London, UK, 257 pp.

Eckert, J.W. and Ogawa, J.M. (1985) The chemical control of post-harvest diseases: subtropical and tropical fruits. *Annual Review of Phytopathology* 23, 421–454.

Eckert, J.W. and Ogawa, J.M. (1988) Chemical control of post-harvest diseases: deciduous fruits, berries, vegetables and root/tuber crops. *Annual Review of Phytopathology* 26, 433–469.

Harvey, J.M. (1978) Reduction of losses in fresh market fruits and vegetables. *Annual Review of Phytopathology* 16, 321–341.

Mitchell, F.G., Guillon, R. and Parsons, R.A. (1972) Commercial cooling of fruits and vegetables. *University of California Manual* No. 43.

Pantastico, E.B. (ed.) (1975) *Post-harvest Physiology, Handling, and Utilization of Tropical and Sub-tropical Fruits and Vegetables.* AVI Publishing, Westport, Connecticut, USA.

Prusky, D. (1996) Pathogen quiescence and post-harvest diseases. *Annual Review of Phytopathology* 34, 413–434.

Ryall, A.L. and Lipton, W.J. (1979) *Handling, Transportation and Storage of Fruits and Vegetables*, 2nd edn, Vol. 1, *Vegetables and Melons.* AVI Publishing, Westport, Connecticut, USA, 587 pp.

Ryall, A.L. and Pentzner, W.T. (1974) *Handling, Transportation and Storage of Fruits and Vegetables*, Vol. 2, *Fruits.* AVI Publishing, Westport, Connecticut, USA, 475 pp.

Sinha, K.S. and Bhatnagar, D. (1998) *Mycotoxins in Agriculture and Food Safety.* Marcel Dekker, New York, USA, 520 pp.

Snowdon, A.L. (1990) *A Colour Atlas of Post Harvest Disease and Disorders of Fruits and Vegetables.* Vol. 1, *General Introduction and Fruits*, 301 pp. (1992) Vol. 2, *Vegetables*, 416 pp. Wolfe Scientific, London, UK.

Sommer, N.F. (1982) Postharvest handling practices and postharvest diseases of fruit. *Plant Disease* 66, 357–364.

Staden, O.L. (1973) A review of the potential of fruit and vegetable irradiation. *Scientific Horticulture* 1, 291–308.

Tuite, J. and Foster, G.H. (1979) Control of storage diseases of grain. *Annual Review of Phytopathology* 17, 343–366.

Wilson, C.L. and Wisniewsky, M.E. (1989) Biological control of post-harvest diseases of fruits and vegetables: an emerging technology. *Annual Review of Phytopathology* 27, 425–441.

General Bibliography of Plant Pathology

Compiled by J.M. Waller

CABI Bioscience UK Centre, Bakeham Lane, Egham, Surrey TW20 9TY, UK

This bibliography lists some of the major publications in English of use to plant pathologists. As far as possible only material published during the past 20 years is included but a few older publications of particular interest are also mentioned. Items are grouped under the headings: general texts; handbooks and sources of information; tropical plant pathology; methods; soil-borne diseases; epidemiology and ecology; bibliographies and abstract journals; periodicals and serial publications. References to individual crops are in Chapter 7.

For bibliographies and reference lists on other subjects listed in the index see the relevant sections in this *Pocketbook*.

General Texts

Agrios, G.N. (1997) *Plant Pathology*, 4th edn. Academic Press, San Diego, USA, 635 pp.

Fry, W.E. (1982) *Principles of Plant Disease Management*. Academic Press, Orlando, Florida, USA, 378 pp.

Horsfall, J.G. and Cowling, E.B. (eds) (1977–1980) *Plant Disease. An Advanced Treatise*. Vol. 1, *How Disease is Managed*, 465 pp. Vol. 2, *How Disease Develops in Populations*, 436 pp. Vol. 3, *How Plants Suffer from Disease*, 487 pp. Vol. 4, *How Pathogens Induce Disease*, 466 pp. Vol. 5, *How Plants Defend Themselves*, 534 pp. Academic Press, New York, USA.

Lucas, J.A. (1998) *Plant Pathology and Plant Pathogens*, 3rd edn. Blackwell, Oxford, UK, 288 pp.

Manners, J.G. (1993) *Principles of Plant Pathology*, 2nd edn. Cambridge University Press, Cambridge, UK, 343 pp.

Scheffer, R.P. (1997) *The Nature of Disease in Plants*. Cambridge University Press, Cambridge, UK, 325 pp.

Strange, R.M. (1993) *Plant Disease Control*. Chapman & Hall, London, UK, 354 pp.

Handbooks and Sources of Information

Agnihotri, V.P., Om-Prakash, R.K. and Misra, A.K. (1996) *Disease Scenario in Crop Plants*. Vol. 1, *Fruits and Vegetables*, 267 pp. Vol. 2, *Cereals, Pulses, Oil Seeds and Cash Crops*, 276 pp. International Books & Periodicals Supply Service, Delhi, India.

Anon. (2000) *Crop Protection Compendium: Global Module*, 2nd edn. (Integrated electronic information resource on crops and their pests and diseases; available as a CD-ROM product.) CAB International, Wallingford, UK.

Dollet, M. (1988) Plant disease caused by flagellate protozoa. *Annual Review of Phytopathology* 22, 115–132.

FAO-GPPIS, online access to FAO's plant protection information database; www.fao.org

Hamilton, J. (1991) *Pest Management: a Directory of Information Sources*. Vol. 1, *Crop Protection*. CAB International, Wallingford, UK, 331 pp.

Holliday, P. (1998) *A Dictionary of Plant Pathology*, 2nd edn. Cambridge University Press, Cambridge, UK, 536 pp.

Horst, R.K. (ed.) (1979) *Westcott's Plant Disease Handbook*, 4th edn. Van Nostrand Reinhold, New York, USA, 803 pp.

Nyvall, R.F. (1989) *Field Crop Disease Handbook*, 2nd edn. Van Nostrand Reinhold, New York, USA, 817 pp.

PEST CABWeb (online subscription product). CAB International, Wallingford, UK.

Vock, N.T. (Compiler) (1978) *A Handbook of Plant Diseases in Colour*. Vol. 1, *Fruit and Vegetables*, 420 pp. Vol. 2, *Field Crops*, 266 pp. Queensland Department of Primary Industries, Brisbane, Australia.

Western, J.H. (ed.) (1979) *Disease of Crop Plants*. Macmillan, London, UK, 404 pp.

Tropical Plant Pathology

Hill, D.S. and Waller, J.M. (1982) *Pests and Diseases of Tropical Crops*. Vol. 1, *Principles and Methods of Control*. Longman, London, UK, 175 pp.

Hill, D.S. and Waller, J.M. (1982) *Pests and Diseases of Tropical Crops*. Vol. 2, *Field Handbook*. Longman, London, UK, 432 pp.

Holliday, P. (1980) *Fungus Diseases of Tropical Crops*. Cambridge University Press, Cambridge, 607 pp. Reprinted 1995, paperback version, Dover, New York, USA.

Kranz, J., Schmutterer, H. and Koch, W. (eds) (1977) *Diseases, Pests and Weeds in Tropical Crops*. Paul Parey, Berlin, Germany, 666 pp.

Raychaudhuri, S.P. (1977) *A Manual of Virus Diseases of Tropical Plants*. Macmillan, Delhi, India, 299 pp.

Thurston, D.H. (1998) *Tropical Plant Diseases*, 2nd edn. APS Press, St Paul, Minnesota, USA.

Wellman, F.L. (1977) *Dictionary of Tropical American Crops and Their Diseases*. Scarecrow Press, Metuchen, 495 pp.

Methods

Burchill, R.T. (ed.) (1981) *Methods in Plant Pathology*. Phytopathological paper no. 26, Commonwealth Mycological Institute, Kew, UK.

Dhingra, O.D. and Sinclair, J.B. (1985) *Basic Plant Pathology Methods*. CRC Press, Boca Raton, Florida, 355 pp.

Gurr, S.J., McPherson, M.J. and Bowles, D.J. (eds) (1992) *Molecular Plant Pathology. A Practical Approach*, Vol. II. IRL Press, Oxford, UK, 304 pp.

Hampton, R., Ball, E. and de Boer, S. (1990) *Serological Methods for Detection and Identification of Viral and Bacterial Plant Pathogens. A Laboratory Manual.* APS Press, St Paul, Minnesota, USA.

Irwin, J.A.G. (1997) Field and laboratory methods for studying soilborne diseases. In: Hillocks, R.J. and Waller, J.M. (eds) *Soilborne Disease of Tropical Crops.* CAB International, Wallingford, UK, pp. 17–37.

Singleton, L.L., Mihail, J.D. and Rush, C.M. (1992) *Methods for Research on Soilborne Phytopathogenic Fungi.* American Phytopathological Society, St Paul, Minnesota, USA, 265 pp.

Waller, J.M., Ritchie, B.J. and Holdernesss, M. (1998) *Plant Clinic Handbook.* IMI Technical series no. 3. CAB International, Wallingford, UK, 94 pp.

Soil-borne Diseases

Domsch, K.H., Gams, W. and Anderson, T.-H. (1980) *Compendium of Soil Fungi,* Vol. 1, 860 pp. Vol. 2, 406 pp. Academic Press, London, UK.

Hillocks, R.J. and Waller, J.M. (eds) (1998) *Soilborne Disease of Tropical Crops.* CAB International, Wallingford, UK, 453 pp.

Hornby, D. (ed.) (1990) *Biological Control of Soil-borne Plant Pathogens.* CAB International, Wallingford, UK, 479 pp.

Katan, J. (1991) Interactions of roots with soil-borne pathogens. In: Waisel, Y., Eshel, A. and Kafkafi, U. (eds) *Plant Roots: the Hidden Half.* Marcel Dekker, New York, USA, pp. 823–836.

Mace, M.E., Bell, A.A. and Beckman, C.H. (eds) (1981) *Fungal Wilt Diseases of Plants.* Academic Press, New York, USA, 608 pp.

Parker, C.A., Rovira, A.D., Moore, K.J., Wong, P.T.W. and Kollmorgan, J.F. (1985) *Ecology and Management of Soilborne Plant Pathogens.* American Phytopathological Society, St Paul, Minnesota, USA, 385 pp.

Schippers, B. and Gams, W. (eds) (1979) *Soil-borne Plant Pathogens.* Academic Press, London, UK, 686 pp.

Epidemiology and Ecology

Blakeman, J.P. and Williamson, B. (1994) *Ecology of Plant Pathogens.* CAB International, Wallingford, UK, 362 pp.

Bunting, A.H., Trask, A.B., Cooper, J.I., Solomon, M.G., McNamara, D.G., Flegg, J.J.M., Summers, D.D.B. and Thresh, J.M. (1981) *Pests, Pathogens and Vegetation. The Role of Weeds and Wild Plants in the Ecology of Crop Pests and Diseases.* Pitman Books Ltd, London, UK, 517 pp.

Campbell, C.L. and Madden, L.V. (1990) *Introduction to Plant Disease Epidemiology.* John Wiley & Sons, New York, USA, 532 pp.

Gareth Jones, D. (ed.) (1998) *The Epidemiology of Plant Diseases.* Kluwer, Dordrecht, The Netherlands, 460 pp.

Harris, K.F. and Maramorosch, K. (eds) (1980) *Vectors of Plant Pathogens.* Academic Press, New York, USA, 480 pp.

Lynch, J.M. (1990) *The Rhizosphere.* John Wiley & Sons, Chichester, UK, 485 pp.

Maramorosch, K. and Harris, K.F. (eds) (1981) *Plant Diseases and Vectors. Ecology and Epidemiology.* Academic Press, New York, USA, 360 pp.

Palti, J. and Kranz, J. (eds) (1980) *Comparative Epidemiology. A Tool for Better Disease Management.* Centre for Agricultural Publishing and Documentation, Wageningen, The Netherlands, 122 pp.

Scott, P.R. and Bainbridge, A. (eds) (1978) *Plant Disease Epidemiology.* Blackwell, Oxford, UK, 329 pp.

Zadoks, J.C. and Schein, R.D. (1979) *Epidemiology and Plant Disease Management.* Oxford University Press, New York, USA, 427 pp.

Bibliographies and Abstract Journals

Many of these are now available online through the internet. The date given is the date of first publication.

Agrindex (1975) FAO, Rome, Italy.

Bibliographie der Pflanzenschutz-literatur (1914) Paul Parey, Berlin, Germany.

Bibliography of Agriculture (1942) CCM Information Corporation, Washington, DC, USA.

Bibliography of Systematic Mycology (1947) CAB International, Wallingford, UK (formerly CMI, Kew).

Biological Abstracts (BIOSIS) (1926) Biosciences Information Service, Philadelphia, USA.

Bulletin Signalétique, Section 380 (1956) Centre National de la Recherche Scientifique, Paris, France.

CABPESTCD (abstracted literature on pests from 1973 onwards; available as a CD-ROM product or as CAB ABSTRACTS available online), CAB International, Wallingford, UK.

Microbiology Abstracts, Section C (1972) Cambridge Scientific Abstracts, Bethesda, Maryland.

Review of Plant Pathology (1922). CAB International, Wallingford, UK (formerly CMI, Kew).

Virology Abstracts (1967) Cambridge Scientific Abstracts, Bethesda, Maryland.

Periodicals and Serial Publications

The list of over 1000 serial publications regularly scanned in producing the *Review of Plant Pathology* and other CAB *International* publications gives an indication of the spread of journals in which articles on plant pathology are to be found. The list is updated annually, and is normally published in the annual index of the *Review*. A high proportion of the papers in the following journals are relevant to plant pathology.

AAB Descriptions of Plant Viruses (1970) Association of Applied Biologists, Wellesbourne (formerly from CMI, Kew).

Acta Phytopathologica Academiae Scientiarum Hungaricae (1966) Akademiai Kiado, Budapest.

African Journal of Plant Protection (1976) Nairobi.

Annales de l'Institut Phytopathologique Benaki (1935) Greece.

Annales de Phytopathologie (1969) Institut National de la Recherche Agronomique, Paris.

Annals of Applied Biology (1914) Cambridge University Press, London.

Annals of the Phytopathological Society of Japan (1918) Tokyo.

Annual Review of Phytopathology (1963) Annual Reviews Inc., Palo Alto, California.

APP (Australasian Plant Pathology) (1972) Australasian Plant Pathology Society, Sydney.

Archiv für Phytopathologie und Pflanzenschutz (1973) Akademie der Landwirtschaftswissenschaften der Deutschen Demokratischen Republik, Berlin.

Canadian Journal of Plant Pathology (1979) Canadian Phytopathological Society, Ottawa.

Canadian Plant Disease Survey (1920) Research Branch, Department of Agriculture, Ottawa.

Cereal Rusts Bulletin (1973) European and Mediterranean Cereal Rusts Foundation, Wageningen.

Crop Protection (1982), Butterworths, Sevenoaks, UK.

Descriptions of Pathogenic Fungi and Bacteria (1964) CABI Bioscience, Egham.

Distribution Maps of Plant Diseases (1942) CAB International, Wallingford.

Egyptian Journal of Phytopathology (1972) Egyptian Phytopathological Society, Cairo.

European Journal of Forest Pathology (1971) Paul Parey, Hamburg.

European Journal of Plant Pathology (1994) (formerly *Netherlands Journal of Plant Pathology*) European Foundation for Plant Pathology, Kluwer, Dordrecht.

FAO Plant Protection Bulletin (1952) FAO, Rome.

Fitopatologia (1966) Asociacion Latinoamericana de Fitopatologia, Lima, Peru.

Fitopatologia Brasileira (1976) Sociedade Brasileira de Fitopatologia, Brasilia.

Fitopatologia Colombiana (1977) Asociacion Colombiana de Fitopatologia y Ciencias Afines 'Ascolfi', Cali.

Indian Journal of Mycology and Plant Pathology (1971) Indian Society of Mycology and Plant Pathology, Udaipur.

Indian Phytopathology (1948) Indian Phytopathological Society, Delhi.

Informatore Fitopatologico (1951) Edagricole, Bologna.

International Journal of Pest Management (1961) (formerly *Tropical Pest Management* and originally *PANS*, Centre of Overseas Pest Research, London) Taylor & Francis, London.

Iranian Journal of Plant Pathology (1964) Plant Pests and Diseases Research Institute, Tehran.

Journal of Phytopathology (1991) (formerly *Phytopathologische Zeitschrift*) Paul Parey, Berlin and Hamburg.

Journal of Plant Pathology (1997) (formerly *Revista di Patologia Vegetale*) Italian Phytopathology Society.

Journal of Turkish Phytopathology (1972) Turkish Phytopathological Society, Bornova.

Kavaka (1973) Mycological Society of India, Madras.

Mikologiya i Fitopatologiya (1967) Nauka, Leningrad.

Molecular Plant Pathology (2000), British Society for Plant Pathology, Blackwells.

Mycologia (1909) New York Botanical Garden, Lancaster.

Netherlands Journal of Plant Pathology (1890–1994) Netherlands Society of Plant Pathology, Wageningen.

Notiziario sulle Malattie delle Piante (1949) Societa Italiana di Fitoiatria, Pavia.

Philippine Phytopathology (1965) Philippine Phytopathological Society, Laguna.

Physiological Plant Pathology (1971) Academic Press, London and New York.

Phytopathologia Mediterranea (1962) Unione Fitopatologica Mediterranea, Bologna.

Phytopathological Papers (1956) CABI Bioscience, Egham.

Phytopathologische Zeitschrift (1929–1991) Paul Parey, Berlin.

Phytopathology (1911) American Phytopathological Society, Ithaca, New York.

Plant Disease (formerly *Plant Disease Reporter*) (1917) American Phytopathological Society, St Paul, Minnesota.

Plantesygdomme i Danmark (1928) Statens Plantepatologiske Forsøg, Lyngby.

Plant Pathology (1952) British Society for Plant Pathology, London (formerly, MAFF Plant Pathology Laboratory, Harpenden).

Revue de Mycologie (1936) Laboratoire de Cryptogamie du Museum National d'Histoire Naturelle, Paris.

Rivista di Patologia Vegetale (1892–1997) Consiglio Nazionale delle Ricerche, Pavia.

Sugarcane Pathologists' Newsletter (1968) International Society of Sugarcane Technologists, Hawaii.

Transactions of the British Mycological Society (1896) Cambridge University Press, London.

Tropical Pest Management (formerly *PANS*) London, Centre for Overseas Pest Research.

Virology (1955) Academic Press, New York.

Zeitschrift für Pflanzenkrantheiten und Pflanzenschutz (1890) Eugen Ulmer, Stuttgart.

Bibliography of Crop and Plant Diseases

Compiled by J.M. Waller

CABI Bioscience UK Centre, Bakeham Lane, Egham, Surrey TW20 9TY, UK

Listed below is a selection of texts covering individual crops or commodity groups. These range from comprehensive texts, some of which, although old, are still invaluable today, to smaller field handbooks. The American Phytopathological Society's 'Compendium' series are useful laboratory guides. There are many small field guides to identification of diseases (and pests) produced by national and international institutions involved with crop protection and improvement and readers are advised to contact these institutions for information on the latest available.

Tropical Cereals

Maize

Shurtleff, M.C. (ed.) (1980) *Compendium of Corn Diseases*, 2nd edn. American Phytopathological Society, St Paul, Minnesota, USA, 105 pp.

Millets

Pall, B.S., Jan, A.C. and Singh, S.P. (1980) *Diseases of Lesser Millets*. Jawaharlal Nehru Krishi Vishwa Vidyalaya, Jabalpur, 69 pp.

Rice

Ou, S.H. (1985) *Rice Diseases*, 2nd edn. Commonwealth Agricultural Bureau, Farnham Royal, 380 pp.

Mueller, K.E. (1974) *Field Problems of Tropical Rice*. International Rice Research Institute, Los Banos, 95 pp.

Webster, R.K. and Gunnell, P.S. (eds) (1992) *Compendium of Rice Diseases*. APS Press, St Paul, Minnesota, USA, 62 pp.

Sorghum

Frederikson, R.A. (ed.) (1986) *Compendium of Sorghum Diseases*. APS Press, St Paul, Minnesota, USA, 82 pp.

Williams, R.J., Frederiksen, R.A. and Girard, J.C. (1978) *Sorghum and Pearl Millet Disease Identification Handbook*. Information Bulletin ICRISAT No. 2, 88 pp.

Williams, R.J., Frederiksen, R.A., Mughogho, L.K. and Bengston, G.D. (eds) (1980) *Proceedings of the International Workshop on Sorghum Diseases, Hyderabad, India, 11–15 December 1978*. International Crops Research Institute for the Semi-Arid Tropics, Hyderabad, 469 pp.

Temperate Cereals

Gareth Jones, D. and Clifford, B.C. (1978) *Cereal Diseases, Their Pathology and Control*. BASF, Ipswich, 279 pp.

Jenkyn, J.F. and Plumb, R.T. (eds) (1981) *Strategies for the Control of Cereal Diseases*. Blackwell, Oxford, UK, 219 pp.

Hornby, D., Bateman, G.I., Gutteridge, R.J., Lucas, P., Osbourn, A.E., Ward, E. and Yarham, D.J. (1998) *Take-all Disease of Cereals*. CAB International, Wallingford, UK, 384 pp.

Zillinsky, F.J. (1983) *Common Disease of Small Grain Cereals: a Guide to Identification*, CIMMYT, Mexico, 141 pp.

Barley

Mathre, D.E. (ed.) (1982) *Compendium of Barley Diseases*. American Phytopathological Society, St Paul, Minnesota, USA, 100 pp.

Wheat

Wiese, M.V. (1987) *Compendium of Wheat Diseases*, 2nd edn. American Phytopathological Society, St Paul, Minnesota, USA, 112 pp.

Root and Tuber crops

Anon. (1978) *Pest Control in Tropical Root Crops*. PANS Manual No. 4, 235 pp.

Cassava

Brekdbaum, T., Belloti, A. and Lozano, J.C. (eds) (1978) *Proceedings of the Cassava Protection Workshop, CIAT, Cali, Colombia, 7–12 November 1977*. Centro Internacional de Agricultura Tropical, Cali, 244 pp.

Potato

Brenchley, G.H. and Wilcox, H.J. (1979) *Potato Diseases*. HMSO, London, 106 pp.
Hooker, W.J. (ed.) (1981) *Compendium of Potato Diseases*. American Phytopathological Society, St Paul, Minnesota, USA, 125 pp.

Sweet potato

Arene, O.B. and Nwankiti, A.O. (1978) Sweet potato diseases in Nigeria. *PANS* 24, 294–305.
Clark, C.A. and Moyer, J.W. (eds) (1988) *Compendium of Sweet Potato Diseases*. APS Press, St Paul, Minnesota, USA, 74 pp.

Taro

Jackson, G.V.H. (1980) *Disease and Pests of Taro*. South Pacific Commission, Noumea, 52 pp.

Legumes

Allen, D.J. and Lenné, J.M. (eds) (1997) *The Pathology of Food and Pasture Legumes*. CAB International, Wallingford, UK, 768 pp.
Howard, R.J. (Compiler) (1981) *Diseases of Pulse Crops in Western Canada*. Alberta Agriculture, Edmonton, Canada, 98 pp.
Ward, A., Mercer, S.L. and Howe, V. (eds) (1981) *Pest Control in Tropical Grain Legumes*. Centre for Overseas Pest Research, London, UK, 206 pp.

Bean

Hall, R. (1991) *Compendium of Bean Diseases*. American Phytopathological Society, St Paul, Minnesota, USA, 73 pp.
Schwartz, H.F. and Galvez, G.E. (eds) (1980) *Bean Production Problems: Disease, Insect, Soil and Climatic Restraints of* Phaseolus vulgaris. Centro Internacional de Agricultura Tropical, Cali, 424 pp.

Chickpea

Nene, Y.L., Reddy, M.V., Haware, M.P., Ghanekar, A.M. and Amin, K.S. (1991) *Field Diagnosis of Chickpea Diseases and Their Control*. ICRISAT-Information-Bulletin, No. 28, International Crops Research Institute for the Semi-Arid Tropics, Patancheru, 52 pp.

Lentils

Beniwal, S.P.S., Baya', a-B., Weigand, S., Makkouk, K. and Saxena, M.C. (1993) *Field Guide to Lentil Diseases and Insect Pests*. International Center for Agricultural Research in the Dry Areas (ICARDA), Aleppo, 107 pp.
Khare, M.N. (1981) Diseases of lentils. In: Webb, C. and Hawtin, G. (eds) *Lentils*. Commonwealth Agricultural Bureau, Farnham Royal, UK, pp. 163–172.

Pea

Hagedorn, D.J. (ed.) (1984) *Compendium of Pea Diseases*. PS Press, St Paul, Minnesota, USA, 57 pp.

Pigeonpea

Reddy, M.V., Raju, T.N., Sharma, S.B., Nene, Y.L. and McDonald, D. (1993) *Handbook of Pigeonpea Diseases*. ICRISAT-Information-Bulletin, 1993, No. 42, International Crops Research Institute for the Semi-Arid Tropics (ICRISAT), Patancheru, 61 pp.

Soybean

Sinclair, J.B. and Backman, P.A. (eds) (1989) *Compendium of Soybean Diseases*. American Phytopathological Society, St Paul, Minnesota, USA, 106 pp.

Vegetables

Dixon, G.R. (1981) *Vegetable Crop Diseases*. Macmillan, London, UK, 404 pp.
Persley, D. (1994) *Diseases of Vegetable Crops*. Information Series Q193024, Queensland Department of Primary Industries, Brisbane, Australia, 100 pp.
Persley, D.M., O'Brien, R. and Syme, J.R. (eds) (1989) *Vegetable Crops. A Disease Management Guide*. Queensland Department of Primary Industries, Brisbane, Australia, 75 pp.
Sherf, A.E. and Macnab, A.A. (1986) *Vegetable Diseases and Their Control*, 2nd edn. John Wiley & Sons, New York, USA, 728 pp.

Beet

Whitney, E.D. and Duffus, J.E. (1986) *Compendium of Beet Disease and Insects*. APS Press, St Paul, Minnesota, USA, 76 pp.

Cucurbits

Blancard, D., Lecoq, H. and Pitrat, M. (1994) *A Colour Atlas of Cucurbit Diseases, Observation Identification and Control*. Wolfe, London, UK, 209 pp.
Zitter, T.A., Hopkins, D.L. and Thomas, C.E. (1996) *Compendium of Cucurbit Diseases*. APS Press, St Paul, Minnesota, USA, 87 pp.

Lettuce

Michael-Davis, R., Subbarao, K.V., Raid, R.N. and Kurtz, E.A. (eds) (1997) *Compendium of Lettuce Diseases*. APS Press, St Paul, Minnesota, USA, 79 pp.

Onion and garlic

Anon. (1986) *Pest Control in Tropical Onions*. PANS Manual, Tropical Development Research Centre, London, UK, 109 pp.
Schwartz, H.F. and Mohan, S.K. (1995) *Compendium of Onion and Garlic Diseases*. APS Press, St Paul, Minnesota, USA, 54 pp.

Tomato

Anon. (1983) *Pest Control in Tropical Tomatoes*. PANS Manual, Centre for Overseas Pest Research, London, UK, 130 pp.
Jones, J.B., Jones, J.P., Stall, R.E. and Zitter, T.A. (eds) (1991) *Compendium of Tomato Diseases*. APS Press, St Paul, Minnesota, USA, 73 pp.

Mushrooms

Fletcher, J.T. and Atkinson, K. (1976) *Mushrooms. A Guide to the Recognition and Control of Diseases, Weed Moulds, Competitors and Pests*. Agricultural Development and Advisory Service, Leeds, UK, 58 pp.

Tropical Fruits and Nuts

Cook, A.A. (1975) *Diseases of Tropical and Subtropical Fruits and Nuts*. Hafner Press, New York, USA, 317 pp.

Persley, D.M., Pegg, K.G. and Syme, J.R. (eds) (1989) *Fruit and Nut Crop*
  ment Guide. Queensland Department of Primary Industries Inform
  Department of Primary Industries, Queensland Government, Brisbane.
Ploetz, R.C., Zentmyer, G.A., Nishijima, W.T., Rohrbach, K.G. and Ohr,
  Compendium of Tropical Fruit Diseases. APS Press, St Paul, Minnesota, USA, 8υ,
Ploetz, R.C. (ed.) (2000) *Diseases of Tropical and Subtropical Fruit Crops*. CAB Internatio..
  Wallingford, UK, 512 pp.

Avocado

Broadley, R.H. (ed.) (1991) *Avocado Pests and Diseases*. Queensland Department of Primary
  Industries, Brisbane, Australia, 74 pp.

Banana

Feakin, S.D. (ed.) (1977) *Pest Control in Bananas*, 3rd edn. PANS Manual No. 1, 126 pp.
Jones, D.R. (ed.) (1999) *Diseases of Banana, Abaca and Enset*. CAB International, Wallingford,
  UK, 544 pp.

Cashew

Ohler, J.G. (1979) *Cashew*. Koninklijk Instituut voor de Tropen, Amsterdam, 260 pp.

Citrus

Fawcett, H.S. (1936) *Citrus Diseases and Their Control*, 2nd edn. McGraw-Hill, New York,
  656 pp.
Knorr, L.C. (1974) *Citrus Diseases and Disorders*. University of Florida, Gainesville, USA,
  163 pp.
Whiteside, J.O., Garnsey, S.M. and Timmer, L.W. (1988) *Compendium of Citrus Diseases*.
  APS Press, St Paul, Minnesota, USA.

Macadamia

Fitzell, R.D. (1994) *Disease and Disorders of Macadamias*. Queensland Department of Primary
  Industries, Brisbane, Australia, 31 pp.

Pineapple

Broadley, R.H., Wassman, R.C. III and Sinclair, E. (1993) *Pineapple Pests and Disorders*.
  Queensland Department of Primary Industries, Brisbane, Australia, 63 pp.

Temperate Fruits and Nuts

Ogawa, J.M., Zehr, E.I., Bird, G.W., Ritchie, D.F., Uriu, K. and Uyemoto, J.K. (1995) *Compendium of Stone Fruit Diseases*. APS Press, St Paul, Minnesota, USA, 98 pp.

Apple and pear

Jones, A.L. and Aldwinckle, H.S. (eds) (1990) *Compendium of Apple and Pear Diseases*. American Phytopathological Society, St Paul, Minnesota, USA, 100 pp.

Blue- and cranberries

Caruso, F.L. and Ramsdell, D.C. (eds) (1995) *Compendium of Blueberry and Cranberry Diseases*. APS Press, St Paul, Minnesota, USA, 87 pp.

Grape

Pearson, R.C. and Goheen, A.C. (eds) (1988) *Compendium of Grape Diseases*. American Phytopathological Society, St Paul, Minnesota, USA, 93 pp.

Raspberry

Ellis, M.A., Converse, R.H. and Williams, R.N. (eds) (1991) *Compendium of Raspberry and Blackberry Disease and Pests*. APS Press, St Paul, Minnesota, USA.

Oil and Fibre Crops

Raychaudhuri, S.P. and Verma, J.P. (eds) (1989) *Review of Tropical Plant Pathology*. Vol. 5. *Diseases of Fibre and Oilseed Crops*. Today and Tomorrow's Printers and Publishers, New Delhi, India, 316 pp.

Coconut

Child, R. (1974) *Coconuts*, 2nd edn. Longman, London, UK, 335 pp. (Diseases are dealt with in Chapter 13, pp. 213–236.)
Oropeza, C., Howard, F.W. and Ashburner, G.R. (1995) *Lethal Yellowing Research and Practical Aspects*. Kluwer Academic, Dordrecht, The Netherlands, 251 pp.

Cotton

Watkins, G.M. (ed.) (1981) *Compendium of Cotton Diseases*. American Phytopathological Society, St Paul, Minnesota, USA, 87 pp.

Hillocks, R.J. (ed.) (1992) *Cotton Diseases*. CAB International, Wallingford, UK, 415 pp.

Flax

Kolte, S.J. and Fitt, D.L. (1998) *Disease of Linseed and Fibre Flax*. Shipra Publishing, Delhi, India, 247 pp.

Groundnut

Feakin, S.D. (ed.) (1973) *Pest Control in Groundnuts*, 3rd edn. PANS Manual No. 2, 197 pp.

Kokalis-Burelle, N., Poter, D.M., Rodriquez-Cabana, R., Smith, D.H. and Subrahmanyam, P. (eds) (1997) *Compendium of Peanut Diseases*. APS Press, St Paul, Minnesota, USA, 94 pp.

Hemp

McPartland, J.M., Clarke, R.C. and Watson, D.P. (2000) *Hemp Disease and Pests: Management with an Emphasis on Biological Control*. CAB International, Wallingford, UK, 208 pp.

Oil palm

Aderungboye, F.O. (1977) *Diseases of the Oil Palm*. PANS 23, 303–305.

Turner, P.D. (1981) *Oil Palm Disease and Disorders*. Oxford University Press, Kuala Lumpur, Malasia, 280 pp.

Ramie

Sarma, B.K. (1981) Diseases of ramie (*Boehmeria nivea*). *Tropical Pest Management* 27, 370–374.

Sisal

Lock, G.W. (1969) *Sisal*, 2nd edn. Longmans, London, UK, 365 pp. (Deficiency disorders are dealt with in Chapter 8, pp. 147–171 and pests and diseases in Chapter 11, pp. 223–245.)

Beverage Crops

Cacao

Thorold, C.A. (1975) *Diseases of Cocoa*. Clarendon Press, Oxford, UK, 423 pp.

Coffee

Coffee Research Services, Kenya (1961) *An Atlas of Coffee Pests and Diseases*. Coffee Board, Nairobi, 146 pp.

Kushalappa, A.C. and Eskes, A.B. (1989) *Coffee Rust*. CRC Press, Boca Raton, Florida, 345 pp.

Wrigley, G. (1988) *Coffee*. Longman, London, UK, 639 pp. (Diseases are dealt with in Chapter 8, pp. 309–349.)

Tea

Eden, T. (1976) *Tea*, 3rd edn. Longman, London, UK, 236 pp. (Diseases are dealt with in Chapter 10, pp. 113–135.)

Industrial Crops

Rubber

Chee, K.H. (1976) *Micro-organisms Associated with Rubber (*Hevea brasiliensis *Mull. Arg.)*. Rubber Research Institute, Kuala Lumpur, Malaysia, 78 pp.

Rao, B.S. (1975) *Maladies of* Hevea *in Malaysia*. Rubber Research Institute, Kuala Lumpur, Malaysia, 108 pp.

Wastie, R.L. (1975) Diseases of rubber and their control. *PANS* 21, 268–288.

Sugarcane

Hughes, C.G. (1978) Diseases of sugarcane – a review. *PANS* 24, 143–159.

Rao, G.P., Gillaspie, A.G. Jr, Upadhyaya, P.P., Bergamin-Filho, A., Agnihotri, V.P. and Chen, C.T. (1994) *Current Trends in Sugarcane Pathology*. International Books & Periodicals Supply Service, Delhi, India, 449 pp.

Rao, G.P., Bergamin-Filho, A., Magarey, R.C. and Autrey, L.J.C. (1999) *Sugarcane Pathology*, Vol. 1, *Fungal Diseases*. Science Publishers, Enfield, USA, 308 pp.

Rott, P., Bailey, R.A., Comstock, J.C., Croft, B.J. and Saumtally, A.S. (2000) *A Guide to Sugarcane Diseases*. CIRAD/ISSCT, Montpellier, France, 339 pp.

Sivanesan, A. and Waller, J.M. (1986) *Sugarcane Diseases*. Phytopathological paper no. 29, CAB International, Wallingford, UK, 88 pp.

Trees

Boa, E. and Lenné, J.M. (1994) *Diseases of Nitrogen Fixing Trees in Developing Countries*. Natural Resources Institute, Chatham, UK, 82 pp.

Browne, F.G. (1968) *Pests and Diseases of Forest Plantation Trees*. Clarendon Press, Oxford, UK, 1330 pp.

Gibson, L.A.S. (1975, 1979) *Diseases of Forest Trees Widely Planted as Exotics in the Tropics and Southern Hemisphere*, Part I. *Important Members of the Myrtaceae, Leguminoseae, Verbenaceae and Meliaceae*, 51 pp. Part 2. *The genus Pinus*, 135 pp. Commonwealth Mycological Institute, Kew and Commonwealth Forestry Institute, Oxford, UK.

Innes, J. (1995) *Forest Health: Its Assessment and Status*. CAB International, Wallingford, UK, 656 pp.

Sincalir, W.A., Lyon, H.H. and Johnson, W.T. (1989) *Disease of Trees and Shrubs*. Cornell University Press, New York, USA, 575 pp.

Tainter, F.H. and Baker, F.A. (1996) *Principles of Forest Pathology*. John Wiley & Sons, New York, USA, 805 pp.

Tattar, T.A. (1989) *Diseases of Shade Trees*. Academic Press, San Diego, USA, 391 pp.

Conifers

Hansen, E. and Lewis, K.J. (1997) *Compendium of Conifer Diseases*. APS Press, St Paul, Minnesota, USA, 101 pp.

Ivory, M.H. (1987) *Diseases and Disorders of Pines in the Tropics. A Field and Laboratory Manual*. Oxford Forestry Institute Research Publication no. 31.

Elm

Stipes, R.J. and Campana, R.J. (eds) (1981) *Compendium of Elm Diseases*. American Phytopathological Society, St Paul, Minnesota, USA, 96 pp.

Pastures

Lenné, J.M. and Trutman, P. (1994) *Diseases of Tropical Pasture Plants*. CAB International, Wallingford, UK, 404 pp.

O'Rourke, C.J. (1976) *Diseases of Grasses and Forage Legumes in Ireland*. An Foras Taluntais, Dublin, UK, 115 pp.

Lucerne

Stuteville, D.L. and Erwin, D.C. (1990) *Compendium of Alfalfa Diseases*, 2nd edn. American Phytopathological Society, St Paul, Minnesota, USA, 84 pp.

Ornamentals and Garden Plants

Buczacki, S.T. and Harris, K.M. (1981) *Collins Guide to Pests, Diseases and Disorders of Garden Plants*. Collins, London, UK, 512 pp.

Chase, A.R. (ed.) (1987) *Compendium of Ornamental Foliage Plants*. APS Press, St Paul, Minnesota, USA.

Chase, A.R. and Broschat, T.K. (eds) (1991) *Disease and Disorders of Ornamental Palms*. APS Press, St Paul, Minnesota, USA, 56 pp.

Pirone, P.P. (1978) *Diseases and Pests of Ornamental Plants*, 5th edn. John Wiley & Sons, Chichester, UK, 566 pp

Bulbs

Moore, W.C. (1979) *Diseases of Bulbs*, 2nd edn. HMSO, London, 205 pp.

Chrysanthemum

Horst, R.K. and Nelson, P.E. (1997) *Compendium of Chrysanthemum Diseases*. APS Press, St Paul, Minnesota, USA, 62 pp.

Rose

Horst, R.K. (ed.) (1983) *Compendium of Rose Diseases*. APS Press, St Paul, Minnesota, USA, 49 pp.

Turf

Couch, H.B. (2000) *Diseases of Turfgrasses*, 3rd edn. Kreiger, Malabar, 421 pp.

Smiley, R.W., Dernoeden, P.H. and Clarke, B.B. (1992) *Compendium of Turfgrass Diseases*, 2nd edn. APS Press, St Paul, Minnesota, USA, 98 pp.

Herbs, Spices and Other Plants

Coca

Lentz, P.L., Lipscomb, B.R. and Farr, D.F. (1975) Fungi and diseases of *Erythroxylum*. *Phytologia* 30, 350–368.

Pepper, black

Kueh Tiong Kheng (1979) *Pests, Diseases and Disorders of Black Pepper in Sarawak*. Department of Agriculture, Sarawak, 68 pp.

Poppy

Schmitt, C.G. and Lipscomb, B. (1975) Pathogens of selected members of the Papaveraceae – an annotated bibliography. *United States Department of Agriculture Agricultural Research Service* No. ARS-NE-62, 186 pp.

Tobacco

Gayed, S.K. (1978) Tobacco diseases. *Canada Department of Agriculture Publications*, No. 1641, 56 pp.
Shew, H.D. and Lucas, G.B. (1991) *Compendium of Tobacco Diseases*. American Phytopathological Society, St Paul, Minnesota, USA, 68 pp.

Collection and Dispatch of Plant Material

J.M. Waller and B.J. Ritchie

CABI Bioscience UK Centre, Bakeham Lane, Egham, Surrey TW20 9TY, UK

Samples of diseased plant material often need to be collected from the field for subsequent laboratory examination and may require dispatch to laboratories in other regions or countries for specialist diagnosis or other forms of investigation. Careful collection and transport of plant material is essential. In the field, brown paper bags are the best containers for most purposes and can easily be written on using a pencil or waterproof pen to avoid smudging. Polythene bags are only suitable for short-term storage of plant material as they encourage condensation which promotes growth of surface saprophytes especially in the tropics. Specimen tubes, e.g. plastic 'universal bottles' are ideal for smaller specimens or those which are very delicate (most seeds, insects, small sporophores, etc.). Bagged specimens need to be placed in a stable rigid container such as a cardboard box to avoid damage during transport; in very hot conditions a cool box is ideal. Care of specimens is often neglected and many are ruined as they are bounced about in the back of a hot dusty vehicle!

The best samples to collect are those in the early to middle stages of disease. Here the pathogen should be active, making examination and any subsequent isolations easier. Grossly diseased samples are normally unusable as the pathogen will probably no longer be viable and saprophytic organisms will have colonized necrotic tissues. Choice of material is vital and a basic knowledge of symptoms and how they are caused is needed in order to ensure that the part of the plant collected is likely to be where the pathogen is located. This may not always be where the symptom occurs, e.g. with wilt diseases symptoms occur on leaves and young shoots but the pathogen is usually located in the roots or mature stems. Fungi for identification should ideally be in good condition and sporulating well on the material. Often a few days' wait after observing developing fructifications will allow the production of spores, which will ensure that the specimen is identifiable. Many fungi will not

©CAB *International* 2002. *Plant Pathologist's Pocketbook*
(eds J.M. Waller, J.M. Lenné and S.J. Waller)

complete their development once the leaf or stem is removed from the host. For identification of *Peronosporales*, portions of stems as well as leaves should be collected; the stems often contain oospores not always present with the conidiophores on leaves.

Leaves should not be collected when wet. If this is unavoidable, the leaves must be blotted to remove excess moisture and then placed between layers of newspaper or other absorbent paper (avoid using thin absorbent paper such as facial tissues as these disintegrate and are difficult to remove from the sample). The leaves need to be spread out so that they are flat and do not overlap. Use of a standard plant press is recommended to prepare dried herbarium specimens of leaves and shoots. The paper may need changing daily. Dried specimens should be packed in envelopes (ungummed) about 15 × 10 cm, made from stout good-quality paper. Larger portions of stems or roots may be partially dried and each portion wrapped separately in newspaper.

Samples with suspected bacterial infection are more difficult to collect for dispatch to a distant laboratory as tissue disintegration is often rapid and they need to arrive in as fresh a condition as possible. In some plant hosts, e.g. *Gramineae*, bacteria die within 12–24 h and cannot be isolated. However, in other plant hosts, e.g. cotton, the bacteria can remain viable for some time.

Material with suspected virus infection needs to be temporarily preserved for dispatch and this can be done using the following method.

Soak filter papers in 50% glycerol so that they are thoroughly wet but without free liquid. Sandwich leaves between the papers and place in a plastic bag. Other kinds of paper, such as facial tissue or paper towels, can be used but they disintegrate more easily. When the glycerol has been rinsed from the specimens they can be prepared for electron microscopy in the normal way (leaf dips) to observe the presence of any virus particles. The virus should also have retained infectivity. Alternatively, specimens may be preserved in 50% glycerol in bottles but these are more expensive to send by post.

Roots should have soil shaken off and be wrapped separately in newspaper. However, when nematodes are suspected, roots with adhering soil should be collected. Stem portions should be cut so that the sample shows clearly healthy and diseased areas of tissue; again these should be dried and carefully wrapped in newspaper. Stems should be wrapped separately, as they are easily damaged if bundled together.

Microscope slides sent with the specimens are useful in drawing attention to the specific fungus of interest in a mixed collection but slides alone are inadequate for accurate identification.

Collection details are also necessary. The identity of the host is often necessary for the correct identification of many obligate pathogens. Geographical area details enable scientists to chart disease progress, etc. Relevant collection details include:

1. Serial number (for your own identification purposes);
2. Host (preferably the scientific name);
3. Locality (country, state, map references, if known);
4. Collection date (necessary if isolations are to be attempted);
5. Collector's name and address;
6. Disease symptoms;

7. Disease severity and distribution, e.g. number and pattern of plants affected.

Samples need to be packed carefully to avoid damage by crushing and exposure to extremes of temperature; saprophytic organisms present on plant surfaces will utilize any excess moisture to germinate and colonize the material. Layers of newspaper and cardboard, or polystyrene granules, will help cushion the samples. Herbaceous plant material should never be wrapped in plastic; this causes the sample to sweat and saprophytic organisms rapidly to colonize and decompose the plant tissues. Packages should be able to withstand the rigours of the postal services; strong cardboard boxes are suitable and the samples should be packed in such a way that they are cushioned from outside pressures. Substrata with delicate fructifications should be stuck down inside a suitable container before dispatch. A representative part of all collections sent for identification should be retained, as specimens cannot normally be returned to the sender.

Samples should be sent to their destination as quickly as possible, using the most rapid method available to avoid progressive deterioration. Except when essential, international airfreight should be avoided as this can cause delay and complicates customs clearance and delivery.

Packages should be clearly marked with their destination and the following:

Perishable biological material – keep in shade
Keep material cool but do not refrigerate

To safeguard plant health and minimize quarantine risks, national and international regulations governing carriage of biological material need to be observed. National post offices can provide information on local regulations including the requirement for permits and where these may be obtained. When sending material to another country, national import permits or other documentation may be needed to comply with local plant health regulations.

It is a good idea to take photographs (colour slides) of samples when collecting from the field. A good pictorial collection of symptoms can be built up fairly quickly. If equipment is available, it is also worthwhile photographing fungal structures when seen either through a dissecting or a compound microscope (see Chapter 39).

Fungi as Plant Pathogens

J.M. Waller and P.F. Cannon

CABI Bioscience UK Centre, Bakeham Lane, Egham, Surrey TW20 9TY, UK

Introduction

References to plant diseases can be traced back to the most ancient written records of crop cultivation and the association of fungi with certain diseases was noticed early in the history of mycology (see Chapter 1). It was not until the mid-19th century, however, that the significance of some fungi as causal agents of infection began to be appreciated. Since then progress in investigation has been rapid, with the accumulation of a vast amount of literature, and the major importance of fungus diseases is now well established.

Plant Pathogenic Fungi

Fungal pathogens of plants are many and diverse. They occur in most taxonomic groups. Only a few, such as rusts (*Uredinales*) and powdery and downy mildews (*Erysiphaceae* and *Peronosporaceae*), are obligate parasites (biotrophs), obtaining their nutrition directly from living plant tissue. Most plant pathogens are necrotrophs, killing plant tissues for their nutrition. These often grow readily in culture and may survive saprobically in soil or dead plant material. The active pathogen may produce spores to disseminate the disease. While the fungus is associated with the living plant, these are commonly conidia, produced mitotically and frequently in large numbers. Meiospores (sexually produced spores, i.e. ascospores or basidiospores) are more usually produced from plant tissues after their death, when the fungus has altered from necrotrophy to saprobic nutrition. Traditionally, the conidium-bearing morphs (anamorphs) of fungi were referred to as the 'imperfect' stages, whereas the meiospore-forming stages (teleomorphs) were considered as the 'perfect' stage. For information on the classification, nomenclature and identification of fungi see p. 80 *et seq.*

Symptoms

A preliminary diagnosis of a fungus disease is sometimes possible from the symptoms, though non-related pathogens and abiotic environmental effects may produce similar types of damage. Infection of the root and collar by soil-inhabiting fungi may lead to damping off of seedlings, sometimes pre-emergent, or wilt or sudden death of mature plants. Stem necrosis may take the form of dieback, cankers or limited anthracnose lesions. On leaves a shot-hole effect is produced when lesions become necrotic and the dead parts fall away. Extensive leaf necrosis is often referred to as blight or blast. The term mildew is used when the pathogen produces a visible growth over the host surface.

Some infections cause growth modifications, either hyperplasia, i.e. over-development (galls, witches' brooms, scab lesions), or hypoplasia (dwarfing, rosetting). Transformations of host tissue are characteristic of smut diseases and ergot. Symptoms may appear on parts of a plant remote from that actually attacked, as in silver leaf (*Chondrostereum purpureum*) of fruit trees and certain wilts, when toxic substances are secreted by the pathogen and transported in the host. When roots or stem bases are diseased, symptoms may be most obvious on the aerial part of the plant as wilting, dieback or shrivelling.

Considerable variability in type or severity of symptoms or in parts attacked is commonly observed, for instance in *Phytophthora* diseases of rubber and cacao, and it must be decided whether this is due to local environment, host variety, or different strains of the pathogen.

Reproduction of the original symptoms upon inoculation of a healthy plant is one of the requirements for proving the pathogenicity of any fungus suspected of causing a disease.

Host–Parasite Relationships

Initial infection depends on successful penetration of the host surface. Germination and growth of fungus spores on aerial plant surfaces depends on moisture and temperature, requirements that vary for different pathogens, and substances diffusing from the host may have a stimulatory or inhibitory effect. The cuticle and epidermis form an efficient barrier to penetration. Some pathogens penetrate directly. Others, notably the germinating uredospores of many rusts, attack the plant through the stomata, whereas wound parasites can invade only damaged tissue.

Many facultative unspecialized pathogens kill the host tissues by enzyme action in advance of the growing hyphae, which are usually intracellular and without haustoria. Obligate parasites are generally relatively specialized in their host range and cause little cell destruction in the early stages, though death of the plant may ensue later when the fungus has produced resistant spores. The mycelium is intercellular with haustoria, or completely intracellular. Invasion by an obligate parasite may be checked by hypersensitivity of the host, death of invaded cells preventing further hyphal growth. Other reactions limiting spread are the development of cork barriers or gum deposits. Resistance may also be expressed by the formation in the host of antagonistic substances, or by inhibition of the pathogen's enzyme system.

The genetic and biochemical aspects of host–parasite interactions have received much attention particularly as they relate to disease resistance. The gene-for-gene relationship (Flor, 1971) between host resistance and pathogen virulence formed the basis of earlier work but there is now a greater understanding of the genetics of pathogenicity and host resistance, particularly in relation to resistance breeding. The biochemical mechanisms involved in host recognition of pathogens and the initiation of defence reactions has also been widely studied (see Chapter 31).

Development and Spread of Plant Diseases

Identification of the source of a fungal pathogen is a key step in devising efficient disease control as this may permit elimination or reduction of the source before disease epidemics can begin. Seed or other planting material, alternative host plants (wild plants or 'volunteers'), crop debris and soil are the most readily treated sources, but pathogens can arrive from more distant diseased crops.

Transmission of fungal spores or other propagules within and between plant populations is an important process in the development of plant diseases, and knowledge of the means of transmission of a pathogen is needed for the selection of appropriate control measures. Wind transport of fungal spores can occur over long distances and at high altitudes; this is responsible for the widespread dissemination of rust pathogens and development of many other plant disease epidemics (Gregory, 1973). Many fungal spores are dispersed in water – either in rain splash or irrigation water – and may give rise to locally intense disease outbreaks. Fungi commonly produce dormant resting spores or other resistant structures that provide a means of interseasonal survival. These may be transmitted on seed, crop debris or through soil or water movement. Insects may also carry fungus spores and this means of transmission is important with several wilt diseases of trees (Dutch elm disease, *Ophiostoma ulmi* and *Ophiostoma novo-ulmi*, and oak wilt, *Ceratocystis fagacearum*). Other major vectors of fungal pathogens are humans and machinery.

See Chapter 25 for further information on plant disease epidemiology.

Control

Choice of the most effective means of control of a fungus disease depends on a detailed knowledge of the particular crop and pathogen and a consideration of economic factors. A number of principles are involved.

Exclusion of a disease from areas not already affected may be possible by quarantine, port inspections, restriction of movement of plants and plant parts, certification and other legislatory measures. Fungi spread on seed and other propagating material are particularly involved.

Eradication of a disease after it has become established involves destruction of alternate and alternative hosts, roguing, excision of infected plant parts, other sanitary measures such as, rotation, sterilization, or disinfection to eliminate soil infestation, and fungicidal treatment of affected growing plants or plant parts.

Protective measures include spraying or dusting plants exposed to disease outbreaks, which may be predictable by forecasting, seed treatment against soil-borne pathogens, treatment of stored products and the use of timber preservatives. Cultural practices such as timing of planting and harvest to reduce the chances of infection and modification of soil conditions to discourage the pathogen and favour antagonists may be effective.

Biological control of plant diseases is becoming better established especially against soil-borne pathogens where preparations of antagonistic fungi such as *Trichoderma* species are available commercially in some countries.

The best means of defence against many diseases, and the only one against some, is by planting resistant cultivars. It is by this means that control of the major diseases of the world staple food crops are achieved.

For more information on disease control see Chapters 27–34.

Classification of Fungi

Although no practising plant pathologist can hope to become an expert on all groups of fungi, some basic understanding of their systematic arrangement is necessary. Various schemes have been proposed which have developed over time as new information becomes available. Cladistic and molecular approaches are currently revolutionizing our understanding of fungal phylogeny and species concepts, and it is clear that much more can be learnt using these approaches. This means that classification schemes are likely to continue to evolve for the forseeable future. This is unfortunate for applied mycologists such as practising plant pathologists, but fossilization of taxonomic systems would be as deleterious in the long term as banning further research in plant pathology.

One of the most significant changes in systematic arrangement in recent years has resulted in the redefinition of the *Fungi* themselves. Traditionally fungi were considered as 'lower plants', largely because they did not generally move and could therefore not be treated as animals. Now, five or more major categories (kingdoms) of organisms are accepted by systematic biologists, with the *Fungi* given their own kingdom. Indeed, evidence from many quarters suggests that the *Fungi* are more closely related to animals than plants.

Many of the organisms previously grouped for convenience under the general heading of 'Fungi' are now known to belong to the kingdoms *Protozoa* and *Chromista*, including important plant pathogens such as *Phytophthora*, *Pythium* and *Polymyxa*. Most of these have cellulose-containing cell walls. The true fungi include four major phyla with chitin as the primary structural component of cell walls, which are distinguished primarily by their separate sexual reproductive forms.

1. *Ascomycota* (meiotic spores contained in asci, mitotic spores exogenous).
2. *Basidiomycota* (meiotic spores borne exogenously on basidia, mitotic spores also exogenous).
3. *Chytridiomycota* (with a motile (flagellate zoospore) phase, but with chitinous cell walls distinguishing them from the *Chromista*).
4. *Zygomycota* (meiotic spores (zygospores) formed by the fusion of morphologically similar gametangia, and endogenous mitotic spores).

Many fungi are known only to produce mitotic spores. As these tend to be more prominent in disease-causing organisms, plant pathogens are particularly well represented, though in many cases the sexual spores are overlooked rather than actually absent. Traditionally, these 'imperfect fungi' were placed in their own class, the '*Deuteromycota*', with three major subdivisions, the hyphomycetes (producing mitotic spores directly from more or less specialized hyphae), the coelomycetes (where the spores are produced in or on fruit bodies) and the agonomycetes (where no recognizable spores are produced at all). It is now universally recognized that the *Deuteromycota* is a completely artificial assemblage, and that its constituent species should properly be assigned to one or other of the phyla of *Fungi*. There are many ways of establishing connections between fungi that only produce mitotic spores and those with meiospores as propagules. These range from direct observation of the two spore-bearing forms developing on the same mycelium, or culture of one morph to produce the other, to various techniques involving nucleic acid (molecular) analysis. If direct connections at species level are not possible, we can obtain clear indications of relationships by morphological comparison; these procedures are well established where some species of a conidium-forming genus also produce sexual spores, such as with *Colletotrichum* where some species are also known to form *Glomerella* teleomorphs. Where morphological comparisons are not possible, we can normally at least place mitotic species as *Ascomycota* or *Basidiomycota* due to differences in cell wall and septum construction. The vast majority of anamorphic fungi have affinities to the *Ascomycota*.

Classification within the major phyla of *Fungi* is complex and is changing rapidly as new information is obtained. Within the *Ascomycota*, the traditional classification system recognized five major subgroups based primarily on fruit body structure, with the hemiascomycetes lacking fruit bodies completely, the discomycetes with asci exposed on flat or cup-shaped structures, the pyrenomycetes with flask-shaped, almost closed fruit bodies, the plecto-mycetes with thin-walled completely closed bodies and the loculoasco-mycetes, which formed fruit bodies without separate walls as locules within stromatic tissues. The system used at present recognizes 46 orders of *Ascomycota*, as it was realized that the traditional classification did not properly reflect their phylogeny. It is clearly not realistic to expect practising plant pathologists to become familiar with all of these subgroups, but a relatively small number contain important plant pathogens. These are summarized below. The classification of the *Basidiomycota* has also undergone radical change in the last 20 years, with the gasteromycetes subsumed within the major group, the *Basidiomycetes*. However, the impact on the plant pathologist in practical terms is minimal.

A General Purpose Classification of the Fungi

Phyla (or divisions) end in '-mycota'; subdivisions in '-mycotina'; classes in '-mycetes'; subclasses in '-mycetidae', orders in '-ales'; and families in '-aceae'.

Families and orders are excluded from this overall scheme (Table 9.1), but indicators of numbers and examples of plant pathogenic genera are given.

A few common names and alternative names have been inserted and examples of major plant pathogenic genera are given.

Systematics and Nomenclature of Fungi

These are distinct processes. Systematics is concerned with deciding where the limits between organisms should be drawn and the level at which they should

Table 9.1. Organisms traditionally treated as fungi.

Kingdom	Phylum	Class	Plant pathogenic genera
Protozoa	Acrasiomycota		
	Dictyosteliomycota		
	Myxomycota	Myxomycetes	
		Protosteliomycetes	
	Plasmodiophoromycota		Plasmodiophora, Polymyxa, Spongospora
Chromista	Hypochytriomycota		
	Labyrinthulomycota		
	Oomycota		19 genera (e.g. Pythium, Phytophthora, Sclerospora)
Fungi	Ascomycota		Many genera in at least 16 orders of which Dothideales, Diaporthales, Erysiphales, Hypocreales and Ophiostomatales are most important as plant pathogens
	Basidiomycota	Basidiomycetes	Many genera in about 10 orders, e.g. Armillaria, Ganoderma, Corticium
		Teliomycetes	The rusts, e.g. Puccinia, Gymnosporangium, Hemileia
		Ustomycetes	The smuts, e.g. Entyloma, Tillettia, Ustilago
	Chytridiomycota		Three genera (e.g. Synchytrium)
	Zygomycota	Trichomycetes	
		Zygomycetes	Six genera (e.g. Rhizopus, Choanephora)

be distinguished (i.e. as families, genera, species, varieties, etc.), and their position in the overall classification. Nomenclature is the mechanism by which names are provided for these units.

Systematics

The basic systematic category is the individual, though even this is controversial when the organism is recognized as comprising mycelium as well as fruit bodies. Recent work has suggested that individual mycelia are among the largest and heaviest organisms in the world, but as new colonies can frequently be formed from uninucleate spores or mycelial fragments, it could alternatively be argued that nuclei should be considered as individuals.

Populations permanently separated from others by discontinuities (morphological, geographical and/or behavioural) are traditionally grouped into species. The types of discontinuities used for species separation vary between different groups of fungi, in practice according to the needs of users as well as differences in attributes. Spore characters are probably most widely employed as species criteria in the fungi as a whole, but there are many exceptions, as in the yeasts, where the ability to utilize particular nutrients assumes paramount importance. Species are grouped into 'genera', all species in a single genus sharing some characters in common, with each genus separated from others by discontinuities in some assumed fundamental characters.

The development of a systematic arrangement reflecting the actual relationships between organisms requires an extensive knowledge of their attributes. In addition to anatomical and morphological criteria, physiological, behavioural and molecular parameters are increasingly used in defining the characteristics of taxonomic groupings. These include growth characteristics, substrate utilization and host range, and nucleic acid sequences. The molecular sequences of certain regions of ribosomal RNA (or the DNA coding for ribosomal RNA) is particularly valuable in determining relationships of fungi, and will increasingly be used in molecular identification systems such as fingerprinting. Different parts of the ribosomal genome are considered to be useful for classifications at different levels of the hierarchy; for example, the small subunit (18S) is valuable at genus level and above (e.g. families, orders), whereas the internal transcribed spacer region is variable especially below the species level. The most useful information is obtained from sequencing these regions, but the process is still expensive and much recent systematic work has been done using less sophisticated techniques such as random amplified polymorphic DNA and restriction fragment length polymorphism analysis. These amplify short segments of DNA or cut it into fragments that can be sorted into different sizes using gel electrophoresis. Such methods can often distinguish well between organisms with different genomes, but give no indication of the extent of the differences, or the mechanisms by which the differences have arisen. This means they are of limited value in phylogenetic studies, but can be of great value in identification systems.

Fungal species concepts are currently poorly understood. The traditional morphologically based systems are increasingly being shown to be unsophisticated and rely too much on subjective decisions from experts who need many years of experience in order to provide reliable opinions of identification or relationship. DNA-based classifications are currently not well enough researched to be used in isolation, although the new knowledge is rapidly becoming more valuable, especially in well-known plant pathogenic organism groups. A particular difficulty is that the traditional hierarchical classifications (variety, subspecies, species, genus, etc.) cannot easily be applied to the complex dendrograms produced by numerical analysis of molecular data. Biological species concepts are well studied for some groups (including plant pathogens such as *Ophiostoma* and *Phytophthora*) but are difficult to apply to fungi that do not undergo sexual recombination, and identification procedures are very cumbersome.

For the pathologists, it is the behavioural attributes reflecting the pathogenicity and ecology of the organism that are often critical, which are sometimes not well reflected using current methods of nucleic acid analysis. Nevertheless, these are real and relevant phenomena, and there are several examples where close considerations of these characteristics have enabled a biologically realistic separation at the species level. Variation within species and populations is of particular concern to pathologists, and here the use of biochemical and molecular techniques linked with sexual or vegetative compatibility groupings and pathogenicity criteria can provide new insights into disease epidemiology and control (see Chapter 23).

Nomenclature

The names given to species are latinized (but need not be derived from Latin) and have two parts. The first is the genus name (with a capital letter) and the second the specific name (or epithet; without a capital letter) as in *Phytophthora infestans*. The names of the people originally naming the fungus are given, often in an abbreviated form, after the name as a method of indicating the sense in which it is used and as a much reduced guide to where to look for its original description in the literature (i.e. in whose publications). In the example given (*Phytophthora infestans* (Mont.) de Bary), the inclusion of a name in brackets shows that the person indicated (Montagne) used the epithet '*infestans*' but in a genus other than *Phytophthora*, and that de Bary transferred it (combined it) into *Phytophthora* from the genus in which it was originally used (which in this case was *Botrytis*). A list of common authors' names and their abbreviations can be found in the *Dictionary of the Fungi* (Kirk *et al.*, 2001).

Genera are grouped into families (the names of which end in '*aceae*' and may be italicized; e.g. *Venturiaceae*), families into orders (names end in '*ales*'; e.g. *Dothideales*), orders into classes (names for *Fungi* end in '*mycetes*'; e.g. *Ascomycetes*), etc. These higher categories (see Classification) are almost all based on fundamental differences in methods of spore production or the development and structure of the spore-bearing organs (sporocarps). Genotypically distinct categories below the rank of species are also available and can be given latinized names. Subspecies, varieties and forms (not to be confused with 'special forms', see below) may be distinguished anatomically and morphologically but by features not regarded as sufficiently important to merit separation at the species level. These are currently rarely used by mycologists, and there is considerable confusion as to the circumstances in which they should be used. Plant pathologists use a further special category, the special form or '*forma specialis*', which is outside the formal nomenclatural system. This is used to differentiate populations separated only by their physiological reactions to particular hosts, though in some cases minor morphological variations correlate with these. For example, *Puccinia graminis* Pers. f.sp. *avenae* occurs on oats, whereas f.sp. *tritici* occurs on wheat.

Any population meriting recognition is termed a 'taxon' (plural 'taxa'), regardless of its rank. For further discussion of the characteristics used in the separation of taxa of different ranks see Hawksworth (1974).

Although experience and subjective judgements are involved in taxonomy, nomenclature follows internationally agreed rules termed 'Codes'. The nomenclature of *Fungi* is controlled by the International Code of Botanical Nomenclature, which is revised periodically by successive International Botanical Congresses, the latest of which was held in 1999 in St Louis, USA, though the Code in force at the time of writing is derived from the previous Congress. Other Codes exist for animals, bacteria, viruses and cultivated plants. A valuable introduction to nomenclature which includes comparative information on the various Codes is provided by Jeffrey (1989).

The International Code of Botanical Nomenclature is a complex legalistic document, the correct interpretation of which requires considerable experience. In general, plant pathologists would be best advised to avoid becoming involved with the intricacies of nomenclature and to seek the views of taxonomists well versed in the use of the Code. It is, however, important for plant pathologists to understand some of the underlying provisions of the Code to appreciate why names change and how they are arrived at. It is not possible to consider all the eventualities covered in the Code here, and only some of those most frequently encountered will be mentioned.

The rules in the Code (termed 'Articles') may most conveniently be viewed as a series of filters, enabling the correct name for a particular taxon to be selected from the competing names proposed as shown in Table 9.2. First (A), names have to be 'effectively published' in journals, books or other items available to botanists generally, and not merely mentioned in written reports for limited circulation, newspapers, or written on herbarium specimens. If effectively published names are to be considered further they must be 'validly published' (B), with the provision of a diagnosis or description in Latin, and deposition of a nomenclatural type in a recognized collection. A type is a permanently preserved herbarium specimen, or dried or lyophilized culture to

Table 9.2. The nomenclatural filter (simplified).

	All names	
(A)	Effective publication	Excludes those unpublished
(B)	Validly published names	Excludes those invalidly published
(C)	Names applicable to taxon	Excludes those not applicable (e.g. belonging to other taxa)
(D)	Legitimate names	Excludes illegitimate names
(E)	Priority	Selects names based on priority
(F)	Correct name	

which the name is irrevocably fixed, which means that the application of the name itself remains unambiguous even when circumscriptions of taxa alter. If name changes are proposed by movement from one genus to another, or by changes in rank (e.g. variety to species), the author making the new transfer (combination) must give full details of the original name (basionym) and its place of publication. If a name does not meet all of the requirements for valid publication, nomenclaturally it is considered not to exist and need not be considered further. The systematist must then decide which of the validly published names (if any) may be applicable to the taxon he wishes to recognize (C).

Valid names may also contravene other Articles in the Code, when they must be rejected as 'illegitimate' (D). Names may be illegitimate for a number of reasons, of which the commonest are that the same name has previously been used for a different taxon in the same rank (a later homonym), or that the author introducing it did not adopt the name he should have under the Code (a superfluous name). The correct name is then selected from the legitimate names, according to the principle of priority of publication (E). The first published legitimate epithet (after 1 May 1753, when the system of binomial nomenclature was first used) must normally be adopted, newly combined with the appropriate generic name if necessary. Names other than the correct one belonging to the same taxon are termed 'synonyms'.

A further special provision of the Code deals with the nomenclature of fungi with more than one distinct state (morph) in their life cycle. This allows different names to be used for the meiospore-forming stage (teleomorph) and one or more mitospore-forming stages (anamorphs). Names of such pleomorphic fungi fall into three categories. Names of anamorphs apply to mitotic asexual forms of propagation, names of teleomorphs apply to meiotic sexual forms and names of holomorphs cover the taxon in all its forms. The correct name for the holomorph is the earliest legitimate name typified by the teleomorph, i.e. the form characterized by production of asci/ascospores, basidia/basidiospores, teliospores, or other basidium-bearing organs. Currently, teleomorph names have priority over anamorph names when taxa are considered to be connected (i.e. part of the same life cycle). This dual nomenclature system is now almost unworkable and gives the misleading impression that different morphs are different organisms. Advances in molecular systematics are allowing more and more reliable links between anamorph and teleomorph, and providing unequivocal evidence of the true relationships of fungi that do not produce sexual spores. However, its replacement by a one name–one fungus system has many practical difficulties. For the moment, the plant pathologist should use the name most familiar (often the anamorph as that stage manifests itself during active disease expression), and leave systematics experts to sort out the details. That does not absolve pathologists from the responsibility of understanding and assimilating name changes necessary for taxonomic reasons (see below).

Why names change

Plant pathologists, in common with most users of names of organisms, often resent changes in the names of taxa with which they are familiar. With an internationally agreed Code of Nomenclature that aims to provide a mechanism for stability in names, they expect stability to increase, but all too often find it does not appear to do so. The reasons why names are changed may be nomenclatural or taxonomic.

Nomenclatural changes are due to the application of the Code to the names that exist. In an ideal situation they would not arise but in practice there are so many names in mycological literature, species often being inadequately described in quite inappropriate genera, that taxonomists are frequently not aware that a name they have not considered belongs to their taxon. If such a name was published earlier than the name currently being employed for a taxon, it must normally be taken up for it under the rules of priority. However, the Code now provides for the system of priority to be overridden in cases where name changes would cause unnecessary confusion, and of course those of prominent plant pathogens will fall into this category. It is now most unlikely that pathogens of significant economic importance will have their names changed for purely nomenclatural reasons.

Changes due to taxonomic decisions are frequently beneficial to plant pathologists, as pathogen species and their interrelationships become more clearly defined. Such changes may include: the union of previously separate species; the realization that a species once thought to be homogeneous comprises several distinct groups; a change in rank (e.g. variety to species); or a remodelling of generic limits.

How to Describe a Fungus

Most fungi are still differentiated at the generic, specific and varietal levels by morphology, although these characters are increasingly being supplemented by molecular data, physiological and biochemical characters, and pathological and behavioural information. Despite recent advances in systematic mycology, most identifications of plant pathogens for the forseeable future will be carried out using a microscope, so it is important that pathologists can analyse and describe specimens to compare with published work or communicate with others. The following procedure is recommended.

Describe the appearance, whether on a natural substratum or in culture, as seen by the naked eye and by examination with a hand lens, a binocular dissecting microscope and at higher magnification. Note details of the substratum type and apparent effect of the fungus on the host material. For cultures, note the colour and texture of the colony and the colour of the underside (the reverse). Note the culture medium and growth conditions and the time of incubation.

Describe and illustrate the microscopic details of mycelium, any special structures, fruit bodies and spores (particularly spore development); use camera lucida drawings and/or photomicrographs at standard magnifications so that quick comparisons can be made. Record the size of microscopic details.

Examine as many specimens and/or cultures as possible from as many different collections as possible to establish the range of variation.

Note also the geographical origin of the specimen, the name of the collector or mycologist who made the original collection or isolate, date of collection and the accession number of the specimen or culture.

Determine any nutritional peculiarities including growth factor requirements and the minimum, optimum and maximum temperatures for growth and for a second state, whether homothallic or heterothallic.

Synonyms of some common plant pathogens including their respective teleomorphs or anamorphs can be found in the CAB *International Descriptions of Fungi and Bacteria* (CAB International, 2001), which currently cover more than 1000 species. Teleomorphic states (in Roman type) of some common plant pathogens are as follows:

Ascochyta cucumis Fautr. & Roum., Didymella bryoniae (Auersw.) Rehm
Ascochyta chrysanthemi Stev., Didymella chrysanthemi (Tassi) Garibaldi & Gullino
Ascochyta lycopersici (Plowr.) Brunaud, Didymella lycopersici Kleb.
Ascochyta pinodes Jones, Mycosphaerella pinodes (Berk. & Blox.) Vestergren
Asteromella brassicae (Chev.) Boerema & van Kest., Mycosphaerella brassicicola (Duby) Lindau

Botrytis convoluta Whetzel & Drayton, Botryotinia convoluta Whetzel & Drayton
Botrytis gladiolorum Timmerm., Botryotinia draytonii Buddin & Wakef.
Botrytis narcissicola Kleb., Botryotinia narcissicola (Gregory) N.F. Buchwald
Botrytis ricini Godfr., Botryotinia ricini (G.H. Godfrey) Whetzel
Botrytis squamosa Walker, Botryotinia squamosa Viennot-Bourgin

Cercospora arachidicola Hori, Mycosphaerella arachidis Deighton
Cercosporidium personatum (Berk. & Curtis) Deighton, Mycosphaerella berkeleyi W.A. Jenkins
Chalara quercina Henry Ceratocystis fagacearum (Bretz) Hunt
Chalaropsis thielavioides Peyron., Ceratocystis radicicola (Bliss) Moreau
Cladosporium echinulatum (Berk.) de Vries, Mycosphaerella dianthi (Burt) J rstad
Cladosporium iridis (Fautr. & Roum.) de Vries, Mycosphaerella macrospora (Kleb.) J rstad
Colletotrichum falcatum Went, Glomerella tucumanensis (Speg.) Arx & Müller
Colletotrichum gloeosporioides (Penz.) Sacc., Glomerella cingulata (Stoneman) Spaulding & von Schrenk
Coniothyrium fuckelii Sacc., Leptosphaeria coniothyrium (Fuckel) Sacc.
Cryptosporiopsis malicorticis (Cordl.) Nannf., Pezicula malicorticis (Jacks.) Nannf.
Curvularia geniculata (Tracy & Earle) Boedijn, Cochliobolus geniculatus Nelson
Cylindrocarpon destructans (Zins.) Scholten, Nectria radicicola Gerlach & Nilsson
Cylindrocarpon mali (Allesch.) Wollenw., Nectria galligena Bres.
Cylindrocladium crotalariae (Loos) Bell & Sobers, Calonectria crotalariae (Loos) Bell & Sobers

Cylindrosporium concentricum Grev., Pyrenopeziza brassicae Sutton & Rawlinson

Dematophora necatrix Hartig, Rosellinia necatrix Prill.
Dothistroma pini Hulbary, Scirrhia pini Funk & Parker
Drechslera avenae (Eidam) Scharif, Pyrenophora avenae Ito & Kuribay.
Drechslera bromi (Died.) Shoemaker, Pyrenophora bromi (Died.) Drechsler
Drechslera cynodontis (Marignoni) Subram. & Jain, Cochliobolus cynodontis Nelson
Drechslera graminea (Rabenh. ex Schlecht.) Shoemaker, Pyrenophora graminea Ito & Kuribay.
Drechslera maydis (Nisikado) Subram. & Jain, Cochliobolus heterostrophus (Drechsler) Drechsler
Drechslera oryzae (Breda de Haan) Subram. & Jain, Cochliobolus miyabeanus (Ito & Kuribay.) Drechsler ex Dastur
Drechslera setariae (Sawada) Subram. & Jain, Cochliobolus setariae (Ito & Kuribay.) Drechsler
Drechslera sorokiniana (Sacc.) Subram. & Jain, Cochliobolus sativus (Ito & Kuribay.) Drechsler ex Dastur
Drechslera teres (Sacc.) Shoemaker, Pyrenophora teres Drechsler
Drechslera tritici-repentis (Died.) Shoemaker, Pyrenophora tritici-repentis (Died.) Drechsler
Drechslera turcica (Pass.) Subram. & Jain, Setosphaeria turcica (Luttr.) Leonard & Suggs
Drechslera verticillata (O'Gara) Shoemaker, Pyrenophora semeniperda (Brittleb. & Adams) Shoemaker
Drechslera victoriae (Mehan & Murphy) Subram. & Jain, Cochliobolus victoriae Nelson

Fusarium decemcellulare Brick, Calonectria rigidiuscula (Berk. & Broome) Sacc.
Fusarium graminearum Schwabe, Gibberella zeae (Schw.) Petch
Fusarium lateritium Fr. em. Snyder & Hansen, Gibberella baccata (Wallr.) Sacc.
Fusarium moniliforme Sheld., Gibberella fujikuroi (Saw.) Wollenw.
Fusarium xylarioides Steyaert, Gibberella xylarioides Heim & Sacc.
Fusicladium carpophilum (Thüm.) Oudem., Venturia carpophila Fisher
Fusicladium cerasi (Rabenh.) Sacc., Venturia cerasi Aderh.
Fusicladium macrosporum Kuyper, Microcyclus ulei (Henn.) v. Arx
Fusicladium pyrorum Fuckel, Venturia pirina Aderh.

Gloeosporium nervisequum (Fuckel) Sacc., Gnomonia platani Kleb.
Gloeosporium ribis (Lib.) Mont. & Desm., Pseudopeziza ribis Kleb.
Graphium ulmi Schwartz, Ophiostoma ulmi (Buism.) Nannf.

Lecanosticta acicola (Thum.) Syd., Scirrhia acicola (Desm.) Siggers

Macrophoma zeae Tehon & Daniels, Botryosphaeria zeae (Stout.) v. Arx & E. Miller
Marssonina fragariae (Lib.) Kleb., Diplocarpon earlianum (Ell. & Ev.) Wolf
Marssonina rosae (Lib.) Died., Diplocarpon rosae Wolf
Monilia cinerea Bon, Monilinia laxa (Aderh. & Rubl.) Honey

Monilia fructigena (Fr.) Westend., Monilinia fructigena (Aderh. & Ruhl.) Honey

Nakataea sigmoidea Hara, Magnaporthe salvinii (Cattaneo) Krause & Webster
Nigrospora oryzae (Berk. & Broome) Petch, Khuskia oryzae Hudson

Oidiopsis taurica (Lev.) Arnaud, Leveillula taurica (Lev.) Arnaud
Oidium erysiphoides Fr., Erysiphe betae (Varha) Weltz.
Oidium tuckeri Berk., Uncinula necator (Schw.) Burr.

Paracercospora fijiensis (Morelet) Deighton, Mycosphaerella fijiensis Morelet
Phacidiopycnis pseudotsugae (M. Wilson) Hahn, Phacidium coniferarum
    (Hahn) DiCosmo
Phloeosporella padi (Lib.) von Arx, Blumeriella jaapii (Rehm) von Arx
Phoma betae Frank, Pleospora betae (Berl.) Nevodovsky
Phoma lingam (Tode ex Fr.) Desm., Leptosphaeria maculans (Desm.) Ces. & de
    Not.
Phyllosticta citricarpa (McAlp.) van der Aa, Guignardia citricarpa Kiely
Pseudocercospora cruenta (Sacc.) Deighton, Mycosphaerella cruenta Latham
Pseudocercospora musae (Zimm.) Deighton, Mycosphaerella musicola Mulder
Pseudocercospora puerariicola (Yamam.) Deighton, Mycosphaerella puerari-
    icola Weimer & Luttr.
Phomopsis citri Fawcett, Diaporthe citri Wolf
Phomopsis phaseoli (Desm.) Grove, Diaporthe phaseolorum (Cook & Ellis)
    Sacc.
Pyricularia grisea Sacc., Magnaporthe grisea (Hebert) M.E. Barr

Ramularia brunnea Peck, Mycosphaerella fragariae (Tul.) Lindau
Rhizoctonia solani Kuhn, Thanatephorus cucumeris (Frank) Donk

Sclerotium gladioli Massey, Stromatinia gladioli (Drayt.) Whetzel
Sclerotium rolfsii Sacc., Corticium rolfsii Curzi
Septoria linicola (Speg.) Garassini, Mycosphaerella linicola Naumov
Septoria pyricola (Desm.) Desm., Mycosphaerella pyri (Auersw.) Boerema
Septoria ribis Desm., Mycosphaerella ribis (Fuckel) Feltgen
Septoria rosae Desm., Sphaerulina rehmiana Jaap
Septoria tritici Rob. ex Desm., Mycosphaerella graminicola (Fuckel) Sanderson
Sphacelia segetum Lév., Claviceps purpurea (Fr.) Tul.
Sphacelia typhina Sacc., Epichloë typhina (Pers. ex Fr.) Tul.
Sphaceloma ampelinum de Bary, Elsinoë ampelina Shear
Sphaceloma australis Bitanc. & Jenkins, Elsinoë australis Bitanc. & Jenkins
Sphaceloma fawcettii Jenkins, Elsinoë fawcettii Bitanc. & Jenkins
Sphaceloma necator (Ell. & Ev.) Jenkins and Shear, Elsinoë veneta (Burkh.)
    Jenkins
Spilocaea pomi Fr., Venturia inaequalis (Cooke) Winter
Stagonospora avenae (A.B. Frank) Bissett, Phaeosphaeria avenaria (Weber) O.
    Erikss.
Stagonospora nodorum (Berk.) E. Castell. & Germano, Phaeosphaeria nodorum
    (E. Müller) Hedjaroude
Stemphylium botryosum Wallr., Pleospora herbarum (Pers. ex Fr.) Rabenh.

Thielaviopsis paradoxa (de Seynes) Höhnel, Ceratocystis paradoxa (Dade)
    Moreau

Identification

Texts that will assist the identification of fungi are listed in the further reading list, but expert advice should be sort for confirmation or where identification is critical, e.g. for new geographic or host records of fungi. This can be obtained from reputable institutions. BioNet is a network of regional institutions that can be of assistance for identification of pest organisms and advice can also be obtained directly from CAB *International*. Details of the identification, information and documentary services provided by CAB *International* may be obtained from The Director, CABI Bioscience, Bakeham Lane, Egham, Surrey, TW20 9TY, UK; fax: +44 (0)1491 829100, email: bioscience@cabi.org. Details of new information products of relevance may be found on the CAB *International* web site at www.cabi.org and on the British Crop Protection Council (BCPC) web site at www.BCPC.org.

References

Flor, H.H. (1971) Current status of the gene-for-gene theory. *Annual Review of Phytopathology* 9, 275–296.

Gregory, P.H. (1973) *Microbiology of the Atmosphere*. Leonard Hill, Aylesbury, UK, 377 pp.

Kirk, P.M., Cannon, P.F., David, J.C. and Stalpers, J.A. (eds) (2001) *Ainsworth and Bisby's Dictionary of the Fungi*, 9th edn. CAB International, Wallingford, UK.

Further Reading

General mycology

Alexopoulos, C.J., Mims, C.W. and Blackwell, M. (1996) *Introductory Mycology*. John Wiley & Sons, New York, USA, 868 pp.

Carlisle, M.J. and Watkinson, S.C. (1994) *The Fungi*. Academic Press, London, UK, 482 pp.

Gravesen, S., Frisvad, J.C. and Samson R.A. (1994) *Microfungi*. Munksgaard, Copenhagen, 168 pp.

Hudson, H.J. (1986) *Fungal Biology*. Edward Arnold, London, UK, 297 pp.

Kendrick (1992) *The Fifth Kingdom*, 2nd edn. Mycologue Publications, Waterloo, Ontario, Canada.

Kirk, P.M., David, J.C. and Stalpers, J.A. (eds) (2001) *Ainsworth and Bisby's Dictionary of the Fungi*, 9th edn. CABI Publishing, Wallingford, UK.

Ulloa, M. and Hanlin, R.T. (2000) *Illustrated Dictionary of Mycology*. APS Press, St Paul, Minnesota, USA, 200 pp.

Webster, J. (1988) *Introduction to Fungi*, 3rd edn. Cambridge University Press, Cambridge, UK.

Taxonomy and classification

Hawksworth, D.L. (1974) *Mycologist's Handbook*. Commonwealth Mycological Institute, Kew. 231 pp. (The nomenclatural sections are now outdated due to changes in the Code, but many of the basic principles remain relevant.)

Jeffrey, C. (1989) *Biological Nomenclature*, 3rd edn. Arnold, London, UK.

Identification of fungi

General

von Arx, J.A. (1981) *The Genera of Fungi Sporulating in Pure Culture*, 3rd edn. J. Cramer,Vaduz, 424 pp.

von Arx, J.A. (1987) *Plant Pathogenic Fungi*. J. Cramer, Berlin & Stuttgart, Germany, 288 pp.

CAB International (2001) *Descriptions of Fungi and Bacteria*. CAB International, Wallingford, UK.

Ellis, M.B. and Ellis, J.P. (1997) *Microfungi on Land Plants*, 2nd edn. Richmond Publishing, Slough, UK, 868 pp.

Farr, D.F., Bills, G.F., Chamuris, C.P. and Rossman, A.Y. (1989) *Fungi on Plant and Plant Products in the United States*. APS Press, St Paul, Minnesota, USA, 1252 pp.

Rossman, A.Y., Palm, M.E. and Spielman, L.J. (1987) *A Literature Guide to the Identification of Plant Pathogenic Fungi*. APS Press, St Paul, Minnesota, USA, 252 pp.

Schots, A., Dewey, F.M. and Oliver, R. (eds) (1994) *Modern Assays for Plant Pathogenic Fungi*. CAB International, Wallingford, UK, 267 pp.

Plasmodiophoromycota

Karling, J.S. (1968) *The Plasmodiophorales*, 2nd edn. Hafner, New York, USA, 256 pp.

Margulis, L., Corliss, J.O., Melkonian, M. and Chapman, D.J. (1990) *Handbook of the Protoctista*. Jones and Bartlett, Boston, USA, 914 pp.

Chytridiomycota

Karling, J.S. (1964) *Synchytrium*. Academic Press, New York and London, 470 pp.

Oomycota

PERONOSPORALES

Constantinescu, O. (1991) An annotated list of Peronospora names. *Thunbergia* 15, 110 pp.

Francis, G.M. and Waterhouse, G.M. (1988) List of Peronosporaceae reported from the British Isles. *Transactions of the British Mycological Society* 91, 1–62.

Spencer, D.M. (ed.) (1981) *The Downy Mildews*. Academic Press, London and New York, 636 pp.

PYTHIALES

Buczaki, S.T. (1983) *Zoosporic Plant Pathogens – a Modern Perspective*. Academic Press, London, UK, 352 pp.

Erwin, D.C. and Robeiro, O.K. (1996) *Phytophthora Diseases Worldwide*. APS Press, St Paul, Minnesota, USA,

Lucas, J.A., Shattock, R.C., Shaw, D.S. and Cooke, L.R. (eds) (1991) Phytophthora. Cambridge University Press, Cambridge, UK, 447 pp.

van der Plaats-Niterink, A.J. (1981) Monograph of the genus *Pythium*. *Studies in Mycology* 21, 244 pp.

Stamps, D.J., Waterhouse, G.M., Newhook, F.J. and Hall, G.S. (1990) Revised tabular key to the species of *Phytophthora*. *Mycological Papers* 162, 28 pp.

Zygomycota

MUCORALES

Kirk, P.M. (1984) A monograph of the *Choanephoraceae*. *Mycological Papers* 152, 61 pp.

Ascomycota

GENERAL TEXTS

Barnett, H.L. and Hunter, B.B. (1998) *Illustrated Genera of the Imperfect Fungi*, 4th edn. APS Press, St Paul, Minnesota, USA, 218 pp.

Barr, M.E. (1990) Prodromus to nonlichenized, pyrenomycetous members of class Hymenoascomycetes. *Mycotaxon* 39, 43–184.

Carmichael, J.W., Kendrick, W.B., Connors, I.L. and Sigler, L. (1980) *Genera of Hyphomycetes*. University of Alberta Press, Edmonton, Canada, 386 pp.

Dennis, R.W.G. (1978) *British Ascomycetes*. J. Cramer, Lehre, 585 pp.

Ellis, M.B. (1971) *Dematiaceous Hyphomycetes*. Commonwealth Mycological Institute, Kew, UK, 608 pp.

Ellis, M.B. (1976) *More Dematiaceous Hyphomycetes*. Commonwealth Mycological Institute, Kew, UK, 507 pp.

Hanlin, R.T. (1998) *Illustrated Genera of the Ascomycetes*. APS Press, St Paul, Minnesota, USA, 258 pp.

Sutton, B.C. (1980) *The Coelomycetes*. Commonwealth Mycological Institute, Kew, UK, 696 pp.

DIAPORTHALES

Barr, M.E. (1978) *The Diaporthales in North America*. J. Cramer, Lehre, 232 pp.

Uecker, F.A. (1988) A world list of Phomopsis names with notes on nomenclature, morphology and biology. *Mycologia Memoir* 13, 231 pp.

DOTHIDEALES

Aptroot, A. (1995) A monograph of Didymosphaeria. *Studies in Mycology* 37, 160 pp.

von Arx, J.A. and Müller, E. (1975) A re-evaluation of the bitunicate ascomycetes with keys to families and genera. *Studies in Mycology* 9, 1–159.

Braun, U. (1998) *A Monograph of Cercosporella, Ramularia and Allied Genera (Phytopathogenic Hyphomycetes)* 2. IHW-Verlag, Eching, 493 pp.

Corlett, M. (1991) An annotated list of the published names in *Mycosphaerella* and *Sphaerella*. *Mycologia Memoires* 18, APS Press, St Paul, Minnesota, USA, 328 pp.

Hughes, S.J. (1976) Sooty moulds. *Mycologia* 68, 693–820.

Shoemaker, R.A. (1985) Canadian and some extralimital *Leptosphaeria* species. *Canadian Journal of Botany* 62, 2688–2729.

Shoemaker, R.A. and Babcock, C.E. (1989) *Phaeosphaeria*. *Canadian Journal of Botany* 67, 1500–1599.

Sivanesan, A. (1984) *The Bitunicate Ascomycetes and Their Anamorphs*. J. Cramer, Vaduz, 701 pp.

ERYSIPHALES

Amano (Hirata), K. (1987) *Host Range and Geographic Distribution of the Powdery Mildew Fungi*. Japan Scientific Press, Tokyo, 74 pp.

Braun, U. (1995) *The Powdery Mildews (Erysiphales) of Europe*. Gustav Fischer Verlag, Jena, 337 pp.

Spencer, D.M. (ed.) (1978) *The Powdery Mildews*. Academic Press, London and New York, 636 pp.

HYPOCREALES

Rossman, A.Y., Samuels, G.J., Rogerson, C.T. and Lowen, R. (1999) Genera of Bionectriaceae, Hypocreaceae and Nectriaceae (Hypocreales, Ascomycetes). *Studies in Mycology* 42, 248 pp.

LEOTIALES

Kohn, L.M. (1979) A monographic revision of the genus *Sclerotinia*. *Mycotaxon* 9, 365–444.
Verhoeff, K., Malathrakis, N.E. and Williamson, B. (eds) (1992) *Recent Advances in Botrytis Research*. PUDOC, Wageningen, The Netherlands, 294 pp.

MELIOLALES

Hosagoudar, V.B., Abraham, T.K. and Pushpangadan, P. (1997) *The Meliolineae. A Supplement*. Tropical Botanic Garden & Research Institute, Kerala, 201 pp.

OPHIOSTOMATALES

Wingfield, M.J., Seifert, K.A. and Webber, J.F. (eds) (1993) *Ceratocystis and Ophiostoma. Taxonomy, Ecology and Pathogenicity*. APS Press, St Paul, Minnesota, USA, 293 pp.

PHYLLACHORALES

Bailey, J.A. and Jeger, M.J. (eds) (1992) *Colletotrichum: Biology, Pathology and Control*. CAB International, Wallingford, UK, 388 pp.
Prusky, D., Freeman, S. and Dickman, M.B. (eds) (2000) *Colletotrichum: Host Specificity and Host–Pathogen Interaction*. APS Press, St Paul, Minnesota, USA, 393 pp.

Basidiomycota

UREDINALES

Bushnell, W.A. and Roelfs, A.P. (eds) (1984–85) *The Cereal Rusts. 1. Origin, Specificity, Structure and Physiology*, 516 pp; 2. *Disease, distribution, epidemiology, control*, 606 pp. Academic Press, New York, USA.
Cummins, G.B. (1971) *The Rust Fungi of Cereals, Grasses and Bamboos*. Springer-Verlag, New York, USA, 570 pp.
Cummins, G.B. (1978) *Rust Fungi on Legumes and Composites in North America*. University of Arizona Press, Tucson, USA, 424 pp.
Cummins, G.B. and Hiratsuka, Y. (1983) *Illustrated Genera of Rust Fungi*, 2nd edn. APS Press, St Paul, Minnesota, USA, 152 pp.
Scott, K.J. and Chakravorty, A.K. (eds) (1982) *The Rust Fungi*. Academic Press, London, UK, 288 pp.
Wilson, M. and Henderson, D.M. (1966) *British Rust Fungi*. Cambridge University Press, Cambridge, UK, 384 pp.

USTILAGINALES

Fischer, G.W. and Holton, C.S. (1957) *Biology and Control of the Smut Fungi*. Ronald Press, New York, USA, 622 pp.
Vanky, K. (1987) *Illustrated Genera of Smut Fungi*. Gustav Fischer Verla, Stuttgart, Germany, 154 pp.
Zundel, G.L. (1953) *The Ustilaginales of the World*. Pennsylvania State College Contribution no. 176, 410 pp.

HYMENOMYCETES

Flood, J., Bridge, P. and Holderness, M. (eds) (2000) Ganoderma *Diseases of Perennial Crops*. CABI Publishing, Wallingford, UK, 350 pp.
Roberts, P. (1999) *Rhizoctonia-forming Fungi. A Taxonomic Guide*. Royal Botanic Gardens, Kew, UK, 239 pp.
Shaw, C.G. and Kile, G.A. (eds) (1991) *Armillaria Root Disease*. Agriculture handbook no. 691, USDA Forest Service, Washington, DC, USA.

Sneh, B., Burpee, L. and Ogoshi, A. (1991) *Identification of Rhizoctonia Species*. APS Press, St Paul, Minnesota, USA, 133 pp.

Stalpers, J.A. (1978) Identification of wood-inhabiting Aphyllophorales in pure culture. *Studies in Mycology* 16, 248 pp.

Woodward, S., Stenlid, J., Karjlaimen, R. and Hutterman. A. (eds) (1998) Heterobasidion annosum: *Biology, Ecology, Impact and Control*. CAB International Wallingford, UK, 589 pp.

Bacteria and Plant Disease

10

G. Saddler

CABI Bioscience UK Centre, Bakeham Lane, Egham, Surrey TW20 9TY, UK

Introduction

Bacteria are ubiquitous and physiologically diverse; those that are found in association with plants exist as epiphytes, endophytes and pathogens. Phytopathogens, members of the latter group, are relatively few in both type and number (Kado, 1992; Sigee, 1993). All bacterial phytopathogens described to date fall within the Domain *Bacteria* (formerly known as the *Eubacteria*). Most possess common morphological characteristics, namely, straight or slightly curved rods with rigid cell walls and aerobic or facultatively anaerobic metabolism. An exception to this rule are members of the Class *Mollicutes*, which, though related to Gram-positive bacteria such as clostridia and bacilli, lack a rigid cell wall. These organisms are dealt with separately in Chapter 12.

Bacterial phytopathogens possessing a cell wall can be subdivided into Gram-positive and Gram-negative. These groupings, based on a well-established microscopy stain, roughly equate with the more recent phylogenetic treatment of bacteria. In this regard, all Gram-negative bacterial plant pathogens are encompassed within the Class *Proteobacteria* (Stackebrandt *et al.*, 1988) whilst the Gram-positives are recovered in Class *Actinobacteria* (Stackebrandt *et al.*, 1997). Most phytopathogenic bacteria are Gram-negative, the majority of which belong to the genera *Acidovorax*, *Agrobacterium*, *Burkholderia*, *Enterobacter*, *Erwinia*, *Pantoea*, *Pseudomonas*, *Ralstonia* and *Xanthomonas* (Table 1). The few Gram-positive pathogens are mainly contained within the genera *Clavibacter*, *Curtobacterium*, *Rathayibacter* and *Streptomyces*. For detailed information on many bacterial phytopathogens see Bradbury (1986).

Table 10.1. The major types of plant-pathogenic bacteria.

Genus/species	General disease symptoms
Gram-negative bacteria	
Acetobacter spp.	Pink disease of pineapple fruit
Acidovorax spp.	Leaf blight, leaf spots/streak
Agrobacterium spp.	Crown gall, hairy root formation
Burkholderia spp.	Vascular wilts, rots
Enterobacter spp.	Cankers, leaf spots and rots
Erwinia spp.	Vascular wilts, dry necroses, leaf spots and soft rots
Gluconobacter oxydans	Pink disease of pineapple fruit
Pantoea spp.	Vascular wilts, rots
Pseudomonas spp.	Leaf spots, vascular wilts, soft rots
Ralstonia spp.	Vascular wilts
Rhizobacter daucus	Bacterial gall of carrot
Serratia marcescens	Crown and root rot of lucerne
Xanthomonas spp.	Leaf spots, vascular wilts, stem cankers
Xylella fastidiosa	Pierce's disease of grape
Xylophilus ampelinus	Bacterial blight of grape
Gram-positive bacteria	
Arthrobacter ilicis	Holly bacterial blight
Clavibacter spp.	Vascular wilts, cankers
Curtobacterium spp.	Silvering disease, vascular wilts
Nocardia vaccinii	Blueberry gall
Rathayibacter spp.	Gumming disease
Rhodococcus fascians	Leafy gall
Streptomyces spp. (*S. scabies*)	Potato scab

Significant Bacterial Plant Pathogens

Gram-negative genera

Acidovorax

This is a member of the family *Comamonadaceae* in the β-subdivision of the class *Proteobacteria*. Members of this genus were formerly assigned to the genus *Pseudomonas*. The genus *Acidovorax* was created after extensive taxonomic studies indicated a number of species were quite distinct from the genus *Pseudomonas sensu stricto* (Palleroni, 1993; Kersters *et al.*, 1996). Phytopathogens contained within this genus were all previously assigned to the genus *Pseudomonas* and include: *A. avenae* subsp. *avenae*, *A. avenae* subsp. *cattleyae*, *A. avenae* subsp. *citrulli* and *A. konjaci* (Willems *et al.*, 1992). In addition there is a newly described pathogen of anthurium, *A. anthurii* (Gardan *et al.*, 2000).

Agrobacterium

This is a member of family *Rhizobiaceae* in the α-subdivision of the class *Proteobacteria*. It is commonly found in soil and in association with plant roots, and relatively few strains belonging to this genus are phytopathogenic (Shaw *et al.*, 1991). Pathogenicity is plasmid borne on tumour-inducing (Ti) or root-inducing (Ri) plasmids, so called for their Ti or Ri ability. Pathogenic strains invade the crown, roots and stem of a great variety of dicotyledonous and some gymnospermous plants, via wounds. This causes the transformation of plant cells into autonomously proliferating tumour cells. Diseases include crown gall, hairy root and cane gall. Saprophytic agrobacteria can become pathogenic through the acquisition of either of the Ti or Ri plasmids (Kado, 1992).

Burkholderia

This is a member of the family *Burkholderiaceae* in the β-subdivision of the class *Proteobacteria*. The majority of strains belonging to this genus are pathogenic to plants or animals. Members of this genus were formerly assigned to the genus *Pseudomonas* (Palleroni, 1993; Kersters *et al.*, 1996) and include: *B. andropogonis*, *B. caryophylli*, *B. cepacia*, *B. gladioli* and *B. glumae* (Yabuuchi *et al.*, 1992). Originally this genus was described encompassing *B. solanacearum*, but this species has since been reassigned to the recently established genus of *Ralstonia*.

Enterobacter

This is a member of the family *Enterobacteriaceae* in the γ-*Proteobacteria*. Widely distributed in nature, only four species are considered phytopathogenic: *E. cancerogenus*, *E. dissolvens*, *E. nimipressuralis* and *E. pyrinus*. These are mostly minor tree pathogens causing cankers, leaf-spots and rots, the exception being *E. dissolvens* which causes stalk rot in maize.

Erwinia

This is a member of the family *Enterobacteriaceae* in the γ-subdivision of the class *Proteobacteria*. The majority of strains belonging to this genus are phytopathogenic. Diseases caused by members of the genus *Erwinia* can be broadly divided into three main types. Pectolytic erwinias cause soft rot in a wide variety of plants, with *E. carotovora* subsp. *carotovora* and *E. chrysanthemi* being the most important. Necrotic diseases are also found, e.g. fireblight of pear and apple caused by *E. amylovora*. Finally, wilt disease can occur, such as the wilt of cucumber caused by *E. tracheiphila*.

Pantoea

This is a member of the family *Enterobacteriaceae* in the γ-subdivision of the class *Proteobacteria*. The genus *Pantoea* was created after extensive taxonomic studies indicated a number of species were distinct from the genus *Erwinia* *sensu stricto* (Mergaert *et al.*, 1993). Phytopathogens contained within this genus were all previously assigned to the genus *Erwinia* and include: *P. ananas*, *P. stewartii* subsp. *indologenes* and *P. stewartii* subsp. *stewartii*. The latter is possibly the most significant pathogen causing Stewart's disease of maize. In addition the non-pathogenic but plant-associated *Erwinia herbicola* has been included in this genus and is now referred to as *P. agglomerans*.

Pseudomonas

This is a member of the family *Pseudomonadaceae* in the γ-subdivision of the class *Proteobacteria*. A definitive genus description is difficult as pseudomonads are very heterogeneous. Saprophytes, autotrophs and human and plant pathogens are all encompassed within the genus *Pseudomonas*. Traditionally, plant-pathogenic pseudomonads could be conveniently split into two groups, the first of which produce a yellow–green diffusible, fluorescent pigment on iron-deficient media and do not accumulate poly-β-hydroxybutyrate (PHB). The second group produce no fluorescent pigments but do accumulate PHB, and most of these species are now contained within the genera *Acidovorax*, *Burkholderia* and *Ralstonia*. As a consequence, the genus description for *Pseudomonas* is in a state of flux; however, in general terms the genus encompasses all of the fluorescent pseudomonad phytopathogens. The phytopathogenic pseudomonads cause an array of diseases in plants ranging from necrotic lesions and spots in fruit, stems and leaves to hyperplasias, tissue macerations, cankers, blights and vascular infections.

Ralstonia

This is a member of the family *Burkholderiaceae* in the β-subdivision of the class *Proteobacteria*. *Ralstonia* shares many similarities with *Burkholderia*, but can be distinguished on the assimilation of galactose, mannitol, mannose and sorbitol (Yabuuchi *et al.*, 1995). Currently there is only one plant pathogen included in this genus, *R. solanacearum*, which is possibly the most significant bacterial phytopathogen worldwide, due to the wide host range and severity of the bacterial wilt diseases it causes. Current thinking suggests that *Pseudomonas syzygii*, the causative organism of Sumatra disease of cloves (Roberts *et al.*, 1990), and the bacterium which causes blood disease of banana in South-East Asia, should be included in this genus (Seal *et al.*, 1999).

Xanthomonas

This is a member of the family *Xanthomonadaceae* in the γ-subdivision of the class *Proteobacteria*. The majority of members of this genus are plant pathogenic. Plant diseases caused by *Xanthomonas* spp. occur worldwide. Practically all major groups of higher plants suffer, to a greater or lesser extent, from one or more *Xanthomonas*-induced disease. Predominantly, symptoms include: necrotic lesions on foliage, stems or fruit, wilts and tissue macerations. The genus has been reclassified on the basis of molecular and chemotaxonomic data and now encompasses 20 species (Vauterin *et al.*, 1995).

Miscellaneous Gram-negative plant pathogens

Acetobacter aceti, *Acetobacter pasteurianus* and *Gluconobacter oxydans* cause pink disease of pineapple fruit. The near ripe fruit becomes discoloured and sours, causing market losses rather than damage to the plant. *Rhizobacter daucus* causes bacterial gall of carrot, a disease first described in 1988, which was observed in the Aomori prefecture of Japan. *Serratia marcescens* causes crown and root rot of lucerne. This organism was not considered to be plant pathogenic until relatively recently, when '*Erwinia amylovora* var. *alfalfae*' was reassigned to the species *S. marcescens*. Like the genus *Erwinia*, *Serratia* belongs to the family *Enterobacteriaceae*. *Xylophilus ampelinus*, causing bacterial blight of grape, was previously classified as *Xanthomonas ampelina*, but is more closely related to *Pseudomonas*. *Xylella fastidiosa* causes Pierce's disease of grape, phony disease of peach, periwinkle wilt and leaf scorches of almond, plum and mulberry. This organism is xylem limited, nutritionally fastidious and related to xanthomonads.

Gram-positive genera

Clavibacter

This is a member of the family *Microbacteriaceae* in the class *Actinobacteria*. All members of this genus are plant pathogenic. Diseases caused by *Clavibacter* spp. include Gumming disease, wilts and cankers. Generally, members of this genus are host specific. Perhaps the most serious disease, ring rot of potatoes, is caused by *C. michiganensis* subsp. *sepedonicus*.

Curtobacterium

This is a member of the family *Microbacteriaceae* in the class *Actinobacteria*. Within the genus only *C. flaccumfaciens* is plant pathogenic. The different pathovars of *C. flaccumfaciens* tend to be host specific. Diseases include silvering disease of sugar beet, vascular wilts of certain members of the *Leguminosae* and leaf spots on poinsettia and tulips. It is found mostly in Europe and North America.

Rathayibacter

This is a member of the family *Microbacteriaceae* in the class *Actinobacteria*. All members of this genus are plant pathogenic. Members of this genus were formerly assigned to the genus *Clavibacter*. The genus *Rathayibacter* was created after extensive taxonomic studies indicated a number of species were quite distinct from the genus *Clavibacter sensu stricto* (Zgurskaya *et al.*, 1993) and comprise: *R. iranicus*, *R. rathayi* and *R. tritici*.

Streptomyces

This is a member of the family *Streptomycetaceae* in the class *Actinobacteria*. It is one of the exceptionally few bacterial phytopathogens that produce vegetative hyphae and extensively branching mycelia. Streptomycetes are distributed widely in nature, and the vast majority are saprophytic and relatively few can cause disease in plants, *S. ipomoeae* causing scab in sweet potato and *S. scabies* potato scab.

Miscellaneous Gram-positive plant pathogens

Arthrobacter ilicis causing bacterial blight of holly is found only in the USA. This yellow-pigmented motile coryneform is the only plant pathogen in this genus. Similar to *A. ilicis*, *Nocardia vaccinii*, which causes blueberry gall, has only been found in the USA. Growth is mycelial to produce red-pigmented colonies. *Rhodococcus fascians*, which causes leafy gall on a wide variety of hosts, is found mainly in Europe and North America.

Nomenclature of Bacterial Plant Pathogens

Nomenclature is basically similar to plants, fungi, etc., but is governed by the *International Code of Nomenclature of Bacteria* (Lapage *et al.*, 1975). New bacterial names or combinations must be validly published in accordance with these rules. The starting date for current bacterial nomenclature was 1 January 1980, coinciding with the publication of the *Approved List of Bacterial Names* (Skerman *et al.*, 1980). This work contains all names that were acceptable at that time and since then there has been one major update (Skerman *et al.*, 1989). All the changes between 1980 and 1989 are listed in Moore and Moore (1989). The purpose of the Approved List is to ensure stability and consistency in bacterial taxonomy. Names not appearing in the List can still be used but normally are enclosed within inverted commas (i.e. ' ') when appearing in written text.

At the introduction of the Approved List many old bacterial names were abandoned and strict rules for publication of new ones came into force. Many distinct plant pathogens were rejected from the Approved List a physiologi-cally too similar to warrant separation as species or subspecies. Plant patholo-gists therefore adopted a 'pathovar nomenclature' outside the jurisdiction of the Code. Lists of pathovars and standards for publication of new pathovars have been published (Dye *et al.*, 1980; Young *et al.*, 1991, 1996). The list is updated annually and is available on www.bspp.org.uk/ispp/nppb.html

Ecology and Spread

Bacterial diseases of plants occur worldwide. However, as they favour moist or warm conditions, they are of greater importance in tropical, subtropical and warm-temperate areas. It is likely that plants from all the major groups in the plant kingdom can suffer from bacterial disease. Crop losses that result from the actions of bacteria can be high; however, in relation to fungi, the potential impact of bacteria as plant pathogens is reduced by the relatively thin bacterial cell wall and their inability to produce resting structures or to penetrate the host other than through natural openings or wounds (Sigee, 1993).

The distribution of some bacterial pathogens mirrors that of the host plant: this is true of *S. scabies* on potato, *Xanthomonas axonopodis* pv. *malvacearum* on cotton and *Pseudomonas savastanoi* pv. *phaseolicola* on bean. Others are of more sporadic occurrence. The reasons for discontinuous distributions are not always clear. Bacterial wilt of groundnut caused by *R. solanacearum* can cause severe problems in China and Indonesia yet is non-existent in other countries, such as India, in which the bacterium is endemic and groundnut is also widely grown (Hayward, 1994). It is possible that these differences may result from the localized development of a population of specialized pathogens, but the impact of a complex mixture of biotic and abiotic factors coinciding to allow disease expression should not be ruled out. Clearly the discontinuous distribution of pathogens may result from geographical barriers preventing their spread. In these instances effective quarantine measures are essential. Failure can have disastrous consequences as is evident from the introduction of fireblight, caused by *E. amylovora*, into the UK and Europe (Lelliot, 1959; Meijneke, 1974) and the more recent potential threat posed by the introduction of bacterial wilt disease on potatoes in Europe (van der Wolf *et al.*, 1998).

Transmission of disease includes two rather different processes: the pathogen may be carried from a diseased to a healthy plant, or infected material may be carried from an area with disease to one without. The first process includes transmission by wind and rain, insects, nematodes, other animals, and humans and their cultural practices. The bacterium must escape from the infected plant, pass to the healthy one and be deposited in a suitable place for infection to take place. The second process includes transmission in seed and vegetative planting material such as tubers, bulbs and graft material.

Transmission over long distances and across geographical barriers usually takes place because of the activities of humans. Infected seed and planting material are of prime importance, but other materials with biological connections, such as earth, wood and plant products, may also spread disease. Even very low rates of infection in both seed and planting material can be a threat, as foci of infection can be established, which may lead to epiphytotics if conditions are favourable. For this reason various sophisticated methods have been devised to detect low levels of contamination of seed; these are reviewed in Chapter 29. Infected plants that show no symptoms, or only very slight ones, are especially dangerous when planting material is being selected.

In transmission over intermediate and short distances insects and other small fauna may be important (Harrison *et al.*, 1980). Insects that feed on plants by sucking or biting frequently transmit bacteria on their mouthparts. Those that visit flowers or leaf scars and other wounds may also transmit bacteria. However, for the vast majority of bacterial plant diseases local transmission by wind and rain together is very important. The rain mobilizes the bacteria, provides droplets for dispersal and frequently helps to waterlog the tissues of the prospective host plant, making infection easier.

Host Range

The host ranges of individual bacterial pathogens vary greatly. Some are very wide, e.g. *Agrobacterium tumefaciens*, *Pseudomonas syringae* pv. *syringae* and *R. solanacearum*, which all affect many genera and various plant families. Some are more restricted, such as *E. amylovora*, which affects a number of genera, nearly all in the family *Rosaceae*. Others have very narrow host ranges, often a single species, or a few species in a single genus, e.g. most pathovars in the genus *Xanthomonas* or the species *P. syringae*. In most instances, as might be expected, the species showing wide host ranges have, on closer examination, been found to be heterogeneous, showing divisions into strains of differing laboratory characteristics (biovars) or pathogenicity (pathovars, races, etc.).

Symptoms and Pathogenesis

Bacterial infections can cause a wide variety of disease symptoms found in plants, namely chlorosis, stunting, wilting, necrosis (either local as in leaf spots or more general as in blights), rots, cankers, scabs, galls and fasciations. Sometimes only one type of symptom is produced, e.g. leaf spots, but increased severity will frequently lead to other symptoms. Leaf spots may coalesce and defoliation occur. This may further result in stunting of the whole plant. The type and severity of symptom is usually the result of several factors including climatic conditions, resistance of the host, site and method of infection and possibly the amount of inoculum.

The mechanisms involved in symptom production are complex, and only relatively recently with developments in cellular and molecular techniques has our understanding improved (Dangl, 1994). Toxins may be produced to cause necrosis or chlorosis, as with *P. syringae* pv. *tabaci* and *P. syringae* pv. *phaseolicola*. Enzymes may be involved, as with soft-rotting *Erwinia* spp. that produce pectinases.

Control

The control of phytopathogenic bacteria is complex, and a detailed knowledge of the bacterial life cycle and its interactions with the host plant is a prerequisite for control. Control at the weakest point(s) of the bacterial life cycle is usually the best form of attack. Use can also be made of resistant varieties, seed or seed-piece selection, seed treatment (dusts and hot water), changes of cultural practices (e.g. pruning, irrigation, composting, rotation), and sanitation (in field, farm and market); sometimes biological control is possible (e.g. *A. tumefaciens* using *Agrobacterium radiobacter* and other antagonists (Cooksey and Moore, 1980; see also Chapter 34)).

It is often stated that bacterial diseases of plants are best controlled by the use of resistant varieties. When successful resistant varieties are widely available this is probably true, but such varieties are not always widely available and attempts to breed them often fail. Care must be taken to select for the best type of resistance according to the circumstances (see Chapter 31).

It must also be remembered when using resistant varieties that they may support growth of the pathogen without showing symptoms. If this is so, the pathogen may be spread to neighbouring susceptible crops, causing an overall increase in the disease and its associated losses. This could be particularly damaging where resistant varieties are not generally available.

Diagnosis of Bacterial Plant Disease

Diagnosis involves careful examination of all symptoms and consideration of other factors that may be important. Bacterial disease may be indicated by absence of more visible pathogens and pests, such as fungi, insects, etc., the presence of ooze, or bacterial streaming may be observed by microscopic examination of infected material. Only by isolation and identification of the causal bacterium can the cause of the symptoms be confirmed. The use of dilution methods to isolate the causative organism, whenever possible, is recommended. The aim is to obtain single colonies of the suspect pathogen, thus ensuring the isolation of pure cultures containing only one species of bacterium.

A range of methods can be used for the identification of phytopathogenic bacteria. Perhaps one of the simplest is based on a dichotomous key built around a small number of biochemical tests (Bradbury, 1970). This simple system has been further refined and packaged to produce the BACTID identification system (Black *et al.*, 1996), comprising miniaturized methods and a simple computer program. More detailed treatments of the use of biochemical or physiological tests are covered by Lelliott and Stead (1987), Schaad (1988) and Klement *et al.* (1990). In addition, a growing number of more sophisticated and sensitive methods are also available, many specifically designed for a particular pathogen and these are reviewed in Chapter 23.

Methods for the Isolation and Preliminary Study of Bacterial Plant Pathogens (see also Chapter 20)

Examination of infected plant material

The first step in diagnosis should be a thorough examination of the symptoms seen on the diseased plant in the field. To confirm whether the observed symptoms are bacterial in origin, a small section from the boundary between diseased and healthy tissue should be removed and mounted in a drop of water on a microscope slide, and examined under a high-power objective (40×) of a compound microscope. Great care must be taken not to confuse particulate material such as latex, plastids or starch granules with bacteria. Large masses of bacteria observed streaming from the plant tissue will generally ensure a successful subsequent isolation.

Isolation of suspected bacterial pathogen

A small piece of diseased tissue is removed aseptically from an area where bacterial activity is suspected (usually the advancing edge of a necrotic lesion or canker, or a piece of discoloured vascular tissue). This is placed in a little sterile water and teased apart with flamed needles. A loopful of the resultant suspension is then removed and streaked over the surface of a suitable agar medium. The aim is to spread the suspension so that single bacterial cells are separated, enabling growth into single colonies. Nutrient agar is usually satisfactory for isolation work, but Medium B of King *et al.* (1954) is often better as many bacteria grow well and information about fluorescence may be obtained. A few bacteria need special media to grow; thus *Xanthomonas albilineans* and *Xanthomonas populi* grow on media containing sucrose or glucose and peptone; some *Erwinia* species belonging to the amylovora group grow much better on nutrient agar supplemented with glucose (1%, w/v). A few plant pathogenic bacteria are highly fastidious, such as *Clavibacter xyli* which requires a very complex medium containing haemin and serum albumen.

After streaking, the dishes are incubated at 28°C and examined daily for 4–5 days, or up to 10–14 days if a slower-growing pathogen is suspected.

Microscopic examination and staining of suspected bacterial plant pathogens

Gram stain and cell morphology

Reagents: 1. Crystal violet (0.5% aq., w/v):
Dissolve in distilled H_2O; filter through Whatman No.1 filter paper.
2. Lugol's iodine:
Dissolve 2 g of potassium iodide in 25 ml distilled H_2O, add 1 g iodine; when dissolved make up to 100 ml with water.
3. Safranin O (0.5% aq., w/v).

Young, actively growing cultures (usually 24–48 h old) should be used. Place a small drop of preferably sterile water on a clean microscope slide. Using a cooled sterile loop remove a part of a single, well-separated colony from the agar surface and smear the bacteria on to the slide. If liquid cultures are to be observed then a loopful of the bacterial suspension will suffice. The smear should be approximately 1 cm^2 and just visible. Air dry the slide then 'heat-fix' by passing the slide, organism up, several times through the flame of a Bunsen burner. DO NOT OVERHEAT THE SLIDE. Fixing ensures the bacteria will adhere to the slide surface during the subsequent staining procedure. Flood the slide with crystal violet and leave for 30 s. Pour off excess fluid and wash with water. Flood the slide with Lugol's iodine and leave for 30 s, wash with ethanol until the colour ceases to come away from the smear (a few seconds is enough), wash with water, blot dry and counterstain with Safranin O for about 1–2 min, wash again with water and dry. Examine the slide under 100× with oil immersion.

Gram-positive cells appear as dark purplish, whereas Gram-negative cells are red. An alternative method (Gregersen, 1978; Suslow *et al.*, 1982) relies on the use of potassium hydroxide (3% aq., w/v). Using a clean Pasteur pipette place a drop of KOH on to a clean microscope slide. Then with a cooled sterile loop remove a part of a single, well-separated colony from a young actively growing agar culture. Mix the bacterial culture into the KOH until an even suspension is obtained. Lift the loop from the slide. If a string of slime is lifted with the loop (ca. 5–20 mm in length), the bacterium is Gram-negative. If a watery suspension is produced and no string of slime observed, the culture is Gram-positive. The destruction of the cell wall of Gram-negative organisms and subsequent liberation of DNA, which is very viscid in water, produces the string of slime. The Gram-positive wall is more resistant to KOH and remains intact, thus no DNA is released. This test is useful in cases of doubtful stain results.

Bacteriological media

Nutrient agar (NA)

The constituents of this medium can be purchased singly or premixed, i.e. Oxoid; CM3 or Difco; Bacto NA.

Lab-Lemco Powder (Oxoid; L29)	1.0 g
Yeast extract (Oxoid; L21)	2.0 g
Peptone Bacteriological (Oxoid; L34)	5.0 g
NaCl	5.0 g
Agar No. 3 (Oxoid; L13)	15.0 g
Distilled H_2O	1000 ml
pH	7.2–7.4

Note: most plant-pathogenic bacteria prefer a neutral or slightly alkaline pH and will grow sufficiently well on this medium. The following organisms do not grow on nutrient agar, or make unsatisfactory growth: sugar cane pathogens, *X. albilineans*, *X. populi*, some *Clavibacter* spp., *Erwinia* spp. and some of the *E. amylovora* group. For these organisms one or more of the following media should be tried: GP, YDC or KB.

Glucose peptone agar (GP)

Peptone Bacteriological (Oxoid; L34)	5.0 g
Glucose	20.0 g
K_2HPO_4 (anhydrous)	0.5 g
$MgSO_4.7H_2O$	0.25 g
Agar No. 3 (Oxoid; L13)	15.0 g
Distilled H_2O	1000 ml
pH	7.2–7.4

Note: suitable for the growth of *X. albilineans*, and has also been used successfully to grow *X. populi*, although a more specialized medium is recommended.

Yeast dextrose chalk agar (YDC)

Yeast extract (Oxoid; L21)	10.0 g
D-Glucose	20.0 g
$CaCO_3$	20.0 g
Agar No. 3 (Oxoid; L13)	12.0 g
Distilled H_2O	1000 ml
pH	7.2

Note: suitable for maintenance and isolation of *Xanthomonas* spp. Calcium carbonate buffers the medium and ensures the pH does not become acidic. *Xanthomonas* spp. are not tolerant of acid pH.

King's medium B (KB)

Proteose peptone	20.0 g
(Difco No.3/ Oxoid; L46)	
Glycerol	10.0 g
K_2HPO_4 (anhydrous)	1.5 g
$MgSO_4.7H_2O$	1.5 g
Agar	15.0 g
Distilled H_2O	1000 ml
pH	7.2

Note: good for fluorescent pseudomonads and *Erwinia* spp.

References

Black, R., Holt, J. and Sweetmore, A. (1996) *BACTID: Bacteriological Identification System for Resource-poor Plant Pathology Laboratories*. Natural Resources Institute, Chatham, UK.

Bradbury, J.F. (1970) Isolation and preliminary study of bacteria from plants. *Review of Plant Pathology* 49, 213–218.

Bradbury, J.F. (1986) *Guide to Plant Pathogenic Bacteria*. CAB International Mycological Institute, Kew, UK.

Cooksey, D.A. and Moore, L.W. (1980) Biological control of crown gall with fungal and bacterial antagonists. *Phytopathology* 70(6), 506–509.

Dangl, J.L. (1994) *Bacterial Pathogenesis of Plants and Animals: Molecular and Cellular Mechanisms*. Springer-Verlag, Berlin, Germany.

Dye, D.W., Bradbury, J.F., Goto, M., Hayward, A.C., Lelliot, R.A. and Schroth, M.N. (1980) International standards for naming pathovars of phytopathogenic bacteria and a list of pathovar names and pathotypes. *Review of Plant Pathology* 59, 153–168.

Gardan, L., Prior, P., Gillis, M. and Saddler, G.S. (2000) Description of *Acidovorax anthurii* sp. nov., a new phytopathogenic bacterium which causes bacterial leaf-spot on *Anthurium*. *International Journal of Systematic Bacteriology* 50, 235–246.

Gregersen, T. (1978) Rapid method for distinction of Gram-negative from Gram-positive bacteria. *European Journal of Applied Microbiology and Biotechnology* 5, 123–127.

Harrison, M.D., Brewer, J.W. and Merrill, L.D. (1980) Insect involvement in the transmission of bacterial pathogens. In: Harris, K.F. and Marramorosch, K. (eds) *Vectors of Plant Pathogens*. Academic Press, New York, USA.

Hayward, A.C. (1994) The hosts of *Pseudomonas solanacearum*. In: Hayward, A.C. and Hartman, G.L. (eds) *Bacterial Wilt: the Disease and its Causative Agent*, Pseudomonas solanacearum. CAB International, Wallingford, UK, pp. 9–24.

Kado, C.I. (1992) Plant pathogenic bacteria. In: Balowes, A., Trüper, H.G., Dworkin, M., Harder, W. and Schleifer, K.-H. (eds) *The Prokaryotes. A Handbook on the Biology of Bacteria: Ecophysiology, Isolation, Identification, Applications*, 2nd edn. Springer-Verlag, New York and Berlin, pp. 659–674.

Kersters, K., Ludwig, W., Vancanneyt, M., Vos, P.D., Gillis, M., Schleifer, K.H. and De Vos, P. (1996) Recent changes in the classification of the pseudomonads: an overview. *Systematic and Applied Microbiology* 19, 465–477.

King, E.O., Ward, M.K. and Raney, D.E. (1954) Two simple media for the demonstration of pyocyanin and fluoroscein. *Journal of Laboratory and Clinical Medicine* 44, 301–307.

Klement, Z., Rudolph, K. and Sands, D.C. (1990) *Methods in Phytobacteriology*. Akadémiai Kiadó Budapest, Hungary.

Lapage, S.P., Sneath, P.H.A., Lessel, E.F., Skerman, V.B.D., Seeliger, H.P.R. and Clark, W.A. (1975) *International Code of Nomenclature of Bacteria*. American Society for Microbiology, Washington, DC, USA.

Lelliot, R.A. (1959) Fireblight of pears in England. *Agriculture* 65, 564–568.

Lelliot, R.A. and Stead, D.E. (1987) *Methods for the Diagnosis of Bacterial Diseases of Plants*. Blackwell Scientific Publications, London, UK.

Meijneke, C.A.R. (1974) The 1971 outbreak of Fireblight in The Netherlands. In: *Proceedings of the 19th International Horticultural Congress*, Vol. 2, pp. 373–382.

Mergaert, J., Verdonck, L. and Kersters, K. (1993) Transfer of *Erwinia ananas* (synonym, *Erwinia uredovora*) and *Erwinia stewartii* to the genus *Pantoea* emend. as *Pantoea ananas* (Serrano, 1928) comb. nov. and *Pantoea stewartii* (Smith, 1898) comb. nov., respectively, and description of *Pantoea stewartii* subsp. *indologenes* subsp. nov. *International Journal of Systematic Bacteriology* 43, 162–173.

Moore, W.E.C. and Moore, L.V.H. (1989) *Index of Bacterial and Yeast Nomenclatural Changes*. American Society for Microbiology, Washington, DC, USA.

Palleroni, N.J. (1993) *Pseudomonas* classification: a new case history in the taxonomy of Gram-negative bacteria. *Antonie van Leeuwenhoek* 64, 231–251.

Roberts, S.J., Eden-Green, S.J., Jones, P. and Ambler, D.J. (1990) *Pseudomonas syzygii*, sp. nov., the cause of Sumatra disease of cloves. *Systematic and Applied Microbiology* 13, 34–43.

Schaad, N.W. (1988) *Laboratory Guide for Identification of Plant Pathogenic Bacteria*, 2nd edn. American Phytopathological Society, St Paul, Minnesota, USA.

Seal, S.E., Taghavi, M., Fegan, N., Hayward, A.C. and Fegan, M. (1999) Determination of *Ralstonia* (*Pseudomonas*) *solanacearum* rDNA subgroups by PCR tests. *Plant Pathology* 48, 115–120.

Shaw, C.H., Loake, G.J., Brown, A.P. and Garrett, C.S. (1991) The early events in *Agrobacterium* infection. In: Smith, C.J. (ed.). *Biochemistry and Molecular Biology of Plant–Pathogen Interactions. Proceedings of the Phytochemical Society of Europe no. 32*. Clarendon Press, Oxford, UK, pp. 197–209.

Sigee, D.C. (1993) *Bacterial Plant Pathology: Cell and Molecular Aspects*. Cambridge University Press, Cambridge, UK.

Skerman, V.B.D., McGowan, V. and Sneath, P.H.A. (1980) Approved list of bacterial names. *International Journal of Systematic Bacteriology* 30, 225–420.

Skerman, V.B.D., McGowan, V. and Sneath, P.H.A. (1989) *Approved List of Bacterial Names; Amended Edition*. American Society of Microbiology, Washington, DC, USA.

Stackebrandt, E., Murray, R.G.E. and Trüper, H.G. (1988) *Proteobacteria classis* nov., a name for the phylogenetic taxon that includes the 'purple bacteria and their relatives'. *International Journal of Systematic Bacteriology* 38, 321–325.

Stackebrandt, E., Rainey, F.A. and Ward-Rainey, N.L. (1997) Proposal for a heirarchic classification system, *Actinobacteria classis* nov. *International Journal of Systematic Bacteriology* 47, 479–491.

Suslow, T.V., Schroth, M.N. and Isaka, M. (1982) Application of a rapid method for Gram differentiation of plant pathogenic and saprophytic bacteria without staining. *Phytopathology* 72, 917–918.

Vauterin, L., Hoste, B., Kersters, K. and Swings, J. (1995) Reclassification of *Xanthomonas*. *International Journal of Systematic Bacteriology* 45, 472–489.

Willems, A., Goor, M., Thielemans, S., Gillis, M., Kersters, K. and De Ley, J. (1992) Transfer of several phytopathogenic *Pseudomonas* species to *Acidovorax* as *Acidovorax avenae* subsp. *avenae* subsp. nov., comb. nov., *Acidovorax avenae* subsp. *citrulli*, *Acidovorax avenae* subsp. *cattleyae*, and *Acidovorax konjaci*. *International Journal of Systematic Bacteriology* 42, 107–119.

van der Wolf, J.M., Bonants, P.J.M., Smith, J.J., Hagenaar, M., Nijhuis, E., van Beckhoven, J.R.C.M., Feulliade, R., Trigalet, A. and Saddler, G.S. (1998) Genetic diversity of *Ralstonia solanacearum* race 3 in Western Europe determined by AFLP, RC-PFGE and PCR with repetitive sequences. In: Prior, P., Allen, C. and Elphinstone, J. (eds) *Bacterial Wilt Disease: Molecular and Ecological Aspects.* Springer-Verlag, Berlin, Germany, pp. 44–49.

Yabuuchi, E., Kosako, Y., Oyaizu, H., Yano, I., Hotta, H., Hashimoto, Y., Ezaki, T. and Arakawa, M. (1992) Proposal of *Burkholderia* gen. nov. and transfer of seven species of the genus *Pseudomonas* homology group II to the new genus, with the type species *Burkholderia cepacia* (Palleroni & Holmes 1981) comb. nov. *Microbiology and Immunology* 36, 1251–1275.

Yabuuchi, E., Kosako, Y., Yano, I., Hotta, H. and Nishiuchi, Y. (1995) Transfer of two *Burkholderia* and an *Alcaligenes* species to *Ralstonia* gen. nov.: proposal of *Ralstonia pickettii* (Ralston, Palleroni and Doudoroff 1973) comb. nov., *Ralstonia solanacearum* (Smith 1896) comb. nov. and *Ralstonia eutropha* (Davis 1969) comb. nov. *Microbiology and Immunology* 39, 897–904.

Young, J.M., Bradbury, J.F., Davis, R.E., Dickey, R.S., Ercolani, G.L., Hayward, A.C. and Vidaver, A.K. (1991) Nomenclatural revisions of plant pathogenic bacteria and list of names 1980–1988. *Review of Plant Pathology* 70, 211–221.

Young, J.M., Saddler, G.S., Takikawa, Y., De Boer, S.H., Vauterin, L., Gardan, L., Gvozdyak, R.I. and Stead, D.E. (1996) Names of plant pathogenic bacteria 1864–1995. *Review of Plant Pathology* 75, 721–763.

Zgurskaya, H.I., Evtushenko, L.I., Akimov, V.N. and Kalakoutskii, L.V. (1993) *Rathayibacter* gen. nov., including the species *Rathayibacter rathayi* comb. nov., *Rathayibacter tritici* comb. nov., *Rathayibacter iranicus* comb. nov., and six strains from annual grasses. *International Journal of Systematic Bacteriology* 43, 143–149.

Virus Diseases

11

Revised by J.M. Waller

CABI Bioscience UK Centre, Bakeham Lane, Egham, Surrey TW20 9TY, UK

Introduction

Viruses are obligate parasites of submicroscopic size, with one dimension smaller than 200 nm. Virus particles, or virions, consist of segments of double- or single-stranded RNA or DNA encased in protein structures, in some cases with lipid and additional substances. Most plant viruses have single-stranded RNA genomes. Identification is based principally on morphology, biochemical and biological properties, and serological reactions of the virus particles.

Viruses, unlike microbial parasites, utilize the processes of the living cell to effect their own multiplication. Viruses also lack the machinery for the production of energy through respiration, and for at least some viruses the isolated nucleic acid genome is infective.

Originally a viral aetiology was inferred for diseases that are transmissible in the absence of a pathogen visible by light microscopy. In many instances the inference has been confirmed through the use of an electron microscope and techniques for concentrating and purifying virus particles. However, some diseases that were previously considered to be caused by viruses have now been shown to be due to other pathogens. These include mycoplasma-like organisms (MLOs) now referred to as phytoplasmas (see Chapter 12). Moreover, biochemical studies have distinguished viruses from viroids. Viroids are uncoated RNA species, which apparently never produce recognized nucleoprotein virus particles and contain insufficient genetic information to code for their own replication.

©CAB *International* 2002. *Plant Pathologist's Pocketbook*
(eds J.M. Waller, J.M. Lenné and S.J. Waller)

Disease Symptoms

In nature, serious disease symptoms normally occur only when a virus has invaded the host plant systemically. The 'local lesions' frequently seen following experimental inoculation of the leaves of test plants have little impact on growth or yield. Plant viruses differ from some other pathogens in that they do not induce disease through the production of a translocatable toxin from a localized point of infection. Virus infection disturbs host cell metabolism; commonly occurring changes include a decrease in photosynthesis, an increased respiration rate, higher phenoloxidase activity, and frequently the abnormal accumulation of metabolic products.

Visible symptoms vary with the interaction of the particular strain of the virus and the genotype of the host plant, and may be modified by environmental conditions. Two strains of the same virus may induce very different symptoms in the same plant cultivar, whereas the same virus isolate may produce very different reactions in different plant species or cultivars. Symptoms are often most intense in the early 'acute' stages of the disease, and subsequently become less obvious in the later 'chronic' phase. Many viruses typically cause light and dark green mosaic or mottling symptoms in affected plants, and there may be slight or severe puckering, distortion or tissue proliferation. The latter may assume the form of superficial leafy outgrowths or 'enations'. Ringspots, and ring and line or 'oak-leaf' patterns of chlorotic or necrotic tissue are commonly seen with many different viruses. Necrotic flecks or areas of dead tissue, sometimes killing the plant, is another symptom. Flowers may also show pale streaks or flecks (colour breaking). Roots seldom show obvious symptoms, even though certain groups of viruses characteristically invade the plant through the roots. Occasionally root necrosis is produced, e.g. with *Cacao swollen shoot virus* and *Cassava brown streak virus*. Certain genetic abnormalities such as leaf variegation can closely resemble virus symptoms.

Virus-infected plants are usually smaller and less vigorous than their healthy counterparts and, especially with mixed infections of two or more viruses, very severe stunting can occur. Fruits may be mottled, misshapen or necrotic; they may drop prematurely, and yield is often considerably reduced.

The latent period between infection and appearance of symptoms, during which the virus may or may not be detectable, is usually between 1 and 6 weeks for herbaceous hosts but can be as short as 24 h for *Tobacco rattle virus* in *Phaseolus vulgaris*. Latent periods are much longer and up to a year or more for viruses of woody perennials such as blackcurrant and other deciduous fruit crops.

Symptoms can be greatly affected by environmental conditions, especially temperature and light intensity. Many viruses are more damaging under conditions of low light and temperature; others (for example *Chrysanthemum stunt viroid*) seldom cause symptoms under such conditions but are clearly manifest under high light and temperature. Many viruses induce microscopic inclusion bodies within infected cells; the bodies may be amorphous ('X-bodies') or crystalline, and are readily visible by light microscopy of epidermal peelings prepared with vital stains such as trypan blue. Electron microscopy of ultrathin sections demonstrates other types of inclusion body, such as the 'pin-wheels' characteristic of the family *Potyviridae*.

A comprehensive definition of symptom types and terms to describe them is given by Bos (1978).

Viruses of Fungi

Viruses or virus-like particles have been reported in well over 100 species of fungi, from all the main taxonomic groups. The effects of the viruses on the fungal hosts are difficult to establish, for very few of the viruses can be transmitted experimentally using cell-free preparations of the virus. In *Agaricus bisporus*, however, it has been well established that the complex of at least five viruses can cause disease and loss of crop; and viruses are involved in the 'killer systems' of brewer's yeast. In some instances virus infection may increase antibiotic production, or other secondary metabolites such as mycotoxins. However, in most cases virus infection has little effect on the production of these substances.

Many plant-pathogenic fungi have been reported to be infected with one or more viruses, but the presence of virus has rarely been associated with fungus pathogenicity. However, hypovirulence in certain isolates of the chestnut blight fungus (*Endothia parasitica*) is associated with virus infection and is transmissible by hyphal anastomosis. Severely affected trees implanted at the sites of cankers with hypovirulent cultures of *E. parasitica* show subsequent regression of the cankers and this has been used in attempted biological control of the disease.

Transmission and Spread of Viruses

Virtually all viruses that infect their host systemically can be transmitted by vegetative propagation including grafting. Viruses are thus frequently disseminated in infected planting material. Some viruses such as those causing cassava mosaic disease may be erratically distributed or localized within the plant, however, and are not necessarily detected in all parts so that a small proportion of the vegetative progeny from, for example cuttings, often escapes infection.

Many viruses can be transmitted by inoculation of sap, although substances present in some plant species may prevent successful transmission by inhibiting infection or inactivating viruses. Relatively few viruses, however, are sufficiently contagious to be transmitted simply by contact between plant tissue or through handling and slight abrasion of the plants during cultural operations. Those that are highly contagious are common in intensive monoculture, and can be difficult to control, e.g. *Potato virus X*, *Tobacco mosaic virus*, *Carnation mottle virus* and the viroids *Potato spindle tuber* and *Chrysanthemum stunt*.

Some viruses may be seed-borne, e.g. many potyviruses (genus *Potyvirus*; family *Potyviridae*) such as *Lettuce mosaic virus*, *Bean common mosaic virus* and *Soybean mosaic virus*. About 10% of the seedlings from an infected parent may be infected, and these play an important part in virus survival and epidemiology. A few viruses, such as *Prunus necrotic ringspot virus*, are also transmitted in pollen to the plant pollinated.

Many viruses are transmitted by insects, especially those with sucking mouthparts (particularly aphids, whiteflies and leafhoppers). Some are transmitted by insects having biting mouthparts (e.g. Chrysomelid beetles). In most instances, there is a specific relationship between the vector and the virus. Some viruses are transmitted in a 'non-persistent' manner when the acquisition and inoculation processes may involve only the terminal part of the stylets. Acquisition and inoculation occur after superficial probes or short periods of feeding, and infectivity is soon lost. These features are characteristic of viruses of the *Carlavirus* and *Potyvirus* genera, as well as cucumoviruses such as *Cucumber mosaic virus* and *Tomato aspermy virus*. Viruses transmitted in a 'persistent' manner are acquired only during a prolonged feed on an infected plant, and an additional latent period may then be necessary before the insect becomes infective and can transmit. Some vectors may remain infective for the rest of their lives, although only in a few instances is there conclusive evidence that the viruses cause disease in the insect. With intermediate 'semi-persistent' viruses, acquisition and inoculation feeds of a few hours are optimal. Most virus–vector relationships are usually highly specific, but some aphid species can transmit several or many different viruses, e.g. *Myzas persicae*. Other arthropod vectors include Eriophyid mites and beetles. The thrips vector of *Tomato spotted wilt virus* must acquire the virus in the larval stage in order to transmit.

A number of free-living ectoparasitic soil nematodes of the genera *Trichodorus*, *Xiphinema* and *Longidorus* are highly specific vectors of viruses of the genera *Nepovirus* and *Tobravirus*. A few viruses are transmitted in soil by zoospores of Chytrid fungi and passive transmission without any living vector occurs in soil with some others.

Although experimental transmission of many viruses has been obtained by training the parasitic plant dodder (*Cuscuta* spp.) from an infected to a healthy plant, it is unlikely that such transmissions occur to any extent in nature.

Naming, Grouping and Classification of Plant Viruses

The International Committee on Nomenclature of Viruses was formed in 1966 and there are separate subcommittees responsible for dealing with the viruses of vertebrates, invertebrates, bacteria and plants. By 1975 the name of the international body had become the International Committee on Taxonomy of Viruses (ICTV), and some 20 groups of plant viruses had been established. Earlier, some use had been made of cryptograms (Gibbs *et al.*, 1966), which indicate in 'short-hand' form certain taxonomically significant properties of the virus. The cryptogram consists of four pairs of symbols indicating: (i) type of nucleic acid; (ii) molecular weight of the nucleic acid and percentage in infective particles; (iii) morphology of the virus particle; and (iv) type of host infected and mode of transmission. Crytograms never became widely used and have been largely abandoned.

Many of the plant viruses that are recognized as distinct taxonomic entities are now grouped into genera and some of these into families, and the Sixth Report of the International Committee for the Taxonomy of Viruses lists nine families and 33 genera. Table 11.1 shows the classification of plant virus genera approved by the ICTV in May 2000. Virus species names are usually based on host plant and symptom.

A database of 864 plant viruses is also given in Viruses of Plants: Descriptions and Lists from the VIDE Database (Brunt *et al.*, 1996).

A summary of the characteristcs of some plant virus genera is as follows.

- *Alfamovirus*: aphid vector, non-persistent; seed, pollen and mechanical transmission. Bacilliform particles; ssRNA; e.g. *Alfalfa mosaic*.
- *Alphacryptovirus*: often no symptoms. Seed and pollen transmission, no vector. Isometric particles; dsRNA; e.g. *White clover cryptic 1*.
- *Badnavirus*: several tropical crops affected. Various insect vectors, semipersistent or persistent. Some may be mechanically, graft- or seed-transmitted. Bacilliform particles; dsRNA; e.g. *Banana streak, Cocoa swollen shoot, Potato leaf curl*.
- *Betacryptovirus*: seed transmission. Isometric particles; dsRNA; e.g. *White clover cryptic 2*.
- *Bromovirus*: transmitted by Coleoptera, mechanical inoculation, grafting and contact. Isometric particles; ssRNA; e.g. *Broad bean mottle, Cowpea chlorotic mottle*.
- *Bymovirus*: mainly cereals affected. Transmitted by fungi (*Plasmodiophorales*) and by mechanical inoculation. Filmentous particles; ssRNA; e.g. *Barley mosaic, Wheat mosaic*.
- *Capillovirus*: some fruit trees. No vector; some transmission by grafting, mechanical inoculation and through seed. Filamentous particles; ssRNA; e.g. *Apple stem grooving*.
- *Carlavirus*: wide host range. Aphid and white-fly vectors, non- or semipersistent, also transmitted mechanically; some by grafting, through seed and pollen. Filmentous particles; ssRNA; e.g. *Potato virus S, Hop mosaic, Cowpea mild mottle*.
- *Carmovirus*: coleopteran or chytrid (fungal) vectors, non-persistent, some may be transmitted by mechanical inoculation, grafting, contact or through seed. Isometric particles; ss RNA; e.g. *Bean mild mosaic, Cucumber leaf spot*.
- *Caulimovirus*: aphid vectors, non- or semipersistent, also transmitted by grafting and some by mechanical inoculation and contact. Isometric particles; dsRNA; e.g. *Cauliflower mosaic, Soybean chlorotic mottle*.
- *Closterovirus*: several insect vectors, non-or semipersistent, also transmitted by grafting and some by mechanical inoculation. Filamentous particles; ssRNA; e.g. *Beet yellows, Citrus tristeza*.
- *Comovirus*: coleopteran vectors, semi- or non-persistent; also transmitted by mechanical inoculation and grafting; some by contact and through seed or pollen. Isometric particles; ssRNA; e.g. *Cowpea mosaic, Squash mosaic*.
- *Cucumovirus*: wide host range. Aphid vector, non-persistent, also transmitted by mechanical inoculation, grafting and through seed. Isometric particles, ssRNA; e.g. *Cucumber mosaic, Peanut stunt*.

Table 11.1. A classification of plant viruses (as approved by ICTV in May 2000).

Genome	Family	Genus	Type species
DNA	*Caulimoviridae*	*Caulimovirus*	*Cauliflower mosaic virus*
		'SoyCMV-like'	*Soybean chlorotic mottle virus*
		'CVMC-like'	*Cassava vein mosaic*
		'PVCV-like'	*Petunia vein clearing virus*
		Badnavirus	*Commelina yellow mottle virus*
		'RTBV-like'	*Rice tungro bacilliform virus*
	Geminiviridae	*Mastrevirus*	*Maize streak virus*
		Curtovirus	*Beet curly top virus*
		Begomovirus	*Bean golden mosaic virus*
		Topocuvirus	*Tomato pseudo-curly top virus*
	–	*Nanovirus*	*Subterranean clover stunt virus*
dsRNA	*Reoviridae*	*Phytoreovirus*	*Wound tumour virus*
		Fijivirus	*Fiji disease virus*
		Oryzavirus	*Rice ragged stunt virus*
	Partitiviridae	*Alphacryptovirus*	*White clover cryptic virus 1*
		Betacryptovirus	*White clover cryptic virus 2*
	–	*Varicosavirus*	*Lettuce big-vein virus*
(–)ssRNA	*Rhabdoviridae*	*Cytorhabdovirus*	*Lettuce necrotic yellows virus*
		Nucleorhabdovirus	*Potato yellow dwarf virus*
	Bunyaviridae	*Tospovirus*	*Tomato spotted wilt virus*
	–	*Tenuivirus*	*Rice stripe virus*
	–	*Ophiovirus*	*Citrus porosis virus*
(+)ssRNA	*Bromoviridae*	*Bromovirus*	*Brome mosaic virus*
		Cucumovirus	*Cucumber mosaic virus*
		Alfamovirus	*Alfalfa mosaic virus*
		Ilarvirus	*Prunus necrotic ringspot virus*
		Oleavirus	*Olive latent virus 2*
	Closteroviridae	*Closterovirus*	*Beet yellows virus*
		Crinivirus	*Lettuce infectious yellows virus*
	Comoviridae	*Comovirus*	*Cowpea mosaic virus*
		Nepovirus	*Tobacco ringspot virus*
	Luteoviridae	*Luteovirus*	*Barley yellow dwarf virus*
		Polerovirus	*Potato leaf roll virus*
		Enamovirus	*Pea enation mosaic virus 1*
	Potyviridae	*Potyvirus*	*Potato virus Y*
		Rymovirus	*Ryegrass mosaic virus*
		Bymovirus	*Barley yellow mosaic virus*
		Macluravirus	*Maclura mosaic virus*
		Ipomovirus	*Sweet potato mild mottle virus*
		Tritimovirus	*Wheat streak mosaic virus*
	Sequiviridae	*Sequivirus*	*Parsnip yellow fleck virus*
		Waikavirus	*Rice tungro spherical virus*
	Tombusviridae	*Tombusvirus*	*Tomato bushy stunt virus*
		Carmovirus	*Carnation mottle virus*
		Necrovirus	*Tobacco necrosis virus A*
		Machlomovirus	*Maize chlorotic mosaic virus*
		Dianthovirus	*Carnation ringspot virus*
		Avenavirus	*Oat chlorotic stunt virus*
		Aureusvirus	*Pothos latent virus*

Continued

Table 11.1.   *Continued*

Genome	Family	Genus	Type species
	Unassigned[a]	Panicovirus	Panicum mosaic virus
		Tobravirus	Tobacco rattle virus
		Tobamovirus	Tobacco mosaic virus
		Hordeivirus	Barley stripe mosaic virus
		Furovirus	Soil-borne wheat mosaic virus
		Pomovirus	Potato mop-top virus
		Pecluvirus	Peanut clump virus
		Benyvirus	Beet necrotic yellow vein virus
		Sobemovirus	Southern bean mosaic virus
		Marafivirus	Maize rayo findo virus
		Umbravirus	Carrot mottle virus
		Tymovirus	Turnip yellow mosaic virus
		Ideovirus	Raspberry bushy dwarf virus
		Ourmiavirus	Ourmia melon virus
		Potexvirus	Potato virus X
		Carlavirus	Carnation latent virus
		Foveavirus	Apple stem pitting virus
		Alexivirus	Shallot virus X
		Capillovirus	Apple stem grooving virus
		Trichovirus	Apple chlorotic leaf spot virus
		Vitivirus	Grapevine virus A

[a]For many reported viruses, there are insufficient data to assign them to a family or genus.

- *Cytorhabdovirus*: aphid or delphacid vectors, persistent, also transmitted by grafting and some by mechanical inoculation. Bullet-shaped particles; ssRNA; e.g. *Cereal northern mosaic, Broccoli necrotic yellows, Lettuce necrotic yellows*.
- *Dianthovirus*: nematode vector, also transmitted by grafting, mechanical inoculation and contact. Isometric particles; ssRNA; e.g. *Carnation ring spot*.
- *Enamovirus*: aphid vector, persistent, also transmitted by mechanical inoculation and through seed. Isometric particles; ssRNA; e.g. *Pea enation mosaic 1*.
- *Fijivirus*: delphacid vector, persistent, some transmission by mechanical inoculation. Isometric particles; dsRNA; e.g. *Sugarcane Fiji disease, Maize rough dwarf*.
- *Furovirus*: fungal vector (*Plasmodiophorales*) or mechanical transmission and by grafting, some by contact and through seed. Rod-shaped particles; ssRNA; e.g. *Beet necrotic yellow vein, Potato mop-top*.
- *Hordeivirus*: no vector; mechanical transmission, by grafting, through pollen and some by seed. Rod-shaped particles; ssRNA; e.g. *Barley stripe mosaic*.
- *Idaeovirus*: no vector; transmission by mechanical inoculation, grafting, through seed and pollen. Isometric particles; ssRNA; e.g. *Raspberry bushy dwarf*.

- *Ilarvirus*: thysanopteran vector, or transmission by mechanical inoculation, grafting, some by contact or through seed and pollen. Isometric particles; ssRNA; e.g. *Apple mosaic.*
- *Ipomovirus*: white-fly vector, persistent; also transmitted by mechanical inoculation and grafting. Filamentous particles; ssRNA; e.g. *Sweet potato mild mottle.*
- *Luteovirus*: aphid vector, persistent, some transmitted by grafting. Isometric particles; ssRNA; e.g. *Barley yellow dwarf, Potato leafroll.*
- *Machlomovirus*: coleopteran or thysanopteran vector, non-persistent, transmitted by mechanical inoculation and through seed. Isometric particles; ssRNA; e.g. *Maize chlorotic mottle.*
- *Macluravirus*: aphid vector, non-persistent; also transmitted by mechanical inoculation and grafting. Filamentous particles; ssRNA; e.g. *Maclura mosaic.*
- *Marafivirus*: cicadellid vectors, persistent. Isometric particles; ssRNA; e.g. *Maize rayo fino.*
- *Mastrevirus*: cicadellid vectors, non- or semipersistent, some transmitted by grafting, mechanical inoculation or through seed. Double (geminate) particles; ssDNA; e.g. *Maize streak.*
- *Nanovirus*: aphid or other insect vectors, semipersistent or persistent; some transmitted by grafting. Isometric particles; ssDNA; e.g. *Banana bunchy top.*
- *Necrovirus*: chytrid (fungal) vectors, also transmitted by mechanical inoculation. Isometric particles; ssRNA; e.g. *Tobacco necrosis.*
- *Nepovirus*: wide host range. Nematode (dorylamid) vector, also transmitted by grafting and mechanical inoculation, some through seed or pollen. Isometric particles; ssRNA; e.g. *Arabis mosaic, Grape fanleaf, Tobacco ringspot, Tomato ringspot.*
- *Nucleorhabdovirus*: mostly aphid, cicadellid or delphacid vectors, persistent, also some transmitted by grafting, mechanical inoculation, contact or through pollen. Bacilliform or bullet-shaped particles; ssRNA; e.g. *Maize mosaic.*
- *Oryzavirus*: delphacid vectors, persistent, some transmitted by mechanical inoculation. Isometric particles; dsRNA; e.g. *Rice ragged stunt.*
- *Ourmiavirus*: no vector; transmission by grafting mechanical inoculation, seed and pollen. Bacilliform particles, ssRNA; e.g. *Ourmia melon.*
- *Phytoreovirus*: cicadellid vectors, persistent. Isometric particles; dsRNA; e.g. *Rice dwarf.*
- *Potexvirus*: wide range of hosts. Aphid or mite vectors, non-persistent; transmitted by grafting, some by mechanical inoculation, contact and through seed. Filamentous particles; ssRNA; e.g. *Papaya mosaic, Potato X.*
- *Potyvirus*: very large range of hosts. Aphid vectors, non-persistent, most also transmitted by grafting, mechanical inoculation, some by contact, through seed and pollen. Filamentous particles; ssRNA; e.g. *Bean common mosaic, Lettuce mosaic, Potato A, Potato V, Potato Y, Tulip breaking, Yam mosaic.*
- *Rymovirus*: mite vector, also transmitted by mechanical inoculation, some through seed. Filamentous particles; ssRNA; e.g. *Ryegrass mosaic.*

- *Sequivirus*: aphid vector, semipersistent; also transmitted by mechanical inoculation. Isometric particles; ssRNA; e.g. *Parsnip yellow fleck*.
- *Sobemovirus*: aphid, cicadellid or coleopteran vectors, non- or semi-persistent, also transmitted by mechanical inoculation, some by grafting, through seed or pollen. Isometric particles; ssRNA; e.g. *Bean southern mosaic, Rice yellow mottle*.
- *Tenuivirus*: cicadellid or delphacid vectors, persistent; one transmitted by mechanical inoculation. Filamentous or bullet-shaped particles; ssRNA; e.g. *Rice grassy stunt, Hoja blanca*.
- *Tobamovirus*: most transmission by mechanical inoculation, grafting, contact or seed; coleopteran vector for one species. Rod-shaped particles; ssRNA; e.g. *Pepper mild mottle, Tobacco mosaic, Tomato mosaic*.
- *Tobravirus*: nematode (trichodorid) vector, also by grafting, mechanical inoculation and through seed. Rod-shaped particles; ssRNA; e.g. *Tobacco rattle, Pepper ring spot*.
- *Tombusvirus*: most transmission by mechanical inoculation, grafting. Chytrid (fungal) vector for one, some contact, seed and pollen trans-mission. Isometric particles; ssRNA; e.g. *Tomato bushy stunt*.
- *Tospovirus*: thysanopteran vectors, semipersistent or persistent, also trans-mitted by mechanical inoculation and by grafting. Isometric particles; ssRNA; e.g. *Tomato spotted wilt*.
- *Trichovirus*: aphid, pseudococcid or mite vectors, semipersistent; also transmitted by grafting and mechanical inoculation, some through seed and pollen. Filamentous particles; ssRNA; e.g. *Apple chlorotic leaf spot*.
- *Tymovirus*: mostly coleopteran vectors, non- or semipersistent; also trans-mitted by mechanical inoculation, grafting, some through contact or seed. Isometric particles; ssRNA; e.g. *Eggplant mosaic, Turnip yellow mosaic*.
- *Umbravirus*: aphid vectors, persistent; also transmitted by mechanical inoculation. Isometric particles; ssRNA; e.g. *Groundnut rosette*.
- *Varicosavirus*: chytrid (fungal) vectors; also some transmitted by mechanical inoculation and grafting. Rod-shaped particles; dsRNA; e.g. *Lettuce big-vein*.
- *Waikavirus*: aphid or cicadellid vectors, semipersistent. Isometric particles; ssRNA; e.g. *Rice tungro bacilliform*.
- Other virus-like bodies
  Satellite viruses: occur as components of other virus diseases, e.g. tobacco necrosis, cucumber mosaic; usually isometric; ssRNA particles; many are viroid-like.
  Plant viroids: are uncoated RNA species which are stable and highly conta-gious and include *Potato spindle tuber, Citrus exocortis, Chrysanthemum chlorotic mottle, Chrysanthemum stunt, Cucumber pale fruit* and *Coconut cadang-cadang*. The disease symptoms include reduced growth or stunting, chlorosis and tissue abnormalities, and the incubation periods in most hosts are often several weeks to several months.

Viroids are most simply detected by sap inoculation or grafting to indicator species, but polyacrylamide gel electrophoresis of nucleic acid extracts can detect the characteristic RNA species not present in healthy plants. See Diener (1979) for a review of viroids.

How to Describe a Virus

Biological or biochemical properties of viruses may differ as much between strains of the same virus as they do between different viruses, whereas other properties may be very similar for a number of different viruses. Properties that are reliable to identify and distinguish between the viruses in one taxonomic group may be of little use with the viruses of other groups. Moreover, full characterization of a virus is often impracticable, either for lack of the facilities and apparatus needed, or because the virus is not amenable to purification using present techniques.

However some characters are much more useful than others for virus identification and for a reasonably full description of a plant virus, as much as possible of the following information should ideally be determined.

- *Tests for experimental transmission*
  By grafting, inoculation of sap, dodder, etc.

- *Host range and symptoms*
  Occurrence in naturally infected plants and tests with experimentally infected plants for propagating the virus, diagnostic reactions and local lesion assay.
  Effects of higher or lower temperature and light regimes on host susceptibility, virus concentration and symptoms produced.

- *Properties in sap* (indicating the limits between which infectivity is lost)
  Dilution end-point (in water).
  Thermal inactivation point (°C for 10 min).
  Duration of survival at laboratory temperature (usually 20°C).
  Duration of survival at 0–2°C.

- *Transmission through seed and pollen*

- *Transmission by vectors*
  Preliminary tests with aphids such as *M. persicae*.
  Tests with other invertebrates, until identity of the vector species is known.
  Times needed by vectors to acquire and inoculate the virus.
  Existence and duration of a latent period.
  Persistence of infectivity in the vector.
  Transmission to vector's progeny and through the moult.

- *Method of purification*
  Effects of different buffer extractants and stabilizing additives.
  Satisfactory precipitation by agents such as ammonium sulphate and polyethylene glycol.
  Suitable centrifugation regimes.

- *Properties in purified preparations*
  Electron microscopy: size and shape of virus particles, including effects of different negative stains. Indications of substructure.

Sedimentation properties and buoyant density, determined by centrifugation (e.g. density-gradient).

Chemical composition: number and molecular weight of capsid proteins, usually determined by polyacrylamide gel electrophoresis.

Type and strandedness of nucleic acid; number and molecular weight of nucleic acid species of the virus genome. Percentage nucleic acid of the virion.

Serological properties: type of precipitate in tube precipitin and micro-drop tests (somatic = granular; flagellar = fluffy). Ability to react in gel-diffusion tests.

- *Tests for relationships with other viruses*
  Serological tests: ability of antiserum to the virus to react with purified preparations of other viruses. Reactions of antisera to other viruses with preparations of the candidate virus.

  Plant protection (pre-immunity) tests: protection of infected plants against experimental infection with other viruses, or protection of plants infected with other viruses from infection with the virus studied.

  Nucleic acid sequencing and comparisons.

- *Cytological examination*
  Presence and morphology of any intracellular inclusion bodies in examination by light microscopy (vital staining) and by thin-section electron microscopy.

- *Elimination of virus*
  Where a vegetatively propagated crop is infected, it is useful to know whether the virus can be eliminated by thermotherapy (usually *c.* 4 weeks at 37°C) and/or by meristem-tip culture.

How to Identify a Plant Virus

The information needed to identify a plant virus differs from one virus group to another; predictably, some viruses are very much more difficult to identify than others. In general, a virus that can be transmitted by inoculation of sap to test plants should present few serious problems in identification, provided that adequate facilities are available, including:

1. Insect-proof compartmented glasshouses for raising and holding test plants;
2. A preparative ultracentrifuge, with both angle and swing-out rotors;
3. Access to an electron microscope;
4. Specific antisera;
5. Molecular biological facilities for nucleic acid analysis.

If no electron microscope is available on site, negatively stained specimens can be prepared on carbon-coated electron microscope-grids and mailed for examination elsewhere. Desirable additional facilities include an ultraviolet spectrophotometer, a density-gradient analyser, chromatography columns, simple growth cabinets, and equipment for electrophoresis, DNA extraction, PCR, etc.

Preliminary tests on a fresh plant specimen might involve inoculation of sap, with suitable buffer, additives and abrasives, to a range of test plants, and electron microscopy examination. 'Quick leaf-dip' preparations provide the quickest method of detecting many filamentous and rod-shaped viruses. A few small epidermal peelings from the underside of a leaf of the specimen are crushed in a few drops of a negative stain (such as 2% potassium phosphotungstate, pH 6.5) and a drop of the liquid is placed on a carbon-coated electron microscope grid and dried. This method is usually very reliable for filamentous viruses, and if measurements of particle lengths are made from micrographs, the virus can be assigned to the appropriate taxonomic group. Grids may also be coated with specific antisera to aid detection of specific viruses. If further identification within the group is required, then inoculations to a range of test plants or serological tests will usually be necessary and a process of elimination can then be followed. Isometric viruses are less reliably detected by electron microscopy of 'quick leaf-dip' preparations, unless the virus occurs in high concentrations in the plant.

If no virus can be detected in the preliminary tests, then grafts from the specimens to healthy plants (or to indicator varieties) of the same species may be necessary to demonstrate transmission of a pathogen. Some viruses, such as those of *Rhabdoviridae*, are best identified by electron microscopy of ultrathin sections of fixed material.

Molecular biochemical methods such as the use of cDNA probes and PCR are now becoming widely used and various protocols have been established (see Chapter 23).

Analytical ultracentrifugation is a valuable aid towards grouping and identification of small isometric viruses. The number and sedimentation coefficients of the virus components are usually characteristic of the virus genus and sometimes enable a virus to be identified with little further testing.

Another property useful in the identification of small spherical viruses is the molecular weight of the virus protein. This is determined by polyacrylamide gel electrophoresis. Although the technique is rather time-consuming, it requires relatively inexpensive apparatus and only minimal amounts of purified virus preparation.

Routine Virus Screening

Screening or indexing plant material for viruses usually involves tests for one or more well-recognized viruses that are known to occur in the plant species concerned. The simplest procedure involves mechanical inoculation of sap from the samples with buffer solution and carborundum or other mild abrasive to appropriate test plants, and can readily be done for many viruses. Grafting on to indicator plants is used with viruses of many vegetatively propagated crops such as citrus and temperate fruit crops.

Electron microscopy can be used to detect some viruses, but serological or DNA-based tests are most commonly used. Standard immunodiffusion and precipitation tests have now largely been replaced by ELISA for routine diagnosis.

ELISA is widely used to detect plant viruses and there are several variants of this test. The precipitated globulin fraction of antiserum is adsorbed to the wells in a microhaemagglutination plate and excess antibody washed away. Test samples are added to duplicate wells at a range of dilutions and incubated for several hours, so that virus can bind to its specific antibody. More globulin fraction of the antiserum is conjugated with alkaline phosphatase or other enzyme, and aliquots added to the appropriate wells, then incubated for a further 3–6 h. Aliquots of substrate are then added to each well and incubated for up to 1 h. The development of a coloration significantly more intense than in the various control mixtures indicates a positive reaction. This technique is valuable for screening large numbers of samples for one or two viruses, but is less suitable for individual tests, each for a different virus. The detection of viruses in seeds is often more difficult than in leaf tissues.

Test Plants

The selection of test plants for routine use will depend on the type of plants likely to be tested. The following list indicates some of the indicator species commonly used and the crops to which they relate.

- *Brassica pekinensis*: for viruses from *Cruciferae*.
- *Chenopodium amaranticolor*: susceptible to a wide range of viruses, giving local lesions and in some instances systemic symptoms. *Chenopodium quinoa*: susceptible to infection with a very wide range of viruses from many different crops – probably reacts to more viruses than any other plant.
- *Cucurbita pepo*: for viruses from *Cucurbitaceae*.
- *Gomphrena globosa*: susceptible to a wide range of viruses, usually producing local lesions but seldom of diagnostic value.
- *Nicotiana* species are widely used: *Nicotiana glutinosa* has long been used to detect and assay *Tobacco mosaic virus* (TMV), *Tomato bushy stunt virus* and *Tomato spotted wilt virus*, and is useful for checking cultures of many viruses for the presence of TMV contamination. *Nicotiana tabacum* reacts with many viruses to produce local lesions, and develops systemic infection with many others.
- *Phaseolus vulgaris*: a most valuable indicator plant, with different cultivars giving differential responses to many viruses. Provides a reliable local lesion assay for assaying the infectivity of many viruses, as well as diagnostic reactions for identification of many legume viruses.
- *Pisum sativum*: for legume viruses and as 'bait plants' for detection of some nepoviruses in soil samples.
- *Physalis floridana*: for potato viruses.
- *Trifolium incarnatum*: for viruses from clovers, legumes and some flower crops.
- *Vicia faba*: susceptible to many legume and clover viruses.
- *Vigna sinensis*: a useful local lesion assay host for many viruses and of diagnostic value with some viruses from tropical *Leguminosae*. Blackeye is a widely used cultivar.

Simple growth cabinets, able to provide controlled conditions of temperature, light intensity and day length, are needed for year-round work with many viruses, as reactions to inoculations are affected by environmental factors.

Tests for viruses in woody plants usually involve grafting to indicator varieties. With tree- and bush-fruits, lists of recommended and available indicator cultivars have been prepared by the International Society for Horticultural Science Working Groups for Virus Diseases of Fruit Trees, and for Virus Diseases of Small Fruits.

Electron microscope grids coated with dilute antiserum can greatly increase the number of particles of specific virus that are retained on the grid.

Maintaining Reference Collections of Plant Viruses

Maintaining a virus in suitable host plants risks incurring changes in the properties of the virus (such as restriction of host range or loss of vector transmissibility) with repeated passage through certain host plants.

Several methods of preserving infective virus have been used. One is to dry small pieces of infective leaf over anhydrous calcium chloride at 0 to –5°C; the dry tissue is stored frozen. This method is useful with a number of viruses of the *Potexvirus* and *Potyvirus* genera, which survive poorly after lyophilization.

Lyophilization, in which 7% (w/v) of peptone and of dextrose are dissolved in freshly expressed infective sap, and 0.25 ml aliquots placed in ampoules, frozen and dried by sublimation of the frozen water under vacuum, has proved very satisfactory with many viruses. The ampoules are sealed under high vacuum (or under N_2) and can be stored at laboratory temperature.

Purified virus preparations can be preserved for serological tests by adding formaldehyde to 2%, or glycerol (1:1 by volume). Treatment with glycerol does not abolish infectivity, and treated preparations retain serological activity for many years.

Control

Some chemicals are known to inhibit virus replication or ameliorate symptoms and they may be used to facilitate production of virus-free plants by meristem-tip therapy. However, there are no therapeutic agents or viricides that can be applied to plants to control virus disease of agricultural or horticultural crops. Consequently, control measures are based mainly on avoiding infection by using host plant resistance or disrupting the epidemic cycle of the disease.

The use of genetically resistant cultivars provides effective control of many virus diseases. Mechanisms of resistance vary; some are related to effects on vectors, whereas others inhibit virus replication. Generally, host plant resistance to virus disease has been more stable than that to fungal pathogens although there have been some notable exceptions where new virus pathotypes have overcome resistance in widely grown cultivars, e.g. *Tomato mosaic virus*. Host plant resistance to some viruses can be induced by inoculation with weekly virulent strains of the same virus, so-called mild strain cross-protection. This has been used for control of *Citrus tristeza*, *Papaya ring spot* and *Zucchini yellow mosaic* viruses. Incorporation of virus coat protein genes into the plant genome is effective in producing resistant plants and this development was the first example of a genetically modified plant. Other strategies are being developed but none is as yet widely used.

Effective control of many virus diseases can be achieved by observing good agricultural practices and crop hygiene measures to eliminate sources of infection and to control or exclude vectors. With vegetatively propagated crops in particular, an effective strategy has been the use of virus-free planting material. Virus-free foundation clones of vegetatively propagated crops such as potato, ornamentals and fruit are usually obtained by selection, thermotherapy and/or meristem-tip culture. The healthy stock is maintained in isolation and progeny are distributed to growers through the various official certification schemes. Unless specially protected, such plants are susceptible to reinfection.

In general, the earlier that infection occurs in the life of the crop, the greater the loss of yield, and measures that delay the entry of virus or the subsequent within-crop spread can minimize losses. With many aphid-transmitted viruses that are seed-borne, such as *Lettuce mosaic virus*, the presence of 1–2% infected seedlings at crop emergence can often result in over 75% of the plants becoming infected before harvesting. Mineral oil sprays or reflective aluminium strip mulches placed beside the plants can repel aphid vectors and somewhat reduce the virus spread. Direct chemical control of the vector can be effective with viruses transmitted in a 'persistent' manner. Insecticides, however, are seldom able to prevent spread of viruses such as those causing cereal yellow dwarf, which are transmitted by aphids in the 'non-persistent' or stylet-borne manner, for the moving aphids can acquire and make several transmission feeds before the chemical takes effect.

Additional control measures may include the isolation and timely removal of any crops retained for seed-production, the removal or ('roguing') of infected plants as soon as they are seen, the elimination of weeds which may act as a reservoir for viruses or their vectors and the spatial or temporal separation of crops susceptible to the same virus.

Select Bibliography

General texts

Bos, L. (1999) *Plant Viruses, Unique and Intriguing Pathogens: a Textbook of Plant Virology.* Backhuys Publishers, Leiden, 358 pp.
Diener, T.O. (1979) *Viroids and Viroid Diseases.* John Wiley & Sons, New York, USA, 252 pp.
Gibbs, A.J., Harrison, B.D., Watson, D.H. and Wildy, P. (1966) What's in a virus name? *Nature, UK* 209, 450–454.
Kurstak, E. (ed.) (1981) *Handbook of Plant Virus Infections.* Elsevier/North Holland, Amsterdam, 944 pp.
Mandahar, C.L. (1989) *Plant Viruses.* Vol. I. *Structure and Replication.* CRC Press, Boca Raton, Florida, 366 pp.
Mandahar, C.L. (1990) *Plant Viruses.* Vol. II. *Pathology.* CRC Press, Boca Raton, Florida, 371 pp.

Matthews, R.E.F. (1991) *Plant Virology*, 3rd edn. Academic Press, New York, USA, 813 pp.
Matthews, R.E.F. (1992) *Fundamentals of Plant Virology*. Academic Press, San Diego, USA, 403 pp.
Smith, K.M. (1977) *Plant viruses*, 6th edn. Chapman & Hall, London, UK, 241 pp.
Walkey, D.G.A. (1991) *Applied Plant Virology*. Chapman & Hall, London, UK, 338 pp.

Virus diseases, host reactions, virus properties and diagnosis

Anon. (1999) *AAB Descriptions of Plant Viruses*. Association of Applied Biology, Wellesborne, CD-ROM.
Bos, L. (1978) *Symptoms of Virus Diseases in Plants*, 3rd edn. Pudoc, Wageningen, The Netherlands, 225 pp.
Brunt, A.A., Crabtree, K., Dallwitz, M.J., Gibbs, A.J. and Wilson, L. (1996) *Viruses of Plants: Descriptions and Lists from the VIDE Database*. CAB International, Wallingford, UK, 1488 pp.
Christie, R.G. and Edwardson, J.R. (1977) Light and electron microscopy of plant virus inclusions. *Monograph, Florida Agricultural Experiment Station* No. 9, 150 pp.
Matthews, R.E.F. (1979) Classification and nomenclature of viruses. *Intervirology* 12, 1–296.
Matthews, R.E.F. (ed.) (1993) *Diagnosis of Plant Viruses*. CRC Press, Boca Raton, Florida, 374 pp.
McDaniel, L.L. and Emerson, E.L. (1990) *Catalogue of Plant Viruses and Antisera*. American Type Culture Collection, Maryland, USA, 51 pp.
Shukla, D.D., Ward, C.W. and Brunt, A.A. (1994) *The Potyviridae*. CAB International, Wallingford, UK, 448 pp.
Smith, G.H. and Barker, H. (eds) (1999) *The Luteoviridae*. CAB International, Wallingford, UK, 320 pp.

Virus diseases of some crop plants

Anjaneyulu, A., Satapathy, M.K. and Shukla, V.D. (1995) *Rice Tungro*. Science Publishers, USA, 228 pp.
Anon. (1976) *Virus Diseases and Non-infectious Disorders of Stone Fruits in North America*. Agriculture, Handbook, United States Department of Agriculture No. 437, 433 pp.
Converse, R.H. (1987) *Virus Diseases of Small Fruits*. USDA Handbook no. 631, 277 pp.
Cooper, J.I. (1993) *Virus Diseases of Trees and Shrubs*, 2nd edn. Chapman & Hall, London, UK, 205 pp.
Desvignes, J.C., Boye, R., Cornaggia, D., Grasseau, N., Hurtt, S. and Waterworth, H. (1999) *Virus Diseases of Fruit Trees. (Diseases Due to Viroids, Viruses, Phytoplasmas and Other Undetermined Infectious Agents.)* Centre Technique Interprofessionnel des Fruits et Legumes (CTIFL), Paris, France, 202 pp.
Edwardson, J.R. and Christie, R.G. (1991) *CRC Handbook of Viruses Affecting Legumes*. CRC Press, Boca Raton, Florida, 505 pp.
Koenig, R. (ed.) (1980) 5th International Symposium on Virus Diseases of Ornamental Plants. *Technical Communication, International Society for Horticultural Science* No. II0, 334 pp.
Kyle, M.M. (1993) *Resistance to Viral Diseases of Vegetables: Genetics and Breeding*. Timber Press Portland, 278 pp.
Loebenstein, G., Lawson, R.H. and Brunt, A.A. (eds) (1995) *Virus and Virus-like Diseases of Bulb and Flower Crops*. John Wiley & Sons, Chichester, UK, 543 pp.
Proceedings of the Conference of the International Organisation of Citrus Virologists (Various volumes).

Salazar, L.F. (1996) *Potato Viruses and Their Control*. International Potato Center (Centro Internacional de la Papa) (CIP), Lima, Peru, 214 pp.

Slykhuis, J.T. (1976) Virus and virus-like diseases of cereals. *Annual Review of Phytopathology* 14, 189–210.

Williams, L.E., Gordon, D.T. and Nault, L.R. (eds) (1977) *Proceedings of the International Maize Virus Disease Colloquium and Workshop*. Ohio Agricultural Research and Development Center, Wooster, 145 pp.

Viruses of fungi

Hollings, M. (1977) Mycoviruses: viruses that infect fungi. *Advances in Virus Research* 22, 1–53.

Molitoris, H.P., Hollings, M. and Wood, H.A. (eds) (1979) *Fungal Viruses*. Springer Verlag, Berlin-Heidelberg, Germany, 194 pp.

Vectors, epidemiology and control

Cooper, J.I. (1995) *Viruses and the Environment*, 2nd edn. Chapman & Hall, London, UK, 210 pp.

Cooper, J.I. and Asher, M.J.C. (eds) (1988) *Viruses with Fungal Vectors*. Association of Applied Biology, Wellesborne, UK, 355 pp.

Cooper, J.I., Kelley, S.E. and Massalski, R. (1988) Virus–pollen interactions. *Advances in Disease Vector Research* 5, 221–249.

Fulton, R.W. (1986) Practices and precautions in the use of cross protection for plant virus control. *Annual Review of Phytopathology* 24, 67–81.

Harrison, B.D. (1977) Ecology and control of viruses with soil-inhabiting vectors. *Annual Review of Phytopathology* 15, 331–360.

McLean, G.D., Garrett, R.G. and Ruesink, W.G. (1986) *Plant Virus Epidemics. Monitoring, Modelling and Predicting Outbreaks*. Academic Press, Sydney, Australia, 550 pp.

Mink. G.I. (1993) Pollen- and seed-transmitted viruses and viroids. *Annual Review of Phytopathology* 31, 375–402.

Pirone, T.P. and Harris, K.F. (1977) Non-persistent transmission of plant viruses by aphids. *Annual Review of Phytopathology* 15, 55–73.

Sturtevant, A.P., Beachy, R.N. and Hiatt, A. (1993) Virus resistance in transgenic plants: coat protein-mediated resistance. In: *Transgenic Plants: Fundamentals and Applications*. Marcel Dekker, New York, USA, pp. 93–112.

Taylor, C.E. and Brown, D.J.F. (1997) *Nematode Vectors of Plant Viruses*. CAB International, Wallingford, UK, 296 pp.

Thresh, J.M. (1974) Temporal patterns of virus spread. *Annual Review of Phytopathology* 12, 111–128.

Timmerman, G.M. (1991) Genetic engineering for resistance to viruses. In: Murray, D.R. (ed.) *Advanced Methods in Plant Breeding and Biotechnology*. Biotechnology in Agriculture No. 4, CAB International, Wallingford, UK, pp. 319–339.

Methods and techniques in plant virology

Clark, M.F. and Adams, A.N. (1977) Characteristics of the microplate method of enzyme-linked immunosorbent assay for the detection of plant viruses. *Journal of General Virology* 34, 475–483.

Foster, G.D. and Taylor, S.C. (1998) *Plant Virology Protocols*. Humana Press, Totowa, 569 pp.

Green, S.K. (1991) *Guidelines for Diagnostic Work in Plant Virology*, 2nd edn. Technical Bulletin Asian Vegetable Research and Development Center, No. 15, Asian Vegetable Research and Development Center, Taipei, 63 pp.

Gugerli, P. (1992) Commercialization of serological tests for plant viruses. In: Duncan, J.M. and Torrance, L. (eds) *Techniques for the Rapid Detection of Plant Pathogens*. Blackwell Scientific Publications, Oxford, UK, pp. 222–229.

Hamilton, R.I. (1992) The use of monoclonal antibodies and cDNA for detection of plant viruses. In: Thottappilly, G., Monti, L.M., Mohan-Raj, D.R. and Moore, A.W. (eds) *Biotechnology: Enhancing Research on Tropical Crops in Africa*. Technical Centre for Agricultural and Rural Cooperation, Wageningen, The Netherlands, pp. 297–303.

Hampton, R., Ball, E. and Boer, S. de (1990) *Serological Methods for Detection and Identification of Viral and Bacterial Plant Pathogens. A Laboratory Manual*. APS Press, St Paul, Minnesota, USA, 389 pp.

Jones, A.T. (1992) Application of double-stranded RNA analysis of plants to detect viruses, virus-like agents, virus satellites and subgenomic viral RNAs. In: Duncan, J.M. and Torrance, L. (eds) *Techniques for the Rapid Detection of Plant Pathogens*. Blackwell Scientific Publications, Oxford, UK, pp. 115–128.

Milne, R.G. and Luisoni, E. (1977) Rapid immune electron microscopy of virus preparations. In: Maramorosch, K. and Koprowski, H. (eds) *Methods in Virology*, Vol. 6. Academic Press, New York, pp. 265–281.

van Regenmortel, M.H.V. (1982) *Serology and Immunochemistry of Plant Viruses*. Academic Press, London, UK, 302 pp.

Sward, R.J. and Eagling, D.R. (1995) Antibody approaches to plant viral diagnostics. In: Skerritt, J.H. and Appels, R. (eds) *New Diagnostics in Crop Sciences*. Biotechnology in Agriculture No. 13, CAB International, Wallingford, UK, pp.171–193.

Torrance, L. (1992) Serological methods to detect plant viruses: production and use of monoclonal antibodies. In: Duncan, J.M. and Torrance, L. (eds) *Techniques for the Rapid Detection of Plant Pathogens*. Blackwell Scientific Publications, Oxford, UK, pp. 7–33.

Phytoplasma Plant Pathogens

P. Jones

Plant Pathogen Interactions (PPI), IACR-Rothamsted, Harpenden, Hertfordshire AL5 2JQ, UK

Introduction

The phytoplasmas, formerly known as mycoplasma-like organisms (MLOs), are prokaryotes that do not possess a cell wall and that cause diseases of over 300 plant species around the world (McCoy *et al.*, 1989). Phytoplasmas belong to the Class *Mollicutes* (Fig. 12.1), which also contains the culturable mycoplasmas, the cause of many diseases in animals and humans, and the spiroplasmas, motile, helical organisms most of which can be cultured *in vitro* without difficulty (Daniels *et al.*, 1973) and which are the cause of at least three plant diseases, corn stunt, citrus stubborn and citrus little leaf (Markham *et al.*, 1974; Liao and Chen, 1977). Members of the *Mollicutes* have some of the smallest genomes of all free-living organisms, ranging in size from about 600 kb to 2000 kb, and lack a number of biosynthetic functions that contribute to the synthesis of certain amino acids, fatty acids and lipids (Sears and Kirkpatrick, 1994; Marcone *et al.*, 1999). Until the 1970s, yellows diseases, which we now know to be caused by phytoplasmas, were thought to be caused by viruses because:

- yellows diseases are graft transmittable;
- yellows diseases could be transmitted by dodder;
- yellows diseases could be transmitted by leaf hopper vectors;
- some yellows diseases could give cross protection to infected hosts.

These are all properties in common with known viruses.

The demonstration that leaf hoppers could transmit a number of yellows diseases led to researchers looking for possible virus-like particles in plant and insect cells by electron microscopy, and it was this that led to the discovery of mycoplasma-like bodies in the sieve elements of plants, such as mulberry affected by dwarf disease and aster affected by aster yellows disease (Doi *et al.*, 1967). Once mycoplasma-like bodies had been found in one plant there were soon reports of them associated with many diseases. Ishiie *et al.* (1967) showed that it was possible to get remission of symptoms by treatment with tetracycline antibiotics and this became accepted practice in establishing a mycoplasma association for these diseases. Studies of the relatedness of plant mycoplasma-like organisms and the culturable mycoplasmas were only possible after recombinant DNA technology became available. The cloning of phytoplasma DNA fragments from the insect vector of Western X Disease (Kirkpatrick *et al.*, 1987) signalled the start of a new era in the study of plant diseases caused by phytoplasmas.

Phytoplasmas can be very difficult to detect in their hosts by light and electron microscopy. This is because the number of cells varies from low or very low in some woody hosts (Braun and Sinclair, 1976; Caudwell and Kuszala, 1992) to high in the experimental host *Catharanthus roseus* (periwinkle). Berges *et al.* (2000) using a competitive PCR approach showed that phytoplasma concentrations varied from 2.2×10^8–1.5×10^9 cells g^{-1} of

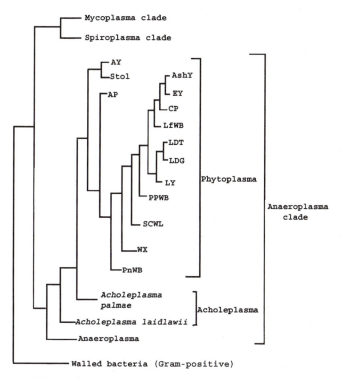

Fig. 12.1. Simplified *Mollicute* phylogenetic tree. AY, aster yellows; Stol, stolbur disease; AP, apple proliferation; Ashy, ash yellows; EY, elm yellows; CP, clover phyllody; LfWB, Loofah witches' broom; LDT, coconut lethal disease, Tanzania; LDG, coconut Cape St Paul Wilt, Ghana; LY, coconut lethal yellowing, Florida; PPWB, pigeon pea witches' broom; SCWL, sugar cane white leaf; WX, western X disease of stone fruits; PnWB, peanut (groundnut) witches' broom.

tissue in periwinkle to 340–37,000 cells g^{-1} of tissue in some apple trees grown on resistant rootstocks and infected with the apple proliferation phytoplasma.

Analysis of the ribosomal RNA gene of phytoplasmas has shown that they are related to but distinct from the animal mycoplasmas (Kirkpatrick, 1989; Lim and Sears, 1989). Sears and Kirkpatrick (1994) proposed the name phytoplasma for the MLOs found in plants because of their similarities with the culturable members of the Class *Mollicutes*.

Symptomatology

Symptoms exhibited by plants infected with phytoplasmas include stunting, floral abnormalities, shoot proliferation, floral necrosis, yellowing of foliage and decline of vigour often leading to death of the plant.

- Floral symptoms consist of virescence or greening of the petals, phyllody in which the petals themselves become leaf-like, and flower production may cease altogether. In tomato big bud disease the plant produces the swollen calyces which give the disease its common name.
- Witches' brooms can develop by the proliferation of shoots from flowers and floral proliferation will cause new flowers to develop from the sexual parts of affected flowers.
- Fruit production often ceases after infection and where fruit had developed they may abort as in the case of coconut palms suffering from premature nut fall, which is one of the earliest signs of infection by the lethal yellows phytoplasma.

There is no evidence that phytoplasma diseases are seed-transmitted. Esau (1965) reports that there are no vascular connections between the embryo and tissues of the parent plant. Where seed is produced by infected plants it is usually sterile (Maramorosch, 1976).

Foliar symptoms often begin with chlorosis and may include vein clearing, which progresses to a systemic yellowing. Sections of infected plants often show phloem necrosis, which may or may not be associated with cell division, resulting in the production of many new sieve elements (Girolami, 1955). Phytoplasma-infected plants may exhibit several symptoms in sequence during disease development, and these can vary from host to host. For example, in coconuts infected with lethal yellowing phytoplasma, premature nut fall is the first sign of infection followed by the appearance of necrosis in immature inflorescences. There then follows a progressive discoloration and shedding of foliage starting at the oldest frond. The spear leaf becomes necrotic. The necrosis spreads down the spear into the meristematic tissues of the bud and the palm dies.

Transmission

Insect vectors of phytoplasmas are restricted to phloem-feeding leaf hopper and plant hopper members of the Cicadellidae, Coccidea, Fulgoroidea and

Psylloidea, which transmit the phytoplasmas in a persistent manner. Although there have been reports of transmission by other insects such as aphids, subsequent re-examination of the disease has shown a viral rather than a phytoplasma pathogen. Successful transmission of disease is dependent on the acquisition of the phytoplasma by the vector during feeding (acquisition access period), the passage of the phytoplasma from the insect gut through the haemocoel and passage across the salivary gland membranes (latent period) and the inoculation of a healthy plant during subsequent feeding (inoculation access period). There have been many reviews of the relationships between phytoplasmas and their vectors (Tsai, 1979) but in recent times vector studies seem to have been abandoned in favour of molecular biological studies.

Other methods of transmission of phytoplasmas include dodder and grafting. Dodders (*Cuscuta* and *Cassytha* spp.) are parasitic vines that develop vascular connections with their hosts through haustoria. When a bridge is established between healthy and phytoplasma-infected plants the phytoplasma will transfer via the connecting phloem elements. Graft transmission can be an easy way of maintaining cultures of phytoplasmas among graft-compatible plants. It has often been used as a rapid means of screening plants for phytoplasmas and for disease-resistance screening of herbaceous perennials (McCoy, 1979).

Experimentally determined host ranges using vectors were considered to be of limited value by Chiykowski and Sinha (1990) from work involving the transmission of the Western aster yellows phytoplasma by the vector *Aphrodes bicinctus* and a clover phyllody phytoplasma by the vector *Macrosteles fascifrons*. They concluded that the differences in host range could be accounted for by differences in vector feeding rather than any differences in pathogenicity.

Chemotherapy

Ishiie *et al.* (1967) reported the first clear results of the therapeutic effects of the tetracycline antibiotics on a *Mollicute* disease. Tetracycline therapy became an essential step in proving the association between a phytoplasma and a disease. Although many antibiotics have been tried only tetracycline and to a lesser extent chloramphenicol antibiotics have been able to alleviate symptoms. The degree of recovery depends on the way in which the antibiotic is applied to the plant and the state of the disease development at the time of application. Typical methods of application include foliar sprays, root immersion, soil drench and trunk injection, but in all cases symptoms reoccur once the antibiotic is removed. Antibiotic therapy is rarely used as a crop protection method, although it has been used in the case of palms of a high landscape value in the urban areas of southern Florida.

Ultrastructure and Cell Biology

In plants, phytoplasmas are confined to the sieve elements of the phloem, unless the host has little or no photosynthetic activity (Sears and Klomparens,

1989). Phytoplasmas have a single plasma membrane and a variable shape but often appear rounded. The internal structure of phytoplasmas consists of ribosomes and strands of DNA (Fig. 12.2). In section they appear more or less circular, up to 1.2 μm in diameter. Some sieve elements can be packed with organisms whereas others contain few. Small circular membranes without contents are often seen and in some hosts the organisms stain so darkly that internal structure cannot be distinguished. Serial sections reveal phytoplasmas as branching filaments extending throughout the length of sieve elements. All electron microscope studies of phytoplasmas in plant sieve tubes show organisms of essentially the same shape and it has been impossible to differentiate between phytoplasmas on the basis of morphology and ultra-structure. In spite of this, electron microscopy has been an essential step for proving the link between a phytoplasma and a disease of unknown aetiology (McCoy, 1979; Cousin *et al.*, 1986).

Little is known about the cell biology of phytoplasmas. Phytoplasmas have an AT-rich genome ranging in size from 500 kb to 1180 kb and are the smallest self-replicating prokaryotes found. By comparison with the spiroplasmas and culturable mycoplasmas, we know that they lack genes to aid DNA repair and to synthesize cell walls and certain amino acids, sterols and lipids, but until they can successfully be isolated and cultured *in vitro* our knowledge of their growth and replication remains incomplete. Sears and Klomparens (1989) reported that phytoplasmas grew to higher titres in leaf-tip cultures of *Oenothera* in which the internal production of oxygen was diminished by mutation, indicating that phytoplasmas may be especially sensitive to oxygen.

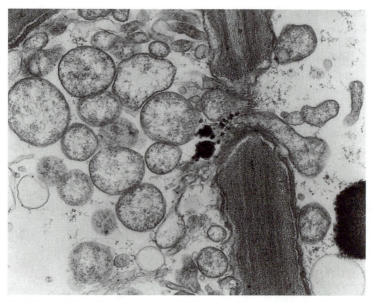

Fig. 12.2.   Transmission electron micrograph of phytoplasmas in the sieve tube of sesame plants infected with sesame phyllody.

The number of phytoplasma genes identified continues to grow. Jarausch *et al.* (1994) identified a nitroreductase gene and Berg *et al.* (1999) identified a gene coding for a major membrane protein from the apple proliferation phytoplasma. Barbara *et al.* (1998) identified a major membrane protein gene from chlorante (aster yellows) phytoplasma and Yu *et al.* (1998) also identified an antigenic protein gene associated with sweet potato witches' broom phytoplasma. The first chromosome map of a phytoplasma was published for the sweet potato little leaf phytoplasma, a phytoplasma with one of the smallest genomes (622 kb), by Padovan *et al.* (2000). They located two RNA operons, the *fus/tuf* genes encoding the G and Tu elongation factors, a *gid* gene coding for a glucose-inhibited division protein and sixteen cleavage sites on the map.

Diagnostics

The demonstration of phytoplasmas in the sieve elements of infected plants by electron microscopy has been an essential step in phytoplasma disease diagnosis but it is expensive in both time and resources. Light microscopy has also been used to detect mollicutes using both chromatic (Diene's stain) and fluorescent (using DAPI (4'6-diamidino-2-phenylindole)) stains to confirm infection of plants by phytoplasmas (Seemüller, 1976; Deely *et al.*, 1979).

Phytoplasmas cannot be grown in culture and it is this one step that has made any progress with detection and diagnosis difficult. Serological assays are routinely used for culturable prokaryotes but difficulties with the purification of phytoplasmas meant that first attempts at purification resulted in cross-reacting polyclonal antisera. Monoclonal antibodies have been raised to a number of phytoplasmas including aster yellows, clover phyllody, Flavescence dorée and elm yellows (Boudon-Padieu *et al.*, 1989; Errampalli *et al.*, 1989; Chang and Chen, 1991). Monoclonal antibodies are useful for differentiating closely related strains of phytoplasmas, whereas polyclonal antibodies can differentiate the more distantly related phytoplasmas (Lee and Davis, 1992). Serological diagnostic tests developed include ELISA, tissue blot and immunosorbent electron microscopy and immunofluoresence (Sinha and Chiykowski, 1984; Boudon-Padieu *et al.*, 1989; Sarindu and Clark, 1993; Minucci *et al.*, 1996).

The application of molecular biological techniques to the study of phytoplasmas has opened up new avenues for diagnostics. DNA hybridization studies using cloned random fragments of a phytoplasma DNA extracted from plants as probes has made it possible reliably and specifically to detect phytoplasmas in their plant and insect hosts (Lee and Davies, 1992).

The use of cloned phytoplasma DNA fragments as probes in hybridization tests allows the detection of these phytoplasmas in infected plant and insect tissues. In addition, these DNA probes have been used to study the genetic relatedness among the phytoplasmas (Harrison *et al.*, 1992, 1994) and for the first time to produce a phytoplasma classification based on their genomes. However, in recent years the use of dot hybridization assays for the detection of phytoplasmas has been superseded by the use of the PCR assay.

Amplification of 16S rDNA from phytoplasmas by PCR provides a much more sensitive detection method than any other yet described (Davis and Eyre, 1996; Minucci *et al.*, 1996). However, the choice of oligonucleotide primers can affect the specificity of the test. There have been a number of publications in which the authors describe the sequence of 'universal' and specific phytoplasma primers (Deng and Hiruki, 1991; Ahrens and Seemüller, 1992; Namba *et al.*, 1993; Smart *et al.*, 1996; Tymon *et al.*, 1998). These primers are based on sequences of the conserved 16S rRNA gene but experience has shown that universal primers may not detect all known phytoplasmas. For various reasons, including low titres of the phytoplasma within the host or the presence of inhibitors in the host, which are extracted along with the phytoplasma DNA, a single series of PCR with as many as 35–40 cycles may not provide a reliable detection method. In such cases the nested PCR, in which a double round of amplifications is performed, can be at least 100 times more sensitive than a single PCR. Gundersen and Lee (1996) have designed nested universal primers based on sequences published by Lee *et al.* (1993a, b). Restriction fragment length polymorphism (RFLP) analyses of the PCR products allows identification of the phytoplasmas associated with the diseased tissues (Kuske *et al.*, 1991; Gundersen *et al.*, 1994). Sequences for some of the commonly used primers are given in Table 12.1.

Protocols for PCR Detection of Phytoplasmas

Phytoplasma enrichment suitable for soft wood tissues, meristematic tissues of palms and leaf and stem tissues of Graminaceous hosts (based on Dellaporta *et al.*, 1983)

1. Grind 100–200 g of tissue in ice-cold EB. Use 3–4 ml buffer per g of tissue and add 0.03 M ascorbic acid just prior to mixing (check pH 7.4: adjust with solid NaOH) and mix well in Waring blender for 10 min stopping every 2 min to scrape down the slurry and to prevent the blender warming up.

2. Leave on ice in cold room for 15–30 min.

Table 12.1. Primers used for PCR amplification of the 16S rRNA genes of phytoplasmas.

Primer	Sequence (5'-3')	Reference
P1	AAGAGTTTGATCCTGGCTCAGGATT	Deng and Hiruki (1991)
520F	GTGCCAGCAGCCGCGG	Namba *et al.* (1993)
P4	GAAGTCTGCAACTCGACTTC	Smart *et al.* (1996)
920R	GTCAATTCCTTTAAGTTT	Namba *et al.* (1993)
P6	CGGTAGGGATACCTTGTTACGACTTA	Deng and Hiruki (1991)
P7 (23S)	CGTCCTTCATCGGCTCTT	Smart *et al.* (1996)
R16mF2	CATGCAAGTCGAACGGA	Gundersen and Lee (1996)
R16mR1	CTTAACCCCAATCATCGAC	Gundersen and Lee (1996)
R16F2n	GAAACGACTGCTAAGACTGG	Gundersen and Lee (1996)
R16R2	TGACGGGCGGTGTGTACAAACCCCG	Gundersen and Lee (1996)

3. Squeeze brie through cheesecloth and transfer to 2×250 ml centrifuge pots per sample.

4. Centrifuge at 3000 *g*, 10 min, 4°C.

5. Transfer supernatant to clean centrifuge pots. Discard pellet.

6. Centrifuge at 20,000 *g*, 30 min, 4°C. Discard supernatant.

7. Thoroughly resuspend pellets in 15 ml DEB (add 7.5 ml to each duplicate, then pool).

8. Add 1 ml 20% sodium dodecyl sulfate (SDS) (to lyse phytoplasma). Mix gently (becomes slightly viscous).

9. Incubate at 65°C with gentle swirling for 10–15 min.

10. Add 5 ml potassium acetate, mix well (NB use acidified 5 M potassium acetate).

11. Leave on ice for 20 min or overnight at 4°C.

12. Centrifuge at 25,000 *g*, 20 min, 4°C.

13. Pour supernatant through miracloth or tissue into 30 ml tubes. Discard pellet.

14. Precipitate DNA by adding 10 ml cold isopropanol (~0.6 vol.).

15. Leave for at least 30 min on ice or overnight at −20°C.

16. Centrifuge at 20,000 *g*, 15 min, 4°C, or spool out DNA depending on amount obtained.

17. Resuspend pellet in 3 ml TE (50 mM Tris-HCl; 20 mM EDTA), leave overnight at 4°C.

18. Add equal volume of PCI (phenol:chloroform:isoamyl alcohol (25:24:1)).

19. Vortex to mix.

20. Separate phases by centrifuging at 10,000 *g*, 10 min, 4°C.

21. Remove upper aqueous phase to clean tube using Pasteur pipette.

22. Add an equal volume of PCI.

23. Vortex to mix.

24. Centrifuge at 10,000 *g*, 10 min, 4°C.

25. Remove upper aqueous phase to clean tube using Pasteur pipette.

26. Chloroform extract once using an equal volume of $CHCl_3$, mix well.

27. Centrifuge at 12,000 *g*, 5 min, 4°C. Remove aqueous phase to clean tube.

28. Remove RNA from samples by addition of 10 µg ml^{-1} of RNase A (stock 10 mg ml^{-1}). Incubate at 37°C for 30–60 min.

29. Reprecipitate DNA by addition of NaCl to a final concentration of 1 M plus 2.5 volumes cold ethanol. Leave at −20°C for 1 h or overnight.

30. If large DNA pellets are obtained, may be able to spool them. If DNA pellets are small, centrifuge at 20,000 *g* for 10 min, 4°C.

31. Wash DNA pellets with cold 70% ethanol to remove salt. Air dry pellets.

32. Resuspend pellets in TE pH 8.0 (10 mM Tris; 1 mM EDTA) using as small a volume as possible.

33. Estimate DNA concentrations using spectrophotometer. Check $A_{260/280}$ readings for purity.

Reagents

EB

$K_2HPO_4.3H_2O$	21.7 g
KH_2PO_4	4.1 g

Sucrose	100 g
Polyvinyl pyrrolidone (PVP) – 40T (or 10)	20 g
Bovine serum albumen (BSA) fraction V	1.5 g

Adjust to pH 7.4 and bring to 1000 ml. Add 0.03 M ascorbic acid just prior to use; recheck the pH and adjust if necessary using solid NaOH.

DEB

	to make 100 ml:
100 mM Tris-HCl, pH 8.0	10 ml of 1 M stock
50 mM EDTA	10 ml of 0.5 M stock
500 mM NaCl	10 ml of 5 M stock
10 mM 2-mercaptoethanol	69.8 µl

TE

	to make 50 ml:
10 mM Tris	0.5 ml 1 M Tris
1 mM EDTA	0.1 ml 0.5 M EDTA

ACIDIFIED POTASSIUM ACETATE (FOR ALKALINE LYSIS). To 60 ml of 5 M potassium acetate add 11.5 ml glacial acetic acid and 28.5 ml H_2O (final vol. 100 ml).

CTAB extraction method suitable for host material where enrichment is not required (after Doyle and Doyle, 1990)

1. Grind 5 g of leaf in 15 ml hot (65°C) CTAB buffer.
2. Incubate at 65°C for 30 min.
3. Add an equal volume of chloroform:isoamyl alcohol (24:1) and mix.
4. Centrifuge at 4000 *g*, 10 min.
5. Remove aqueous (upper) layer to clean tube.
6. To aqueous layer add 2/3 volume cold isopropanol or 2.5 volumes of cold absolute ethanol and one tenth volume of 3 M sodium acetate. Leave overnight at –20°C.
7. Centrifuge at 10,000 *g* for 30 min. Pour off supernatant and dry pellet (you may not see anything).
8. Resuspend in 0.5 ml TE buffer and transfer to clean tube. This is the stock template.

Reagents

CTAB (MIXED ALKYL TRIMETHYL AMMONIUM BROMIDE (SIGMA M-7635)).
CTAB BUFFER

to make 100 ml solution:		
CTAB	2%	2 g
NaCl	1.4 M	28 ml of 5 M solution
EDTA pH 8.0	20 mM	4 ml of 0.5 M solution
Tris-HCl pH 8.0	100 mM	10 ml of 1 M solution
2-mercaptoethanol	0.2%	200 µl

DNA amplifications

A pair of oligonucleotide primers is used to prime the amplification of 16S rDNA sequence from infected tissue. Template DNAs from healthy and diseased plants are quantified by spectrophotometry and diluted to a final concentration of 50 µg ml^{-1}. Amplifications are performed in either 25 µl or 50 µl reaction volumes, each containing 50–100 ng of template DNA; 100 ng each primer; 150 µM of each dNTP; 0.5 U *Taq* DNA polymerase and recommended PCR buffer. Reactions are overlaid with 50 µl of mineral oil and 35 cycles of PCR are performed in a thermal cycler. Reaction conditions are typically: 95°C for 5 min followed by 35 cycles of denaturation (95°C) for 30 s; annealing (58°C) for 1 min; extension (72°C) for 1 min; followed by a final extension step (72°C) for 10 min. Reaction mixtures containing only healthy plant DNA or sterile distilled water substituted for template DNA serve as negative controls in each experiment.

Nested PCR

Following a first round PCR using primers such as P1/P7, a 1 µl aliquot of the PCR mix is taken to act as template for the nested PCR reaction. Some workers advise dilution of the template between 40 and 100 times before adding to the nested reaction mixture. This has the advantage of diluting out any inhibitory substances and avoiding too high a concentration of template DNA in the nested reaction. Choice of primer for the nested PCR will depend on the experimental aims but we have found the following to work well (see Table 12.1 for primer sequences):

First PCR pair	Nested primer pair
P1/P7	P4/P7
P1/P7	R16F2n/R16R2
R16mF2/R16mR1	R16F2n/R16R2

The nested PCR products are electrophoresed through a 1% agarose gel using 1 × TAE (40 mM Tris-acetate; 1 mM EDTA) followed by staining with ethidium bromide and visual examination by UV illumination.

Analysis of PCR amplification products

An aliquot (8–10 µl) of each reaction can be analysed by electrophoresis through 1% agarose gels (Gibco BRL) using 1 × TAE followed by staining with ethidium bromide. Further aliquots (10–15 µl) can then be digested by the addition of 0.5 U of restriction enzyme followed by incubation at the appropriate temperature for a minimum of 4 h. The enzymes, *Alu*I, *Bcl*I, *Eco*RI, *Rsa*I and *Tru*9I have all proved useful for phytoplasma identification. Products are analysed by electrophoresis through 3% NuSieve GTG agarose gels (FMC BioProducts, supplied by Flowgen Instruments Ltd, UK) with 1 × TBE (90 mM Tris-borate; 2 mM EDTA) as running buffer.

Note: *Taq* DNA polymerase is a highly labile enzyme. Amersham Pharmacia have recently introduced a stable enzyme mix called Ready-To-Go® PCR beads. We have found these to be very convenient and to give highly reproducible results in difficult environments.

Phytoplasma Collections

Phytoplasma reference collections are maintained by the following institutions.

Plant Pathogen Interactions (PPI),
IACR-Rothamsted,
Harpenden,
Herts AL5 2JO,
UK
Contact: P. Jones

Laboratoire de Biologie Cellulaire et Moleculaire,
Institut National de la Recherche Agronomique,
BP 81,
33883 Villenave d'Ornon 59,
France
Contact: M. Garnier

Instituto di Patologia Vegetale,
University of Bologna,
V.F. Re 8,
40126 Bologna,
Italy
Contact: A. Bertaccini

References

Ahrens, U. and Seemüller, E. (1992) Detection of DNA of plant pathogenic mycoplasma-like organisms by a polymerase chain reaction that amplifies a sequence of the 16S rRNA gene. *Phytopathology* 82, 828–832.

Barbara, D.J., Davies, D.L. and Clark, M.F. (1998) Cloning and sequencing of a major membrane protein from chlorante (aster yellows) phytoplasma. *12th International Organisation of Mycoplasmology Conference*, July 1998, Sydney, Australia, Poster G04.

Berg, M., Davies, D.L., Clark, M.F., Vetten, H.J., Maier, G., Marcone, C. and Seemüller, E. (1999) Isolation of the gene encoding an immunodominant membrane protein of the apple proliferation phytoplasma, and expression and characterisation of the gene product. *Microbiology* 145, 1937–1943.

Berges, R., Rott, M. and Seemüller, E. (2000) Range of phytoplasma concentration in various plant hosts as determined by competitive polymerase chain reaction. *Phytopathology* 90, 1145–1152.

Boudon-Padieu, E., Larrue, J. and Caudwell, A. (1989) ELISA and dot-blot detection of flavescence doree-MLO in individual leafhopper vectors during latency and inoculative state. *Current Microbiology* 19, 357–364.

Braun, E.J. and Sinclair, W.A. (1976) Histopathology of phloem necrosis in *Ulmus americana*. *Phytopathology* 66(5), 598–607.

Caudwell, A. and Kuszala, K. (1992) Mise au point d'un test ELISA sur les tissus des vignes atteintes de flavescence dorée. *Research in Microbiology* 143, 791–806.

Chiykowski, L.N. and Sinha, R.C. (1990) Differentiation of MLO diseases by means of symptomatology and vector transmission. *Zentrallblatt Bakteriologische Supplement* 20, 280–287.

Cousin, M.T., Sharma, A.J. and Misra, S. (1986) Correlation between light and electron microscopic observations and identification of mycoplasma-like organisms. *Phytopathologische Zeitschrift* 115, 368–374.

Daniels, M.J., Markham, P.G., Meddins, B.M., Townsend, R., Plaskitt, K. and Bar-Joseph, M. (1973) Axenic culture of a plant-pathogenic spiroplasma. *Nature* 244, 523–524.

Davis, D.L. and Eyre, S. (1996) Detection of phytoplasmas associated with pear decline in pear psyllids by polymerase chain reaction. *British Crop Protection Council Symposium Proceedings* No. 65, 67–72.

Deely, J., Stevens, W.A. and Fox, R.T.V. (1979) Uses of Dienes stains to detect plant diseases induced by mycoplasmalike organisms. *Phytopathology* 69, 1169–1171.

Dellaporta, S.L., Wood, J. and Hicks, J.B. (1983) A plant DNA minipreparation, version II. *Plant Molecular Biology Reporter* 1, 19–21.

Deng, S. and Hiruki, C. (1991) Amplification of 16S rRNA genes from culturable and nonculturable Mollicutes. *Journal of Microbiological Methods* 14, 53–61.

Doi, Y.M., Teranaka, M., Yora, K. and Asuyama, H. (1967) Mycoplasma or PLT-group like micro-organisms found in the phloem elements of plants infected with mulberry dwarf, potato witches' broom, aster yellows and paulownia witches' broom. *Annals of the Phytopathological Society of Japan* 33, 259–266.

Doyle, J.J. and Doyle, J.L. (1990) Isolation of plant DNA from fresh tissue. *Focus* (Life Technologies Inc.) 12, 13–15.

Errampalli, D., Fletcher, J. and Sherwood, J.L. (1989) Production of monospecific polyclonal antibodies against the aster yellows mycoplasmalike organisms (AY MLO) of Oklahoma. *Phytopathology* 79, 1137.

Esau, K. (1965) *Plant Anatomy*, 2nd edn. John Wiley & Sons, New York, USA.

Girolami, G. (1955) Comparative anatomical effects of the curly top and aster yellows viruses on the flax plant. *Botanical Gazette* 116, 305–322.

Gundersen, D.E. and Lee, I.M. (1996) Ultrasensitive detection of phytoplasmas by nested PCR using two universal primers. *Phytopathologia Mediterranea* 35, 144–151.

Gundersen, D.E., Lee, I.M., Rehner, S.A., Davis, R.E. and Kingsbury, D.T. (1994) Phylogeny of mycoplasmalike organisms/phytoplasmas: a basis for their classification. *Journal of Bacteriology* 176, 5244–5254.

Harrison, N.A., Bourne, C.M., Cox, R.L., Tsai, J.H. and Richardson, P.A. (1992) DNA probes for detection of mycoplasmalike organisms associated with lethal yellowing disease of palms in Florida. *Phytopathology* 82, 216–224.

Harrison, N.A., Richardson, P.A., Jones, P. and Tymon, A.M. (1994) Comparative investigation of MLOs associated with Caribbean and African coconut lethal decline diseases by DNA hybridization and PCR assays. *Plant Disease* 78, 507–511.

Ishiie, T., Doi, Y., Yora, K. and Asuyama, H. (1967) Suppressive effects of antibiotics of the tetracycline group on symptom development in mulberry dwarf disease. *Annals of the Phytopathological Society of Japan* 33, 267–275.

Jarausch, W., Saillard, C., Dosba, F. and Bové, J.M. (1994) Differentiation of mycoplasma-like organisms (MLOs) in European fruit trees by PCR using specific primers derived from the sequence of a chromosomal fragment of the apple proliferation MLO. *Applied Environmental Microbiology* 60, 2916–2923.

Kirkpatrick, B.C. (1989) Strategies for characterising plant pathogenic mycoplasmalike organisms and their effects on plants. In: Kosuge, T. and Nester, E.W. (eds) *Plant–Microbe*

Interactions, Molecular and Genetic Perspectives, Vol. 3. McGraw-Hill, New York, USA, pp. 241–293.

Kirkpatrick, B.C., Stenger, D.C., Morris, T.J. and Purcell, A.H. (1987) Cloning and detection of DNA from a non-culturable plant pathogenic mycoplasmalike organism. *Science* 238, 197–200.

Kuske, C.R., Kirkpatrick, B.C. and Seemüller, E. (1991) Differentiation of virescence mycoplasmalike organisms using western aster yellows MLO chromosomal DNA probes and restriction fragment length polymorphism analysis. *Journal of General Microbiology* 137, 153–159.

Lee, I.M. and Davis, R.G. (1992) Mycoplasmas which infect insects and plants. In: Maniloff, J., McElmansey, R.N., Finch, L.R. and Baseman, J.B. (eds) *Mycoplasmas; Molecular Biology and Pathogenesis*. American Society for Microbiology, Washington, DC, USA, pp. 379–390.

Lee, I.M., Davis, R.E., Sinclair, W.A., DeWitt, N.D. and Conti, M. (1993a) Genetic relatedness of mycoplasmalike organisms detected in *Ulmus* spp. in the United States and Italy by means of DNA probes and polymerase chain reactions. *Phytopathology* 83, 829–833.

Lee, I.M., Hammond, R.W., Davis, R.E. and Gundersen, D.E. (1993b) Universal amplification and analysis of pathogen 16S rDNA for classification and identification of mycoplasmalike organisms. *Phytopathology* 83, 834–842.

Liao, C.H. and Chen, T.A. (1977) Culture of corn stunt spiroplasma in a simple medium. *Phytopathology* 67, 802–807.

Lim, P.-O. and Sears, B.B. (1989) 16S rRNA sequence indicates that plant-pathogenic mycoplasmalike organisms are evolutionarily distinct from animal mycoplasmas. *Journal of Bacteriology* 171, 5901–5906.

Maramorosch, K. (1976) Plant mycoplasma disease. In: Heitefuss, R. and Williams, P.H. (eds) *Physiological Plant Pathology*, Vol. 4 of *The Encyclopaedia of Plant Physiology* New Series, Pirson, A. and Zimmerman, M.H. (eds). Springer-Verlag, Berlin, Germany, pp. 150–171.

Marcone, C., Neimark, H., Ragozzino, A., Lauer, U. and Seemüller, E. (1999) Chromosome sizes of phytoplasmas comparing major phylogenetic groups and subgroups. *Phytopathology* 89, 805–810.

Markham, P.G., Townsend, R., Bar-Joseph, M., Daniels, M.J., Plaskitt, K. and Meddins, B.M. (1974) Spiroplasmas are the causal agents of citrus little leaf disease. *Annals of Applied Biology* 78, 49–57.

McCoy, R.E. (1979) Mycoplasmas and yellows diseases. In: Whitcomb, R.F. and Tully, J.G. (eds) *The Mycoplasmas*, Vol. 3. *Plants and Insect Mycoplasmas*. Academic Press, New York, USA, 351 pp.

McCoy, R.E., Cauldwell, A., Chang, C.J., Chen, T.A., Chiykowski, L.N., Cousin M.T., Dale, J.L., de Leeuw, G.T.N., Golino, D.A., Hackett, K.J., Kirkpatrick, B.C., Marwitz, R., Petzold, H., Sinha, R.C., Sugiura, M., Whitcomb, R.F., Yang, I.L., Zhu, B.M. and Semµller, E. (1989) Plant diseases associated with mycoplasmalike organisms. In: Whitcomb, R.F. and Tully, J.G. (eds) *The Mycoplasmas*, Vol. 5. *Spiroplasma, Acholeplasmas and Mycoplasmas of Plants and Arthropods*. Academic Press, San Diego, USA.

Minucci, C., Rajan, J. and Clark, M.F. (1996) Differentiation of phytoplasmas associated with sweet potato little leaf disease and other phytoplasmas in the Faba Bean Phyllody cluster. *British Crop Protection Council Symposium Proceedings* No. 65, 61–66.

Namba, S., Oyaizu, H., Kato, S., Iwanami, S. and Tsuchizaki, T. (1993) Phylogenetic diversity of phytopathogenic mycoplasmalike organisms. *International Journal of Systematic Bacteriology* 43, 461–467.

Padovan, A.C., Firrao, G., Schneider, B. and Gibb, K.S. (2000) Chromosome mapping of the sweet potato little leaf phytoplasma reveals genome heterogeneity within the phytoplasmas. *Microbiology* 146, 893–902.

Sarindu, N. and Clark, M.F. (1993) Antibody production and identity of MLOs associated with sugar-cane whiteleaf disease and bermuda-grass whiteleaf disease from Thailand. *Plant Pathology* 42, 396–402.

Sears, B.B. and Kirkpatrick, B.C. (1994) Unveiling the evolutionary relationships of plant pathogenic mycoplasma-like organisms. *ASM News* 60, 307–312.

Sears, B.B. and Klomparens, K.L. (1989) Leaf tip cultures of the evening primrose allow stable, aseptic culture of mycoplasma-like organism. *Canadian Journal of Plant Pathology* 11, 343–348.

Seemüller, E. (1976) Fluorescence optical demonstration of mycoplasma-like organisms in the phloem of trees with pear decline or proliferation disease. *Phytopathologische Zeitschrift* 85, 368–372.

Seemüller, E., Schneider, B., Maurer, R., Ahrens, U., Daire, X., Kison, H., Lorenz, K.-H., Firrao, G., Avinent, L., Sears, B.B. and Steckebrandt, E. (1994) Phylogenetic classification of phytopathogenic mollicutes by sequence analysis of 16S ribosomal DNA. *International Journal of Systematic Bacteriology* 44, 440–446.

Sinha, R.C. and Chiykowski, L.N. (1984) Purification and serological detection of mycoplasma-like organisms from plants affected by peach eastern X-disease. *Canadian Journal of Plant Pathology* 6, 200–205.

Smart, C.D., Schneider, B., Blomquist, D.L., Guerra, L.J., Harrison, N.A., Ahrens, U., Lorenz, K.-H., Seemüller, E. and Kirkpatrick, B.C. (1996) Phytoplasma-specific PCR primers based on sequences of the 16S–23S rRNA spacer region. *Applied and Environmental Microbiology* 62, 2988–2993.

Tsai, J.H. (1979) Vector transmission of mycoplasmal agents of plant diseases. In: Whitcomb, R.F. and Tully, J.G. (eds) *The Mycoplasmas*, Vol. 3. *Plant and Insect Mycoplasmas*. Academic Press, New York, USA, pp. 265–307.

Tymon, A.M., Jones, P. and Harrison, N.A. (1998) Phylogenetic relationships of coconut phytoplasmas and the development of specific oligonucleotide PCR primers. *Annals of Applied Biology* 132, 437–452.

Yu, Y.L., Yeh, K.W. and Lin, C.P. (1998) An antigenic protein gene of a phytoplasma associated with sweet potato witches' broom. *Microbiology* 144, 1257–1262.

Plant Parasitic Nematodes

J. Bridge[1] and T.D. Williams[2]

[1]CABI Bioscience, Bakeham Lane, Egham, Surrey TW20 9TY, UK; [2]4 Warren Park, West Hill, Ottery St Mary, Devon EX11 1TN, UK

General

Nematodes that feed on higher plants are known as 'plant parasitic nematodes' or simply 'plant nematodes' to distinguish them from the many other 'free-living' nematodes found in soils and plant material that are saprophytes or feed on bacteria or fungi. Plant parasitic nematodes are also referred to as 'eelworms' in the earlier literature. Many different genera and species of nematodes are known to be crop pests (Table 13.1), some more damaging than others.

Taxonomy

Plant parasitic nematodes belong to the orders Tylenchida and Dorylaimida in the phylum Nematoda.

Morphology

Most plant parasitic nematodes are minute, vermiform animals ranging in size from less than 0.2 mm (*Paratylenchus* spp.) to over 12 mm (*Longidorus* spp.). They generally show few morphological differences between the sexes except for sexual organs, although in some genera the females become comparatively large and swollen, as in the gall-forming root-knot nematodes (*Meloidogyne* spp.) and the cyst-forming genera *Heterodera*, *Globodera* and *Punctodera* or saccate as in *Achlysiella*, *Nacobbus*, *Tylenchulus* and *Rotylenchulus*. All known plant parasitic nematodes possess a stylet or mouth-spear that is

©CAB International 2002. Plant Pathologist's Pocketbook
(eds J.M. Waller, J.M. Lenné and S.J. Waller)

Table 13.1. Distribution and crop hosts of nematode genera and species known to be important pests worldwide.

Nematodes and damage symptoms	Main crops affected	World and climatic distribution
Achlysiella		
Necrosis of roots		Tropical
A. williamsi	Sugarcane	Australasia
Anguina	Cereals and grasses	Temperate: worldwide
Seed and leaf galls, distortion of leaves		
A. tritici (Ear cockle nematode)	Temperate cereals, mainly wheat	Temperate: North Africa, India, China, eastern Europe, eastern Mediterranean, Middle East
Aphasmatylenchus		Tropical: West Africa
Poor root growth, leaf chlorosis		
A. straturatus	Groundnuts	Tropical: West Africa
Aphelenchoides		Temperate and tropical
Necrosis and distortion of leaves, buds and seeds; destruction of fungal mycelium		
A. arachidis (Groundnut testa nematode)	Groundnut	Tropical: West Africa
A. besseyi (Rice white tip nematode)	Rice	Tropical: rice-growing areas worldwide
A. fragariae (Strawberry crimp nematode)	Strawberry	Temperate: Europe, North America, Japan
A. ritzemabosi (Leaf nematode)	Chrysanthemum	Temperate: Europe. North and South America, East and South Africa, Australia
A. composticola (Mushroom nematode)	Mushrooms	In mushroom cultivation areas worldwide
Belonolaimus (Sting nematodes)		Subtropical
Necrosis of roots, leaf chlorosis, wilt		
B. longicaudatus	Sweetcorn, vegetables, groundnut, citrus, cotton, grasses	Subtropical: south-eastern USA
Bursaphelenchus		Temperate
Chlorosis and tree death		
B. xylophilus (Pine wilt nematode)	Pine	Temperate: North America, China, Taiwan, Korea

Continued

Table 13.1. *Continued*

Nematodes and damage symptoms	Main crops affected	World and climatic distribution
Criconemella (Ring nematodes) Chlorosis, necrosis of roots and pods, wilting	Common on tree crops	Temperate and tropical
C. onoensis	Rice	Tropical: USA, West Africa, Central and South America
C. ornata	Groundnut	Subtropical: USA
C. xenoplax	Fruit trees	Subtropical: USA
Ditylenchus (Stem/bulb nematodes) Lesions of stems and leaves, distortion of flowers and foliage, bulb and tuber rot		Temperate and tropical
D. africanus (Groundnut pod nematode)	Groundnut	Subtropical: South Africa
D. angustus (Ufra nematode)	Rice	Tropical: Bangladesh, India, Burma, Vietnam
D. destructor	Potatoes	Temperate: Europe, North America
D. dipsaci	Field beans, onions, daffodils and other bulb crops, cereals	Temperate: Europe, North and South America, Australasia
D. myceliophagus	Mushrooms	Temperate: mushroom-cultivating areas worldwide
Dolichodorus Necrotic and shortened (stubby) roots		Warm temperate
D. heterocephalus (Awl nematode)	Vegetables (celery, tomato, beans)	Warm temperate: USA, South Africa
Globodera (Round cyst nematodes) Cysts on roots, poor root growth, leaf chlorosis, wilting		Temperate
G. pallida, G. rostochiensis (Potato cyst nematodes)	Potato and other species of Solanaceae	Temperate: potato-growing areas worldwide
Helicotylenchus (Spiral nematodes) Necrosis of roots	Common on many crops but damage largely unknown for most species	Temperate and tropical: worldwide
H. multicinctus	Bananas and plantains	Tropical/subtropical: banana-growing areas worldwide

Nematode	Host	Distribution
Hemicriconemoides Root destruction, chlorosis, twig dieback		Temperate and tropical
H. mangiferae	Fruit trees	Subtropical: South Asia, Africa, South and Central America, Caribbean
Hemicycliophora (Sheath nematodes) Root-tip galls, stunting		Temperate and tropical
H. arenaria	Citrus	Subtropical: USA
Heterodera (Cyst nematodes) Cysts on roots, poor root growth, leaf chlorosis, wilting		Temperate and tropical
H. avenae (Cereal cyst nematode)	Temperate cereals	Temperate: cereal-growing areas worldwide
H. cajani (Pigeon pea cyst nematode)	Pigeon pea	India
H. ciceri (Chickpea cyst nematode)	Chickpea, lentils	Mediterranean
H. glycines (Soybean cyst nematode)	Soybean, beans	Subtropical: North and South America, Japan, China
H. goettingiana (Pea cyst nematode)	Peas, beans, lentils	Temperate: Europe, North America, Mediterranean
H. oryzae (Rice cyst nematode)	Rice	Tropical: India, Bangladesh
H. sacchari (Sugarcane cyst nematode)	Sugarcane, rice	Tropical: West Africa, India
H. schachtii (Sugarbeet cyst nematode)	Beets, swedes and other brassicas	Temperate/subtropics: Europe, North America, West and South Africa, Australia
H. trifolii (Clover cyst nematode)	Clover	Temperate: Europe
H. zeae (Maize cyst nematode)	Maize	Tropical: India
Hirschmanniella (Rice root nematodes and mitimiti nematode) Root lesions and corm rot (mitimiti disease)		Tropical
H. gracilis, H. imamuri	Rice	Tropical
H. oryzae, H. spinicaudata	Rice	Tropical: Africa, North and South America, South and South-East Asia
H. miticausa	Taro	Tropical: Pacific
Hoplolaimus (Lance nematodes) Necrosis of roots		Tropical
H. columbus	Cotton	Subtropical: USA, Egypt
H. seinhorsti	Cotton, vegetables	Tropical: Africa, South Asia, South America

Continued

Table 13.1. *Continued*

Nematodes and damage symptoms	Main crops affected	World and climatic distribution
Longidorus (Needle nematodes) Root-tip galling. Transmit viruses		Temperate and tropical
L. elongatus	Strawberry, sugarbeet	Temperate: Europe, Canada
Meloidogyne (Root-knot nematodes) Galling of roots and tubers, leaf chlorosis, wilting		Temperate and tropical
M. acronea	Cotton, sorghum	Tropical: South Africa
M. africana	Coffee	Tropical: Africa
M. arenaria	Groundnut	Tropical: worldwide
M. chitwoodi	Potatoes, sugarbeet, cereals	Temperate: North America, Mexico, South Africa, Australia, Europe
M. coffeicola, M. exigua	Coffee	Tropical: South America
M. graminicola	Rice	Tropical: South and South-East Asia
M. hapla	Pyrethrum, vegetables, clover	Temperate: worldwide
M. incognita, M. javanica	Vegetables, cotton, tobacco, ++ very wide host range	Tropical: worldwide
M. naasi	Grasses	Temperate: Europe, USA, New Zealand
M. oryzae	Rice	Tropical: South Asia
Nacobbus (False root-knot nematodes) Root galling, leaf chlorosis, wilting		Temperate/subtropical
N. aberrans	Vegetables, potato, sugarbeet	Temperate/subtropical: South, Central and North America, glasshouses in Europe
Paralongidorus (Needle nematodes) Root-tip galling. Transmit viruses		Temperate and tropical
P. australis	Rice	Subtropical: Australia

Nematode / damage	Hosts	Distribution
Paratrichodorus (Stubby root nematodes) Shortened (stubby) blackened roots. Transmit viruses		
P. minor	Vegetables	Temperate and subtropical
Pratylenchus (Lesion nematodes) Necrosis of roots, pods, corms and tubers, stunting, toppling (bananas), tuber rot (yams)		
P. brachyurus	Groundnuts, pineapple	Temperate: Europe, USA; Subtropical: Asia, Africa, Central and South America, Australia
P. coffeae	Bananas, yams, coffee, citrus, spices ++ very wide host range	Temperate and tropical; Tropical: worldwide
P. goodeyi	Bananas	Tropical: worldwide
P. loosi	Tea	Subtropical: East and West Africa, Canaries
P. neglectus	Cereals, grasses, fruit trees	Subtropical: South Asia
P. penetrans	Fruit and nut trees, soft fruits, vegetables, flower crops	Temperate: worldwide
P. zeae	Maize, upland rice	Temperate: worldwide
P. vulnus	Fruit and nut trees, flower crops	Tropical: South and South-East Asia
Punctodera Stunting and chlorosis, cysts on roots	Cereals and grasses	Temperate: worldwide
P. punctata	Grasses, temperate cereals	Temperate and tropical
P. chalcoensis	Maize	Temperate: Europe, USA, Russia
Radopholus (Burrowing nematodes) Necrosis of roots and tubers, rots, root breakage, toppling		Tropical: Mexico
R. citri	Citrus	Tropical
R. similis	Bananas, citrus, root and tubers, coconut, tea, black pepper and other spices	Tropical: Indonesia; Tropical: worldwide

Continued

Table 13.1. *Continued*

Nematodes and damage symptoms	Main crops affected	World and climatic distribution
Rhadinaphelenchus		
Necrosis (red ring) of stems and inflorescence, nut fall		Tropical
R. cocophilus (Red ring nematode)	Coconut, oil palm	Tropical: South and Central America, Caribbean
Rotylenchulus (Reniform nematodes)		
Poor root growth, leaf chlorosis, stunting		Tropical
R. parvus	Pigeon pea, sweet potato	Tropical: Africa
R. reniformis	Pineapple, vegetables	Tropical: worldwide
R. variabilis	Sweet potato	Tropical: Africa
Rotylenchus (Spiral nematodes)		
Unthrifty growth		Temperate and subtropical
R. robustus	Vegetables, tree seedlings	Temperate: Europe, North America, India
Scutellonema (Spiral nematodes)		
Dry rot of tubers, poor root growth		Mainly in the tropics and Africa
S. bradys (Yam nematode)	Yams	Tropical: West Africa, Caribbean
S. cavenessi	Groundnuts	Tropical: West Africa
Trichodorus (Stubby root nematodes)		
Shortened (stubby) blackened roots. Transmit viruses		Temperate and tropical
T. primitivus	Sugarbeet	Temperate: Europe, North America, New Zealand
T. viruliferus	Sugarbeet, potato	Temperate: Europe, North America
Trophotylenchulus		
Reduced root growth		
T. obscurus	Coffee	Tropical: Africa

Tylenchorhynchus (Stunt nematodes)	Cereals, vegetables	Temperate and tropical
Stunting of roots		
T. annulatus	Cereals, grasses	Tropical/subtropical: worldwide
T. brevilineatus	Groundnuts	Tropical: India
T. claytoni	Cereals, grasses, trees	Temperate: North America, Europe, Japan
T. dubius	Cereals, grasses	Temperate: Europe
Tylenchulus		Tropical/subtropical
Poor root growth, slow decline of trees, dieback		
T. semipenetrans (Citrus nematode)	Citrus	Tropical/subtropical: citrus-growing areas worldwide
Xiphinema (Dagger nematodes)		Temperate and tropical
Root-tip galling. Transmit viruses		
X. americanum (group)	Trees, grape	Temperate/subtropical: worldwide
X. diversicaudatum	Roses, grape, soft fruits	Temperate: Europe, North America, Australia, New Zealand
X. index	Grape, fruit trees, rose	Temperate: Europe, South America, Mediterranean, South Africa, Australia

extruded to pierce plant cells; some stylets may bear basal knobs (Fig. 13.1). In some groups, the stylet is greatly reduced or absent in the males as in some criconematid genera.

Biology and life cycles

Nematodes generally have four moulting juvenile stages between egg and adult. Eggs are laid in soil and/or plant tissues singly, or in large numbers in gelatinous egg-masses, or are retained in the dead females as with the cyst-forming genera *Heterodera*, *Globodera* and *Punctodera*. Life cycles of tropical plant

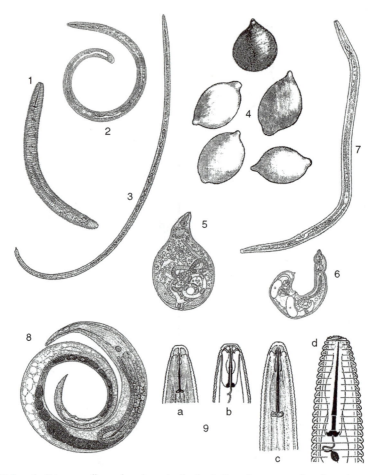

Fig. 13.1. 1. *Criconemella* sp. female entire body. 2. *Scutellonema* sp. female entire body. 3. *Xiphinema* sp. female entire body. 4. Cysts of *Heterodera* and *Globodera* spp. 5. *Meloidogyne* sp. mature swollen female. 6. *Rotylenchulus* sp. mature swollen female. 7. *Pratylenchus* sp. female entire body. 8. *Anguina* sp. female entire body. 9. Heads of nematodes showing mouth stylets: a, *Tylenchorhynchus*; b, *Pratylenchus*; c, *Helicotylenchus*; d, *Criconemella*.

parasitic species are generally shorter than those found in temperate environments. Typically the life cycle from egg to egg of the advanced nematodes such as *Meloidogyne* spp. in the tropics is around one month although it can vary. The shortest life cycles of 10 days or less are found in the relatively simple nematodes such as *Aphelenchoides*. Many cycles can occur in the tropics during one or more cropping seasons but some of the major nematode pests in temperate climates, such as the potato cyst nematodes, may only have one life cycle per year. Males are absent or very rare in some of the parasitic species and these reproduce parthenogenetically. The reproductive capability of plant-parasitic nematodes is considerable and numbers can rapidly increase. A single female of a *Meloidogyne* species may produce as many as 2000 viable eggs although the average is between 200 and 500. There can be an 80-fold increase of the potato cyst nematode *Globodera rostochiensis* in untreated potato fields, and numbers of stem-nematodes *Ditylenchus dipsaci* in soil may increase 50-fold after an oat crop, and much higher rates may occur within individual host plants.

Plant parasitic nematodes feed by inserting their hollow stylet or mouth spear into plant cells and sucking out the contents. Digestion of host cell contents may be partially extracorporeal as salivary fluids can be exuded; these fluids may also have a toxic or growth-modifying effect on the plant tissues. With *Meloidogyne*, for example, hyperplasia accompanies cell hypertrophy, often producing massive galling of roots in which the nematodes develop.

Plant-parasitic nematodes are dependent on a water film, either in the soil or on the plant, for mobility and become inactive or die if this is absent. Some species enter a quiescent phase with the onset of dehydration or extreme low temperatures and can survive for long periods in such conditions. Examples are the pre-adult stage of *D. dipsaci* (stem and bulb nematode), and the juvenile stages of *Anguina* and related species found in abundance in the seed galls of grasses and cereals when seed cleaning is inefficient. In most species the egg is the stage most able to survive adverse conditions.

Parasitic habits of plant nematodes

There are different types of parasitic habits in plant nematodes depending on whether they enter plant tissues and whether they remain migratory or become immobile and sedentary (Table 13.2):

1. *Ectoparasites*: generally the nematodes remain on the surface of the plant tissues feeding by inserting the stylet into cells that are within reach.
2. *Migratory endoparasites*: all stages of the nematodes can completely penetrate the plant tissues remaining mobile and vermiform and feeding as they move through tissues; they often migrate between the soil and roots.
3. *Sedentary endoparasites*: immature female or juvenile nematodes enter the plant tissues, develop a permanent feeding site, become immobile and swell into obese bodies.
4. *Semi-endoparasites*: immature female or juvenile nematodes only partially penetrate the roots leaving the posterior half to two-thirds of the body

projecting into the soil. Nematodes become immobile in a fixed feeding site and the projecting posterior of the body swells. (Some migratory nematodes can also be found in a semi-endoparasitic position on the roots.)

Table 13.2.   Parasitic habits.

Habitat	Nematode genera and species	
Ectoparasites		
Foliar		
Ectoparasites feeding generally on epidermal plant cells of young leaves, stems and flower primordia often enclosed by other foliage.	*Anguina* *Aphelenchoides* *Ditylenchus*	
Root		
Ectoparasites with short stylets feeding mainly on outer root cells and root hairs.	*Helicotylenchus* (some) *Paratrichodorus*	*Trichodorus* *Tylenchorhynchus*
Ectoparasites with long stylets that can be inserted deep into root tissues normally at growing tip. (Some can become relatively immobile.)	*Belonolaimus* *Cacopaurus* *Criconema* *Criconemella* *Dolichodorus* *Hemicriconemoides*	*Hemicycliophora* *Longidorus* *Paralongidorus* *Paratylenchus* *Xiphinema*
Migratory endoparasites		
Foliar		
Endoparasites in stems, leaves, flower primordia or seeds.	*Aphelenchoides* *Bursaphelenchus xylophilus* *Ditylenchus dipsaci* *Rhadinaphelenchus cocophilus*	
Below ground		
All stages of endoparasites found throughout different tissues in roots, corms, bulbs, tubers and seeds (groundnuts).	*Aphasmatylenchus* *Ditylenchus* (some) *Helicotylenchus* (some) *Hirschmanniella* *Hoplolaimus* *Pratylenchoides*	*Pratylenchus* *Radopholus* *Rotylenchoides* *Rotylenchus* (some) *Scutellonema*
Sedentary endoparasites		
Enlarged nematode bodies generally within tissues. Different life stages present; females become sedentary and obese. Expansion of tissues (galling) can occur around nematodes.	*Achlysiella* *Globodera* *Heterodera* *Meloidodera*	*Meloidogyne* *Nacobbus* *Punctodera*
Semi-endoparasites		
Different stages of nematodes partly embedded in root; posterior part of mature females projecting from root becomes swollen.	*Rotylenchulus* *Sphaeronema* *Trophotylenchulus* *Tylenchulus*	

Distribution and Crops Damaged

Many genera and species are cosmopolitan, whereas others are more limited in distribution either by their own ecological requirements, such as temperature, soil moisture and soil type, or by the range restrictions of their preferred hosts (Table 13.1). Climatic conditions can limit distribution; some nematode species are restricted to tropical environments, others to temperate regions. The latter regions include the higher altitudes in tropical countries where nematodes that are normally restricted to cooler latitudes may be found. Some nematodes that are now serious crop pests in the cool temperate latitudes originated from the high altitudes in the tropics; examples are the potato cyst nematodes *G. rostochiensis* and *Globodera pallida* from the Andes in South America. Conversely, nematodes from northern Europe including the same potato cyst nematodes and others such as the stem and leaf nematodes, *Aphelenchoides ritzemabosi* and *D. dipsaci*, are being introduced into the high altitude regions of the tropics.

Alone, or in association with other pathogens, plant parasitic nematodes can be very destructive. When they reach economic threshold levels they can cause very significant mechanical or physiological damage and this can inhibit or prevent the uptake of water and nutrients. Many crops are affected by a range of different nematodes (Table 13.1).

Symptoms of Nematode Damage

Field symptoms

Mechanical injury is caused by nematodes feeding and moving through or between cells; some nematodes secrete pectinases that dissolve the middle cell lamella. Necrosis and other changes in cell contents may be caused by salivary secretions and more complex plant–nematode interactions occur in response to species of genera such as *Meloidogyne*, *Heterodera*, *Globodera* and *Nacobbus*, which produce specific feeding sites. Nematodes feeding ectoparasitically at the root tips suppress cell division in the apical meristem and cause stubby root formation; some root-tip galling may also occur. The symptoms of nematode injury may be very variable depending on the kind of nematode, the type and age of the host plant, the part of the plant affected and general growing conditions.

A clear association between disease or damage symptoms and plant parasitic nematodes is often not obvious, although there are a few exceptions. Nematode injury to roots produces above-ground symptoms similar to mineral deficiencies, inadequate or excessive water, and generally poor soils. Often so called 'sick' or 'tired' soils are a result of a build-up of large populations of parasitic nematodes in the soil. Symptoms are generally more pronounced if the plants are already affected by adverse growing conditions or are being attacked by other pathogens. Plants growing under highly favourable conditions may be heavily attacked by nematodes yet show few, if any, symptoms

Table 13.3. Symptoms of nematode damage.

Above-ground symptoms	Below-ground symptoms
1. Stunting, reduced foliage and dieback	1. Reduced root systems, particularly reduction in secondary feeder roots
2. Poor yield	2. Abnormal roots
3. Yellowing (chlorosis) of leaves	(a) Overall galling of roots, often severe. Large
4. Wilting	uneven galls (most *Meloidogyne* spp.)
5. Early senescence	(b) Overall galling of roots, often severe. Small
6. Fruit drop and poor or malformed fruits	rounded galls (*Nacobbus*, some *Meloidogyne* spp.)
7. Interveinal discoloration and necrosis of leaves (e.g. *Aphelenchoides ritzemabosi* on chrysanthemum leaves.	(c) Root-tip galling, rounded (*Longidorus*, *Hemicycliophora*, coffee *Meloidogyne* spp.)
	(d) Swollen, hooked root tips (*Subanguina* spp., *Xiphinema* spp., *Meloidogyne graminicola*)
8. Stem, leaf and seed galls (e.g. *Anguina* spp. on leaves and seeds of grasses and cereals)	(e) Clumping of lateral roots in a ball (some *Meloidogyne* spp, *Heterodera* spp.)
9. Twisting of leaves and white tips (e.g. *Aphelenchoides besseyi* on rice)	(f) Necrotic root lesions or overall necrosis of roots (*Pratylenchus* spp., *Radopholus* spp., *Hirschmanniella* spp., some *Helicotylenchus*)
10. Twisting of leaves and raised yellow lesions on stems and leaves (e.g. *Ditylenchus dipsaci* on daffodils and onions)	(g) Accumulation of soil particles and debris on roots (*Tylenchulus semipenetrans*)
11. Twisted panicles and empty grains (e.g. *Ditylenchus angustus* on rice)	(h) Stubby roots (*Paratrichodorus*, *Trichodorus*)
12. Yellowing and rapid death of pine trees (*Bursaphelenchus xylophilus*)	3. Cysts (white, yellow or dark-brown specks) on root surface (*Heterodera*, *Globodera*)
13. Yellowing and collapse of palm leaves followed by rapid death. Red necrosis of internal stem tissues usually in a ring (*Rhadinaphelenchus cocophilus* on coconut and oil palm)	4. Internal rotting of tubers, corms and bulbs (*Ditylenchus*, *Pratylenchus*, *Scutellonema*)
	5. Galled or warty tubers (potato, yam) (*Meloidogyne* spp.)
14. Distorted apical growth and 'crimping' of leaves and inflorescence (*A. besseyi* and *Aphelenchoides fragariae* on strawberry)	6. Surface cracking of tubers (potato, sweet potato, yam) (*Ditylenchus destructor*, *Scutellonema bradys*, *Pratylenchus coffeae*, *Meloidogyne* spp.)
	7. Lesions on groundnut pods (*Pratylenchus*, *Criconemella*)
	8. Brown and shrivelled groundnut seeds (*Aphelenchoides arachidis*, *Ditylenchus africanus*)

above ground. Under such circumstances nematodes reproduce better and can represent a severe and hidden threat to succeeding crops.

Symptoms of nematode damage can be divided into two categories, above ground and below ground (Table 13.3). It is usually necessary to examine roots and other plant tissues to establish a connection between damage symptoms and nematodes. To be certain of the association, nematodes have to be extracted from soil, roots or other plant material and identified microscopically.

Seed/planting material: symptoms of nematode damage

The spread of nematodes in planting material can lead to serious yield losses in the mature crop. Where symptoms of damage can be detected in material prior to planting it will be possible to prevent this spread. Many nematodes are spread by this means but mainly in vegetatively propagated material (tubers, corms, bulbs) or seedlings and rootstocks; very few nematodes are actually disseminated in true seeds (Table 13.4).

Interrelationships

Nematodes as vectors of plant viruses

As well as directly damaging plants, nematodes are vectors of several harmful plant viruses. Under very favourable circumstances a nematode can acquire a virus in feeding periods of an hour or less and transmit it equally rapidly. The viruses are widely distributed although few are recorded in the tropics. The nematode vector may remain infective for several months, more so in the genera *Xiphinema* and *Trichodorus* than in *Longidorus* spp., but infectivity does not survive a moult nor is the virus transmitted via the nematode egg. Two major virus groups are carried only by nematodes in the order Dorylaimida:

1. The nematode-transmitted polyhedral viruses (*Nepoviruses*). Species of *Xiphinema* and *Longidorus* are the nematode vectors transmitting the following viruses:
Xiphinema – *Tomato ringspot, Cherry rasp leaf, Peach rosette mosaic, Grapevine yellow vein, Arabis mosaic, Strawberry latent ringspot, Grapevine fanleaf.*
Longidorus – *Artichoke Italian latent, Tomato blackring, Peach rosette mosaic, Raspberry ringspot, Mulberry ringspot.*
2. The nematode-transmitted tubular viruses (*Tobraviruses*). Species of *Trichodorus* and *Paratrichodorus* are the vectors of *Tobacco rattle, Pea early browning* and *Pepper ringspot.*

Interactions with fungi and bacteria

Disease complexes between nematodes and wilt-inducing and root-rot fungi in particular are well documented for a number of fungi and nematode genera. *Fusarium oxysporum* and other *Fusarium* spp., *Verticillium* spp., *Pythium* spp. and *Cylindrocarpon* spp. are examples of fungi that interact with nematodes, mainly *Meloidogyne* spp., *Pratylenchus* spp. and *Rotylenchulus reniformis*, with a synergistic effect on plant damage.

Interactions are also known to occur between the disease-causing bacteria *Clavibacter* spp., *Pseudomonas* spp. and *Agrobacterium* spp., and species of the nematode genera *Meloidogyne, Pratylenchus, Anguina* and *Ditylenchus*.

Table 13.4.   Detection of nematode symptoms in seed or planting material.

Crop	Nematodes	Symptoms of damage
Seeds		
Wheat (*Triticum aestivum*)	*Anguina tritici*	Black, misshapen seed galls.
Beans (*Vicia faba*)	*Ditylenchus dipsaci*	Shrivelled and discoloured seeds, sometimes with 'nematode wool' attached.
Onion (*Allium cepa*)	*Ditylenchus dipsaci*	No observable symptoms. Nematodes emerging from soaked seeds examined microscopically.
Rice (*Oryza sativa*)	*Aphelenchoides besseyi*	No observable symptoms. Nematodes emerging from soaked seeds examined microscopically.
Groundnut (*Arachis hypogaea*)	*Aphelenchoides arachidis* *Ditylenchus africanus*	Shrivelled seeds, dark-brown testas.
Tubers		
Yam (*Dioscorea* spp.)	*Scutellonema bradys* *Pratylenchus coffeae* *Radopholus similis*	Internal dark-brown 'dry rot' in periphery of tuber; observed when cut or when epidermis is scraped away. Whole tubers spongy to touch with surface cracks.
	Meloidogyne spp.	Uneven, knobbly tubers; internal necrotic spots.
Potato (*Solanum tuberosum*)	*Pratylenchus* spp.	Uneven tubers pitted with dark, dry necrotic lesions.
	Meloidogyne spp.	Watery tubers, swellings and lesions on surface.
	Globodera spp.	Cysts on tuber or in adhering soil (difficult to detect).
	Ditylenchus destructor	Surface cracking and internal dark-brown rot.
Bulbs		
Onion, garlic (*Allium* spp.)	*Ditylenchus dipsaci*	Internal necrosis or rot of sets and cloves, sometimes with 'nematode wool' attached.
Rhizomes		
Ginger (*Zingiber officinale*)	*Meloidogyne* spp.	Light-brown, watery, necrotic areas in outer rhizome, often in surface folds.
	Radopholus similis	Small, sunken water lesions on surface.
Turmeric (*Curcuma domestica*)	*Radopholus similis*	Shallow, water-soaked brownish lesions and rotting.
Corms		
Banana, plantain, abaca (*Musa* spp.), Enset (*Ensete*)	*Radopholus similis* *Pratylenchus coffeae* *Pratylenchus goodeyi*	Purple to dark-brown necrotic lesions throughout cortex of attached roots and similar lesions in outer tissues of corm.
	Helicotylenchus multicinctus	Purple to dark-brown necrotic lesions in periphery of cortex of attached roots.

Table 13.4. *Continued*

Crop	Nematodes	Symptoms of damage
Taro (cocoyam) (*Colocasia esculenta*)	*Hirschmanniella miticausa*	Internal red necrotic streaks and peripheral rot.
Seedling transplants		After uprooting, the following symptoms may be observed:
Vegetables, rice, coffee, etc.	*Meloidogyne* spp.	Swellings or galls on roots.
	Nacobbus spp.	Discrete swellings or galls on roots.
	Pratylenchus spp. *Hirschmanniella* spp. *Radopholus* spp.	Yellow or brown lesions on root surface.
	Heterodera spp.	Spidery or distorted root growth sometimes with white specks (females) on roots of older seedlings.
Rootstocks		
Grapevine, peach, etc.	*Meloidogyne* spp.	Galls on roots, sometimes accompanied by rotting.
	Pratylenchus spp.	Brown lesions on root surface.
Citrus	*Tylenchulus semipenetrans*	Irregular, 'dirty' roots with soil adhering to surface.

From Bridge (1987) in Brown and Kerry (1987).

Two well-known examples of nematode–bacteria interactions are that of *Meloidogyne* spp. and *Ralstonia solanacearum* causing bacterial wilt of many crops (tobacco, potato, tomato, aubergine) and that of the ear cockle nematode, *Anguina tritici*, and *Clavibacter tritici* causing a disease in wheat referred to as 'tundu' in India (Khan, 1993).

Management of Nematodes

Many reviews on methods to manage nematodes in field soils and planting material are available and should be consulted for details. The most useful sources of general information are Webster (1972), Decker (1981), Nickle (1984), Brown and Kerry (1987), Luc *et al.* (1990), Bhatti and Walia (1992), Evans *et al.* (1993), Whitehead (1997), Dasgupta (1998) and Marks and Brodie (1998).

Briefly, management of nematodes can be by chemical or non-chemical means. The use of chemical nematicides has declined considerably over the years as a number of the products have been banned on the world market. Both fumigant nematicides, such as 1,3-dichloropropene, and non-fumigants, such as oxime carbamates, can be used. Their toxicity, costs and difficulties of application are their main drawbacks. There are many non-chemical cultural and physical practices that exist that can successfully be used to manage nematodes. The practices used, or the choice of the practices, will vary with the different types of crops or cropping systems. Generally, a combination or integration of a number of practices will have the most beneficial effect on reducing nematode damage. These fall mainly under two categories: (i) preventing the introduction and spread of nematodes by the use of nematode-free planting material; and (ii) using direct, cultural and physical control methods. The possible techniques to reduce nematode populations and the crop damage they cause are:

1. Rotation of crops;
2. Resistant cultivars;
3. Fallows;
4. Adjusting planting time/escape cropping;
5. Antagonistic plants and trap cropping;
6. Burning infested stubble;
7. Surface burning;
8. Flooding: artificial and natural;
9. Postharvest destruction/removal of infected crop residues;
10. Cultivating/turning soil between crops;
11. Grafting;
12. Improved crop husbandry (compensation for damage);
13. Soil solarization;
14. Hot water treatment of planting material;
15. Removal of nematode-infested tissues from planting material;
16. Organic soil amendments and biological control.

(See also Chapter 30.)

Techniques

Sampling

The many, well-documented techniques for sampling and processing soil and plant material all have a part to play in survey work, each one having its advantages and disadvantages. In general, enlightened use of even the simplest techniques will provide some pertinent information on the presence of most nematodes.

Field samples

Assessment of nematodes in fields can be done by taking and analysing individual samples or by taking a large number of samples from similar plants and bulking into one composite sample before extraction and further processing. The *sample size* is only critical for determination of nematode populations in advisory and regulatory surveys. For this work, the number of samples recommended is very variable, although a figure of 20–25 samples per area is usually suitable. The *time of sampling* can be critical; obviously fewer nematodes will be found in the dry season or winter months, outside the growing season. The most representative picture of nematodes parasitic on crop plants will be obtained by sampling the plants from the middle to the end of the growing season, but before plant senescence.

Soils

Although custom-made soil sampling equipment can be used, no special implement is required for collecting standard soil samples as they can be taken with whatever local digging tool is available. Most nematodes will occur around the plant root zone at depths of 5–30 cm, with few occurring in the surface layers. The many extraction methods that can be used to separate nematodes from soil are mainly based on the following techniques: Baermann funnel or tray method; flotation, sedimentation and sieving; elutriation, centrifugal flotation; and flotation, flocculation and sieving. The basic tray method is where active nematodes move from the soil into a dish of water, the soil (100–200 ml) being separated from the water by tissue or muslin cloth supported by various devices such as plastic sieves (Fig. 13.2). Water is added to make the soil moist and the extraction left for 24 h after which the extraction filter is carefully removed and the water–nematode suspension poured off into a beaker. Sieving involves pouring a suspension of soil (100–500 ml in a bucket of water) through a bank of suitable sieves. For most plant parasitic nematodes sieve meshes of 75 μm and 53 μm are adequate, and for maximum

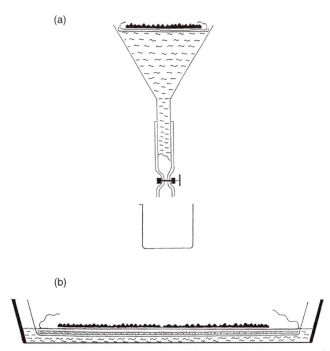

(a)

(b)

Fig. 13.2. Simple equipment for extraction of nematodes: (a) Baermann funnel; (b) tray modification.

retention of nematodes ideally there should be two or three 53 μm sieves in a bank with one or two 75 μm sieves on top. The material trapped on each sieve is then washed into a beaker with a gentle jet of water applied to the back of each sieve.

All extraction methods are fully described in most nematology books including *Plant Parasitic Nematodes in Subtropical and Tropical Agriculture* (Luc *et al.*, 1990), *Plant Parasitic Nematodes in Temperate Agriculture* (Evans *et al.*, 1993) and *Laboratory Methods for Work with Plant and Soil Nematodes* (Southey, 1986).

Roots

The type and amount of roots sampled can be important. It is preferable to examine all roots from plants with comparatively small root systems, but this is not always possible with crop plants in the field. The techniques used for assessing populations of nematodes in roots will depend on whether the nematodes are active or sedentary. Active or migratory nematodes can be extracted in a similar way to those from soils over a period of 72 h or more. Sedentary nematodes can be separated from root tissues by maceration–sieving and maceration–centrifugal flotation techniques. Staining also gives additional information on the direct association between presence of nematodes and tissue damage. Nematodes in plant material can be stained by immersion for 3 min in a boiling solution of equal parts lactic acid, glycerol and distilled water + 0.05% acid fuchsin or cotton blue. Stain is cleared from the plant tissues in a 50:50 mixture of glycerol and distilled water. Nematodes should be stained bright pink or blue, depending on the stain used.

Direct assessment of visible root damage symptoms induced by nematodes can be a great help in estimating actual crop damage and nematode infestation levels. For root-galling nematodes, e.g. *Meloidogyne* spp., a quantitative analysis of the extent of galling can be expressed as the percentage of roots galled, number of galls, size of galls, or types of roots galled. Similarly with the migratory endoparasites, e.g. *Pratylenchus*, *Radopholus*, the amount of visible necrosis or browning can again be determined on a percentage basis.

Foliage

Most leaf, stem and seed nematodes are active and can be extracted by methods used for root nematode extraction. The main consideration is to select those parts of the plant where the nematodes are either known to be, or likely to occur, for extraction. Populations are normally quantified on numbers of nematodes per weight of stems and leaves, and weight or numbers of seeds. Infested seeds can be left in a shallow dish of water and nematodes will emerge over a period of 72 h.

Direct assessment of foliar damage symptoms is also possible, either in the field or laboratory. The number of plants with seed galls or number of seed galls per head caused by *A. tritici*, and leaf twisting, stem and inflorescence distortion and necrosis caused by *Anguina*, *Aphelenchoides* and *Ditylenchus* can be visually assessed.

Killing and fixing

Nematodes are best killed by gentle heat (55–60°C) in water. After killing, the sample is left to cool and then the whole nematode suspension is fixed by adding an equal volume of 'double strength' fixative. Alternatively, individual nematode specimens are picked out of the suspension after heating and cooling and transferred into the cold fixatives at the normal strength shown below.

The main fixatives used are:

	Normal strength	Double strength
1. TAF		
40% Formaldehyde (formalin)	7 ml	7 ml
Triethanolamine	2 ml	2 ml
Distilled water	91 ml	45 ml
2. FA 4:1		
40% Formaldehyde (formalin)	10 ml	10 ml
Glacial acetic acid	1 ml	1 ml
Distilled water	89 ml	45 ml
3. Formaldehyde (formalin)	2% (5%)	4% (10%)

Nematodes in fixative should be left for 12 h or overnight before processing further.

CABI Bioscience, plant nematode identification service

Precise identification to species level is essential to determine the potential economic importance and host range of a plant parasitic nematode. Such identifications can be made by the Plant Disease and Diagnostic Services, CABI Bioscience UK Centre, Egham, Surrey, UK. Plant material, soil or extracted nematodes are accepted by this Centre and information on pathogenicity, host ranges, bionomics and management of the relevant nematode species can be provided. For further information on methods of extraction and dispatching through the post contact the Director of the Centre. The identification service may be free for material originating from priority countries for UK Department for International Development (DFID); a charge is made for commercial and other material (see also Chapter 8).

Samples for identification may be sent to the Centre in the form of soil samples, plant material such as diseased roots, stems or leaves, and extracted nematode suspensions.

• *Soil*: About 200 g. Most of the plant parasitic nematodes are found in the soil around roots. Care should be taken to ensure that the soil is moist during transit. Soil samples with or without roots included travel well in polythene bags.

- *Roots*: 5–20 g, including healthy, lightly infested and severely infested roots. These are best sent in polythene bags to keep them damp but they should not be very wet.
- *Stems and leaves*: Plant tops usually decompose faster than roots and should be wrapped in paper in separate bags if they are stored for more than a day or two.
- *Extracted ·nematodes* fixed in TAF or FA 4:1 should be sent in glass or polythene vials with caps or stoppers but not corks.

Acknowledgements

We wish to thank our friends and colleagues particularly in IARC-Rothamsted and CABI *Bioscience* for their valuable help and use of material in preparing this chapter with especial thanks to Dr David Hunt who also compiled the drawings in Fig. 13.1.

General References

Select Bibliography

CIH Descriptions of Plant-parasitic Nematodes. Descriptions, in a standardized format, of plant parasitic nematodes, 15 species per set, in a folder. Set I (1972) to Set 8 (1985). List distribution, host ranges and any interactions with fungi, bacteria or viruses. Published by CAB International, Wallingford, UK.
Nematological Abstracts (Continues *Helminthological Abstracts Series B – Plant Nematology*). Published by CAB International, Wallingford, UK.

Journals

Devoted entirely to plant and soil nematodes

Indian Journal of Nematology (1971) Nematological Society of India.
International Journal of Nematology (1991) (formerly *Afro-Asian Journal of Nematology*).
Japanese Journal of Nematology (1972) Japanese Nematological Society, Tokyo.
Journal of Nematology (1969) Society of Nematologists, USA.
Nematologia Mediterranea (1973) Laboratorio di Nematologia Agraria de CNR, Bari, Italy.
Nematology (1998) (formerly *Nematologica* (1956) and *Fundamental and Applied Nematology* (1992)).
Nematropica (1971) Organization of Tropical American Nematologists.
Pakistan Journal of Nematology (1983) Pakistan Society of Nematologists.

Include articles on plant and soil nematodes

Annals of Applied Biology (1914) Cambridge University Press, London, UK.

International Journal of Pest Management (formerly *Tropical Pest Management* and *PANS*) (1961) Taylor and Francis, London.
Journal of Helminthology (1923) CAB International, Wallingford, UK.
Phytopathology (1911) American Phytopathological Society, Ithaca, New York, USA.
Plant Disease (formerly *Plant Disease Reporter* (1917)) American Phytopathological Society, Ithaca, New York, USA.
Plant Pathology (1952) Blackwell (formerly HMSO), London, UK.
Proceedings of the Helminthological Society of Washington (1934).
Systematic Parasitology (1979) W. Junk, The Hague.

Nematology Books

Since 1980

Barker, K.R., Carter, C.C. and Sasser, J.N. (1985) *An advanced treatise on* Meloidogyne, Vol. II, *Methodology*. North Carolina State University, Raleigh, USA, 223 pp.
Bhatti, D.S. and Walia, R.K. (1992) *Nematode Pests of Crops*. CBS Publishers, Delhi, 381pp.
Brown, R.H. and Kerry, B.R. (1987) *Principles and Practice of Nematode Control in Crops*. Academic Press, London, UK, 447 pp.
Dasgupta, M.K. (1998) *Phytonematology*. Naya Prokash, Calcutta, India, 846 pp.
Decker, H. (1981) *Plant Nematodes and Their Control (Phytonematology)*. Amerind Publishing, New Delhi, India, 540 pp.
Decraemer, W. (1995) *The Family Trichodoridae: Stubby Root and Virus Vector Nematodes*. Kluwer Academic Publishers, Dordrecht, The Netherlands, 360 pp.
Evans, K., Trudgill, D.L. and Webster, J.M. (eds) (1993) *Plant Parasitic Nematodes in Temperate Agriculture*. CAB International, Wallingford, UK, 648 pp.
Hall, G.S. (ed.) (1996) *Methods for the Examination of Organismal Diversity in Soils and Sediments*. CAB International, Wallingford, UK, 307 pp.
Hunt, D.J. (1993) *Aphelenchida, Longidoridae and Trichodoridae: Their Systematics and Bionomics*. CAB International, Wallingford, UK, 372 pp.
Jepson, S.B. (1987) *Identification of Root-knot Nematodes (*Meloidogyne *Species)*. CAB International, Wallingford, UK, 265 pp.
Khan, M.W. (1993) *Nematode Interactions*. Chapman & Hall, London, UK, 377 pp.
Kishi, Y. (1995) *Pine Wood Nematode and the Japanese Pine Sawer*. Thomas Company Ltd, Minato-ku, Japan, 302 pp.
Lamberti, F. and Taylor, C.E. (1986) *Cyst Nematodes*. Plenum Press, New York, USA, 467 pp.
Luc, M., Sikora, R.A. and Bridge, J. (eds) (1990) *Plant Parasitic Nematodes in Subtropical and Tropical Agriculture*. CAB International, Wallingford, UK, 629 pp.
Maggenti, A. (1981) *General Nematology*. Springer-Verlag, New York, USA, 372 pp.
Marks, R.J. and Brodie, B.B. (eds) (1998) *Potato Cyst Nematodes. Biology, Distribution and Control*. CAB International, Wallingford, UK, 408 pp.
Nickle, W.R. (1984) *Plant and Insect Nematodes*. Marcel Dekker, New York, USA, 372 pp.
Perry, R.N. and Wright, D.J. (eds) (1998) *The Physiology and Biochemistry of Free-living and Plant-parasitic Nematodes*. CAB International, Wallingford, UK, 438 pp.
Santos, M.S.N. de A., Abrantes, I.M. de O., Brown, D.J.F. and Lemos, R.M. (1997) *An Introduction to Virus Vector Nematodes and Their Associated Viruses*. Universidade de Coimbra, Portugal, 535 pp.
Sasser, J.N. and Carter, C.C. (1985) *An Advanced Treatise on* Meloidogyne. Vol. I. Biology and Control, North Carolina State University, Raleigh, USA, 422pp.
Sharma, S.B. (1998) The Cyst Nematodes. Kluwer Academic Publishers, Dordrecht, The Netherlands, 452 pp.

Siddiqi, M.R. (2001) *Tylenchida: Parasites of Plants and Insects*, 2nd edn. CAB International, Wallingford, UK, 833 pp.
Southey, J.F. (1986) *Laboratory Methods for Work with Plant and Soil Nematodes*. Reference Book 402. Ministry of Agriculture, Fisheries and Food, HMSO, London, UK, 202 pp.
Taylor, C.E. and Brown, D.J.F. (1997) *Nematode Vectors of Plant Viruses*. CAB International, Wallingford, UK, 296 pp.
Whitehead, A.G. (1997) *Plant Nematode Control*. CAB International, Wallingford, UK, 448 pp.

Selected texts published before 1980

Ayoub, S.M. (1980) *Plant Nematology. An Agricultural Training Aid*. Nemaid, Sacramento, California, USA, 195 pp.
Caveness, F.E. (1974) *A Glossary of Nematological Terms*. International Institute of Tropical Agriculture, Ibadan, Nigeria, 68 pp.
Chitwood, B.G. and Chitwood, M.B. (1940, 1974) *Introduction to Nematology*. University Park Press, Baltimore, Maryland, USA, 372 pp., 325 pp.
Dropkin, V.H. (1980) *Introduction to Plant Nematology*. John Wiley & Sons, New York, USA, 293 pp.
Goodey, J.B., Franklin, M.T. and Hooper, D.J. (1965) *T. Goodey's The Nematode Parasites of Plants Catalogued under Their Hosts*, 3rd edn. John Wylie, Commonwealth Agricultural Bureau, Farnham Royal, UK, 214 pp.
Jenkins, W.R. and Taylor, D.P. (1967) *Plant Nematology*. Reinhold, New York, 270 pp.
Lamberti, F. and Taylor, C.E. (1979) *Root-knot Nematodes (Meloidogyne Species). Systematics, Biology and Control*. Academic Press, London, 477 pp.
Lamberti, F., Taylor, C.E. and Seinhorst, J.W. (1975) *Nematode Vectors of Plant Viruses*. Plenum Press, London, 460 pp.
Mai, W.F. and Lyon, H.H. (1975) *Pictorial Key to Genera of Plant-parasitic Nematodes*, 4th edn. Cornell University Press, Ithaca, New York, USA, 219 pp.
Norton, D.C. (1978) *Ecology of Plant-parasitic Nematodes*. John Wiley & Sons, New York, USA, 268 pp.
Southey, J.F. (1978) *Plant Nematology*. HMSO, London, 440pp. (MAFF/ADAS GD1; replaces *Technical Bulletin* No. 7.).
Tarjan, A.C. and Hopper, B.E. (1974) *Nomenclatorial Compilation of Plant and Soil Nematodes*. Society of Nematologists, Deheon Springs, Florida, USA, 419 pp.
Taylor, A.L. (1971) *Introduction to Research on Plant Nematology. An FAO Guide to the Study and Control of Plant Parasitic Nematodes*. FAO, Rome, Italy, 133 pp.
Taylor, A.L. and Sasser, J.N. (1978) *Biology, Identification and Control of Root-knot Nematodes (Meloidogyne Species)*. North Carolina State University, Raleigh and USAID, 111 pp.
Thomas, P.R. and Taylor, C.E. (1968) Plant nematology in Africa south of the Sahara. *Technical Communication, Commonwealth Institute of Helminthology* No. 39, 83 pp.
Thorne, G. (1961) *Principles of Nematology*. McGraw-Hill, New York, 553 pp.
Wallace, H.R. (1963) *The Biology of Plant Parasitic Nematodes*. Edward Arnold, London, UK, 280 pp.
Wallace, H.R. (1973) *Nematode Ecology and Plant Disease*. Edward Arnold, London, UK, 228 pp.
Webster, J.M. (1972) *Economic Nematology*. Academic Press, London, 563 pp.
Zuckerman, B.M., Mai, W.F. and Rohde, R.A. (1971–1981) *Plant Parasitic Nematodes*, Vol. 1, *Morphology, Anatomy, Taxonomy and Ecology*, 345 pp. Vol. 2, *Cytogenetics, Host–parasite Interactions and Physiology*, 347 pp. Vol. 3, 528 pp. Academic Press, New York and London.

Insect and Other Arthropod Pests 14

G.W. Watson

*Entomology Department, The Natural History Museum,
Cromwell Road, London SW7 5BD, UK*

Introduction

These notes provide information on how to process and identify arthropod pests. Further information can be found in the publications listed in the bibliography, and in the electronic products and literature abstracts produced by CAB *International*. The latter are available either as a monthly journal, *Review of Agricultural Entomology* (*RAE*), or as CAB ABSTRACTS on CD-ROM; the CAB ABSTRACTS database is also available online through computer networks. Details of the identification, information and documentary services provided by CAB *International* may be obtained from The Director, CABI Bioscience UK Centre, Bakeham Lane, Egham, Surrey TW20 9TY, UK; fax +44 (0) 1491 829100, email: bioscience@cabi.org. Details of new information products of relevance may be found on the CAB *International* web site at www.cabi.org/ and on the British Crop Protection Council (BCPC) web site at www.BCPC.org/

Diagnosis

The first step in dealing with a plant health problem is diagnosis of the cause, based on the symptoms. Where arthropods are responsible for damage, the type of injury is often distinctive, e.g. faeces, holes in leaves, and the pest may still be present. However, some damage inflicted by invertebrates does not immediately suggest the cause, and the plant pathologist should keep this in mind when confronted with an unfamiliar situation. Sometimes the causal agent is no longer present, and the symptoms may be misleading. The following are examples.

1.   The damage may imitate symptoms of a different complaint. Examples are: flower-shed of tomatoes (apparently due to a physiological problem) may be caused by larvae of the midge *Contarinia lycopersici*; poor seed production in sorghum (often attributed to genetic sterility or bird attack) caused by *Contarinia sorghicola*; leaf-puckering of citrus caused by aphid attack of the young growth; and russeting of fruit caused by mites.
2.   The primary damage may be concealed. Examples are: dieback of citrus caused by root-feeding weevil larvae; and 'dead heart' of sugarcane caused by stem-boring caterpillars.
3.   Secondary damage may be present; for example, stem-boring larvae may infest the plant via pruning wounds.
4.   Secondary organisms may be present; for example, drosophilid fruit fly and carpophilid beetle larvae in citrus fruit previously damaged by fruit-piercing moths.

Identification

Once an arthropod has been found to be responsible for the problem, the next step is to obtain an accurate identification. An identification to species provides access to information on its biology, economic and quarantine importance, distribution and control. Sometimes biological information is available in treatises on the crop attacked. More frequently, information needs to be gathered from several sources, and the most rapid way to access it is to consult CAB ABSTRACTS, either in *RAE* or on CD-ROM or online.

Often only immature stages of the pest are present. Immature phytophagous mites can be identified relatively easily but most insects can only be authoritatively identified to species from adults. If the sample contains only immature specimens, it may be advisable to rear some through to adults before attempting identification. More information on rearing is provided in the section on Collection below.

If an authoritatively identified insect reference collection is available for comparison, or if the specimens agree in all particulars with a textbook account of a pest on the same crop, tentative identification may be possible. However, it is prudent to have the identification confirmed by an expert, to ensure that any biological or chemical control measures planned will be appropriate and effective.

For authoritative identification, specimens should be sent to an entomologist or acarologist in a local Department of Agriculture or university or museum; alternatively, the Network Co-ordinating Institute (NECI) of the BioNET LOOP for your region may be able to provide a list of experts willing to make identifications. The identification service provided by the Natural History Museum, London, UK may also be used; a charge is made for this service, the rate being lower for CABI member countries. Details are available from The Insect Identification Service, Department of Entomology, The Natural

History Museum, Cromwell Road, London SW7 5BD, UK; fax +44 207942 5229; email: insect-enquiries@nhm.ac.uk

Control

Action against an arthropod pest is only appropriate if damage is still in progress. If the pest activity is drawing to its end, e.g. if caterpillars are pupating, control may be unnecessary. Selection of the most appropriate control method requires that the identity of the pest and the nature and area of the crop be considered. Application of pesticides is widely used against many pests, but more sustainable options may be effective. Biocontrol agents can be used against larval stages of several foliage pests and those under glass; cultural methods to disrupt the life cycle of the pest, for example by the removal of crop remains that might harbour eggs or pupae, are also effective against many pests. The cheapest method consistent with effective control and environmental acceptability should be selected. Choice may sometimes be affected by additional considerations, such as a need for low phytotoxicity effects, likely impact on pollinators and natural enemies, compatibility with fungicides, low persistence of toxic residues or avoidance of taint in produce. Integrated pest management seeks a combination of methods which will give optimum efficiency at minimum environmental cost.

Different control methods and new pesticides are constantly being devised and tested. The most rapid way of obtaining up-to-date information on this field is to consult the literature abstracts produced by CAB *International* and the latest information published by BCPC.

Collection and Processing of Arthropod Specimens

The following notes cover the main requirements for collecting, rearing, killing, preserving, mounting and storing insects and mites. Additional information is available in the references in the bibliography.

Collection

When collecting arthropod specimens for identification, it is important that collection data are recorded at the time. Relevant information includes: place and date of collection, name of the substrate (host-plant or -animal); any specific details, e.g. what part of the plant the specimen was found on, what it was doing (e.g. resting or feeding or ovipositing); collector's name; the appearance of any damage caused; what life cycle stages were present; and any association with other organisms observed. Associated details like distinctive damage or galls may be photographed, preferably against a ruler to give an idea of size. Such information greatly facilitates identification and may add to what is known about the pest species. Full collection data should accompany any specimens sent away for authoritative identification.

The usefulness of specimens collected by light-, suction-, pitfall or other traps is limited because the biological data associated with trap catches is minimal. Small, inactive insects should be collected on pieces of host-plant material; more active arthropods may be caught by sweeping and brushing over vegetation with a net, or be dislodged from bushes by beating so that they fall on a dark sheet spread below, from which they may be removed individually (using an aspirator/paintbrush/forceps/tweezers) into a box or vial. In dense vegetation it is often best to stalk insects singly, so as not to disturb nearby specimens that may be caught subsequently. A large specimen tube, slowly lowered over the insect from above, is fairly effective for the capture of wary species. Insects that jump may be persuaded to jump into a tube.

Most arthropods are identified from adults, but a few groups are identified from immature stages (see Table 14.1). If adults are required for identification but the sample contains only immature stages, several specimens should be reared through to adult by caging them with their food in appropriate environmental conditions; all natural enemies must be removed. Some knowledge of the organism concerned may be required, as it may be necessary to provide suitable places for pupation, e.g. leaf litter or soil, or to use an appropriate level of illumination. The atmosphere in the rearing cage should not be extremely dry, as this may cause difficulties with moulting and expansion of the wings. Reared adults should be allowed time to expand, harden and darken fully before being killed.

Killing

Immature stages and many small species can be killed by dropping them into 80% ethanol, in which they can subsequently be stored (although colour and wax secretions are not preserved well in alcohol). Small, sessile insects should be killed in alcohol on small pieces of the host-plant, as forcible removal from the plant often damages the specimens. Enzyme activity sometimes makes larvae turn black if they are killed in alcohol at room temperature; for a speedy kill and enzyme denaturation, the larva should be dropped into recently boiled water for about 30 s, before being transferred into alcohol for preservation.

Adults of larger species can be killed by confinement in a clean, wide-mouthed glass jar containing fumes of ethyl acetate or chloroform, which can be applied to a piece of absorbent material pinned to the underside of the stopper. Most insects will die within 30 min. In the tropics, an alternative is to put samples in a freezer for several days, but some species survive this. Very large insects, e.g. glossy beetles, may be killed quickly in very hot water but this method should not be used with dull, hairy or scaly species, as wetting may change their appearance.

Specimens collected for molecular analysis should be killed rapidly by a method considered appropriate by the molecular biologists concerned, e.g. deep freezing in sterile tubes, in liquid nitrogen or on dry ice (solid carbon dioxide). They must not be allowed to thaw until they are being prepared for analysis.

Preservation

There are two main methods of preservation: wet in alcohol (usually small species and all immature stages); or dry on stainless-steel pins (usually large/hairy/scaly species). The most appropriate preservation method(s) for each arthropod group are summarized in Table 14.1. In some groups (e.g. Hemiptera), the best preservation method may depend on the family.

Preservation of specimens in alcohol is best if the sample is heated to about 75°C (e.g. by standing the tube in recently boiled water) for 20 min immediately after killing, to denature enzymes in the body. Since material containing water will dilute the alcohol, large samples or those containing plant tissues should be bottled with plenty of alcohol, or the liquid should be replaced with fresh 80% ethanol after 24 h; care may be necessary to avoid accidentally discarding very small specimens with the waste fluid.

Once killed, specimens that are to be preserved dry should be processed quickly to avoid them going greasy or developing mould. The aim is to preserve complete specimens in good condition; they should therefore be handled very gently to avoid breakage or loss of setae, scales or other features that may be essential for identification.

Specimens for molecular analysis must be preserved in a sterile environment maintained continuously at an appropriately low temperature; this may require liquid or solid gases, and special containers (e.g. Dewar flasks) and transport arrangements.

Storage

Specimens can be damaged by water, high humidity, heat, fungi, mites or insects. Even undamaged specimens become worthless if the labels bearing the collection data are damaged (e.g. by termites or water) or lost. In hot, humid climates the store-room should be air-conditioned if possible, and large specimens should be dried quickly in a drying cabinet to prevent decay.

Dry specimens should be stored in boxes with tight-fitting lids; a liberal supply of naphthalene (immobilized by pins or a partition) should be included to deter pests. Each specimen/sample should be accompanied by a label bearing the collection data, written in ink or pencil insoluble in water or alcohol. Specimens may be kept in paper envelopes or mounted on pins; pins actually passing through specimens should be specialist entomological pins made of stainless steel, to avoid corrosion problems.

If a box of dry specimens is attacked by mites or insects, these can be killed by completely sealing the box with polythene and sticky tape and deep freezing it for one week. On removal from the freezer, the seal should not be broken until the box and its contents have reached room temperature, to avoid problems with condensation on the specimens.

Material preserved in alcohol should be kept at the bottom of the vial by loosely crumpled tissue paper (not cotton wool, which tangles with the specimens' claws). Small vials intended for prolonged storage should be stored in alcohol in lidded jars, which should be topped up regularly.

Table 14.1. Selection and treatment of specimens in particular groups.

Group	Stage, form or sex	Method of preservation	Appropriate container or treatment
Diplura	Adult	80% alcohol	In tubes
Protura	Adult	80% alcohol	In tubes
Collembola	Adult	80% alcohol	In tubes
Thysanura	Adult	80% alcohol	In tubes
Ephemeroptera	Adult: males desirable	80% alcohol	In tubes
Odonata	Adult: males desirable	Dry	In paper
Plecoptera	Adult: males desirable	80% alcohol	In tubes
Orthoptera	Adult	Dry rapidly to prevent decay	Pinned through rear of prothorax, slightly to right of mid-line
Phasmida	Adult	Dry rapidly to prevent decay	Pinned through thorax or in paper
Dictyoptera	Adult	Dry	Pinned through rear of prothorax or in paper
Dermaptera	Adult	Dry or in 80% alcohol	Pinned through right tegmen or in tubes
Embioptera	Adult: males desirable	80% alcohol	In tubes
Isoptera	Adult workers, soldiers (and alates if possible)	80% alcohol	In tubes
Psocoptera	Adult: males desirable	80% alcohol or (if scaly winged) dry	In tubes or in paper
Mallophaga	Adult	80% alcohol	In tubes
Anoplura	Adult	80% alcohol	In tubes
Thysanoptera	Adult	60% alcohol (preferably mixed with glycerine and acetic acid in the ratio 10:1:1)	In tubes
Hemiptera			
Coleorrhyncha	Adult: both sexes desirable	80% alcohol mixed with glycerine in the ratio 20 : 1	In tubes
Auchenorrhyncha	Adult: males desirable	Dry; or in 80% alcohol mixed with glycerine in the ratio 20:1	Pinned through prothorax or in tubes

Sternorrhyncha			
Psylloidea	Adult: males desirable	95% alcohol; galls dry	In tubes; galls in paper
Aleyrodoidea	Pupal cases, with associated adults if possible	95% alcohol (pupal cases *in situ* on host tissue)	In tubes
Aphidoidea	Adult: apterae desirable	95% alcohol	In tubes
Coccoidea	Adult female (if possible, not old), preferably *in situ* on plant material	80–95% alcohol	In tubes
Heteroptera			
Cimicoidea	Adult	80% alcohol	In tubes
Other Heteroptera	Adult: males desirable	Dry; if small, in 80% alcohol mixed with glycerine in the ratio 20:1	Pinned through prothorax, or in tubes
Neuroptera	Adult	Dry; in 80% alcohol if small	Pinned through thorax or in tubes
Mecoptera	Adult: males desirable	Dry or in 80% alcohol	Pinned through thorax or in tubes
Trichoptera	Adult	Dry; in 80% alcohol if small	Pinned through thorax or in tubes
Lepidoptera	Adult: males usually desirable	Dry	Pinned through thorax with wings spread out
Diptera	Adult: males desirable: provide host name or sample if relevant	Dry if large; in 80% alcohol if small	Pinned through thorax just to one side of mid-line with a micropin or in tubes
Siphonaptera	Adult	80% alcohol	In tubes
Hymenoptera			
Sawflies, bees and wasps	Adult	Dry	Pinned through thorax
Ants	Workers	80% alcohol	In tubes
Parasitica	Adult: females desirable: provide host name or sample	80% alcohol (90–95% for large ichneumonids)	In tubes
Coleoptera	Adult: both sexes especially males desirable (plant-boring forms with associated larvae if possible)	Dry; larvae or very small adults in 80% alcohol	Pinned through anterior third of right elytron or in tubes
Strepsiptera	Adult male	80% alcohol	In tubes
Acarina	Adult: both sexes desirable	80% alcohol	In tubes
Other Arthropoda	Adult: both sexes desirable	80% alcohol	In tubes

Packing specimens for dispatch

Arthropods, especially dry specimens, are very fragile and easily damaged. Inadequate packing may result in severe damage or total loss of specimens. Specimens should be packed in a way that prevents them moving about. Weighty objects such as tubes and pieces of twig should be packed separately from pinned insects, and large specimens should not be included in the same box as small ones if possible.

For dry specimens, pins should be firmly inserted into the base of the box, which should be lined with a thick layer of cork or Pastazote (*not* expanded polystyrene, as this has insufficient grip). Objects likely to swivel about, such as large specimens or long labels, should be cross-pinned obliquely to secure them in position. There should be nothing liable to work loose in the box, and glass tubes or balls of naphthalene should not be included with pinned specimens.

Specimens preserved in alcohol should be placed in small, tightly sealed vials completely filled with alcohol (as movement of a bubble may damage the specimens). It is advisable to restrict the movement of specimens with a plug of tissue paper (not cotton wool) large enough not to move about in the tube. Tubes should be separated from each other with soft packing material like bubble-wrap, crumpled paper or expanded polystyrene chips. The type of alcohol used should always be fully miscible with water (i.e. it should remain transparent after dilution); this is not true of some commercial brands.

The insect box should be placed in a larger box of thin plywood or robust cardboard and surrounded on all sides by shock-absorbent material (e.g. expanded polystyrene chips or wood shavings or crumpled paper), not too tightly packed. The parcel should be clearly labelled with both the destination and originating addresses in permanent ink.

Accompanying data

Within a parcel of specimens, each specimen/sample should be labelled clearly with its own reference number and collection data in insoluble ink or pencil (see Collection section above). The reference number is essential, but if the collection data are too lengthy to fit entirely on a label or the specimen envelope, they can be listed separately against the reference number in a covering letter. This letter also should give the name and address of the sender and any other relevant information. It is better to provide too much information, than too little.

Dispatch

The shorter the transit, the less likely the material is to be damaged by rough handling. Airmail is preferable to surface mail. Large parcels are best sent by air parcel post and small parcels, weighing less than 500 g, by second-class airmail. Air freight should *not* be used because there are large charges for customs clearance and delivery, and delivery may be greatly delayed.

Select Bibliography

General textbooks and information sources

Borror, D.J., DeLong, D.M. and Johnson, N.F. (1989) *Introduction to the Study of Insects*, 6th edn. Saunders College Publishing, Philadelphia, USA, 875 pp.

CAB ABSTRACTS (abstracted literature from 1973 onwards; available online or as a CD-ROM product) CAB International, Wallingford, UK.

Chapman, R.F. (1998) *The Insects – Structure and Function*, 4th edn. Cambridge University Press, Cambridge, 770 pp.

Chinery, M. (1993) *Insects of Britain and Northern Europe*, 3rd edn. Harper Collins, London, UK, 320 pp.

Elzinga, R.J. (1997) *Fundamentals of Entomology*, 4th edn. Prentice-Hall, Englewood Cliffs, New Jersey, USA, 475 pp.

Evans, G.O. (1992) *Principles of Acarology*. CAB International, Wallingford, UK, 563 pp.

Gullan, P.J. and Cranston, P.S. (1999) *The Insects – an Outline of Entomology*, 2nd edn. Chapman & Hall, London, 491 pp.

Naumann, I.D. (1991) *The Insects of Australia*, 2nd edn. Melbourne University Press, Carlton, Victoria, 1137 pp.

Pearce, M.J. (1997) *Termites: Biology and Pest Management*. CAB International, Wallingford, UK, 190 pp.

Scholtz, C.H. and Holm, E. (1985) *Insects of Southern Africa*. 4th impression, 1996, Butterworths Publishers, University of Pretoria, Durban, 502 pp.

Walter, D.E. (1999) *Mites: Ecology, Evolution and Behaviour*. CAB International, Wallingford, UK, 352 pp.

Pests of cultivated plants

Anon. (2000) Crop Protection Compendium: Global Module, 2nd edn (Integrated electronic information resource on crops and their pests and diseases; available as a CD-ROM product.) CAB International, Wallingford, UK.

Buczacki, S.T. and Harris, K.M. (1981) *Collins Guide to Pests, Diseases and Disorders of Garden Plants*. William Collins, London, 512 pp.

Bohlen, E. (1978) *Crop Pests in Tanzania and Their Control*, 2nd revised edn. Verlag Paul Parey, Berlin, Germany, 142 pp.

CAB ABSTRACTS (abstracted literature from 1973 onwards; available online or as a CD-ROM product) CAB International, Wallingford, UK.

CABPESTCD® (abstracted literature on pests from 1973 onwards; available as a CD-ROM product) CAB International, Wallingford, UK.

Caresche, L., Cotterell, G.S., Peachey, J.E., Trayner, R.W. and Jaques-Felix, H. (1969) *Handbook for Phytosanitary Inspectors in Africa*. OAU/STRC, Lagos, 444 pp.

Distribution Maps of Plant Pests. Compiled by CAB *International* in association with the European and Mediterranean Plant Protection Organization (EPPO). CAB International, Wallingford, UK.

Forestry Compendium: module 1. (Integrated electronic information resource on forestry trees, pests and diseases; available as a CD-ROM product.) CAB International, Wallingford, UK.

Hill, D.S. (1983) *Agricultural Insect Pests of the Tropics and Their Control*, 2nd edn. Cambridge University Press, Cambridge, 746 pp.

Hill, D.S. (1994) *Agricultural Entomology*. Timber Press, Portland, Oregon, USA, 635 pp.

Hill, D.S. and Waller, J.M. (1988) *Pests and Diseases of Tropical Crops*, Vol. 2. *Field Handbook*. Intermediate Tropical Agriculture Series. Longman Scientific & Technical, Harlow, 432 pp.

Jeppson, L.R., Keifer, H.H. and Baker, E.W. (1975) *Mites Injurious to Economic Plants*. University of California Press, Berkeley, 614 pp.

Kalshoven, L.G.E. (1981) *Pests of Crops in Indonesia*. Ichtiar Baru – Van Hoeve, Jakarta, 701 pp.

Kranz, J., Schmutterer, H. and Koch, W. (1977) *Diseases, Pests and Weeds in Tropical Crops*. Verlag Paul Parey, Berlin, Germany, 666 pp.

Lewis, T. (1997) *Thrips as Crop Pests*. CAB International, Wallingford, UK, 736 pp.

Matthews, G. and Tunstall, J. (1994) *Insect Pests of Cotton*. CAB International, Wallingford, UK, 592 pp.

PEST CABWeb® (online subscription product) CAB International, Wallingford, UK.

Polazek, A. (1998) *African Cereal Stem Borers: Economic Importance, Natural Enemies and Control*. CAB International, Wallingford, UK, 592 pp.

Smith, I.M. and Charles, L.M.F. (1999) *Distribution Maps of Quarantine Pests of Europe*. CAB International, Wallingford, UK, 768 pp.

Smith, I.M. and McNamara, D.G. (1996) *Quarantine Pests for Europe*, 2nd edn. CAB International, Wallingford, UK, 1440 pp.

Smith, I.M. and Roy, A.S. (1996) *Illustrations of Quarantine Pests for Europe*. CAB International, Wallingford, UK, 220 pp.

Collecting, rearing, preserving and studying

Carter, D. and Walker, A.K. (1999) *The Care and Conservation of Natural History Collections*. Butterworth-Heinemann, Oxford, UK, 226 pp.

Martin, J.E.H. (1978) *The Insects and Arachnids of Canada*, Part 1. *Collecting, Preparing and Preserving Insects, Mites and Spiders*. Publication no. 1643, Canada Department of Agriculture, Québec, 182 pp.

Sánchez, J.V. and Stirnemann, E.G. (1995) *Captura, Preparacion y Conservacion de Insectos*. Editorial universitaria; Argumentos 9. Universidad Nacional de Misiones, Misiones, Tucuman, Argentina, 43 pp.

Steyskal, G.C., Murphy, W.L. and Hoover, E.M. (1986) *Insects and Mites. Techniques for Collection and Preservation*. Miscellaneous Publication No. 1443. United States Department of Agriculture, Washington, DC, USA, 103 pp.

Upton, M.S. (1991) *Methods for Collecting, Preserving, and Studying Insects and Allied Forms*, 4th edn. Miscellaneous Publications No. 3. The Australian Entomological Society, Indooroopilly, Queensland, Australia, 86 pp.

Uys, V.M. and Urban, R.P. (1996) *How to Collect and Preserve Insects and Arachnids*. Plant Protection Research Institute Handbook No. 7. Plant Protection Research Institute, Pretoria, 73 pp.

Identification and nomenclature

ANI-CD (1995) (a checklist of preferred names for insects of economic importance, based on Wood, A. (1989) below) CAB International, Wallingford, UK.

Blackman, R.L. and Eastop, V.F. (1997) *TAXAKEY to Aphids on the World's Crops*. (Electronic identification system on CD-ROM.) CAB International, Wallingford, UK.

Blackman, R.L. and Eastop, V.F. (2000) *Aphids on the World's Crops*, 2nd edn. John Wiley & Sons, Chichester.

Booth, R.G., Cox, M.L. and Madge, R.B. (1990) *IIE Guides to Insects of Importance to Man. 3. Coleoptera*. CAB International, Wallingford, UK, 392 pp.

Booth, R.G., Madge, R.B. and Cox, M.L. (1994) *CABIKEY to Major Beetle Families*. (Expert identification system on computer disks.) CAB International, Wallingford, UK.

Chinery, M. (1993) *Insects of Britain and Northern Europe*, 3rd edn. Harper Collins, London, UK, 320 pp.

Gorham, J.R. (1991) *Insect and Mite Pests in Food; an Illustrated Key*. United States Department of Agriculture, Agriculture Handbook No. 655. United States Department of Agriculture, Washington, DC, USA, 767 pp.

Hollis, D. (1980) *Animal Identification: a Reference Guide*. Vol. 3. *Insects*. British Museum (Natural History), London, and John Wiley & Sons, Chichester, UK, 160 pp.

Holloway, J.D., Bradley, J.D. and Carter, D.J. (1987) *IIE Guides to Insects of Importance to Man. 3. Lepidoptera*. CAB International, Wallingford, UK, 262 pp.

Martin, J.H. (1987) An identification guide to common whitefly pest species of the world (Homoptera, Aleyrodidae). *Tropical Pest Management* 33(4), 298–322.

Miller, D.R., Ben-Dov, Y. and Gibson, G. (1998) *Scalenet: a Searchable Information System on Scale Insects*. http://www.sel.barc.usda.gov/scalenet/scalenet.htm

Mound, L.A. (1989) *Common Insect Pests of Stored Products*, 7th edn. British Museum (Natural History), London, UK, 68 pp.

Oldroyd, H. (1970) *Diptera. 1 Introduction and Key to Families*, 3rd edn. *Handbooks for the Classification of British Insects* 9(1). Royal Entomological Society of London, London, UK, 104 pp.

Scoble, M.J. (1995) *The Lepidoptera: Form, Function and Diversity*. The Natural History Museum in association with Oxford University Press, Oxford, UK, 404 pp.

White, I.M. and Elson-Harris, M.M. (1992) *Fruit Flies of Economic Significance: Their Identification and Bionomics*. CAB International, Wallingford, UK, 600 pp.

White, I.M. and Hancock, D.L. (1997) *CABIKEY to the Dacini (Diptera, Tephritidae) of the Asia-Pacific-Australasian regions*. (Expert identification system on CD-ROM.) CAB International, Wallingford, UK.

Wilson, M.R. and Claridge, M.F. (1991) *Handbook for the Identification of Leafhoppers and Planthoppers of Rice*. CAB International, Wallingford, UK, 150 pp.

Wood, A. (1989) *Insects of Economic Importance: a Checklist of Preferred Names*. CAB International, Wallingford, UK, (available on CD-ROM as ANI-CD, 1995), 160 pp.

Zhang, B.-C. (1994) *Index of Economically Important Lepidoptera*. CAB International, Wallingford, UK, 600 pp.

Terminology

Nichols, S.W. (1989) *The Torre-Bueno Glossary of Entomology*. New York Entomological Society in cooperation with American Museum of Natural History, New York, USA, 840 pp.

Uys, V.M. and Urban, R.P. (1996) *How to Collect and Preserve Insects and Arachnids*. Plant Protection Research Institute Handbook No. 7, Plant Protection Research Institute, Pretoria, 73 pp.

Control methods

Copping, L. (1999) *The Biopesticides Manual*. British Crop Protection Council.

Debach, P. and Rosen, D. (1991) *Biological Control by Natural Enemies*, 2nd edn. Cambridge University Press, Cambridge, 440 pp.

Denholm, I., Pickett, J.A. and Devonshire, A.L. (1999) *Insecticide Resistance: from Mechanisms to Management*. CAB International, Wallingford, UK, 144 pp.

Dent, D. (2000) *Insect Pest Management*, 2nd edn. CAB International, Wallingford, UK, 480 pp.

Dent, D.R. and Walton, M.P. (eds) (1997) *Methods in Ecological and Agricultural Entomology*. CAB International, Wallingford, UK, 400 pp.

Highley, E., Wright, E.J., Banks, H.J. and Champ, B.R. (1994) *Stored Product Protection*. CAB International, Wallingford, UK, 1312 pp.

Hill, D.S. and Waller, J.M. (1982) *Pests and Diseases of Tropical Crops*. Vol. 1, *Principles and Methods of Control*. Intermediate Tropical Agriculture Series. Longman Group, Harlow, London, UK, 175 pp.

Kranz, J., Schmutterer, H. and Koch, W. (1977) *Diseases, Pests and Weeds in Tropical Crops*. Verlag Paul Parey, Berlin, Germany, 666 pp.

Panda, N. and Kush, S.G. (1995) *Host Plant Resistance to Insects*. CAB International, Wallingford, UK, 448 pp.

Paull, R.E. and Armstrong, J.W. (1994) *Insect Pests and Fresh Horticultural Products: Treatments and Responses*. CAB International, Wallingford, UK, 368 pp.

Polazek, A. (1998) *African Cereal Stem Borers: Economic Importance, Natural Enemies and Control*. CAB International, Wallingford, UK, 592 pp.

Whitehead, R. (Annual publication) *The UK Pesticide Guide* (also available on CD-ROM). Joint publication by British Crop Protection Council and CAB International, Wallingford, UK, 736 pp.

Weeds

P.J. Terry[1] and C. Parker[2]

[1]Long Ashton Research Station, University of Bristol, Long Ashton, Bristol BS18 9AF, UK; [2]5 Royal York Crescent, Bristol BS8 4JZ, UK

Introduction

Weeds, like pathogens, are pests of crops and some of the worst are parasites (see Chapter 16), which are closely analogous to some fungal diseases in their mode of attachment to the host and in the production of phytotoxins. Most weeds have a less close association with the crop but can cause severe losses by competition for the available nutrients or water in the soil, or for light. A further cause of damage is allelopathy, whereby toxic exudates from weeds have an inhibitory effect on crop germination or growth. Weed competition is generally most serious during the first third of the crop's life and it is therefore important for an annual crop to be kept clean during the period from 2 to 6 weeks after sowing. The overall direct losses due to weeds are often overlooked or underestimated as the first 10–20% of crop loss may be accompanied by no visible damage symptoms. Furthermore, in developing countries, labour required for weed control can be the limiting factor in the area that can be farmed by a family unit. There are also indirect effects of weed growth which include effects on crop quality, increased costs of harvesting and influence on pests and diseases.

Interactions of Weeds with Plant Pathogens

Weeds act as alternative hosts to many crop pests, including insects, nematodes, viruses, bacteria and fungal pathogens. Some insects and nematodes are pests in their own right but they can also be vectors of disease organisms. Examples of weeds as alternative hosts are given in Table 15.1, but many others

Table 15.1. Examples of weeds as alternate hosts of crop pests.

Weed (and type)	Type of pest	Species	Crops affected
Senecio	Insect	*Myzus persicae*	Potato, beet, lettuce
vulgaris	Nematode	*Pratylenchus penetrans*	Cereals, clover, beet, potato
(annual	Fungus	*Coleosporium tussilagenis*	Pine
broadleaf)	Virus	*Cucumber mosaic*	Cucurbits, celery, tobacco
Chenopodium	Insect	*Aphis fabae*	Beans, beet
album (annual	Nematode	*Heterodera schachtii*	Sugarbeet, mangolds
broadleaf)	Fungus	*Verticillium albo-atrum*	Hops
	Virus	*Beet mosaic*	Beet
Capsella	Insect	*Brachycaudatus helichrysi*	Chrysanthemum, plum, peach
bursa-pastoris			
(annual broadleaf)	Nematode	*Ditylenchus dipsaci*	Oats, beet, onion, bean, clover
	Fungus	*Verticillium albo-atrum*	Lucerne
	Virus	*Turnip yellow mosaic*	Brassicas
Avena fatua	Insect	*Macrosiphum avenae*	Cereals and grasses
(annual grass)	Fungus	*Erysiphe graminis*	Oats
	Virus	*Barley yellow dwarf*	Cereals, grasses
Alopecurus	Fungus	*Claviceps purpurea*	Wheat
myosuroides	Fungus	*Gaeumannomyces graminis*	Cereals
(annual grass)			
Elymus repens	Fungus	*Gaeumannomyces graminis*	Cereals
(perennial grass)	Fungus	*Rhynchosporium secalis*	Cereals
Sorghum	Virus	*Maize dwarf*	Maize, grasses
halepense			
(perennial grass)			

are known. Almost all viruses that infect crop plants also have weed hosts, which may remain without symptoms, but the importance of weeds as sources of infection varies greatly. Although most types of weed can be alternative hosts, perennial species can be a problem where they enable vectors and pathogens to survive in the interval between cropping seasons. Land management practices, such as fallowing or retaining vegetation cover along irrigation canals, can allow pests and diseases to survive on weed refuges.

In many cases, the removal of weeds is an appropriate practice for management of crop pests and vectors of diseases. However, there is a possibility that weed removal can aggravate pest problems because: (i) pests and vectors are diverted on to the crop; (ii) natural enemies of insect pests and vectors lose the weeds on which they shelter; and (iii) insect-repelling weeds are eliminated. Though total weed eradication is generally desirable for removing competition with crops, the merits of keeping some weeds should be considered as part of an overall pest management strategy.

Methods of Weed Control

Non-chemical methods

Non-chemical methods of weed control are widely practised in developing countries where herbicides are not available or beyond the resources of the farmers. They are the preferred methods in organic agriculture and can be divided into three broad categories.

Cultural weed control

Cultural methods include various techniques for the prevention or avoidance of weeds. Sowing clean seed is an obvious example, as well as avoiding the introduction of new weeds through manure, fodder, soil on transplanted crop plants, contaminated equipment or irrigation water. Time of planting, such as immediately after preparation of a seed bed, reduces early competition from weeds. Competitive crop varieties, such as those that establish quickly to produce a shady canopy and extensive root system, help to suppress weeds. Seed advancement, where crop seeds are soaked in water to advance the germination process before sowing, can help speedy establishment. Crop rotations can be manipulated to prevent or avoid the build-up of problems from particular weed species.

Physical methods of weed control

Primary tillage breaks and loosens soil to depths of 15–90 cm using mould-boards, discs, chisels or hand tools. Secondary tillage works the soil to depths of about 15 cm or less and uses equipment such as harrows, cultivators, rotary hoes or hand equipment, the objective being to produce a good seedbed. Most annual weeds are readily controlled by a combination of tillage methods but control of perennials with deep roots or rhizome systems may be particularly dependent on the primary cultivation. Where this is reduced, as in some minimum tillage systems, it is the perennials that tend to increase and which may then require chemical treatment. In general, reduced tillage has no detrimental effect on crop growth and any beneficial effects of cultivation are largely due to weed control. Tillage after crop establishment requires care to avoid damage, especially to young plants and root systems, and requires crops to be planted in rows where mechanized tillage with tractors or draught animals is used.

Cutting of weeds, whether by mechanized equipment or hand tools, has the advantage of conserving soil but weeds can quickly regrow, and prostrate weeds are encouraged, especially some perennial grasses. Hand pulling is a laborious exercise, often done when weeds are large enough to grasp, by which time they have harmed the crop. Pulling or roguing can be desirable for removing weeds that have survived a herbicide treatment, thereby helping to prevent the build-up of resistant weeds, but also for collecting weeds that are useful (e.g. for food, animal fodder and medicines).

Flooding is an effective method of physically controlling weeds by covering the soil surface, usually a rice paddy, to a water depth of 5–10 cm. Timely land preparation and water management can reduce the need for other methods of weed control.

Biological control of weeds

Classical biological control has been successfully introduced for certain important weed species, particularly perennial aquatic and rangeland species, where there is opportunity for the steady build-up and spread of phytophagous insects or pathogenic fungi. It has generally been less successful on annual weeds in annual crops but augmentative forms of biological control, such as mycoherbicides (fungal pathogens formulated to enable them to be sprayed on to weeds), have proved effective in these situations.

Chemical control

In developed countries and increasingly in developing countries, weeds are now very largely controlled by chemical means. Because weeds are almost invariably present, it is quite normal for herbicides to be applied on a routine basis, often as prophylactic, pre-emergence treatments.

Well over 200 different herbicides and many more formulated products are now available. Herbicides can be used in all crops, with the exception of some very minor or specialized commodities, and there are usually five to ten possible compounds for use in any major crop, sometimes as many as 20–30. The choice of herbicide will depend, to some extent, on the crop growing method, soil type, climate, etc., but mainly on the combination of weeds present. Although some crops have good physiological tolerance to some herbicides, all herbicides are intrinsically phytotoxic and their selectivity usually depends on accurate dosing. Concentrations of spray are not necessarily critical but dose per unit area must be carefully controlled if the treated crop is not to be damaged. Even if the treated crop is tolerant, an excessive dose could in some cases lead to more prolonged soil residues and damage to a following crop. Such carry-over of residues is not a common problem with correct usage but can occur with some compounds, especially when cold or dry conditions slow the natural breakdown processes in the soil.

There are three main classes of herbicide: (i) the soil-acting or residual herbicides generally used as presowing or pre-emergence treatments to kill weeds as they germinate; (ii) contact herbicides which scorch or kill the foliage that is wetted but have little or no translocated effect; and (iii) translocated herbicides, which can be applied to foliage and are translocated to the stems and underground parts of the plant to give a systemic kill. Individual herbicides can show combinations of two or more of these types of action.

Some classes of compounds have very broad-spectrum activity on both broad-leaved and monocotyledonous weeds, e.g. the substituted triazines, ureas, uracils, imidazilinones, paraquat and glyphosate. Others, including

2,4-D (2,4-dichlorophenoxyacetic acid) and related compounds, are pre-dominantly active on broad-leaved weeds and sedges (though with some useful pre-emergence activity against annual grasses too). For control of problem species, such as broad-leaved weeds in broad-leaved crops or wild oats (*Avena* spp.) and other grass weeds in cereals, many herbicides have been developed with quite narrow specificity. Some, indeed, control only *Avena* spp. and no other weeds. The mechanism of the selectivity of herbicides between crops and weeds is not always clearly understood. In most cases, the herbicide is known to be successfully degraded within the resistant crop but not within the susceptible weeds. In a few cases (MCPB (4-(4-chloro-2-methylphenoxy) butanoic acid) and 2,4-DB (4-(2,4-dichlorophenoxy) butyric acid)), the complete opposite occurs; the product itself is not active but is converted into an active compound in the weed and not in the crop. In other cases, there is less distinct physiological immunity and selectivity depends on differential uptake of the herbicide, either as a result of different retention on the foliage or different exposure of the crop to the herbicide. For instance, a deep-rooted crop may escape damage from a relatively immobile herbicide applied to the soil surface or an established crop may be treated with a non-selective contact herbicide by carefully directed application between the rows.

The specificity of herbicides (in insect control the term selectivity would be used, but this is generally reserved for the selectivity between crop and weeds rather than between weed species) can result in quite closely related or similar-looking weed species having very different susceptibilities to a herbicide, meaning that *identification* can be a very important preliminary to a decision on what compound to use. Guides to weed identification are available for many regions of the world and many species are illustrated on the Internet. Where it is necessary to refer weeds to a herbarium, the whole plant should be collected, if possible, including underground parts, and notes made of the size and habit of the plant, the colour and shape of flowers and the locality, crop and soil conditions in which it was found.

Because of their basic phytotoxicity, herbicides have to be applied with greater care than fungicides. In particular, it is usual for spraying pressures to be low or equipment otherwise designed to produce relatively large droplets so that wind will not drift spray on to neighbouring areas where other crop species might be severely damaged. It is important for spray equipment to be very thor-oughly cleaned before it is used on a different crop, or it may even be necessary to reserve particular equipment for particular crops. Controlled drop applica-tion (CDA) using a spinning disc principle is a way of reducing spray volumes of 200–400 l ha^{-1} commonly applied with pneumatic sprayers to 20 l ha^{-1}. This technique is suitable for most residual and translocated herbicides but may not be acceptable for some contact herbicides. Even lower volume rates can be obtained with weed wipers where a concentrated solution of herbicide (usually glyphosate) is smeared on to the target weeds. Such a method can be used where the weeds grow well above the crop canopy. Granular formulations of herbicides are not widely used but are particularly suitable in flooded rice.

A risk from continued use of any herbicide is that the weed flora in a crop changes from one with several generally susceptible weeds to one that is dominated by a few tolerant species. Such a situation occurs in cereals where use of 2,4-D and similar products leads to the proliferation of annual grass weeds; likewise, perennial sedges and grasses tend to become a problem where paraquat is used in minimum tillage. Perhaps more alarming is the development of resistance in weeds that hitherto were susceptible to a herbicide. So far, over 150 weed species are known to have developed resistance to herbicides and nine new cases, on average, are being found each year. Herbicide resistance is being exploited where it has been bred or genetically engineered into a crop, enabling broad-spectrum herbicides, such as glyphosate and glufosinate, to be used without risk to the crop. The danger, of course, is that the high selection pressure arising from use of a single product will lead to the development of resistant weeds. An even greater fear is that the genes for herbicide resistance in crops will escape into the weed community, probably through near relatives, to produce 'super weeds' that are very difficult to control.

Herbicides generally have a lower mammalian toxicity than other pesticides but they can, in some instances, persist in the soil and contaminate groundwater. There is little doubt that the public would prefer pesticide use to be curtailed, if not prohibited. However, the developed world and, to some extent, developing countries have acquired a herbicide dependency without which there could be a serious decline in crop production. There is, consequently, increasing awareness of the need to minimize herbicide use by the integration of non-chemical methods into the weed management system wherever possible.

Interactions of Herbicides with Plant Diseases and Disease Control

Many crop protection chemicals, particularly herbicides and growth regulators, produce changes in the physiology and development of crop plants, which may result in substantial alteration to disease susceptibility. Two opposing side effects on disease may result from any herbicidal regime – one increasing and the other decreasing the same disease. In general, four major herbicidal effects lead to increased plant disease: (i) reduction of structural defences of the host (e.g. cell and membrane disruption); (ii) stimulation of increased exudation from host plants (e.g. sugars, amino acids, other carbon compounds and minerals); (iii) stimulation of pathogen growth; and (iv) inhibition of microflora competing with potential pathogens. By contrast, major effects of herbicides leading to decreased disease incidence and/or severity are: (i) increased host structural defences; (ii) increased host biochemical defences; and (iii) decreased growth of potential pathogens. Some examples of the effects of herbicides on pathogens are given in Tables 15.2 and 15.3.

Herbicide damage has sometimes been wrongly identified as symptoms of a disease. Herbicides related to 2,4-D cause various deformities, including narrowing of leaves. Others cause chlorosis, stunting, scorch or other abnormalities, which could perhaps be mistaken for disease symptoms. Some of these symptoms are illustrated in Anon. (1977).

Table 15.2. Increased damage by pathogens resulting from herbicide use.

Herbicide	Diseases increased by herbicide use	Possible causes of increased pathogenicity
2,4-D	Southern corn leaf blight	Protein content of corn is increased, favouring growth of the pathogen, *Cochliobolus heterostrophus*
Atrazine	Root rot of navy bean	Growth of *Fusarium solanii* is stimulated by atrazine
Cycloate	Damping-off of sugarbeet	Glucose and mineral ions leak from sugarbeet hypocotyls, providing a nutrient source for *Rhizoctonia solani*, the causal organism of damping-off disease
Diphenamid	Damping-off of peppers	*R. solani* increases due to damage by the herbicide to microorganisms that are antagonistic to this pathogen
Mecoprop	Take-all disease of spring wheat	Herbicide causes malformed root tips in spring wheat, enabling easier penetration by *Gaeumannomyces graminis*
Picloram	Root rot of wheat and maize	Growth of soil-borne pathogens increased due to exudation of carbohydrates from the cereals
TCA (trichloroacetic acid)	Downy mildew of peas	TCA reduces leaf wax of peas, allowing easier penetration of *Peronospora viciae*
Trifluralin	Damping-off of soybean	Seedling emergence of soybeans delayed, increasing susceptibility to damping-off

Table 15.3. Decreased damage by pathogens resulting from herbicide use.

Herbicide	Diseases decreased by herbicide use	Possible causes of decreased pathogenicity
Atrazine	Damping-off of soybean	Atrazine has a fungicidal effect on *Pythium*
Diallate	Root rot of maize	Fungicidal effect of diallate on *Fusarium moniliforme* and prevention of the fungus from penetrating the stele of maize roots
Diuron	Root rot of winter wheat	Host resistance to *Pseudocercosporella herpotrichoides* is increased by diuron
Monolinuron	Powdery mildew of wheat	Decreased sugar content and other biochemical changes in the crop give protection against infection by *Erisyphe graminis*

References and Selected Bibliography

Journals

Weed Abstracts (1954) CAB International, Wallingford, UK.
Weed Research (1961) Blackwell Science, Edinburgh, UK.
Weed Science (1968) Weed Science Society of America, Lawrence, Kansas, USA.
Weed Technology (1987) Weed Science Society of America, Lawrence, Kansas, USA.

Books and reviews

Akobundu, I.O. (1987) *Weed Science in the Tropics. Principles and Practices.* John Wiley & Sons, New York, USA.
Aldrich, R.J. and Kremer, R.J. (1997) *Principles in Weed Management*, 2nd edn. Iowa State University Press, Ames, Iowa, USA.
Altman, J. and Campbell, C.L. (1977) Effect of herbicides on plant diseases. *Annual Review of Phytopathology* 15, 361–385.
Altman, J. and Rovira, A.D. (1989) Herbicide–pathogen interactions in soil-borne root diseases. *Canadian Journal of Plant Pathology* 11, 166–172.
Anon. (1977) *Herbicide Injury Symptoms and Diagnosis.* Bulletin, North Carolina Agricultural Extension Service No. AG-85, North Carolina, USA.
Holm, L., Doll, J., Holm, E., Pancho, J. and Herberger, J. (1997) *World Weeds. Natural Histories and Distributions.* John Wiley & Sons, New York, USA.
Holm, L.G., Plucknett, D.L., Pancho, J.V. and Herberger, J.P. (1977) *The World's Worst Weeds.* Hawaii University Press, Honolulu, USA.
Labrada, R., Caseley, J.C. and Parker, C. (1994) *Weed Management for Developing Countries.* Food and Agriculture Organization, Rome, Italy.
Leela, D. and Ganeshan, G. (1993) Herbicides and plant disease interactions: a review. *Advances in Horticulture and Forestry* 3, 1–48.
Powles, S.B. and Holtum, J.A.M. (1994) *Herbicide Resistance in Plants: Biology and Biochemistry.* Lewis Publishers, Boca Raton, Florida, USA.
Thurston, J.M., Moore, F.J., Franklin, M.T., Heathcote, G.D. and van Emden, H.F. (1970) Some examples of weeds carrying pests and diseases of crops. In: *Proceedings 10th British Weed Control Conference*, Brighton, UK, pp. 953–957.

Parasitic Higher Plants

C. Parker

5 Royal York Crescent, Bristol BS8 4JZ, UK

Introduction

Parasitic weeds constitute a small but very important group of plant pathogens which can be divided conveniently (but not always precisely) into two groups showing total and partial parasitism, respectively.

Total (holo-) parasites include the *Orobanchaceae*, which completely lack chlorophyll, have leaves reduced to scales and are totally dependent on the host for nutrition. Some genera of the *Convolvulaceae* (*Cuscuta*) and *Lauraceae* (*Cassytha*) also are virtually devoid of chlorophyll.

By contrast, the partial (or hemi-) parasites are green with mostly normal-looking leaves, which have varying ability to photosynthesise. Many have less chlorophyll than fully autotrophic plants and may still depend on the host for photosynthate as well as water and minerals. All of the *Loranthaceae*, *Viscaceae* and *Santalaceae* and the parasitic members of the *Scrophulariaceae* (e.g. *Striga*, *Alectra*, *Rhinanthus*, *Seymeria*) are partial parasites. Some of these, e.g. *Osyris* in the *Santalaceae* and *Rhinanthus* in the *Scrophulariaceae*, are facultative parasites, able to establish and mature without a host. Others, including *Loranthaceae*, *Viscaceae* and most *Striga* and *Alectra* species are obligate parasites, unable to establish and grow to maturity on their own.

In all species, contact is established by haustorial systems, which arise from stem or root tissues of the parasite to penetrate the host tissues. These unions may occur between roots (*Striga*, *Santalum*, *Orobanche*, *Lathraea* and others), stems (*Cuscuta*, *Cassytha*) or haustorial structures replacing the root of the parasite which invade the shoot of the host (*Loranthaceae*, *Viscaceae*). The group includes annuals and perennials, and the range of hosts parasitized by any single parasite can vary from a single genus to a wide range of species.

Within certain species, such as *Viscum album* and *Striga gesnerioides*, physiologically specialized forms have developed with their own narrow host range.

Many parasitic higher plants are only of academic interest but the mistletoes (*Loranthaceae, Viscaceae*), witchweeds (*Striga* and *Alectra* spp.), broomrapes (*Orobanche* spp.) and the dodders (*Cuscuta* spp.) cause economically significant losses to annual and perennial crops in many parts of the world, while *Santalum album*, the source of sandal wood, is a valuable crop plant in its own right.

Mistletoes

The economically important species all belong to the *Loranthaceae* and *Viscaceae*. They separate morphologically into the leafy or green mistletoes (*Viscum, Loranthus, Phoradendron, Tapinanthus, Phragmanthera* spp., etc.) with well-developed shoots and seeds adapted for distribution by animals and birds, and the dwarf mistletoes (*Arceuthobium* spp.) with greatly reduced foliage and chlorophyll, and seeds dispersed by an explosive mechanism. In both groups the seeds are coated with a sticky substance, viscin, by which they adhere to the host.

The leafy mistletoes colonize many species of hardwoods and conifers and are distributed throughout the five continents. Hundreds of species in various genera of *Loranthaceae* occur in Africa, Asia, Australasia and (to a limited extent) in Europe; in the *Viscaceae, Phoradendron* spp. are found throughout the temperate and tropical Americas and *Viscum* spp. in more temperate regions of Europe, Africa, Asia and Australasia. By contrast, *Arceuthobium* spp. are found only on conifers and are distributed mainly in North and Central America (about 30 species) with only four species known in the Old World.

All mistletoe infections are perennial with a span of 6–7 years between infection of the host and the first production of seeds. The seed germinates without special stimulation from the host (unlike witchweeds and broomrapes) and the radicle, after a brief period of exploratory growth, penetrates the host bark to the cambial layer. Successive layers of host wood become invaded as the haustorium is engulfed by the lateral growth of the host stem tissues. The region of the host colonized by a mistletoe shows a hypertrophic reaction, varying from moderate swelling to the formation of gall tissue, burrs or witches' brooms (the latter are particularly characteristic of infection by *Arceuthobium* spp.). Bark cracking and gum or resin exudation may also occur, which, in turn, may predispose the host to invasion by insects or secondary fungal pathogens.

The leafy mistletoes obtain water, mineral salts and possibly some photosynthate from their hosts, but they return nothing. Light infections may have no discernible effect on host tree growth, but there can be gradual (or more dramatic) debilitation of the host due to competition for water in times of moisture stress, or the gradual replacement of the crown by an increasing mass of unproductive foliage. Well-established infections may also girdle large branches killing the distal parts, so further reducing the photosynthetic area of the host. Although the leafy mistletoes are not usually regarded as a serious problem in natural forests, they can be of concern in tree and other plantation crops, as with *Dendrophthoe falcata* in teak in India, *Tapinanthus* spp. in cocoa in Ghana, and various species in citrus in Central America.

The dwarf mistletoes have little chlorophyll and cause much greater damage to their hosts by depleting them of photosynthate, as well as causing severe deformity, abnormal branching, etc. They are considered to be the most important of all forest disease organisms in certain regions of North America. Their potential menace to forestry in other parts of the world is recognized in the fact that *Arceuthobium* spp. are the only weeds to be completely prohibited as imports to the European Community.

Control of leafy mistletoes in the shorter tree crops is usually feasible by timely pruning, though this must involve the whole host branch from below the point of infection, not just the mistletoe shoot itself, or regrowth will occur. Chemical control of mistletoes by injecting toxic salts, such as copper sulphate solutions, or herbicides, such as 2,4-D (2,4-dichlorophenoxyacetic acid) or metribuzin, into the host has been effective at times but has not been adopted on any significant scale. Maintaining optimum crop vigour is important as an indirect means of suppressing the leafy mistletoes, as they require light for rapid growth and can be effectively shaded out by vigorous tree growth. Severe mistletoe infestations are more often the *result* of weak tree growth, than its *cause*. The use of isolation strips has been recommended for the protection of regenerating or new plantations from mistletoe attack in North America and the West Indies.

There are virtually no direct means of control for the dwarf mistletoes, and the problem has to be managed in such a way as to minimize spread to non-infected or new stands via programmes of selective thinning, clear felling, fire, open corridors, or barrier strips of non-susceptible tree species. Computer programs have been developed to help make appropriate management decisions.

Witchweeds

The *Striga* spp. or witchweeds are a group of annual hemiparasites belonging to the family *Scrophulariaceae* that cause enormous losses to cereal crops in Africa and lesser damage elsewhere. Sorghum, maize and pearl millet (*Pennisetum americanum*) are most seriously affected, but rice, finger millet (*Eleusine coracana*) and sugarcane can also be attacked. *Striga asiatica* is the most widely distributed species, through Africa and Asia into Indonesia, whereas *Striga hermonthica* is confined to tropical Africa and Arabia. Other less important species include *Striga aspera* and *Striga forbesii* on cereals in Africa, *Striga densiflora* and *Striga angustifolia* on cereals in India, *Striga curviflora* in Timor and *S. gesnerioides* on cowpea in West Africa. Also sometimes known as witchweed, the closely related *Alectra vogelii* is a damaging parasite of cowpea and groundnut sporadically across tropical Africa.

All these species produce large numbers of minute seeds (less than 0.5 mm long), which are dispersed by wind or water or on plant debris, eventually becoming incorporated in the soil where they may remain viable for more than ten years. Germination normally occurs only in response to stimulant substances exuded by the roots of potential host plants. This ensures that germination takes place close to crop roots to which the parasite seedling attaches itself by a haustorium. It grows as a total parasite for several weeks until emergence. Normal green foliage is then formed, but it continues to draw a large proportion of its water, minerals and carbohydrate from the host. The flow from host to parasite is enhanced by the open stomata and high transpiration rate of *Striga* foliage. In addition to causing a drain of foodstuffs from the host, there is a very marked influence on the host architecture, such that roots may be stimulated while shoot growth is stunted. The eventual result of infection can be a severe debilitation involving stunting, wilting, chlorosis, scorch and, in extreme cases, crop death.

Conditions favouring serious build-up of the major *Striga* spp. on cereals include low or erratic rainfall, low soil fertility and repeated growing of susceptible crops, leading to a build-up of seed in the soil. Conversely, varying degrees of cultural control can be achieved by combinations of suitable rotations, improvement of soil fertility (especially nitrogen levels), moisture conservation, intercropping and hygiene (see Chapter 30). The rotation should include a minimum of susceptible cereals and the greatest possible proportion of trap crops. These are crops whose roots exude the germination stimulant but are not susceptible to attack. They include cotton, cowpea, groundnut, soybean and many other legumes. Nitrogen fertilizer helps to reduce the early germination of *Striga* and the amount of damage caused, and is particularly useful for protecting maize. The flow of water and nutrients to the parasite is reduced under humid conditions, as in wet weather, or under a dense crop canopy, or with suitable mixed cropping, preferably with the intercrop planted in the cereal row rather than in separate rows.

Hygienic methods include the use of clean crop seed for planting, and hand-pulling or other means of destruction of emerged *Striga* before flowers have time to set seed. Benefits are not immediate owing to the seed reserve in the soil but such hygiene is essential for a long-term control programme.

Other more direct control methods include catch crops, i.e. susceptible cereals grown for a few weeks to stimulate seed germination and then ploughed in prior to the main crop, but the delay in planting date is rarely acceptable.

Herbicides, such as 2,4-D, ametryne, etc., can be used to kill emerged *Striga* and prevent seeding but they may not do much to prevent damage to the current crop. Selective control has been reported with pre- or early post-emergence treatment with some other herbicides, including chlorsulfuron, but selectivity is very narrow. Reliable selective control is currently available only with the use of genetically modified herbicide-resistant crop varieties, such as maize, which is being developed with resistance to imidazolinone herbicides such as imazethapyr, used at time of planting, or to glyphosate which can be applied after *Striga* emergence. Cost and complexity of herbicide use is still an enormous barrier to their use on most *Striga*-affected farms, but trials suggest that the quantity of herbicide, and hence the cost, can be minimized by applying herbicide only to the crop seed before planting. In the USA, where large areas of South and North Carolina were found to be infested with *S. asiatica* in the 1950s, the infestation has now been largely eradicated (at great cost) by strict quarantine, the repeated use of 2,4-D in maize crops, the use of other herbicides in rotational crops to eliminate grass weeds on which the *Striga*

might grow, injection of ethylene gas into the soil to stimulate suicidal germination of the seed and, in small areas, the use of methyl bromide fumigation
(this is no longer permitted).

The possibilities for biological control continue to receive research
attention, with emphasis on the potential for use of fungal pathogens,
especially *Fusarium* species, as mycoherbicides, and on the understanding and
possible exploitation of '*Striga*-sick soils'.

The development of *Striga*-resistant crop varieties has long been the main
hope for resource-poor farmers. In spite of several decades of work by ICRISAT,
there are still very few areas where *Striga*-resistant varieties of sorghum have
been proved sufficiently resistant, reliable and acceptable to the farmers, but
selection of less-susceptible varieties should always be considered. In the
case of maize, some tolerant hybrids (supporting *Striga* development but being
relatively undamaged) have been developed by IITA, but these too have failed
to prove reliable. The one success has been with cowpea in which almost
total resistance has been found to both *S. gesnerioides* and *A. vogelii*. This
resistance, based on the cowpea land-race B.301 from Botswana, has now been
incorporated into agronomically acceptable varieties by IITA.

With increasing population pressure on land in Africa, and corresponding
decreases in fallowing and soil fertility, the *Striga* problem is regrettably
continuing to increase. In the absence of simple direct control measures, there
is a need for farmers to be educated as fully as possible on the biology of *Striga*
and their cooperation sought in the development of integrated programmes of
Striga control suited to their conditions. It is important that such programmes
are seen as long-term, as successful reduction in the problem can only be
expected after a number of years of sustained effort. This effort may need to
be aimed as much towards the improvement of soil fertility as at the *Striga*
problem. Complete success is very unlikely where soil fertility remains low.

Broomrapes

The *Orobanche* spp. or broomrapes (*Orobanchaceae*) are all total root parasites
without any chlorophyll. There are five or six major species causing locally
severe losses including *Orobanche ramosa* and *Orobanche aegyptiaca* on
solanaceous crops such as tobacco, tomato, potato and aubergine and many
legumes and brassica crops, mainly in the Mediterranean basin but also
sporadically elsewhere. *Orobanche crenata* is especially damaging to faba
bean, peas and lentils and has seriously reduced the areas planted to these
crops around the Mediterranean. *Orobanche cernua* exists in two forms, one of
which affects *Solanaceae* in the Middle East, Africa and India, whereas the
other (also known as *Orobanche cumana*) affects large areas of sunflower in
Russia and other countries of eastern and southern Europe.

The biology of the broomrapes is very similar to that of the witchweeds,
starting with a small seed whose germination is triggered by crop root exudates.
The effect on the host, however, is different, mainly involving a simple transfer
of photosynthates from host to parasite, although this loss of resource may
be aggravated by depletion of water. Damage is therefore worst under dry

conditions, but most species can flourish under irrigated conditions. They are not usually reduced by high fertility but cultural control by rotation and hygiene is equally important. The effectiveness of trap crops is not so clearly defined.

The problem in sunflower has been greatly reduced by selection of Orobanche-resistant varieties, but new races of parasite have evolved or been selected, for which new resistance genes have to be found. At least five such races are now known but more are suspected, and a continuous breeding and selection programme is needed. No fully satisfactory resistance has been found in the other major crops affected, with the possible exception of faba bean. Work on this crop in Egypt and at ICARDA in Syria has resulted in some promising material but this is not yet widely available.

A number of herbicides have given good selective control of Orobanche species, including the sulphonyl urea chlorsulfuron, and imidazolinone compounds imazaquin, imazethapyr and imazapyr, applied in various ways pre- or postemergence, or even as dressings on the crop seed, in faba bean, pea, sunflower and tobacco. Low doses of glyphosate have been used postemergence in faba bean, but selectivity is marginal and its use is not widespread. The value of herbicide-resistant crop varieties has already been demonstrated and near-perfect selectivity achieved with glyphosate, sulphonyl urea or imidazolinone compounds in correspondingly modified crops. This approach is likely to become an important weapon in the control of Orobanche species, especially in tobacco and tomato.

In high-value crops, fumigants have been used with success, and although methyl bromide will no longer be available for this use, others including metham-sodium continue to be of value. A further technique for high-value situations is the use of plastic sheeting for solarization.

The agromyzid fly Phytomyza orobanchia has been used as a biocontrol agent in eastern Europe with apparent success in the past. It involves collecting infected capsules and storing them under the correct conditions over the winter before release in the spring. A project is attempting to prove their usefulness in Chile. There is also continuing research into the potential for use of fungal pathogens, especially Fusarium species, as mycoherbicides.

Dodders and Cassytha

Cuscuta spp. (dodders) belong to the Convolvulaceae or are sometimes placed in their own family Cuscutaceae. Cassytha filiformis belongs to the Lauraceae but is very similar in character and often mistaken for a Cuscuta sp. They consist of little more than leafless yellow or orange twining stems, with leaves reduced to small scales, forming a parasitic web over the above-ground parts of both crop and other weeds. They are virtually devoid of chlorophyll, although some is detectable and functional at a very low level in some species. Between 10 and 15 species can occur as significant weed problems, of which the most widespread and damaging is Cuscuta campestris, a North American species now introduced (usually via contaminated crop seed or forage) to most warm and temperate regions of the world. One species, Cuscuta epilinum, has very narrow host range, affecting only flax and linseed, but most others have a broad host range, affecting many vegetable and field crops, such as lucerne, sugarbeet, potato, carrot and niger seed (*Guizotia abyssinica*). The problems in lucerne and niger seed are largely due to contamination of crop seed. Some species, including *C. campestris, Cuscuta reflexa, Cuscuta monogyna* and *C. filiformis* can also attack perennial fruit and tree crops. Grass species are not generally attacked. Dodder seeds germinate spontaneously without stimulation from a host crop. After germination the seedling nutates, i.e. swings in circles, until contact is made with a host stem. The dodder stem forms a tight coil around the host and sinks haustoria into the host tissue, penetrating the vascular bundles and making connections with the host phloem. Physiological studies show that the parasite has an extremely powerful sink effect, diverting much of the host resources to itself, and starving the crop, and especially any developing fruit, of photosynthate. This causes severe debilitation and reduction in crop growth and yields.

Cultural control measures include: rotation with grass and cereal crops that are not attacked; improved seed cleaning; cutting, pulling or spot spraying to prevent new seed formation; and clearing of susceptible species (e.g. *Convolvulus arvensis*) from around field borders. Herbicides can be used selectively in some crops to prevent germination and/or to kill young seedlings. The range of compounds includes trifluralin, pendimethalin, chlorthaldimethyl, propyzamide, thiazopyr and imazethapyr for pre-emergence treatments, and glyphosate and paraquat postemergence. The last two are generally used for non-selective spot-spraying, but glyphosate may give selective control at low doses in lucerne.

Biological control possibilities have been studied but have not yet been successfully exploited.

Select Bibliography

Hawksworth, F.G. and Wiens, D. (1972) *Biology and Classification of Dwarf Mistletoes (Arceuthobium)*. Agricultural Handbook, United States Department of Agriculture Forest Service No. 401, 234 pp.

Kuijt, J. (1969) *The Biology of Parasitic Flowering Plants*. University of California Press, Berkeley, 246 pp.

Moreno, M.T., Cubero, J.I., Berner, D., Joel, D.M., Musselman, L.J. and Parker, C. (1996) *Advances in Parasitic Plant Research. Proceedings of the Sixth International Parasitic Weed Symposium, Cordoba, Spain, 1996*. Junta de Andalucia, Seville, Spain.

Parker, C. and Riches, C.R. (1993) *Parasitic Weeds of the World: Biology and Control*. CAB International, Wallingford, UK.

Press, M.C. and Graves, J.D. (1995) *Parasitic Plants*. Chapman & Hall, London, UK.

Smith, I.M. and Roy, A.S. (1996) *Illustrations of Quarantine Pests for Europe*. EPPO, Paris, and CAB International, Wallingford, UK, 241 pp.

Non-infectious Disorders

17

J.M. Lenné

ICRISAT, Patancheru, Andhra Pradesh 502 324, India

General

Non-infectious disorders are conditions in which no primary parasite is involved but which are brought about by abnormal conditions of temperature, light or the atmosphere, disturbance of water relationship, nutritional imbalance, the toxic action of fungicides or other applied chemicals, or by injury from such physical causes as lightning and wind. The symptoms of non-infectious disorders often resemble and are frequently confused with those caused by pathogens, but as they are non-infectious epidemic development of the condition does not take place. Often non-parasitic disturbances predispose plants to infection, either through physiological impairment or through injury, permitting pathogens to enter and damage the plant.

Much economic loss of crops is brought about by adverse environmental factors. The severity and type of injury vary with the plant, its stage of maturity, when the disturbance occurs, and the part of the plant involved. Certain of the better-known types of non-infectious disorder will be dealt with briefly.

Low-temperature Effects

Plants differ notably in their sensitivity to low temperatures and frost. Low temperatures not only reduce growth, but can cause plants to remain sterile or produce deformed flowering structures. Low night temperatures at high altitudes in the tropics are associated with tissue malformation as in 'hot and cold' disease of coffee. Most species in temperate zones are subject at one time or another to low temperatures and thus to freezing injury. Small grains, maize

©CAB *International 2002. Plant Pathologist's Pocketbook*
(eds J.M. Waller, J.M. Lenné and S.J. Waller)

and other crops often fail to reach proper maturity before being killed by frost in some seasons in northern latitudes. Freezing injury directly kills plant cells and causes necrosis of plant tissues. This may appear initially as a wilting of shoot tissues, which then become blackened. Plants may also suffer death of twigs and branches, splitting of trunks, and the loss of fruit crops when flowers are killed.

Low temperatures are also associated with various forms of damage to fruits and vegetables held in cold storage. A familiar effect of low temperature is the sweetening of potato tubers in storage, caused by the transformation of starch into sugar at temperatures near freezing point. The fruits and vegetables that require high temperatures for growth generally are the ones most subject to injury by chilling or low temperature. Among them are bananas, citrus fruits, cucurbits, aubergines, peppers, sweet potatoes and tomatoes.

High-temperature Effects

Plants may be injured on days of high temperature and bright sunshine. Unduly high temperatures are responsible for such damage as sunscald of fruit and foliage, or heat cankers of the stem at soil level. Seeds may fail to germinate if soil temperatures are too high. High levels of solar radiation can induce high surface temperatures, which can kill plant tissues, especially if accompanied by water stress, which limits the cooling effect of transpiration. The scalding effects of high temperatures are most commonly seen on fruit, but can affect whole plants.

Excessive temperatures may induce black heart of potato in storage by increasing the rate of respiration of tubers to such an extent that the oxygen within the tissues is used up more rapidly than it can be replenished. Black heart has also been found occasionally in tubers in the field, particularly in those areas where high temperatures occur during periods of tuber growth and maturation.

Soil-moisture Disturbances

Most crop plants grow well on relatively well-drained soil that may be subject to leaching or temporary flooding, but most plants will not survive persistent flooding, which destroys the root system primarily by depriving the tissues of oxygen but also predisposing it to infection by pathogens such as *Pythium* spp. Temporary wilting is often evident and the plant may fail to recover. Wilting or other signs of water stress may also be a result of damage to the plant's roots or vascular system caused by pests or pathogens.

Deficiencies in soil moisture also cause temporary wilting and during critical growth periods may result in stunted growth or more serious tissue damage. Maize, for example, first shows rolling of leaves; if the drought continues it may suffer to the degree that the upper part of the plant, including the male inflorescence, dries up and fertilization cannot take place. A major effect of soil-moisture limitations is to predispose plants to infection. Charcoal root rot (*Macrophomina phaseolina*), a disease of several herbaceous crops in the semiarid tropics, is often associated with low soil moisture. Widely fluctuating soil-moisture availability is also associated with root disease problems.

Effects of Atmospheric Pollution

These are dealt with in Chapter 18.

Losses in storage occurring as a result of extraneous gases or the volatile by-products of the stored plant tissues themselves are a form of pollution. Apple scald is a serious physiological disorder of some apple varieties and is caused by one or more of the volatile constituents that are associated with the characteristic odour of apple fruits. Discoloration of yellow or red onions in storage is brought about when leakage in the cooling plant permits small amounts of ammonia gas to enter the atmosphere. Similarly, certain discoloration occurs when apples, peaches, pears and bananas are exposed to ammonia fumes.

Lightning and Hail Injury

Lightening frequently damages trees and causes splitting of bark and the death of tissues beneath as the high potential discharge runs to earth. Trees or their branches may be killed but often the only resultant damage may be the development of a lateral stripe canker down the affected side of the tree. Lightning striking the ground causes the spread of high electric potential in a roughly circular direction. The extent of such spread before it becomes dissipated will depend on the nature of the soil and the vegetation covering it. Grasses and cereals are apparently quite resistant to damage since few reports of injury to such crops are on record. More succulent plants such as potato, tomato, celery and brassicas are very readily damaged. The occurrence of lightning injury is seldom noticed until some weeks after the strike. It has then become evident by a roughly circular bare spot, in which most or all of the plants are completely dead, having been killed rapidly. At the periphery plants show various degrees of retardation in growth.

Damage due to hail results in holes and shedding of leaves and bruising to stems. Such wounding facilitates entry of pathogens. If a substantial amount of leaf is lost, growth is retarded and yield reduced.

Mineral Nutritional Disorders

Nutritional disorders may result from the toxic effects of salts present in excess, or from unduly acid or alkaline soils, but are more commonly induced by deficiencies of one or more of the elements necessary for plant growth. This may either be due to their lack in the soil or to their being unavailable to the plant because of chemical reactions in the soil or injury to the plant inhibiting nutrient uptake. The symptoms induced by such factors include yellowing and other colour changes in the leaves, of varying pattern, marginal scorching and other forms of necrosis. On alkaline soils, iron may be unavailable to some plants, inducing chlorosis, especially of young tissues. Nitrogen deficiency causes a general chlorosis most marked in older leaves which senesce early. Stunting or deformation of foliage and/or fruit can occur when some of the minor elements such as boron, zinc and manganese are deficient. Symptoms of disease are also evident when some of the major fertilizer elements such as potassium, phosphorus or nitrogen are not present in sufficient quantities in the soil or are rendered unavailable.

Many nutrient elements in excess may cause symptoms of toxicity. For example, boron in any considerable amount results in a marginal necrosis of the older leaves often followed by stunting and death. These symptoms have frequently been seen following the use of irrigation water carrying toxic amounts of boron. The use of potash salts in fertilizers containing excessive amounts of boron has also caused losses.

It is not possible to distinguish clearly between the disorders that are due to a deficiency as such and those that are due to too much of another element. A deficient supply of one element often implies an excess of others. A mass-action effect may arise when too much of one element may interfere with the solubility, absorption and utilization of another element to the extent of developing acute deficiency effects. Furthermore, symptoms of pathogenic disease are often the result, at least in part, of an imbalance in plant nutrition induced by the pathogen.

Effects of Agricultural Chemicals

Chemicals applied as sprays or used for fruit set or colouring may cause serious damage to the fruit or foliage and soil-applied fertilizers may cause scorching of tender plant tissues. All plant parts may be injured by some pesticides, particularly by older compounds such as Bordeaux mixture and those containing sulphur. Cucurbits are particularly 'sulphur-shy'. Some of the effects are: blemishes on the fruit, as in apple; foliage injury, as in peach; retarded growth, as in cucumber; excessive transpiration resulting in drought injury, as in some vegetables; blossom drop, resulting in delayed production, as in tomato. Lime-sulphur also may cause lesions on foliage or fruit and premature fruit drop. The most common injury is a dull-brown spotting of the leaves or burning of margins and tips.

Many injuries have followed the improper use of herbicides. These can range from occasional spotting to deformation and necrosis of leaves and stems. Minute amounts of growth-regulating herbicides drifting from their site of application are enough to injure plants.

Climate Change

The effects of global changes in climate are considered in detail in Chapter 19. Direct effects of small changes in temperature and CO_2 levels over time on plant pathogens are difficult to predict. Periodic monitoring is needed to understand the possible effects.

Select Bibliography

Altman, J. and Campbell, C.L. (1977) Effect of herbicides on plant diseases. *Annual Review of Phytopathology* 15, 361–385.

Anon. (1976) *Fertilizer use and plant health. Proceedings, 12th Colloquium of the International Potash Institute.* International Potash Institute, Berne, 330 pp.

Anon. (1978) Plant production under stress. 70th Annual meeting of the American Phytopathological Society. *Phytopathology News* 12, 105–228. (Abstracts of 475 papers.)

Ayres, P.G. and Body, L. (1986). *Water, Fungi and Plants.* BMS Symposium no. 11. Cambridge University Press, Cambridge, UK, 413 pp.

Basra, A.S. and Basra, R.K. (1997) *Mechanisms of Environmental Stress Resistance in Plants.* Harwood Academic Publishers, Amsterdam, 407 pp.

Bennett, W.F. (1993) *Nutrient Deficiencies and Toxicities in Crop Plants.* APS Press, St Paul, Minnesota, USA, 202 pp.

Chicchio, F.P.B. and Santos Filho, H.P. (1980) Non-infectious gummosis caused by an insecticide on citrus. *Fitopatologia-Brasileira* 5(2), 207–209.

Eagle, D.J. and Caverly, D.J. (1981) *Diagnosis of Herbicide Damage to Crops.* HMSO, London, UK, 76 pp.

Edward, W. (1996) Measurement methods and strategies for non-infectious microbial components in bioaerosols at the workplace. *Analyst* 121(9), 1197–1201.

Griffiths, E. (1981) Iatrogenic plant diseases. *Annual Review of Phytopathology* 19, 69–82.

Kozlowski, T.T. (1978) *Water and Plant Disease.* Academic Press, New York, USA, 323 pp.

Nriagu, J.O. (1978) *Sulfur in the Environment. Part 2: Ecological Impacts.* John Wiley & Sons, New York, USA, 482 pp.

Shurtleff, M.C. and Averre, C.W. (1997) *The Plant Disease Clinic and Field Diagnosis of Abiotic Diseases.* American Phytopathological Society, St Paul, Minnesota, USA, 245 pp.

Tattar T.A. (1980) Non-infectious diseases of trees. *Journal of Agriculture* 6, 1–4.

Tickell, Sir C., Bell, J.N.B., McNeill, S., Houlden, G., Brown, V.C., Mansfield, P.J., Bucke, D., Bentham, G., Bell, E.A., Walsh, J.F., Molyneux, D.H., Birley, M.H., Rogers, D.J., Williams, B.G., Phillips, D.R., Hominick, W. and Chappell, L.H. (1993) The impact of global change on disease. *Parasitology* 106 (Suppl.), S1–S107.

Weir, R.G. and Cresswell, G.C. (1993) *Plant Nutrient Disorders 1: Temperate and Subtropical Fruit and Nut Crops.* Australia Inkata Press, North Ryde, Australia, 93 pp.

Weir, R.G. and Cresswell, G.C. (1993) *Plant Nutrient Disorders 3: Vegetable Crops.* Inkata Press, North Ryde, Australia, 105 pp.

Air Pollution Effects and Injury

A.R. Wellburn (Deceased)

Formerly of Division of Biological Sciences, Lancaster University, Lancaster LA1 4YQ, UK

Visible or Invisible Injury

There is now abundant evidence that the major effect of most air pollutants on plants is invisible rather than visible and due to long-term chronic exposures rather than isolated and often accidental acute exposures. This physiological invisible injury often causes significant losses of yield and reductions in biomass formation and has been traced to a variety of changes. Principal among these are changes in stomatal behaviour, which can disturb both photosynthesis and transpiration, direct losses in photosynthetic capacity, failure to provide sufficient antioxidant protection, imbalances and accumulations of certain metabolites or nutrients (e.g. S and N) and disturbances to the translocation of photosynthate (Wellburn, 1994).

The losses of yield and biomass can be considerable (up to 50%) without visible injury appearing, which poses many problems for the plant pathologist in terms of evaluation of field conditions. It is only when unpolluted controls are available alongside during experimental fumigations that these changes can be fully appreciated. Even then it is extremely difficult to design such fumigations so that they accurately reflect the surrounding environment, although much improvement has been made in experimental design and practice in recent years. Basically these are of two types. One possibility is the exclusion of all air pollutants, which is not easy because activated charcoal does not adsorb all air pollutants. This provides clean-air-grown plants, which, in turn, may not reflect true ambient conditions. Far more common is a design where fumigations are done at rural background ambient levels as controls and certain air pollutants are added in an episodic and modulated (i.e. updated minute by minute) manner over and above the ambient levels to a prescribed target addition so as to achieve a certain dosage over a particular period.

Adverse yield responses to ozone (O_3) are measured differently in Europe and North America. The European Open-Top Chambers Programme (EOTCP) and other studies in Europe selected the AOT40 (accumulated exposure over a threshold of 40 nl l^{-1}) as a critical exposure index to assess crop growth. This demonstrated that for a range of crop species concentrations of O_3 in the range 35–60 nl l^{-1}, a frequent occurrence in Europe in the 1990s (European Environment Agency, 1997), are capable of affecting crop yield at AOT40s > 5300 nl l^{-1} h from May to July (Ashmore, 1993; Legge et al., 1995; Fuhrer et al., 1997). The USA NCLAN (National Crop Loss Assessment Network) by contrast uses the SUMO6 (sum of all hourly concentrations above 60 nl l^{-1}), which suggests that O_3 concentrations in the range 50–87 nl l^{-1} are the best predictors of crop yield in the USA (Legge et al., 1995). Further European studies have shown that natural vegetation and trees are often insufficiently protected from O_3 by similar AOT40s and that equivalent AOT30s are now being developed. So far, these concepts have only been used to describe losses arising from O_3, the most important crop-loss pollutant, but such critical level concepts could equally well be used to describe and predict losses from oxides of nitrogen, NH_3 and perhaps SO_2.

It is very difficult to ascribe the current balance between the extents of visible and invisible injury, but it is probably more than 98% in favour of the latter if O_3, NH_3 and oxides of nitrogen are included. In the case of all the other gases, which are far more likely to be highly localized and caused by accidental acute spillages and emissions, the balance may be slightly shifted towards a greater likelihood of visible symptoms. This now includes SO_2. Over the last 20 years, there has been a dramatic decline in ambient levels of this gas over Europe, so much so that NO_2 has long replaced it as the major component of acid rain. Indeed, symptoms of S deficiency are now being reported for crops requiring high S inputs (e.g. oilseed rape which produces white flowers when deficient).

Interactions

A major factor in assessing air pollution response is the possibility of an interaction between two air pollutants or an air pollutant and another abiotic or biotic factor. These interactions are of two types: synergistic or more-than-additive and antagonistic or less-than-additive. Interactions between SO_2 and either NO_2 or O_3 are strongly synergistic in terms of depression of growth and yield and have been studied in some detail (Mansfield and McCune, 1988; Barnes and Wellburn, 1998). Surprisingly, little study has yet been made of the commonest combination by far – O_3 in the presence of oxides of nitrogen. Combinations of O_3 with drought or insect predation also show strong synergism but this is more easily explained by poor root growth and the ability of insects to detect weakened plants (see below).

Influencing Factors

Air pollution generally affects roots more than foliage. A large number of studies of chronic low level fumigations of plants with different air pollutants have shown that poor root growth is a common outcome (Darrell, 1989). As a consequence, root:shoot ratios are lower and such exposed plants, although usually not showing visible injury, are highly susceptible to subsequent drought. This effect has often been traced to pronounced detrimental effects of the pollutants on phloem loading, which in certain circumstances can produce bigger shoot masses than in ambient controls. Different parts of the foliage are affected by acute exposures to particular air pollutants. Younger tissues are affected by HCl, SO_2 and oxides of nitrogen and earlier immature tissues are sensitive to HF, whereas ethene causes epinasty to expanding leaves. Ozone generally affects intermediate and older leaves.

Dose responses to air pollutants are sigmoidal in shape. Generally, very low dosages do not produce visible symptoms and low doses very little. However, all relationships differ between species and often within species. Plants are usually more sensitive at higher temperatures but conifers, cabbages and leeks exposed to NH_3 in winter are the exception. Low light levels also increase the effects of NH_3 and oxides of nitrogen but generally higher light levels increase sensitivity towards the other air pollutants. Seasonal changes in sensitivity are also possible.

Excesses and deficiencies of soil nutrition can also affect the sensitivity of vegetation to air pollution injury but this depends on the air pollutant, the species and the nutrient as to whether the sensitivity is magnified or lessened. Higher relative humidities generally increase the extent of injury but lower soil moisture content reduces sensitivity because the stomata tend to close. By stripping away the protective boundary layer of still air around leaves, wind has a pronounced effect on air pollutant uptake and hence sensitivity.

There is a pronounced genetic element within species which determines sensitivity towards air pollutants. This has been evaluated in detail in a number of species particularly in the context of O_3. This has resulted in the availability of both sensitive and tolerant cultivars of the same species, which can be used in area mapping of air pollution injury by setting out the relevant pairs in selected locations over the area of study at regular intervals. The best-known examples of this type of survey have used Bel-W3 (sensitive) and Bel-B (tolerant) tobaccos (e.g. Ashmore *et al.*, 1978) but this is not always convenient because of the chilling sensitivity of tobacco. Other pairs are now available for birch, loblolly pine, poplar, plantains, clover and radish (Wellburn and Wellburn, 1996).

Types of Visible Injury

There is a wide range of visible injury caused by acute concentrations of differ-ent air pollutants on various species and there is no substitute for a colour photograph in each case to describe the symptoms. Such a compendium does not exist and useful photographs are scattered around the literature. The best examples, in terms of quality, are provided by Treshow (1984) and Wellburn (1994), but the most comprehensive are those of Taylor *et al.* (1989) although the reproduction is inferior. They do, however, have the advantage of being published alongside examples of possible confusions with other causative factors (see below). Taylor *et al.* (1989) also provide comprehensive tables of species' response to each type of air pollutant, the symptoms found, the concentrations known to cause injury to different species and the relative sensitivities of various species to SO_2, fluoride, HF, oxides of nitrogen, Cl_2, HCl, NH_3, ethene, dusts and particulates, acid mist, H_2S, CO, Hg vapour, Br_2, I_2, O_3 and peroxyacetylnitrate (PAN). The nature and range of the different types of injury on different species is so disparate that it is not profitable to make any generalizations here and direct reference must be made to Taylor *et al.* (1989) for detailed information. However, the relative sensitivities and symptoms of easily recognized species to acute concentrations of different gases is an important first step towards identification and elimination of possible air pollutants causing problems in the field (Table 18.1).

Special mention, however, must be made of the visible injury caused by chronic exposures of certain plants to O_3, which have enabled surveys of this gas to be made over large areas. Current-year needles of certain sensitive long-needle pines show chlorotic mottling or banding, which form as the needles elongate during episodes of high O_3. These symptoms were first observed on *Pinus ponderosa* and *Pinus jeffrii* in the San Bernardino mountains around Los Angeles but are also now shown by *Pinus halepensis* around the Mediterranean region. Broad-leaved plants also show visible injury after long-term chronic exposures to O_3. The best-known example is that of black cherry (*Prunus serotina*) in the Great Smoky Mountains of the eastern USA.

Mention has already been made of bioindicator surveys using sensitive and tolerant pairs but specific injury responses of certain species have been made for much longer. Table 18.2 shows a listing of suitable species for surveys in Europe.

Possible Confusions

Many similar symptoms to those caused by air pollutants can be caused by biotic and abiotic factors. Again these are dealt with in detail by Taylor *et al.* (1989) who attempt to list the most commonly mistaken attributions and show a number of illustrative figures to emphasize possible confusions.

In the case of biotic agents, fungi, bacteria, viruses and mycoplasmas, nematodes, insects and mites can all produce similar injury to that caused by air pollution. Fungi cause chlorosis, necrosis, distortion, discoloration and abscission but a hand-lens examination will often eliminate confusion with that caused by air pollutants. Effects of the following are found to be most similar: rose black and bean chocolate tar spots, potato blight, tomato wilt, *Alternaria*, honey fungus, barley leaf stripe and sooty mould.

Table 18.1.   Relative sensitivity and responses of some common, easily identifiable species to acute concentrations of several air pollutants. (Adapted from Taylor *et al.*, 1989.)

Species	SO_2	F⁻,HF	NO_2	Cl_2	HCl	NH_3	C_2H_4	O_3	PAN
Apple	I	I	S n	S n	–	S d	T	–	–
Ash	T	S nt	–	–	–	S d	–	S s	–
Aspen	I	S n	–	–	S c	S b	–	S n	–
Barley	S n	S nt	S n	S n	S c	S c	T	S c	T
Bean	S nc	I	S n	I	S n	S d	S a	S n	S g
Birch (Silver)	I	T	S n	–	I	S n	–	T	–
Blackberry	I	T	–	S nc	–	–	–	–	–
Carrot	I	I	S n	–	S c	I	I	I	I
Cherry	I	I	–	–	S c	S n	–	S b	–
Chickweed	S n	I	T	–	–	S d	–	–	S g
Clover	S n	S c	S n	–	–	I	T	S c	S g
Cocksfoot	I	S c	I	–	–	I	–	S b	–
Dandelion	S n	T	I	I	I	I	T	–	–
Fathen	I	I	T	S c	I	S d	S a	T	I
Fir (Douglas)	I	S nc	S a	S n	S tb	I	T	T	–
Iris	I	S t	–	T	I	–	–	–	–
Larch	S c	I	S na	–	S t	T	–	I	–
Lettuce	S f	I	S a	S n	–	S d	–	T	S g
Lilac	T	S d	–	S n	–	S n	–	S l	–
Lucerne	S b	I	S n	S n	S n	I	S a	S n	I
Oak	T	S n	T	–	S c	T	–	T	–
Oat	I	I	S n	S n	–	–	T	S c	S y
Onion	T	S gt	T	S n	–	I	T	S f	T
Pea	S n	I	S n	–	–	S d	S e	I	–
Pear	I	T	S n	–	I	S d	T	–	–
Petunia	I	T	I	I	–	I	S i	S b	S g
Potato	T	I	T	–	S c	S n	S a	S b	–
Privet	T	T	I	I	–	S d	S a	I	–
Radish	S n	–	S n	S n	S n	S d	T	S cn	T
Raspberry	I	T	–	–	S c	S d	–	–	–
Rose	I	I	S n	S s	S n	T	S cn	–	–
Ryegrass	S n	T	I	–	–	I	T	–	–
Smooth-stalked meadow grass	S n	I	I	S n	–	T	T	–	–
Spruce	I	S nt	I	S n	I	S d	T	T	–
Sunflower	S n	I	S n	S n	–	S d	S a	–	–
Sweet pea	S n	T	S n	–	–	S d	S e	–	–
Wheat	I	I	I	–	–	–	–	S c	I
Willow	I	T	–	–	S a	S b	–	–	–

S, Sensitive; I, Intermediate; T, Tolerant; –, No Information.

a, abscission; b, bronzing; c, chlorosis; d, discoloration; e, epinasty; f, shot holing; g, glazing; i, inhibition of flowering; l, leaf curling; n, necrosis; s, stippling; t, tip burning; y, yellow or white banding.

Table 18.2.   List of plant species suitable for use as air pollution bioindicators in Europe. (After Steubing and Jäger, 1982.)

Pollutant	Species and variety
SO$_2$	Lucerne, *Medicago sativa* L.cv. Du Puits Clover, *Trifolium incarnatum* L. Pea, *Pisum sativum* L. Buckwheat, *Fagopyrum esculentum* Moench. Great plantain, *Plantago major* L.
NO$_2$	Wild celery, *Apium graveolens* L. *Petunia* sp. Ornamental tobacco, *Nicotiana glutinosa* L.
O$_3$	Tobacco, *Nicotiana tabacum* L. cv. Bel W3
PAN	Small nettle, *Urtica urens* L. Annual meadow grass, *Poa annua* L.
HF, fluorides	*Gladiolus gandavensis* L.cv. Snow Princess Tulip, *Tulipa gesneriana* L.cv. Blue Parrot
General accumulators	Italian rye grass, *Lolium multiflorum* Lam. ssp. *italicum* Cabbage, *Brassica oleracea* L.cv. Acephala
Bark accumulators	Rose, *Rosa rugosa* Thunb. *Thuja orientalis* L.

Bacteria may cause chlorotic mottle, distortion and reduced growth. The most likely confusions are with lucerne wilt, fireblight and bean halo blight. Similarly viruses cause chlorosis and necrosis. The most common confusion is between fluoride injury and *cherry mottle leaf virus*. Nematodes cause very similar effects to both bacteria and viruses and again confusion with fluoride injury is possible.

Bark beetles, spider mites, leaf hoppers, aphids, thrips, sawflys and leaf miners can also cause various discolorations similar to that caused by air pollution which may be discriminated using a hand lens. However, there is a wealth of data in the literature to demonstrate that insect and mite attack on plants often follows exposure of plants to air pollution. There is no doubt that insects can recognize such weakened plants at a distance by mechanisms little understood but highly efficient and very much better than any diagnostic system available to ourselves. Once in close vicinity they are able to gain access more easily because plant defences are simultaneously weakened by air pollution. The advantages to the insects are often nutritional. For example, extra S may be gained from plants exposed to SO$_2$ or additional N from those experiencing NH$_3$, NO and NO$_2$.

Abiotic factors which can mimic air pollution effects include nutrient deficiencies, mineral excesses, drought, waterlogging, extremes of temperature, herbicides and pesticides and genetic disorders. Full lists of the most common confusions are given by Taylor *et al.* (1989) along with selected photographs.

Diagnosis Procedure

Careful attention to procedure is the key to accurate diagnosis of air pollution injury. The following questions given by Taylor *et al.* (1989) are a useful guide.

1. Are many plant species affected? *Air pollution usually affects a wide range of species.*

2. What symptoms are shown? *Chlorosis and necrosis are most common.*

3. Which part of the plant is most affected? *Different air pollutants and mimicking effects affect different ages of tissues.*

4. Are the affected plants in the same small area? *Often a patch of the same poor soil type, particular microclimate or frost hollow, etc., can be responsible.*

5. Is there a gradation in injury symptoms with area? *Plumes from point sources often touch ground level at certain points with lesser effects towards the edges.*

6. Is an organism present? (*See 'Possible confusions' section above.*)

7. Is there a history of such symptoms in the area? *Season, wind direction and extremes of temperature and moisture may be correlated with the cause.*

8. What management practice has been previously undertaken? *A variety of abiotic factors, such as salting, treatment with pesticides, soil sterilants, etc., could be a possible explanation.*

9. Did the symptoms appear at a specific time? *If they were quick then frost or air pollution may be the cause.*

10. Are local pollution sources present? *With the exception of ozone, most acute air pollution incidents occur close to sources.*

11. Have analyses be done? *Analysis of leaf and soil samples can be of great assistance in diagnosis. Taylor et al. (1989) have set out an appropriate sampling and analysis procedure.*

12. If air pollution is still suspected, are some of the well-known sensitive species affected? (See Table 18.1).

References

Ashmore, M.R. (1993) Critical levels and agriculture in Europe. In: Jäger H.J., Unsworth, M., Temmerman, L. De. and Mathy, P. (eds) *Effects of Air Pollution on Agricultural Crops in Europe.* Air Pollution Research Report No. 46, CEC, Brussels, pp. 105–130.

Ashmore, M.R., Bell, J.N.B. and Reily, C.L. (1978) A survey of ozone levels in the British Isles using indicator plants. *Nature* 276, 813–815.

Barnes, J.D. and Wellburn, A.R. (1998) Air pollutant combinations. In: DeKok, L. and Stulen, I. (eds) *Responses of Plant Metabolism to Air Pollution and Global Change.* SBS Publishers, The Hague.

Darrall, N.M. (1989) The effect of air pollutants on physiological processes in plants. *Plant, Cell and Environment* 12, 1–30.

European Environment Agency (1997) *Air Pollution in Europe 1997.* Environment Monograph No. 4, EEA, Copenhagen.

Fuhrer, J., Skarby, L. and Ashmore, M.R. (1997) Critical levels for ozone effects on vegetation in Europe. *Environmental Pollution* 97, 91–106.

Legge, A.H., Grunhage, L., Nosal, M., Jäger, H.J. and Krupa, S.V. (1995) Ambient ozone and adverse crop response – an evaluation of North American and European data as they relate to exposure indexes and critical levels. *Angewandte Botanik* 69, 192–205.

Mansfield, T.A. and McCune, D.C. (1988) Problems of crop loss assessment when there is exposure to two or more gaseous pollutants. In: Heck, W.W., Taylor, O.C. and Tingey, D.T. (eds) *Assessment of Crop Loss from Air Pollutants*. Elsevier Applied Science, London, UK, pp. 317–344.

Steubing, L. and Jäger, H.J. (1982) *Monitoring of Air Pollutants by Plants: Methods and Problems*. Dr W. Junk, The Hague.

Taylor, H.J., Ashmore, M.R. and Bell, J.N.B. (1989) *Air Pollution Injury to Vegetation*. HM Health and Safety Executive, IEHO, London, UK.

Treshow, M. (1984) *Air Pollution and Plant Life*. John Wiley & Sons, Chichester, UK.

Wellburn, A.R. (1994) *Air Pollution and Climate Change: the Biological Impact*. Longmans, Harlow, UK.

Wellburn, F.A.M. and Wellburn, A.R. (1996) Variable patterns of antioxidant protection but similar ethene differences in several ozone-sensitive and -tolerant selections. *Plant, Cell and Environment* 19, 754–760.

Effects of Climate Change

S. Chakraborty

CSIRO Plant Industry, CRC Tropical Plant Protection,
University of Queensland, Queensland 4072, Australia

Climate Change

Solar radiation drives global climate as it interacts with land, water, vegetation and the atmosphere. Net radiation reaching the earth's surface causes the earth to emit thermal radiation. Radiatively active gases such as water vapour, CO_2, ozone, methane and nitrous oxide (N_2O), naturally present in the atmosphere, partially trap this outgoing radiation, raising the mean surface temperature to about 15°C, which causes the 'greenhouse effect'. Without these gases in the atmosphere the mean surface temperature of the earth would be −18°C. Although palaeoclimatic records indicate changes in climate in the past, human activities are increasingly modifying global climate. For instance, burning of fossil fuels and the large-scale clearing of forests have increased the atmospheric concentration of CO_2 by 30%, methane by 145% and N_2O by 15% since pre-industrial times. This increasing concentration of the radiatively active gases has enhanced the greenhouse effect. Global warming of 0.3–0.6°C has occurred since the late 19th century and the earth's surface temperature is projected to rise between 1 and 3°C by 2100.

Climate change scenarios are developed using general circulation models (GCMs), which simulate climate using mathematical formulations of the processes that comprise the climate system. All GCMs show warming due to increased greenhouse gas concentrations but uncertainties remain due to incomplete understanding of ocean circulation patterns, lack of knowledge concerning the formation and feedback from clouds, simplistic simulation of hydrological processes and coarse resolution. With current spatial resolution of approximately 250 km by 350 km, GCMs are better at simulating temperature than precipitation, which is spatially and temporally discontinuous. Down-scaling of GCM outputs using mesoscale models can generate more realistic outputs, which represent conditions at the surface of a geographic zone with 10–50 km grid resolution (Russo and Zack, 1997).

Approaches Used to Study Effects on Plant Diseases

Although no disease can develop in the absence of conducive weather conditions, research on how changing climates can influence plant diseases has only started in recent years. Experimental studies in controlled environments have been used to determine effects of individual weather factors and/or varying atmospheric composition on the physiology of host–pathogen interaction. Polycyclic epidemics can not be experimentally studied in growth cabinets. Although open-top chambers with CO_2 and temperature controls are useful (Norby et al., 1997), modelling approaches are more suitable to simultaneously examine multiple climate change scenarios and interacting factors. Most models have considered changes in mean temperature, although change in variability is equally relevant. A modest warming can cause a significant increase in heat sums above a critical temperature threshold. This will influence host physiology and resistance to a pathogen. Diseases may also respond to slow long-term changes in climate. One analysis has shown a significant association between interannual variations in wheat rust severity and El Niño–Southern Oscillation events, which are driven by variation in sea surface temperatures in the Pacific Ocean (Scherm and Yang, 1995). Moisture is important for dispersal, infection, survival and other events in pathogen life cycles, but there has been little work on how pathogens respond to changes in moisture variables such as precipitation, dew or surface wetness.

Two types of models are commonly used to study effects of climate change on plant diseases. In the empirical 'climate-matching' models, a profile of climatic preferences is developed from meteorological data and observed distribution of the pathogen. This profile is used to explore potential distribution under a changed climate. Climate-matching models are useful as a first-pass analysis but they do not consider other interacting factors such as management interventions or direct effects of CO_2 enrichments. 'Process-based' models are more flexible and have a broad range of application beyond determining the effects of climate change. Damage mechanisms, which explain the quantitative effect of a disease on crop growth and development, are linked to pathogen (e.g. severity) and crop (e.g. leaf area) variables and the model is run for current and changed climate scenarios.

Effects on Plant Diseases

The most likely effects of climate change are altered geographical distribution of host and pathogens, and changes in crop loss and the efficacy of control measures. These effects are due to altered physiology of host–pathogen interactions and changes in stages and/or rates of pathogen development.

Geographical distribution

Warming will cause a poleward shift of agroclimatic zones and crops that grow in these zones. Pathogens will follow the migrating hosts. For example, the oak decline pathogen *Phytophthora cinnamomi* and plant parasitic nematodes *Xiphinema* and *Longidorus* are predicted to spread to the north of Europe (Boag *et al.*, 1991; Brasier and Scott, 1994). Migration may alter the type, relative importance and spectrum of diseases. The thermophilic *Melampsora alli-populina* appears only sporadically in northern Europe; a change to warmer climate will have serious implications for the large areas where susceptible poplar clones are grown as an alternative to producing agricultural surpluses. A crop may continue to be grown for agroecological or economic reasons despite climate change altering its suitability for certain locations. Marginal climates would impose chronic stress and may increase susceptibility to diseases.

Physiology of host–pathogen interaction

Changes due to elevated CO₂

If other resources are non-limiting, a doubling of CO_2 consistently increases yield by about 33% due to enhanced rate of photosynthesis and water use efficiency; realization of this potential net increase in production will depend on damage from disease, weed competition and herbivory. Associated physiological and anatomical changes which influence host–pathogen interactions include: reduction in stomatal density and conductance; greater accumulation of carbohydrates in leaves; more waxes, extra layers of epidermal cells and increased fibre content; production of papillae and accumulation of silicon at the sites of appressorial penetration; and greater number of mesophyll cells. In most diseases caused by necrotrophic pathogens severity is increased at high CO_2, and increased canopy size and density enhance inoculum survival on crop residues. The severity of other diseases caused by necrotrophic and biotrophic pathogens may reduce or remain unchanged at high CO_2.

Elevated CO_2 changes the onset and duration of stages in pathogen life cycles. The latent period, i.e. the time between inoculation and sporulation, is extended under high CO_2 in all pathogens studied so far and fecundity in some pathogens increases by up to 20-fold. Host resistances may be overcome more rapidly as a result of accelerated evolution of pathogen populations due to increased fecundity.

Changes due to increased ozone and ultraviolet B

About 90% of ozone occurs in the stratosphere, which extends from 15 to 50 km, with the remainder in the troposphere, which extends some 15 km above the ground. Depletion of stratospheric ozone by chlorine-containing compounds such as chlorofluorocarbons (CFCs) allows more UV radiation in the 288–320 nm range (UV-B) to reach the earth's surface. In most cases, exposure to UV-B predisposes the host to increased disease severity (Manning and Tiedemann, 1995). Plants become stunted, increase branching, reduce leaf area, and accelerate ripening and reproduction when exposed to enhanced UV-B; continued exposure can lower the production of antifungal compounds. Despite the stimulatory effect of near-UV on spore production, many fungi are damaged by UV-B.

The tropospheric ozone concentration has increased around major urban centres due to vehicle emissions. Ozone induces stress reactions in plants, which can either enhance tolerance or increase susceptibility to a second stressor, such as a pathogen. The effects of elevated ozone on host–pathogen interaction and crop yield have been extensively studied (Sandermann, 1996).

Conclusion

Climate change is a gradual process and its impact on a production system is modified by interactions with other changes such as the introduction of a resistant cultivar. Consequently, it becomes difficult to isolate effects due to climate change alone. Uncertainties about future emissions of greenhouse gases and weaknesses in GCMs limit our ability to project future changes in climate. There is a paucity of knowledge on the effect of climate change on plant diseases, and the lack of spatial resolution in GCMs does not allow prediction of location-specific effects on plant diseases. In rice blast, the effect of temperature changes varied with the agroecological zone. The risk of serious rice blast epidemics was high in the cooler subtropics but low in the humid tropics and warm humid subtropics (Luo *et al.*, 1995). Thus, examining the effects of climate change at a regional level may uncover variables not readily identified at a paddock/crop level. A challenge for the future will be to develop landscape-scale disease models, which can be integrated with regional climate models for realistic appraisal of crop loss under a changed climate. The economic impact can be positive, negative or neutral as climate change may reduce, increase or have no effect on some diseases in some regions; mitigation strategies need to consider this. For hosts most at risk, pre-emptive breeding strategies will need to start early due to the long time required for development and release of cultivars. Through its influence on the efficacy of biological and chemical control options, climate change will impact strongly on disease management and research policy.

References and Further Reading

Boag, B., Crawford, J.W. and Neilson, R. (1991) The effect of potential climatic changes on the geographical distribution of the plant parasitic nematodes *Xiphinema* and *Longidorus* in Europe. *Nematologia* 37, 312–323.

Brasier, C.M. and Scott, J.K. (1994) European oak declines and global warming: a theoretical assessment with special reference to the activity of *Phytophthora cinnamomi*. *EPPO Bulletin* 24, 221–232.

Chakraborty, S., Murray, G.M., Magarey, P.A., Yonow, T., O'Brien, R., Croft, B.J., Barbetti, M.J., Sivasithamparam, K., Old, K.M., Dudzinski, M.J., Sutherst, R.W., Penrose, L.J., Archer, C.

and Emmett, R.W. (1998) Potential impact of climate change on plant diseases of economic significance to Australia. *Australasian Plant Pathology* 27, 15–35.

Chakraborty, S., Tiedmann, A.V. and Teng, P.S. (2000) Climate change and air pollution: potential impact on plant diseases. *Environmental Pollution* 108, 317–328.

Cheddadi, R., Yu, G., Guiot, J., Harrison, S.P. and Prentice, I.C. (1996) The climate of Europe 6000 years ago. *Climate Dynamics* 13, 1–9.

Coakley, S.M. and Scherm, H. (1996) Plant disease in a changing global environment. *Aspects of Applied Biology* 45, 227–237.

Coakley, S., Scherm, H. and Chakraborty, S. (1999) Climate change and disease management. *Annual Review of Phytopathology* 37, 399–426.

Frankland, J.C., Magan, N. and Gadd, G.M. (1996) *Fungi and Environmental Change*. Cambridge University Press, Cambridge, UK, 351 pp.

Goudriaan, J. and Zadoks, J.C. (1995) Global climate change: modeling the potential responses of agro-ecosystems with special reference to crop protection. *Environmental Pollution* 87, 215–224.

Hibberd, J.M., Whitbread, R. and Farrar, J.F. (1996) Effect of elevated concentrations of CO_2 on infection of barley by *Erysiphe graminis*. *Physiological and Molecular Plant Pathology* 48, 37–53.

IPCC (1996) Contribution of Working Group 1 to the second assessment report of the Intergovernmental Panel on Climate Change. In: Houghton, J.T., Meira Filho, L.G., Callander, B.A., Harris, N., Kattenberg, A. and Maskell, K. (eds) *Climate Change 1995: the Science of Climate Change*. Cambridge University Press, Cambridge, UK, 572 pp.

Luo, Y., TeBeest, D.O., Teng, P.S. and Fabellar, N.G. (1995) Simulation studies on risk analysis of rice blast epidemics associated with global climate in several Asian countries. *Journal of Biogeography* 22, 673–678.

Manning, W.J. and Tiedemann, A.V. (1995) Climate change: potential effects of increased atmospheric carbon dioxide (CO_2), ozone (O_3), and ultraviolet-B (UVB) radiation on plant diseases. *Environmental Pollution* 88, 219–245.

Norby, R.J., Edwards, N.T., Riggs, J.S., Abner, C.H., Wullschleger, S.D. and Gunderson, C.A. (1997) Temperature-controlled open-top chambers for global change research. *Global Change Biology* 3, 259–267.

Prestidge, R.A. and Pottinger, R.P. (1990) *The Impact of Climate Change on Pests, Diseases, Weeds and Beneficial Organisms Present in New Zealand Agricultural and Horticultural Systems*. MAF Technology, Ruakura Agricultural Centre, Hamilton, New Zealand.

Rosenzweig, C. and Hillel, D. (1998) *Climate Change and the Global Harvest*. Oxford University Press, New York, USA, 324 pp.

Russo, J.M. and Zack, J.W. (1997) Downscaling GCM output with a mesoscale model. *Journal of Environmental Management* 49, 19–29.

Sandermann, H. Jr (1996) Ozone and plant health. *Annual Review of Phytopathology* 34, 347–366.

Scherm, H. and Yang, X.B. (1995) Interannual variations in wheat rust development in China and the United States in relation to the El Niño/Southern Oscillation. *Phytopathology* 85, 970–976.

Sutherst, R.W., Yonow, T., Chakraborty, S., O'Donnell, C. and White, N. (1996) A generic approach to defining impacts of climate change on pests, weeds and diseases in Australasia. In: Bouma, W.J., Pearman, G.I. and Manning, M.R. (eds) *Greenhouse, Coping with Climate Change*. CSIRO, Australia, pp. 281–307.

Teng, P.S., Heong, K.L., Kropff, M.J., Nutter, F.W. and Sutherst, R.W. (1996) Linked pest-crop models under global change. In: Walker, B. and Steffen, W. (eds) *Global Change and Terrestrial Ecosystems*. Cambridge University Press, Cambridge, pp. 291–316.

Detection and Isolation of Fungal and Bacterial Pathogens

J.M. Waller

CABI Bioscience UK Centre, Bakeham Lane, Egham, Surrey TW20 9TY, UK

General

Some diseases can be recognized from symptoms alone especially where the characteristic appearance of the pathogen, as in mildew, rust and smut diseases, is an integral part of the symptom or where the observer is very familiar with the disease. Even then, identification of the species involved may require microscopy of the sporulating structures. Confirmation of the presence of a pathogen in a diseased plant is an essential step in the diagnosis of disease and this often requires procedures of detection and isolation. However, there are many situations where ill-health of plants may not be clearly caused by obvious pathogenic diseases. Some problems caused by viruses or root pathogens may not produce overt symptoms of disease but nevertheless result in substantial reduction of plant productivity. Mechanical or physiological damage by environmental factors, by pests and by weed competition may contribute to plant ill-health and an appreciation of these factors is required for successful diagnosis.

Detecting a pathogen requires knowledge of where it is in relation to the symptoms. With local lesions or where pathogen sporulation is clearly visible, the pathogen is, or has been, present in the lesion. Where rapid necrosis follows infection, secondary organisms invade and frequently displace the pathogen from older lesions so that the pathogen may only be detected in young active lesions. Invasion of damaged tissue by secondary organisms is a major cause of diagnostic problems. Where symptoms are systemic, such as wilts, stunting and general chlorosis, the pathogen may not be situated where the symptom is most obvious. Roots and vascular systems need to be examined for signs of the pathogen, e.g. brown flecking of the xylem tissue associated with fungal wilt pathogens, exudation from vascular tissues of cut stems associated with bacterial wilts.

Pathogens are most readily detected by microscopy of affected tissues. Low-power stereoscopic microscopy can reveal the presence of sporulating structures of fungal pathogens or evidence of bacterial exudation; further examination under a compound microscope is often needed to identify the pathogen. However, immunological, biochemical or molecular techniques can now be used to detect the presence of specific pathogens, often at very low levels, in diseased tissue (Duncan and Torrence, 1992; Henson and French, 1993; see also Chapters 22 and 23). Where the presence of a specific pathogen is not known, high-power microscopy of diseased tissues may reveal it (see Chapter 21) or isolation and culture of the pathogen may be required.

Isolation

Pathogens often have to be isolated and cultured from diseased specimens before they can be identified. Cultures of pathogens may also be needed for other purposes. Only pathogens capable of saprophytic growth (facultative or necrotrophic organisms) can generally be grown in culture and some of these are fastidious in their requirements.

When a fungus is sporulating on the surface of a lesion or other substratum, a pure culture can frequently be obtained by direct transfer to a growth medium. This can be done by merely touching the spores (under a hand lens or a low-power binocular microscope if necessary) with a sterile fine gauge inoculating needle, either dry, or moistened by first stabbing it into the sterile medium, and then streaking on to a plate or slope.

Moist chambers

Surface sporulation is often encouraged by keeping the specimen in a moist chamber overnight but this treatment also encourages quick-growing saprobic fungi. Containers used as moist chambers should be clean and preferably sterile. Sterile filter papers or absorbent laboratory tissues (autoclaved) are added to the containers and moistened with sterile distilled water. A few drops of glycerol are added to slow down drying out. Specimens are best kept above the damp paper by supporting on small pieces of plastic grid or similar supports; these are sterilized by dipping in alcohol and quickly flaming. Moist chambers should not be incubated at high temperatures and are best kept in the light, e.g. on the laboratory bench, but out of direct sunlight and at a fairly constant temperature to avoid condensation. They need to be checked daily for sporulation under a low-power stereoscopic microscope.

Isolation from tissues

Isolation of fungi from plant material is usually achieved by placing small portions of relevant tissue on to a suitable agar growth medium in sterile Petri dishes. Aseptic conditions are necessary to avoid contamination, and surface sterilization of excised tissue is often necessary before plating out to remove saprophytic organisms, which commonly grow over plant surfaces, but in some situations described below this may not be necessary or advisable. Working surfaces for aseptic dissection of plant material should be hard and non-porous; a glazed tile or plate glass sheet can be used if the bench surface is porous and cleaned by wiping with industrial spirit. Instruments can be sterilized with alcohol and/or heating. As a general rule, isolations from plant material should be undertaken on plain agar (tap-water agar, TWA). This favours the growth of the pathogen, which uses the plant material as a food base, rather than the growth of contaminant saprophytic fungi. Nutrient agars favour the growth of faster-growing saprophytes. When plating pieces of tissue on to agar, Petri dish lids should be carefully lifted and replaced to avoid entry of airborne contaminants.

For surface lesions on leaves and stems small pieces of diseased tissue of a few cubic millimeters excised from the lesion margin are surface sterilized (1–3 min in 10% NaOCl), washed in sterile water and then placed on to the agar surface. Where there are internal lesions of the vascular tissue, careful excision may avoid the need for surface sterilization. The sample is split longitudinally from healthy to diseased area using a clean instrument and small slivers of tissues from the edges of the lesion on the newly exposed internal surface removed and plated out using sterile instruments.

Examination of root systems can be done after washing off adhering soil. Small and fine roots are best examined in water over a white surface, e.g. in a white dish to assist the detection of lesions. Approximately 1 cm lengths of roots can be excised and plated out. A very light surface sterilization (dip in alcohol or 10% NaOCl and wash in sterile water) can be undertaken, but with such delicate material, which may be infected by pathogens particularly sensitive to surface-sterilizing chemicals, a prolonged washing procedure may be best. This can be achieved by retaining root pieces in a fine sieve placed under a gentle stream of clean running tap water for 30 min–2 h.

Surface sterilizing agents and their uses

Alcohol (70% ethanol, industrial or methylated spirit)

Used as a surface swab or a dip, this is a useful cleansing and wetting agent especially on hard surfaces. There is no need for post-treatment washing as it can be flamed off or left to evaporate. It is not effective against many fungi.

Sodium or calcium hypochlorite (NaOCl or CaOCl)

Sodium hypochlorite is widely used and very effective. A stable 'stock' solution (commercial bleach) contains 10–14% available chlorine. It is usually used at 10% dilution (1–1.4% available chlorine) with immersion times of 1–5 min.

The free chlorine has good penetrating power and evaporates so that post-treatment washes can be omitted especially with CaOCl. It should be stored in a refrigerator as it loses potency with age especially on a laboratory bench; a fresh solution should be made up every 2–3 weeks.

Hydrogen peroxide (H$_2$O$_2$)

This is used at 30% (100 volumes) strength for robust tissues (seeds, wood, etc.). Dilutions down to 10% can be used for more delicate tissues. It is gentler in action than hypochlorite and immersion times are longer – up to 10 min. It should be kept cool and in darkness.

Other chemicals such as mercuric chloride (0.1% HgCl$_2$) are not recommended for general use as they require several post-sterilization washes to remove traces of the chemical, they are toxic and their disposal presents an environmental hazard.

Isolation of Bacteria

The presence of bacteria in a lesion can often be seen under the microscope as streaming from the cut edge of the lesion mounted in water. For isolation of bacteria small pieces of diseased tissue are placed in a small amount of sterile water in a watch glass or similar container for about 15 min to enable the bacteria to exude into the water. Using a sterile wire loop, a loopful of the resultant dilute suspension is streaked on to one side of a nutrient agar plate. The loop is resterilized and a succession of streaks drawn from the original inoculation around the sides of the plate to achieve single cell separation.

Isolation of soil fungi

A range of techniques has been devised for the isolation of fungi from soil. Many of these are more appropriate to the study of soil microbial ecology than plant pathogens. Soil fungi have been most frequently isolated by modification of the dilution plate techniques referred to in Chapter 37 in which finely ground soil is used instead of spore suspensions. Direct plating of soil is achieved by taking a small soil sample suspension, dispersing it across the bottom of sterile Petri dishes and then adding molten, cooled agar. The particles are distributed throughout the medium by rotating the dish. Suspension techniques will produce more colonies derived from spores than from hyphae and will not give a true representation of the populations of active soil fungi including plant pathogens. Various refinements have been developed to overcome this problem including microscopic examination of soil suspensions to enable direct observation and removal for culture of fungal hyphae. *In situ* isolation from soil can also be done using agar contained within sterile sealed tubes or plates. These agar 'traps' are placed in the soil after small holes have been made in them. Fungi grow into the agar through the holes and can then be removed for further culture.

Selective Isolation

Media

Selective media have been devised for the isolation of many pathogenic fungi. Many rely on the use of various antibiotics or fungicidal substances to suppress bacteria or groups of fungi (Tsao, 1970). Selective isolation may also be required to remove bacterial contaminants from existing cultures or when plating out spore suspension prepared from plant lesions. Chapter 38 lists the more common substances used for these purposes.

Baiting

Plant material can be used as bait to isolate specific pathogens from the environment or from lesions contaminated with secondary invaders and saprophytic organisms. Seedlings, fruit, etc. can be used to detect the presence of particular pathogens in air, water or soil and may be used in epidemiological investigations. Many methods have been tried and published. Other substances can be used for isolating organisms that metabolize particular substrates. Suitable baits include stems, leaves, surface-sterilized roots, fruits, pollen grains, hemp seed, insect wings, hair, snakeskin casts, feathers, etc. For chitinous material it is customary to place the dry baits on top of the soil, which must be kept moist.

Whole hard fruits, such as apples or pears, can be used as bait by inoculating with a small quantity of soil or diseased plant tissue. The fruit is surface sterilized, a flap is cut in the flesh with a sterile scalpel, a small portion of the test material placed underneath and the flap sealed with adhesive tape. When a lesion develops, the fungus can be recovered by excising a portion from the edge of the lesion some distance away from the original inoculation site and cultured on to a suitable medium. This method is useful for isolating *Phytophthora* and *Pythium* spp.

Baiting techniques are particularly appropriate for zoosporic fungi as germinating sporangia produce zoospores that can actively move towards and infect the bait. Thus suitable baits can be applied to soil suspensions or to water in which pieces of disease roots or leaves have been added in order to trap these fungi. Almost pure cultures of *Phytophthora* spp. have been obtained by burying relevant fruits in orchard soils suspected of harbouring these diseases, e.g. citrus fruits in citrus orchards, avocado fruits in soils with avocado root rot, pineapple leaves near parent diseased plants, etc. For general methods of cultivation of zoosporic fungi see Fuller (1978).

With all baiting and isolation techniques, careful monitoring of progress is necessary. Incubation regimes should be as stable as possible, and extremes of temperature, humidity and light should be avoided. Subsequent subculturing should be on to a suitable medium, avoiding nutrient-sugar-rich media which will promote excessive vegetative growth at the expense of sporulation.

Incubation and Subculturing

Incubating dishes upside-down prevents condensation forming on the lid of the dish. Sealing dishes with adhesive PVC (electrical) tape avoids drying-out of the agar medium and can protect from mites. Inoculated agar plates are usually incubated at the temperature most suitable for growth of the pathogen; 25°C is usual and in tropical areas an incubator that provides incubation temperatures below ambient is useful. Incubated plates should be examined daily. Fungi growing from the plated lesion pieces on to TWA will often sporulate especially if exposed to daylight thus enabling them to be identified directly without subculturing. When the fungus has grown 1–2 cm away from the tissue piece, it should be subcultured from hyphal tips on to a suitable nutrient medium, e.g. potato carrot agar or oat agar. These media do not contain large amounts of added sugar and therefore do not promote excessive vegetative growth at the expense of sporulation. Bacterial isolation plates should be incubated in darkness and examined daily. As soon as individual colonies appear, they should be subcultured by transferring a minute piece of the colony on to a sterile loop or needle and streaking on to the surface of a nutrient agar plate.

Spore Trapping

Methods are best developed for trapping airborne spores, but a variety of techniques have been used to trap water-borne spores (Waller, 1972; Fitt, 1983). For a general discussion see Gregory (1973).

Spore traps have been used mainly for investigating the epidemiology of pathogens. Traps in which live plants are used can detect the presence of particular pathogens in the air (Wolfe *et al.*, 1981). They are more usually used to measure either the concentration of spores in a given volume of air or water or the number of spores deposited on a surface. The two are not related as the amount of deposition from a particular concentration especially for airborne fungi depends on meteorological and crop phenology factors. Methods can be further classified according to whether they give an estimate based on counting morphologically identified spores under the microscope, or on counting colonies developing in culture (both methods have severe limitations).

The simplest deposition method is the sticky horizontal microscope slide (or open Petri dish with medium) exposed to the air (under a shelter to keep off rain if out of doors) but is biased in favour of larger spores; smaller types are under-represented and may be overlooked. A vertical sticky slide (or Petri dish) has also been used. A sticky slide inclined upwards at about 45° and mounted in a vane so that it faces the wind has given useful results with cereal rusts. Quantitatively, the number of spores trapped is doubly dependent on wind speed and spore size.

Dependence on external wind speed is a disadvantage that can be over-come by using some powered method to produce a nearly constant wind speed over the catching surface. For some purposes a 'whirling arm' suffices. Here the sticky surface is moved rapidly through the air to increase the impaction efficiency of spores on the surface, and to standardize the volume of air swept per unit of time. The simplest of these devices is the original rotorod sampler (Perkins, 1957), easily made from a battery-driven record-player motor or obtainable commercially. More usually, power is used to draw air by suction through an orifice behind which the spores are collected on a sticky surface or in liquid. The Andersen Sampler (Andersen, 1958) deposits spores on solid media in Petri dishes, fractionated into six size ranges.

For microscopic examination, the Hirst automatic volumetric suction trap is a robust instrument sampling 10 l air min^{-1}, and is widely used for continuous sampling in the field (Hirst, 1952). One pattern (Casella and Co., Regent House, Britannia Walk, London Nl, UK) collects airborne spores on a slowly moving microscope slide, which is changed daily. A later model (Burkhard Manuf. Co., Rickmansworth, Herts., UK) collects on a moving transparent plastic band, which is changed weekly. Both patterns allow changes in spore concentration to be studied hour by hour. Deposits can be mounted permanently for reference in various mountants; glycerine jelly is excellent optically and entirely satisfactory in practice. For sampling larger quantities of air a cyclone dust extractor may be used. Small models are used for collecting rust and smut spores for use as inoculum (Cherry and Peet, 1966).

References and Further Reading

Andersen, A.A. (1958) New sampler for the collection, sizing and enumeration of viable airborne particles. *Journal of Bacteriology* 76, 471–484.

Ausher, R., Ben-Ze'ev, I.S. and Black, R. (1996) The role of plant clinics in plant disease diagnostics and education in developing countries. *Annual Review of Phytopathology* 34, 51–66.

Booth, C. (ed.) (1971) *Methods in Microbiology*, Vol. 4. Academic Press, London, 795 pp.

Chee, K.H. and Newhook, F.J. (1965) Improved methods for use in studies on *Phytophthora cinnamomi* Rands and other *Phytophthora* spp. *New Zealand Journal of Agricultural Research* 8, 88–95.

Cherry, E. and Peet, C.E. (1966) An efficient device for the rapid collection of fungal spores from infected plants. *Phytopathology* 56, 1102–1103.

Duncan, J.M. and Torrance, L. (eds) (1992) *Techniques for the Rapid Detection of Plant Pathogens*. Blackwell Scientific Publications, Oxford, UK, 235 pp.

Fitt, B.D.L. (1983) Evaluation of samplers for splash dispersal of fungal spores. *EPPO Bulletin* 13, 57–61.

Fox, R.T.V. (1993) Isolation of pathogens and their preliminary identification. In: *Principles of Diagnostic Techniques in Plant Pathology*. CAB International, Wallingford, UK, pp. 37–65.

Fuller, M.S. (ed.) (1978) *Lower Fungi in the Laboratory*. University of Georgia, Athens, USA, 213 pp.

Gregory, P.H. (1973) *The Microbiology of the Atmosphere*, 2nd edn. Leonard Hill, London, UK.

Hansen, H.N. and Snyder, W.C. (1947) Gaseous sterilization of biological materials for use as culture media. *Phytopathology* 37, 369–371.

Henson, J.M. and French, R. (1993) The polymerase chain reaction and plant disease diagnosis. *Annual Review of Phytopathology* 31, 81–109.

Hirst, J.M. (1952) An automatic volumetric spore trap. *Annals of Applied Biology* 39, 257–265.

Miller, S.A. and Martin, R.R. (1988) Molecular diagnosis of plant pathogens. *Annual Review of Phytopathology* 26, 409–432.

Narayanaswamy, P. (1997) *Plant Pathogen Detection and Disease Diagnosis*. Marcel Dekker, New York, USA, 331 pp.

Perkins, W.A. (1957) The rotorod sampler. *2nd Semiannual Report, Aerosol Laboratory, Department of Chemistry and Chemical Engineering, Stanford University* No. CML 186, 66 pp.

Schots, A., Dewey, F.M. and Oliver, R. (1994) *Modern Assays for Plant Pathogenic Fungi: Identification, Detection and Quantification*. CAB International, Wallingford, UK.

Singleton, L.L., Mihail, J.D. and Rush, C.M. (1992) *Methods for Research on Soilborne Phytopathogenic Fungi*. American Phytopathological Society, St Paul, Minnesota, 265 pp.

Strouts, R.G. and Winter, T.G. (1994) *Diagnosis of Ill-Health in Trees*. Forestry Commission, HMSO, London, UK, 307 pp.

Tsao, P.P. (1970) Selective media for the isolation of pathogenic fungi. *Annual Review of Phytopathology* 8, 157–186.

Waller, J.M. (1972) Water-borne spore dispersal in coffee berry disease and its relation to control. *Annals of Applied Biology* 71, 1–18.

Waller, J.M., Ritchie, B.J. and Holderness, M. (1998) *Plant Clinic Handbook*. IMI Technical series no. 3. CAB International, Wallingford, UK, 94 pp.

Wolfe, M.S., Slater, S.E. and Minchin, P.N. (1981) Mildew of barley. In: *Annual Report of the UK Cereal Pathogen Virulence Survey 1980*. Plant Breeding Institute, Cambridge, UK, pp. 42–56.

Microscopy

21

J.M. Waller

CABI Bioscience UK Centre, Bakeham Lane, Egham, Surrey TW20 9TY, UK

Light Microscopy

Low-power stereoscopic microscopy enables the characteristics of disease lesions to be determined and delicate dissections to examine internal lesions to be made. Incident light is used for opaque material but transmitted light from a substage illuminator can also reveal structures in translucent material such as embedded perithecia not visible using direct light. Magnifications of up to ×100 enable fungal structures to be removed and mounted on a slide for more detailed microscopy. Fine root tissues are best examined in water over a white surface. Low-power microscopy is also used to examine pathogen cultures for the presence of sporulation, fruit bodies, mites, etc.

The use of a compound microscope with magnifications up to ×1000 is required to enable detection of pathogens inside plant tissue and for identification of pathogenic fungi on plant surfaces and in culture. Removal of small pieces of plant tissue by careful dissection, surface scraping or sectioning is usually carried out after examination under the stereoscopic microscope. Freehand sections using a razor blade require practice; a freezing microtome is easier but for more exacting work, material may need to be dehydrated, embedded and sectioned using appropriate techniques given in microscopy manuals. Woody materials can be soaked and sectioned directly on a microtome. Pieces or sections of specimens can be placed directly on to a glass microscope slide in a drop of water and covered with a coverslip. This simple technique allows a preliminary scan to be made for the presence of some pathogens and is useful for detecting bacterial streaming from lesions caused by bacteria. More usually the specimen is mounted in a high refractive index clearing medium and warmed slightly to remove air bubbles and clear the tissue.

©CAB *International* 2002. *Plant Pathologist's Pocketbook*
(eds J.M. Waller, J.M. Lenné and S.J. Waller)

Lactic acid and glycerol can be used for simple mounts of fungi but do not clear plant tissues well. Lactophenol is a better mounting and clearing medium for diseased plant tissues but is a carcinogen and must be used with care. A simple combined clearing, staining and mounting medium is lactophenol with 0.01% cotton or trypan blue. Pieces of leaf tissue, fine roots, stem sections, etc. can be cleared, stained and examined as a whole mount. These techniques are useful for examining obligate leaf pathogens, delicate leaf surface fungi or structures embedded in tissues. *Phytophthora* or *Pythium* species, or vascular/arbuscular fungal structures can be detected in root tissues using these methods. Plant tissue may be decolorized by boiling and immersion in alcohol or by various bleaching methods. Soaking overnight or warming in 10% KOH solution will soften tissues, which can then be neutralized, cleared, stained and squashed, e.g. on a microscope slide under a no. 3 coverslip. Other staining and clearing methods are given in Chapter 38.

For culture examination and identification, small pieces of mycelium with spores or fruit bodies are removed from the culture and mounted on the slide in a suitable medium with stain if required. The high-power oil immersion objective ($\times$100), which can allow resolution of structures down to 1.2 μm, may be needed to examine suspected bacteria or finer detail of pathogen conidiogenic structures. Oil must be removed from the objective with lens tissue after use.

Examination of delicate fungal structures on leaf surfaces or on the surface of cultures can often be best achieved by application of transparent sticky tape to the surface, removing and mounting face down over a suitable stain before placing under the microscope. This technique is particularly useful for powdery mildews, *Cercospora* species, etc.

Another method of making slide preparations of delicate microfungi on leaf and other surfaces that retain the arrangement of the structures on the surface is to apply a thin cellulose acetate solution such as 'Necol' or dilute nail varnish to the surface structures, allow to dry, peel off, stain and mount for microscopic examination.

Slide cultures can be used for preparing slides of microfungi in culture; this preserves fragile structures (e.g. chains of spores), which would disintegrate if a mount were made in the usual way. This involves placing a small block of agar on a sterile microscope slide and placing a sterile coverslip on it. The block is inoculated with the fungus on each side and the slide culture placed on pieces of glass rod in a Petri dish containing sterile moist filter paper for incubation. When the fungal culture has grown over part of the coverslip and slide, the agar block is removed and the coverslip and slide with adhering mycelium is stained and mounted separately for microscopic examination.

Various optical techniques can be used to provide improved visibility of microscopic structures. Dark-field illumination enables transparent structures to be examined over a dark background while illuminated from below, usually by inserting an opaque screen in the condenser. Phase-contrast microscopy has a somewhat similar function and involves the use of a special condenser with an annular phase plate inserted above. In Nomarski optics, the beam of light is split then recombined to form an image of the object based on two-wave interference that produces coloured differentiation of the object. Fluorescence microscopy is used for some applications that use fluorescent antibodies for staining or other fluorochromic stains; a UV light source is used, which, because of the shorter wavelength, enables finer resolution of microscopic detail.

Care and Adjustments

Optical elements occasionally need cleaning and great care is needed to avoid damaging the optical surfaces that are specially coated. Light dust and grease on the eyepieces and objectives can be removed by gently wiping with a lens tissue. Avoid applying undue pressure on the lenses as this may force dirt across the surface, leaving a scratch. In warm humid climates, mould fungi may grow on lens surfaces and may damage the coating. Keeping microscopes (and other optical equipment) in a simple 'hot box', such as a box containing a low wattage electric light bulb, can avoid this.

An eyepiece micrometer is essential for measuring microscopic objects; most can be inserted into the eyepiece after unscrewing the top lens. Calibration needs to be done with a stage micrometer; by focusing on to the scale on the stage micrometer, the divisions on the eyepiece scale can be aligned and measured against the stage scale. This method can also be used to calculate the 'field diameter' of the microscope at each magnification; this is a useful measurement to have when assessing spore densities or other frequency observations.

Bright-field illumination (Köhler illumination) adjustments

- Place a specimen slide on the microscope stage and focus on the specimen.
- Close down the field diaphragm (above light source).
- Focus on the field diaphragm orifice by using the condenser.
- Centre the field diaphragm orifice if necessary by using the condenser centring adjustment screws.
- Open the field diaphragm until the edge of the orifice just disappears from view.

Phase-contrast adjustment

- Adjust as for Köhler illumination.
- Position the phase objective.
- Replace the eyepiece with a centring telescope. Both the bright and the dark ring should be in focus.
- With the adjustment controls of the phase-contrast condenser move the bright ring until it lies exactly within the dark ring.
- Replace the centring telescope with the eyepiece and resume viewing.

Electron Microscopy

Electron microscopes are of two basic designs and are referred to as transmission electron microscopes (TEMs) or scanning electron microscopes

(SEMs). Essentially the TEM gives results comparable with those obtained with a light microscope (compound type) whereas results obtained with the SEM are comparable with those obtained with a binocular (stereo) microscope.

The 'illumination' source in both types of electron microscope is a beam of electrons produced by a hot filament (e.g. tungsten) usually located at the top of the instrument. The electrons are accelerated down the microscope by a potential difference (accelerating voltage), which exists between the cathode (the filament) and the anode, and are focused into a narrow beam by means of electromagnetic lenses in a manner comparable to the condenser of a light microscope. Because gas molecules would deflect and disperse the beam of electrons it is essential that the interior of the microscope is maintained in a state of high vacuum.

In the TEM the beam of focused electrons passes through a thin section of the material under investigation and further electromagnetic lenses magnify the image formed in the beam, in a manner comparable to the objective and eyepiece of a light microscope. The final image is made visible to the eye by projecting the electrons on to a fluorescent screen. Such images can be recorded directly by use of special photographic film.

In the SEM the beam of focused electrons is scanned across the specimen in a manner similar to that used to form the picture on a television screen. On striking the specimen the beam of electrons causes the emission of secondary electrons. A proportion of the secondary electrons is attracted towards a positively charged grill and detection system that produces an electronic signal of variable strength. After signal amplification and processing the final image is displayed on the screen of a cathode-ray tube. This may be photographed directly.

Material for the TEM is usually chemically treated and embedded in a high-polymer plastic or resin. Such material is then sectioned with an ultra-microtome using glass or diamond knives to produce sections typically 0.05–0.1 μm thick. These are collected on a small circular copper mesh ('grid'), which supports the sections and holds them in the microscope. Contrast in the specimen can be increased by staining the sections with salts of certain heavy metals (e.g. lead), which have a high electron deflecting potential. For rapid detection of virus particles in plant sap, material is ground in a suitable buffer, filtered, mixed with a negative stain and sprayed directly on to grids with an air brush (see Chapter 11).

Material for the SEM is not sectioned as the SEM provides information on surface topography. Specimens up to 2 cm across can be examined. Since many fungal structures have a high water content and are fragile, a special technique must be employed to avoid shrinkage, distortion and collapse when the specimen is exposed to the rapid desiccating effect of the high vacuum inside the SEM. Damage occurs when the large surface tension forces at the liquid–gas interface travel through the material during desiccation. Two techniques are available: freeze-drying and critical point drying.

Freeze-drying is accomplished by putting specimens in a chamber, which is evacuated; the water in the specimen gradually sublimes off and condenses on the colder refrigeration coils or is removed by a chemical drying agent (e.g. phosphorus pentoxide).

Critical point drying, the more widely used technique, involves the use of a high-pressure vessel with heating facilities. Specimens, dehydrated through an acetone series, are placed in the pressure vessel and the transitional liquid (e.g. liquid CO_2) is infiltrated to replace the acetone. After removal of all the acetone the vessel is isolated and the temperature is raised until the critical point is reached ($31.3°C$ at 73×105 N m^{-2} for CO_2). At the critical point no liquid phase can remain, regardless of pressure. A single phase is present above the critical point and the transitional liquid, now in a gaseous form, is bled off from the pressure vessel. The relatively low critical temperature of CO_2 usually causes no damage to the specimens during processing.

To obtain satisfactory results when examining material with the SEM, specimens must be coated with a thin film of an electrical conductor (e.g. carbon, gold, etc.) to reduce charging and increase electron emission. By far the most efficient method of coating specimens is by use of a sputter coater. Sputtering is accomplished by an electrical discharge passing between two electrodes under low pressure (13.3 N m^{-2}). The basic apparatus consists of a vacuum chamber containing an anode, to which the specimens can be attached, and immediately above this, a circular, usually metal (e.g. gold) cathode. Gold atoms, freed from the cathode by bombardment with ionized argon atoms, undergo successive multiple collisions with the argon and penetrate all irregularities on the surface of the specimen to produce a complete coat of known thickness. Heating of the specimen during coating can be prevented by cooling the specimen holder (e.g. by using a thermoelectric device).

Using modern TEMs, resolution of 1 nm or less with effective magnifications of up to ×200,000 or more can easily be obtained from well-prepared specimens. In the case of modern SEMs resolution of 20 nm or less with effective magnifications of up to ×40,000 or more are possible. This compares with a resolution of 200 nm obtained from good-quality light microscopes. An additional advantage of the SEM is the greater depth of focus obtained, which is usually better than the light microscope by a factor of at least 250 at similar magnifications.

Select Bibliography

Bradbury, S. (1976) *The Optical Microscope in Biology*. Edward Arnold, London, UK.

Fox, R.T.V. and Waller, J.M. (1993) Using microscopy. In: Fox, R.T.V. (ed.) *Principles of Diagnostic Techniques in Plant Pathology*. CAB International, Wallingford, UK, pp. 87–128.

Kinden, D.A. and Brown, M.F. (1975) Techniques for scanning electron microscopy of fungal structures within plant cells. *Phytopathology* 65, 75–76.

Lung, B. (1974) The preparation of small, particulate specimens by critical point drying applications for scanning electron microscopy. *Journal of Microscopy* 101, 77–80.

Mendgen, K. and Lesemann, D.E. (1991) *Electron Microscopy of Plant Pathogens*. Springer-Verlag, Berlin, Germany, 336 pp.

Parsons, E., Bole, B., Hall, D.J. and Thomas, W.D.E. (1974) A comparative survey of techniques for preparing plant surfaces for the scanning electron microscope. *Journal of Microscopy* 101, 59–75.

Rawlins, D.J. (1992) *Light Microscopy*. BIOS Scientific Publishers, Oxford, UK.

Samson, R.A., Stalpers, J.A. and Verkerde, W. (1979) A simplified technique to prepare fungal specimens for scanning electron microscopy. *Cytobios* 24, 7–11.

Immunological Techniques

22

F.M. Dewey

Plant Sciences Department, University of Oxford,
South Parks Road, Oxford OX1 3RB, UK

Immunoassays are used routinely for the detection, 'visualization' and quantification of a number of plant pathogens, particularly viruses. Development of such assays for the detection of complex organisms, such as fungi, has been slow but is now possible. The techniques utilize the exquisite specificity and sensitivity of the mammalian and avian immune systems. When a foreign organism, protein, glycoprotein, lipopolysaccharide or complex carbohydrate is injected into an animal, the animal's immune system responds by making antibodies (Abs) that specifically recognize and bind to specific sites on the foreign molecule, known as epitopes (Harlow and Lane, 1998). The substances or organisms used to induce antibodies are commonly referred to as immunogens and the molecules recognized by the antibodies as antigens. Small molecules, such as mycotoxins, are not immunogenic by themselves, but, if linked covalently to a protein such as keyhole limpet haemocyanin, they will induce antibodies that will bind directly to the small molecule.

Antibodies, which are secreted by lymphocytes that circulate in the blood, are present in the cell-free component of the blood known as the antiserum. When an animal is injected with a purified antigen, a number of antibodies recognizing different binding sites on the same molecule are produced. Some of the binding sites or epitopes may be the same as those that are found on similar or unrelated molecules, in which case the antiserum is said to be cross-reactive. If, however, all the epitopes recognized by the antibodies in the antiserum recognize only the immunogen then the antiserum is said to be monospecific and it is this type of antiserum that is most useful for diagnostic purposes. Viruses, because of their molecular simplicity, generally induce a highly specific response but immunization with more complex organisms, such as fungi, or extracts from these organisms will induce a number of antibodies that are cross-reactive, recognizing both related and unrelated species and host molecules. To obtain an antiserum for complex organisms that is useful diagnostically, i.e. taxonomically specific, it is necessary first to identify and purify specific molecules but, even then, there is no guarantee that such an immunogen will result in an antiserum that is specific, because some of the epitopes on the antigen may also occur on totally unrelated molecules in different organisms. However, the advent of hybridoma technology developed by Kohler and Milstein (1975) has enabled the production and selection of unlimited quantities of immortalized lymphocytes that secrete specific antibodies all with the same specificity, known as monoclonal antibodies (MAbs). Hybridoma cell lines can be raised and selected that secrete MAbs with the required level of taxonomic or chemical specificities. Using specific MAbs, immunoassays have been developed that can discriminate between closely related viruses, fungi, bacteria, nematodes and intractable obligate pathogens such as mycoplasmas.

Antibodies are large glycoprotein molecules known as immunoglobulins (Igs). There are five main classes of antibodies designated IgA, IgD, IgE, IgG and IgM; the class IgG is split into a number of subclasses, the main ones being IgG_1, IgG_{2a}, IgG_{2b} and IgG_3. All antibodies have the same basic structure but differ considerably in size and properties. The basic unit consists of two long or heavy chains (H) and two short or light chains (L) joined by covalent bonds (Fig. 22.1). Each unit has two binding sites composed of amino acids from the ends of both the heavy and light chains. The region of the binding site is known as the variable region because rearrangement of the amino acid sequences composing this region can result in up to 10^{10} potentially different binding sites. The sequence of the amino acids in the Fc region of the antibody is constant and specific to each animal, i.e. the Fc region of antibodies raised in mice (murine antibodies) differs from the Fc region of antibodies raised in rats and horses. The carbohydrate side chains attached to the amino acids in this region form a convenient means of linking reporter enzymes such as

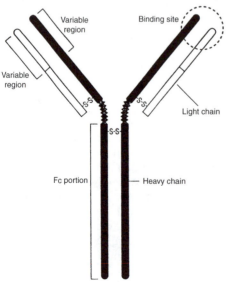

Fig. 22.1. Structure of an antibody molecule.

horseradish peroxidase and alkaline phosphatase and fluorescent conjugates to the antibody molecule. Cleavage by the proteolytic enzymes pepsin or papain just above the Fc region yields either F(ab)2 or Fab and Fc fragments.

When an animal is first injected with a foreign molecule, IgM antibodies are induced. These are large antibodies with five basic units joined by a J-chain. On the second and subsequent exposure to the same immunogen, the immune response is faster and IgM antibodies are generally replaced by IgG antibodies, unless the immunogen is a complex carbohydrate or a glycoprotein, in which case only IgM antibodies continue to be produced. Purification and conjugation of reporter molecules to IgM antibodies is more difficult than with IgG antibodies. In general there is little need for individual plant pathologists to raise their own antisera or monoclonal antibodies because many are now available commercially, as are kits for the detection of economically important pathogens (e.g. Adgen Ltd, Auchincruive, Scotland).

Immunization of Animals

In most countries, the immunization of animals is tightly controlled by legislation. In the UK, both a Home Office project licence and a personal animal experimentation licence are required. Rabbits and rats are commonly used for the small-scale production of antisera, and sheep and goats for large-scale production. Splenocytes from immunized mice and rats are used for the production of monoclonal antibodies. For details of injection procedures, dosage and methods of obtaining test and final bleeds see Harlow and Lane (1998). Most animals require three or more injections at 2–3-week intervals before a good response is obtained. Antisera can be obtained by leaving blood from an immunized animal overnight at 4°C, during which time it will clot and separate. The upper, relatively clear liquid, the antiserum, can then be removed easily by suction. If necessary, low-speed centrifugation can be applied to facilitate the separation.

Purification of Antibodies

IgG antibodies can be purified from antiserum by passing them down a protein A or protein G–Sepharose column to which they will readily bind. The bound antibodies can be eluted with a Tris-glycine buffer of low pH. A number of commercial kits are now available for the purification of IgG antibodies. Purification of IgM antibodies is more tedious but can be done by ammonium sulphate precipitation or loading on to a snowdrop lectin Sepharose column and eluting with a Tris-glycine buffer.

Production of Monoclonal Antibodies

For details of hybridoma technology see Harlow and Lane (1998). The process involves fusing splenocytes (the best source of lymphocytes) from immunized mice or rats, at random, with myeloma cells from a specifically engineered non-antibody-producing lymphatic cancer cell line, using polyethylene glycol. The fusion mixture is diluted extensively into a selective medium that only allows the growth of lymphocytes that have fused with myeloma cells. This mixture of cells is plated out into ten 96-well microtitre plates and incubated in an atmosphere of 5% CO_2 at 37°C. From 7 days onwards, supernatants from those wells that contain colonies of hybridoma cells are screened by immunoassay for the production of antibodies. Supernatants from those cell lines producing antibodies that recognize the target antigen or organism are then further screened against related and unrelated non-target antigens or organisms. Potentially useful cell lines are re-cloned, grown in bulk, preserved by freezing slowly in fetal bovine serum plus dimethylsulphoxide (DMSO) and stored in liquid nitrogen. The MAbs present in hybridoma supernatants can be used directly or purified and concentrated in much the same way as those from antisera.

Molecular methods are now available for the production *in vitro* of recombinant antibodies. The most successful methods to date appear to be those using phage display libraries of antibody genes expressing SCvF fragments. Selection methods involve panning blots of phage plate cultures with the target antigen (Torrance, 1995). The advantages of *in vitro* methods are that they eliminate the need for animals and enable antibodies to be raised to molecules toxic to animals, such as pesticides.

Immunoassays

There are several immunoassay formats, all of which depend on the 'visualization', either directly or indirectly, of the binding of a specific antibody to its respective (homologous) antigen. Earlier methods included immunoprecipitation of the antibody/antigen complex in narrow tubes or diffusion through agar, as in Ouchterlony assays. In both, a visible line can be seen at the point where the antibody meets and forms an insoluble complex with its respective antigen. Electrophoresis of antigen mixtures, followed by immunodiffusion of the antibodies, was also used. Agglutination of microorganisms on exposure to their respective antibody is still used as a means of detecting some bacteria. Agglutination of antigens with their respective antibodies adsorbed on to protein A on the surface of *Staphylococcus aureus* or covalently bound to protein A-coated Sepharose beads is also used for diagnostic purposes. However, most of these assays have now been superseded by ELISAs.

Enzyme-linked immunosorbent assays

ELISAs are highly sensitive assays that can easily be replicated, automated and quantified. The only disadvantage is that they require laboratory facilities. Almost all the assays are conducted in polystyrene microtitre wells to which either the antigen or antibody is passively adsorbed. The same kinds of assays using larger volumes (0.5 ml) and shorter incubation times can now be done in specialized plastic tubes. There are three main types of ELISA format: the plate-trapped antigen-ELISA (PTA-ELISA), the double antibody sandwich-ELISA (DAS-ELISA); and competition-ELISA (C-ELISA). In PTA-ELISAs the soluble antigen or target microorganism is incubated in microtitre wells for a few hours at room temperature or overnight at 4°C, in an appropriate buffer, such as phosphate-buffered saline (PBS) or bicarbonate buffer. Subsequently, the antigen solution is washed out using a saline buffer, and the unbound sites are blocked by incubating with a solution of non-fat dried milk, pure casein or bovine serum albumin. Wells are then separately incubated with the specific antibody solution, secondary antibody–enzyme conjugate and the chromogenic enzyme substrate, with saline washes containing Tween 20 between each incubation step. The conversion of the substrate by the enzyme to a coloured product is finally stopped and the absorbance values read using an ELISA plate reader with a filter of appropriate wavelength. In direct assays, the enzyme is conjugated directly to the primary antibody. In indirect assays, a second antibody, generally a commercial antibody, raised in another animal species, that recognizes and binds to the primary antibody is used. Alkaline phosphatase and horseradish peroxidase are commonly used as reporter enzymes.

In DAS-ELISAs, the antigen is captured between two antibodies by incubating test samples in microtitre wells precoated with one antibody and then incubating with a second antibody that may or may not be directly conjugated to a reporter enzyme. The two antibodies forming the sandwich are commonly different, one a monoclonal antibody and the other a polyclonal antibody. The only time that the same antibody can be used successfully to capture and detect an antigen is when the antigen bears repeat epitopes. Indirect DAS-ELISAs are sometimes called TAS-ELISAs because a third antibody–enzyme conjugate is used as a reporter enzyme.

C-ELISAs are commonly used for the detection of small molecules such as pesticides and mycotoxins and for assays requiring high levels of sensitivity. In these assays, test samples are preincubated with a very dilute solution of the specific antibody before exposure to antigen-coated wells. If the test sample contains the suspect antigen, it will block all the antibody binding sites and prevent any antibody binding to the antigen on the precoated wells. However, if the test sample does not contain the homológous antigen, it will not block the antibody binding sites and, therefore, the antibody will be free to bind to the antigen on the coated wells. Thus the presence of colour at the end of a C-ELISA indicates a negative result and absence of colour is a positive result. In these assays, the amount of antigen needed to inhibit 50% binding of the antibody is usually determined and denoted as the I_{50} value.

Estimation of biomass by any type of ELISA requires comparison with a dilution series of standards. The calibration must be repeated each time that a sample is tested by ELISA because of variability in ELISA response curves not only from day to day but, to a lesser extent, from plate to plate.

The extraction of antigens from plants and soil particles is often difficult. In many cases, simple extraction buffers such as tris-buffered saline (pH 8.2), bicarbonate (pH 9.6), acetate (pH 4.5) and, more commonly, PBS are inadequate. Addition of surfactants such as Tween 20 or Nonidet P40 and polyvinyl pyrolidone (PVP), which reduces interference from phenols, can improve extraction, but the presence of detergents in extracts limits their use to DAS-ELISA formats because detergents inhibit the binding of antigens to microtitre wells. Where extraction of the antigen is difficult, biological amplification of antigens is often used to boost the signal.

Membrane assays

The mechanism of dot-blot, dip-stick or squash-blot assay is essentially the same as that in ELISAs in that the membrane provides a surface to which antigens or antibodies can be passively adsorbed. Some test systems use nitrocellulose membranes or nitrocellulose-coated plastic tags or cards, and others use polyvinylidene difluoride (PVDF, Immobilon P, Millipore). The reporter molecule is generally an enzyme conjugate, commonly alkaline phosphatase, but gold conjugates, which can be silver enhanced, have been used by a number of workers.

More recently an altogether different type of membrane assay has been developed. It is an immunochromatographic assay, sometimes known as a lateral flow assay or device, that can be developed in 3–5 min (Fig. 22.2; Danks and Barker, 2000). Such devices are now being used to confirm the presence of *potato viruses* X and Y in symptomatic leaves and the presence of fungal antigens in grape juice. The devices consist of pathogen-specific antibodies, bound either to coloured latex beads or gold particles preimmobilized on a membrane. Addition of a test sample, such as plant extracts diluted in a carrier buffer, to the preimmobilized membrane allows the tagged antibodies to move, by capillary action, through the membrane. If the homologous antigen is present in the test solution it will form a complex with the tagged antibodies and be carried by them through the membrane. Movement of this complex along the length of the membrane is arrested by a preprinted line of either the same pathogen-specific antibody or another antibody recognizing a specific antigen or epitope (the test line), and as the complexes become arrested the line then becomes visible to the naked eye because of the accumulation of the latex beads or gold particles to which the antibodies are bound. Any tagged antibodies that do not form a complex with the antigen will also move up the membrane and

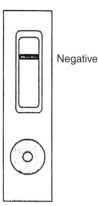 Negative

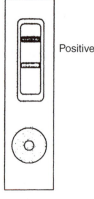 Positive

Fig. 22.2.   Lateral flow device.

are trapped by a second line of antibodies that are antispecies antibodies (the control line). Thus, the appearance of two lines, i.e. both test and control lines, indicates a positive result, i.e. antigens of the pathogen are present in the test sample. The appearance of only one line, the control line, indicates that the assay has run correctly but that the sample is negative, i.e. it does not contain any pathogen-specific antigens. Alternatively, competition assays can be employed in which case one line would be positive and two lines negative.

Magnetic bead assays

Magnetic beads provide a promising alternative solid support system for capturing viruses, bacteria or soluble antigens from larger pathogens. Immunoassays that use small beads are ideally suited to efficiently extracting antigens from samples that contain particulate debris such as soils.

Immunolabelling Techniques

Immunolabelling techniques are most useful as research tools for visualization of pathogens *in vitro* or *in situ* but in general such techniques, because they involve microscopy, are not used for diagnosis and quantification of pathogens. Tissues are immunolabelled by allowing the pathogen-specific antibody to bind directly to the organism either *in situ*, such as in sections of infected plant material, or *in vitro*, e.g. with bacterial smears on glass slides or fungal spores that have been allowed to germinate on a glass slide. Bound antibodies are detected by using a second antibody, an antispecies antibody, that is conjugated either to a fluorescent compound such as fluorescein or rhodamine, a gold particle or an enzyme such as peroxidase or phosphatase, followed by an appropriate chromogenic substrate.

Visualization of fluorescent compounds necessitates the use of a UV microscope. When such molecules are excited by exposure to UV light they fluoresce, emitting light at a different wavelength. Background fluorescence or autofluorescence from soil particles or plant materials, particularly cellulose and lignin, can be partially eliminated with a conventional epifluorescent microscope by using narrow band filters or, more effectively, by using a confocal microscope. Blocking with a non-specific conjugate such as rhodamine/gelatin is also effective in reducing autofluorescence. Double labelling of antigens can be achieved by using specific antibodies raised in different animal species followed by the respective second antispecies antibodies conjugated to different fluorochromes. Fixing of tissue with 3% paraformaldehyde or 70% alcohol before labelling helps the retention of soluble antigens.

The methods used in immunolabelling tissues with gold conjugates are the same as those used for immunofluorescence labelling except that the pathogen-specific antibodies are located using a secondary antibody–gold conjugate. Conjugates with different sized gold particles are commercially available. For light microscopy larger particles are required than for electron microscopy. Amplification of trace levels of antigen can be achieved by silver enhancement of bound gold particles. Viewing of bound immunolabelled gold particles at the light level can also be enhanced by using polarized light. Some antigens are very sensitive to stringent fixation techniques commonly used in electron microscopy such as osmium tetroxide; more gentle fixation techniques such as a combination of paraformaldehyde and gluteraldehyde are recommended. Filtering of all antibody and wash solutions through a microbial membrane and incorporation of a blocker are important when working at the electron microscopic level.

References and Further Reading

Danks, C. and Barker, I. (2000) The on-site detection of plant pathogens using lateral flow devices. *EPPO Conference on Diagnostic Techniques for Plant Pests*, Wageningen, The Netherlands, February 2000.

Harlow, E. and Lane, D. (1998) *Antibodies: a Laboratory Manual.* Cold Spring Harbor, Laboratory Press, Cold Spring Harbor, New York, USA, 726 pp.

Johnstone, A. and Thorpe, R. (1996) *Immunocytochemistry in Practice*, 3rd edn. Blackwell Science, Cambridge, UK, 362 pp.

Kohler, G. and Milstein, C. (1975) Continuous culture of fused cells secreting antibodies of predefined specificity. *Nature* 256, 495–497.

Marshall, G. (1996) *Diagnostics in Crop Production*. BCPC Symposium Proceedings No. 65, Association of Applied Biology, Surrey, UK, 395 pp.

Skerrit, J. and Appels, R. (eds) (1995) *New Diagnostics in Crop Sciences, Biotechnology in Agriculture* No. 13. CAB International, Wallingford, UK, 338 pp.

Torrance, I. (1995) Use of monoclonal antibodies in plant pathology. *European Journal of Plant Pathology* 101, 351–363.

Biochemical and Molecular Techniques

P. Bridge

Mycology Section, Royal Botanic Gardens, Kew, Richmond, Surrey TW9 3AE, UK and School of Biological and Chemical Sciences, Birkbeck, University of London, Malet St, London WC1E 7HX, UK

Historically, laboratory techniques for the classification and identification of plant pathogenic microorganisms have differed considerably, depending on whether the pathogen was a bacterium or a fungus. Bacterial identification has for the most part been based on combinations of tests for physiological and biochemical attributes (Fox, 1993), whereas fungal identifications have generally been made entirely on the basis of morphological features such as spore type, shape and colour (Hawksworth, 1974). There have been considerable advances in the understanding of microbial biochemistry and molecular biology in the last 30+ years. These advances, together with the increased availability and simplification of laboratory techniques, have led to a rapid expansion of diagnostic methods and identification procedures. This expansion in methods has been particularly significant for the identification of closely related pathotypes and in the delineation of pathogen populations (Bridge *et al.*, 1993; Schaad *et al.*, 1994; Brygoo *et al.*, 1998).

Physiological and Biochemical Tests

The use of physiological and biochemical characters to differentiate between microorganisms was first established in the 1930s. Since that time such tests have become one of the primary tools for the systematics of bacteria (Skerman, 1969; Barrow and Feltham, 1993). Among the eukaryotic microbes similar techniques are used for taxonomic studies with yeasts (Barnett *et al.*, 1990), but have only been used occasionally for filamentous fungi (Abe, 1956; Frisvad, 1981). Filamentous fungi in general show greater phenotypic plasticity than bacteria, and physiological and biochemical characteristics are less defined between species and genera.

Physiological and biochemical tests can broadly be divided into four types: resistance/tolerance, assimilation/enzyme production, fermentation and metabolite production. Characters derived from the first three categories are reflections of the organism's abilities to grow and multiply in the environment, and it is often the case that organisms found in specialized environments will have specialized physiological or biochemical properties. Examples of this are the pectinases produced by many plant pathogenic organisms (Mateos *et al.*, 1992) or the specialized proteases produced by entomopathogenic fungi (St Leger, 1995). As physiological and biochemical properties will vary widely between different groups of organisms there is no single standard set of tests for all bacteria, with different test sets used for isolates of different genera. Standard test sets can, however, be used for some groups of genera, and these are utilized in a number of commercially available kits for bacterial and yeast identification. Many of these are designed for specific bacterial groups, e.g. Gram-negative enteric organisms (Holmes *et al.*, 1978), although some kits based on assimilation tests have been produced for all Gram-positive or all Gram-negative bacteria (Biolog GN kit (Biolog Inc., USA)). Similar kits have been designed for yeast identification and can give good identifications for clinical yeast isolates (API 32C (BioMérieux, 1993)), although slower-growing environmental isolates may be more problematic. As mentioned before these types of methods have rarely been used with filamentous fungi, although some studies have given promising results (Bridge and Hawksworth, 1985; Manczinger and Polner, 1985).

The production of metabolites differs from other physiological and bio-chemical tests in that their environmental or ecological significance is not often apparent. The production of particular metabolites can be of significance in the systematics of filamentous fungi, particularly as some metabolites are only produced by a very narrow range of fungal species (e.g. aflatoxin, ochratoxin). Classification schemes have been proposed for a number of fungal groups based largely on the particular metabolites produced (Frisvad and Filtenborg, 1990), but although many hundred fungal metabolites have been described, there is very little information available for many genera, including most of the important plant pathogens.

Methods

The detection of physiological and biochemical activities is largely undertaken on either solid or liquid growth media, with positive reactions being the appearance of growth or a colour change. Such tests involve the inclusion of appropriate substrates within the media, together with a system that allows the detection of either their degradation or one of the products associated with it. For assimilation tests, results are frequently measured as the growth obtained, or as a pH change. Examples of such plate-based tests include casein hydroly-sis, where positive activity is seen as a zone of clearing where the casein has been degraded, or aesculin hydrolysis where a black colour is formed from the reaction between iron in the medium and the hydrolysis product aesculetin

(Rutherford *et al.*, 1993). One other approach to detecting enzymatic activity is to use specific substrates conjugated to easily detectable compounds, such as nitrophenols. One example of this is the *p*-nitrophenyl-β-D-galactopyranoside (PNPG) test for β-galactosidase, where the substrate is PNPG. β-galactosidase activity removes the galactopyranoside leaving *p*-nitrophenol which is coloured. Other applications of this approach have been the inclusion of naphthyl-substituted compounds where the liberated naphthyl groups are detected with one of the Fast stain reagents, or the use of methylumbelliferone, where the free methylumbelliferone is fluorescent (Barth and Bridge, 1989; Goodfellow and James, 1994). Tests based on the use of substituted substrates to give a colour reaction can be miniaturized, and this methodology has been used to develop commercially available kits such as the API ZYM system produced by BioMérieux (Humble *et al.*, 1977). These kits have been used in limited studies for bacterial and fungal identification (Tharagonnet *et al.*, 1977; Bridge and Hawksworth, 1985).

Metabolite detection is of most importance with filamentous fungi, and the detection of metabolites can be undertaken by a variety of methods. Although a small number of metabolites can be detected through colour changes with specific reagents (e.g. metabolite E from *Phoma exigua* (Logan and Khan, 1969)), the most widely used detection methods are chromatographic. Metabolites are most easily separated and detected by thin-layer chromatography (TLC) and high-performance liquid chromatography (HPLC) (Paterson and Bridge, 1994; Frisvad *et al.*, 1998). Fungal metabolites can be broadly described as extracellular or intracellular. Extracellular metabolites can frequently be detected in chromatographic systems directly from agar plugs or culture broths, whereas intracellular metabolites require some extraction procedure prior to analysis (Paterson and Bridge, 1994). Serological methods have been designed that incorporate antibodies to specific metabolites and a number of commercial kits are available for the detection of individual compounds.

Chemotaxonomy

Chemotaxonomy involves the study of the chemical constituents of the cell and cell walls and membranes. Most chemosystematic work has been undertaken with one or more of three classes of compounds: proteins and amino acids, lipids and fatty acids, and carbohydrates (see Priest and Austin, 1993). In addition, some approaches have involved the comparison of the total chemical constituents of cells, such as through Curie point pyrolysis and mass spectrometry (Gutteridge *et al.*, 1980). The introduction of chemotaxonomy was a significant advance in bacterial systematics, and allowed the resolution of many taxonomic problems (Priest and Austin, 1993). Chemotaxonomy has been less used in diagnostics, although protein and fatty acid profiles have been used for the analysis of plant pathogenic bacteria (Stead, 1995). A semiautomated system for fatty acid analysis has been used to identify bacterial plant pathogens at both the specific and subspecific levels (Persson and Sletten, 1995). The first widespread use of strictly chemotaxonomic methods was the use of protein electrophoresis, which became established during the 1960s (Lund, 1965), and techniques related to cell wall and membrane analysis such as lipid and fatty acid analysis were developed in the 1970s (Minikin and Goodfellow, 1980; Collins, 1985). Chemosystematic methods are dependent upon the biochemical pathways which are expressed at the time of analysis under the growth conditions used. As a result some chemosystematic methods can be extremely sensitive to the growth media used, time of growth and growth temperature.

Proteins

The first applications of chemosystematics were the analysis of total protein profiles generated by separating whole cell protein extracts by electrophoresis. These patterns are very complex and analysis by eye can generally only detect gross differences. Subsequent developments in densitometry and computer recognition of patterns were necessary before these patterns could be used extensively in systematics and diagnosis. Several software packages are now available for analysing electrophoresis gel patterns. Simpler patterns can be obtained from the same whole cell extracts by staining the subsequent gel for specific enzymic activities. Under the appropriate conditions multilocus enzyme systems may show between one and 15–20 bands. These bands are in most cases isoforms of the enzymes and the term isoenzyme electrophoresis has been used in these cases. Both of these electrophoretic methods have been used to a wide extent in bacteriology; like many chemotaxonomic approaches they can be considerably influenced by the growth media used and to an extent by the time of incubation used to obtain the initial samples. In fungi the situation is further complicated by the presence of differentiated and differentiating tissues. Protein electrophoresis is obviously dependent on the proteins expressed at a given time, and in fungal cultures, different proteins can be expressed by spores, mycelia and fruiting bodies. Therefore, different proteins will be expressed at different times in a single growth period. Direct electrophoresis of proteins has, therefore, only been used in isolated cases with filamentous fungi and generally from homogenous samples such as spore suspensions. Isoenzyme electrophoresis has been used for defining groups within both bacteria and fungi, and in recent years has been more widely used in mycology than bacteriology. A wide variety of enzymes have proved useful for differentiating at both the specific and subspecific level in both bacteria and fungi; these include esterases, phosphatases and dehydrogenases (Williams and Shah, 1980; Bonde et al., 1986). Extracellular enzymes can also be separated by electrophoresis, and these have been used extensively in fungal systematics. This technique involves growing the fungus in a suitable liquid medium in order to induce the enzymes, and then screening culture fluids directly. The most widely used system of relevence to plant pathology is the characterization of the pectinase enzyme complex (Cruickshank and Wade, 1980). In this procedure isolates are grown in a broth containing pectin, and culture fluids are subsequently loaded on to electrophoresis gels containing pectin. After electrophoresis, gels are incubated in buffer to allow the separated enzymes to react with the pectin, and the pectin in the gel is then stained in ruthenium red. Areas of gel corresponding to pectin-degrading enzymes, such as polygalacturonidases, appear as clear zones in the red gel, and areas corresponding to pectin-modifying enzymes, such pectin lyase and esterases, appear as yellow or dark red zones. In this system isoforms of single enzymes may be detected, together with further enzymes.

Amino acid sequences have been considered in the taxonomy of bacteria, particularly in relation to the amino acid chains that form the backbone of the cell wall peptidoglycan. The peptidoglycan type is very uniform among the Gram-negative bacteria, but amino acid chains and the associated sugars vary among the Gram-positive organisms. This technique has been particularly associated with the coryneform taxa and actinomycetes (Keddie and Cure, 1978).

Lipids

There are many lipids associated with the cell walls and membranes of bacteria and other organisms that can provide significant taxonomic information. The most widely used of these are the isoprenoid quinones and long-chain fatty acids. Isoprenoid quinones are components of the electron transport pathway and are present in all organisms. In microbes, two classes are found, the menaquinones and the ubiquinones (Priest and Austin, 1993). Quinones are extracted from cells with organic solvents and then detected by either HPLC or TLC. The major quinone types present in organisms have been widely surveyed for bacteria, yeasts and fungi. In bacteria there are generally only one or two predominant quinones, and these have been used to determine generic and specific separations (Minikin and Goodfellow, 1980; Collins, 1985). Quinones appear to be more variable in the eukaryotic organisms (Kuraishi *et al.*, 1985); they have been considered extensively in yeasts and they have proved particularly useful in helping to define generic concepts in the heterobasidiomycetes (Boekhout, 1991). The value of quinone types is less clear with filamentous fungi, where there can be a considerable number of quinones present in significant amounts (Paterson and Buddie, 1991).

The long-chain fatty acid components of microbial walls and membranes can be extracted as their methyl esters and separated by gas–liquid chromatography (GLC). Fatty acid types, in terms of chain length, hydroxylation and position and number of double bonds, can be used to generate profiles for bacteria that can in many instances be species or subspecies specific (see earlier). These profiles can therefore be used in routine bacterial identification. The use of automated GLC analysis, combined with computer-based comparison of traces, can result in a system particularly suited for diagnostic work with large numbers of organisms. Fatty acid profiles have also been considered with yeasts and filamentous fungi (Kock and van der Walt, 1986; Kock and Botha, 1998). In these cases the situation is less clear cut as in general terms the eukaryotic organisms show a reduced range of fatty acids. The smaller number of components obtained for eukaryotes has meant that in yeasts, basic profiles may not provide sufficient discriminatory information and further characters have been obtained by considering the ratios of sets of components. Species- and genus-specific profiles can be obtained for some yeasts, and the technique has been used to detect infraspecific groups in filamentous organisms (Johnk and Jones, 1994).

Two other lipid classes that have been used in bacterial taxonomy are the free polar lipids including phospho- and glycolipids and mycolic acids. Polar lipids can be of use at the family and genus level, whereas mycolic acids are restricted to some Gram-positive taxa (Komagata and Suzuki, 1987).

Vegetative Compatibility

One technique that has proved to be useful in differentiating closely related populations of plant pathogenic fungi is vegetative compatibility testing. In this method different strains of a pathogen are 'crossed' to determine if their hyphae can fuse and allow the expression of the nuclei from both strains. This is achieved by selecting mutants of the original strains that are defective in one part of their nitrate utilization pathway. When strains with different (complementary) mutations fuse, the resulting hybrid can express the complete pathway and so compatibility can be assessed by the ability to grow on nitrate. The underlying assumption is that for this to occur, the different strains must be genetically very similar. Strains can then be categorized into groups, each of which consists of vegetatively compatible strains.

Vegetative compatibility groups (VCGs) have been used as epidemiological markers in some fungi, particularly in the special pathogenic forms of *Fusarium oxysporum*. The true significance of these groups, and their relationship to other pathologically defined groups such as races, is uncertain. In *F. oxysporum* f.sp. *cubense* multiple VCGs may occur within a single race, whereas in other groups a single VCG may contain isolates of different pathological races (Bridge *et al.*, 1993; Rutherford *et al.*, 1995). A further complication to interpreting VCG data is that a number of 'bridging' strains have been found that are compatible with more than one VCG (Lodwig *et al.*, 1999).

Molecular Techniques

DNA extraction

In order to examine the molecular composition of an organism it is first necessary to extract the DNA from the cell. The extraction procedure necessary is dependent on the organism being studied and the quantity and purity of DNA required. For DNA probe-based methods or restriction enzyme-based techniques it is necessary to culture the organism, disrupt the cells and extract and purify the DNA. DNA extraction methods have been considerably simplified since the early protocols of the 1960s. Significant (10 μg+) quantities can be obtained from microorganisms in procedures that take 2 h or less (Cenis, 1992). DNA extraction procedures for bacteria and fungi are very similar and are generally based on a number of precipitation reactions from either phenol and/or chloroform extractions, the use of carrier compounds such as cetyl-trimethyl ammonium bromide (CTAB or cetrimide) or hydroxyapatite, or combinations of both. Additional steps such as the inclusion of RNase, β-mercaptoethanol or polyvinyl polypyrrolidine may be included, depending on the original DNA sample and the intended use of the final product. A significant feature of any DNA extraction is the initial disruption of the cells, and this will be very dependent on the organism under study. Simple lysis techniques that are suitable for Gram-negative bacteria (Sambrook *et al.*, 1989) will be less effective or inappropriate for Gram-positive organisms, whereas for filamentous fungi some form of physical disruption will generally be required (Paterson and Bridge, 1994). Physical disruption such as grinding or shaking with glass beads can result in considerable heat being generated locally and so the sample should be either cooled or dehydrated. The most common methods used for filamentous fungi are either to grind liquid nitrogen-frozen material or to use lyophilized material.

Restriction digestion

Restriction endonucleases are naturally occurring enzymes that recognize sites consisting of specific sequences of bases. Double-stranded DNA will be split at these sites when the enzymes are present and the salt and temperature conditions are optimal. Different endonucleases recognize different sequences and these in general range from about 4 to 12 base pairs. The specific activities of restriction enzymes have proved to be fundamental to many aspects of molecular biology. In molecular systematics restriction enzymes have most commonly been used to provide defined fragments of DNA, and differences in fragment size and number have given rise to a range of techniques defined as restriction fragment length polymorphism (RFLP) analysis.

Direct RFLPs

Direct digestion of DNA with restriction enzymes results in the DNA being digested into X fragments, where X is the number of restriction sites +1. DNA samples that contain a small number of sites will therefore produce a small number of fragments and these can be compared directly by electrophoresis. This direct approach is commonly used to study small regions of DNA, such as the digestion of PCR-amplified products or discrete samples such as purified plasmid or mitochondrial DNA (Jacobson and Gordon, 1990; Cooley, 1992; Appel and Gordon, 1995). DNA extracts from whole organisms will in most cases have too many sites to give clear separation of restriction fragments. Two exceptions to this have been the use of 'rare' cutting enzymes with bacterial genomes and GC 4-base cutting enzymes with fungal genomes.

Rare cutting enzymes are restriction enzymes where the site recognized is present only rarely in the genome, and as a result digestion of a complete bacterial genome results in the production of a small number of large fragments. These fragments are too large to be separated in conventional gel electrophoresis, but may be separated by techniques such as pulsed-field gel electrophoresis. RFLPs derived in this way have the advantage of including the entire genome and so can show polymorphisms that are the result of infrequent events. As a result the technique has been particularly useful in discriminating between extremely closely related isolates of a bacterial species or subspecies (Smith *et al.*, 1995).

GC 4-base cutting enzymes are enzymes that recognize short four-base sites consisting primarily of guanine and cytosine. These are particularly useful with fungi, as in general such sites are rare in the mitochondrial DNA and very common in chromosomal DNA. Digestion of total fungal DNA with these enzymes will result in many small fragments from the chromosomal DNA, whereas the mitochondrial DNA will generally be recovered as a small number of large fragments (Spitzer *et al.*, 1989; Typas *et al.*, 1992). This method has been used extensively with fungi, and the resulting band patterns of adenine- and thiamine-rich fragments have been termed 'presumptive' mitochondrial RFLPs. A number of studies with a restricted number of genera have shown equivalence between these and RFLPs obtained from purified mitochondrial DNA (Marriot *et al.*, 1984; Typas *et al.*, 1998). In general, band patterns have been used to subdivide species, and in *F. oxysporum* f.sp. *cubense* there is some correlation reported between such groups and pathogenic race (Thomas *et al.*, 1994).

Probes

The most widely used method of identifying polymorphisms in 'whole organism' DNA samples has been to use probes to detect specific regions of DNA. Probes are short pieces of DNA which are labelled with either a radioactive, or more recently chemical, marker. In the most general application of DNA probes the total DNA extracts are digested with frequent-cutting restriction enzymes, and the digested DNA is separated by electrophoresis to give a 'smear' of fragments of varying sizes. The gel containing the DNA is then treated to denature the double-stranded DNA, and the single-stranded DNA fragments are transferred on to an inert support, commonly a nitrocellulose or nylon membrane. This transfer is achieved by blotting the DNA fragments from the gel either by capillary blotting (also known as Southern blotting) or by one of a number of techniques such as vacuum blotting. Once the DNA has been transferred to the membrane it is fixed in place by heating or exposure to UV (Sambrook *et al.*, 1989). The prepared membrane is then immersed in a buffer containing the labelled probe, which hybridizes to the DNA wherever there is a complementary sequence. As the probe is labelled its presence can be detected, and the presence of the labelled probe indicates the position of that DNA sequence in the original sample. The specificity with which the probe binds to the membrane is determined by the conditions of the hybridization step; the type and strength of the buffer used and the temperature of the reaction must be carefully monitored to ensure specificity. Many different DNA sequences have been used as probes in this way, including particular genes or gene regions, random probes and total mitochondrial DNA (Jacobson and Gordon, 1990; Saylor and Layton, 1990). The most taxonomic information has been obtained from probes to repeated sequences, as these generally form the greatest part of random genomic libraries and they usually generate a greater number of restriction fragments than single copy genes. Regions that have been used extensively include the ribosomal RNA genes for bacteria and fungi, and mitochondrial sequences for eukaryotes (Gardes *et al.*, 1991). Small probes made from both simple and tandemly repeated repetitive sequences have also been used as DNA fingerprinting tools (Schlick *et al.*, 1994), and in some applications with very short synthetic repetitive probes it has been possible to generate fingerprints directly from gels without the need to transfer the target DNA to membranes.

The methodology employed for deriving RFLPs with probes can be adapted for *in situ* hybridization. *In situ* hybridization works on the same principle as any other hybridization where a single DNA strand is bound to a complementary sequence. In *in situ* hybridization this reaction takes place within a cell preparation, and the DNA probe is generally labelled with either a coloured or a fluorescent dye. This technique has been used to locate the position of individual gene sequences on plant chromosomes, and so can be of particular benefit in allowing plant breeders to follow the presence and location of DNA sequences during breeding programmes (Skerritt and Appels, 1995).

The polymerase chain reaction

The PCR is a method by which sections of a genome may be amplified (Saiki *et al.*, 1988). The technique generally requires some knowledge of the target DNA sequence, and relies on the binding of short oligonucleotides (primers) to known sequences on either side of the target region. The first step involves a denaturation of the DNA to separate the two DNA strands. One primer binds to the upper strand on the 5′ (upstream) side of the target, and the second binds to the lower strand on the downstream side (Fig. 23.1). A DNA polymerase then adds the individual phosphorylated nucleotides along the DNA strand from the 5′ end, starting at the primer. A further denaturation to give single-stranded molecules helps to remove any overlapping bases and the process is then repeated for further cycles. The DNA denaturation method used at the beginning of each cycle involves heating the double-stranded DNA to approaching 100°C, and then cooling to a lower temperature (annealing temperature) to allow the primers to bind. The separated DNA strands will also be able to bind at the annealing temperature and this is minimized by having a large excess of primers.

The heating step in each cycle requires that the DNA polymerase used is heat stable and traditionally these have been obtained from thermophilic

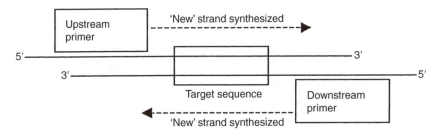

Fig. 23.1. Binding of primers to DNA strands and activity of DNA polymerase during PCR.

bacteria such as *Thermus aquaticus* (*Taq* enzyme) and *Thermus thermophilus* (*Tth* enzyme). These polymerases have relatively high temperature optima and so the addition of the nucleotides (extension) is carried out at 72°C. Each step in the cycle is relatively short, typically 30–90 s, and it is therefore important that heating and cooling between steps are carried out as quickly as possible. The entire reaction is performed in a dedicated machine referred to as a thermal cycler. The number of complete cycles necessary in a PCR reaction will vary depending on the abundance of the original DNA sequence. Each completed cycle leads to a doubling in the amount of the target DNA and programmes generally include 20–40 cycles.

The PCR reaction in itself is merely a method of DNA amplification and its utility in plant pathology depends on the application for which it is used; the two most common applications are for amplifying regions of DNA specific to individual pathogens for their subsequent identification (Seal *et al.*, 1992; Brown *et al.*, 1993), or for providing considerable quantities of selected DNA regions for subsequent analysis (White *et al.*, 1990). An important feature of PCR is that it can be used to amplify from very small quantities of DNA (in theory, a single strand) and so PCR-based identification methods will not necessarily require the isolation of the pathogen.

Probes and primers: species diagnostics

Taxon-specific probes and primers are those that hybridize to or amplify a DNA sequence that is specific to a particular taxon, and this approach has been investigated extensively with bacteria and fungi. Taxon-specific regions may be derived from almost any part of the genome, but the most widely used area is that containing the genes that code for the rRNA subunits. In a eukaryotic organism, such as yeasts and fungi, these genes are arranged in a cluster consisting of the genes for the small and large ribosomal subunits, separated by the 5.8S subunit. Between the 5.8S gene and those for the other subunits there are two internally transcribed spacers (ITS) (White *et al.*, 1990). There are many copies of the gene cluster in the genome, and these are arranged as a long string of repeated units, each separated by an intergenic spacer. In prokaryotic organisms such as bacteria, the genes may be arranged in a single unit with tRNA genes separating the subunits, or the individual subunit genes may be widely separated in the genome (Hillis and Dixon, 1991). In fungi, primers for specific plant pathogens have largely been developed from the ITS regions (Brown *et al.*, 1993). In prokaryotes, signature oligonucleotides have been found for the major groups of bacteria in the 16S DNA, and more variable regions have shown species specificity (Fox and Stackebrandt, 1987). There has also been interest in using the 28S DNA sequences. There has been considerable activity in the development of diagnostic probes and primers and many other genes have been considered to give specificity for organisms with specific properties. In these instances the target genes have been those associated with a particular property such as the enteric toxin gene in *Escherichia coli* (Priest and Austin, 1993) or specific mycotoxin genes in *Aspergillus* and other toxin-producing fungi (Geisen, 1998).

PCR-based fingerprinting techniques

There are many PCR-based fingerprinting techniques for bacteria and fungi. Different names have been applied to these at different times, but they can all be grouped under the general title of direct amplification fingerprinting. The earliest PCR-based fingerprinting technique was a method referred to as either random amplification of polymorphic DNA or arbitrarily primed PCR (Welsh and McClelland, 1990; Williams *et al.*, 1990). In this method short oligonucleotide primers are used at a reduced hybridization temperature, and this, together with an excess of magnesium ions, results in extensive binding of varying specificity. These methods have been used to both group and separate closely related isolates, but there have been some concerns over their reproducibility (Lamboy, 1994; Bridge and Arora, 1998). The random nature of the binding that occurs in this type of fingerprinting approach also makes it difficult to interpret measures of difference between organisms. The more commonly used fingerprinting primers are those based on repetitive DNA sequences, and these are essentially of two types. The first, which have been used extensively with both bacteria and fungi, are relatively complex sequences of around 15–25 base pairs that occur frequently in the genome. These sequences are often considered as 'universal' primers, as those derived from bacteria will also amplify DNA from eukaryotes. Particular examples are the protein II gene sequence from M13 bacteriophage, ERIC and REP primers (Hulton *et al.*, 1991; Versalovic *et al.*, 1991; Meyer *et al.*, 1992). The second type of repetitive sequences used for fingerprinting are the simple sequence repeats derived from microsatellite sequences. These vary in length from around 12 to 20 base pairs, and consist of doublet, triplet and quaternary repeats such as $(GA)_8$, $(CAG)_5$ and $(GACA)_4$. These primers are believed to hybridize to microsatellite regions, which are hypervariable loci consisting of tandem repeats of short core sequences present in all eukaryote genomes (also known as variable number tandem repeats; VNTRs). The number of repeats at these loci will vary between unrelated individuals, and microsatellites have a high mutation rate, so unique fingerprint patterns can be generated (Schlick *et al.*, 1994; Buscot *et al.*, 1996; Bridge *et al.*, 1997).

Another technique that has recently been introduced for fingerprinting microorganisms is amplification fragment length polymorphism. In this method total genomic DNA is extracted and digested with restriction enzymes. Linkers containing short oligonucleotides are then attached to the ends of the restriction fragments. Primers that match part of the oligonucletide sequence are then used to amplify a proportion of the digestion fragments. This technique will produce many more fragments than repetitive sequence fingerprinting (Vos *et al.*, 1995).

Sequencing

All of the above molecular methods contain a presumptive element in that bands obtained from different isolates that appear in the same position on a gel are assumed to be the same. In order to arrive at reliable systematic and phylogenetic classifications, and to be able to design specific probes and primers, it is necessary to sequence regions of the genome. The regions that are sequenced will depend largely on the purpose of the work, but for systematic and diagnostic applications, sequences from both variable and conserved regions of rRNA genes are widely used. The use of DNA sequence information is becoming increasingly common in plant pathology for designing diagnostic tools such as pathogen-specific probes and primers, and for elucidating relationships between organisms. DNA sequencing is also a powerful tool in the identification and characterization of the specific genes involved in pathogenicity and resistance, and is increasingly being incorporated into plant breeding programmes.

There are a number of routes to obtaining DNA sequences, but most are now generated through automatic sequencers with fluorescent dyes and laser detection. In order to sequence a DNA fragment it must first be separated, purified and concentrated. Two main routes are used for this, either by cloning the separated fragment into some form of vector (e.g. phage, plasmid or cosmid) and propagating the vector in a bacterium, or by preparing samples directly from PCR reactions. Cloning has the advantages that large fragments can be more easily sequenced in sections and that the clone provides for continual production and maintenance of the fragment. The PCR method requires less technical input, providing primers are available for the fragment, and the final PCR product should then be cleaned through one of the commercially available DNA purification kits to remove any unincorporated nucleotides, primers, etc. Consideration must be given to the possibility of mixed products, in which case cloning will be required to ensure the final sequence is generated from a single product.

Conclusions

Biochemical and molecular techniques can provide considerable information about the identity and function of plant pathogenic microorganisms. It must, however, be remembered that plant pathogens are a heterogeneous group of organisms, and that characteristics that may prove useful for the study of one pathogen can prove to be inappropriate for others. Modern molecular diagnostics have the potential to greatly improve early disease detection and for the screening of large numbers of plants. However, these methods can be highly specific and so accurate, correct targeting is required for the development and selection of the methodology employed.

References

Abe, S. (1956) Studies on the classification of the *Penicillia*. *Journal of General and Applied Microbiology* 2, 1–344.

Appel, D.J. and Gordon, T.R. (1995) Intraspecific variation within populations of *Fusarium oxysporum* based on RFLP analysis of the intergenic spacer region of the rDNA. *Experimental Mycology* 19, 120–128.

Barnett, J.A., Payne, R.W. and Yarrow, D. (1990) *Yeasts: Characteristics and Identification*, 2nd edn. Cambridge University Press, Cambridge, UK.

Barrow, G.I. and Feltham, R.K.A. (1993) *Cowan and Steel's Manual for the Identification of Medical Bacteria*, 3rd edn. Cambridge University Press, Cambridge, UK.

Barth, M.G.M. and Bridge, P.D. (1989) 4-methylumbelliferyl substituted compounds as fluorogenic substrates for fungal extracellular enzymes. *Letters in Applied Microbiology* 9, 177–179.

BioMérieux (1993) *ID32C Analytical Profile Index*. BioMérieux, Marcy-l'Etoile, France.

Boekhout, T. (1991) A revision of ballistoconidia-forming yeasts and fungi. *Studies in Mycology* 33.

Bonde, M.R., Micales, J.A. and Peterson, G.L. (1986) The use of isozyme analysis in fungal taxonomy and genetics. *Mycotaxon* 27, 405–449.

Bridge, P.D. and Arora, D.K. (1998) Interpretation of PCR methods for species definition. In: Bridge, P.D., Arora, D.K., Elander, R.P. and. Reddy, C.A. (eds) *Applications of PCR in Mycology*. CAB International, Wallingford, UK, pp. 64–83.

Bridge, P.D. and Hawksworth, D.L. (1985) Biochemical tests as an aid to the identification of *Monascus* species. *Letters in Applied Microbiology* 1, 25–29.

Bridge, P.D., Ismail, M.A. and Rutherford, M. (1993) An assessment of aesculin hydrolysis, vegetative compatibility and DNA polymorphism as criteria for characterising pathogenic races within *Fusarium oxysporum* f. sp. *vasinfectum*. *Plant Pathology* 42, 264–269.

Bridge, P.D., Pearce, D., Rutherford, M. and Rivero, A. (1997) VNTR derived oligonucleotides as PCR primers for population studies in filamentous fungi. *Letters in Applied Microbiology* 24, 426–430.

Brown, A.E., Muthumeenakshi, S., Sreenivasaprasad, S., Mills, P.R. and Swinburne, T.R. (1993) A PCR primer-specific to *Cylindrocarpon heteronema* for detection of the pathogen in apple wood. *FEMS Microbiology Letters* 108, 117–120.

Brygoo, Y., Caffier, V., Carlier, J., Fabre, J.-V., Fernandez, D., Giraud, T., Mourichon, X., Neema, C., Notteghem, J.-L., Pope, C., Tharreau, D. and Lebrun, M.-H. (1998) Reproduction and population structure in phytopathogenic fungi. In: Bridge, P.D., Couteaudier, Y. and Clarkson, J.M. (eds) *Molecular Variability of Fungal Pathogens*. CAB International, Wallingford, UK, pp. 133–148.

Buscot, F., Wipf, D., Battista, C.D., Munch, J.-C., Botton, B. and Martin, F. (1996) DNA polymorphism in morels: PCR/RFLP analysis of the ribosomal DNA spacers and microsatellite-primed PCR. *Mycological Research* 100, 63–71.

Cenis, J.L. (1992) Rapid extraction of fungal DNA for PCR amplification. *Nucleic Acids Research* 20, 2380.

Collins, M.D. (1985) Isoprenoid quinone analysis in bacterial classification and identification. In: Goodfellow, M. and Minikin, D.E. (eds) *Chemical Methods in Bacterial Systematics*. Academic Press, London, UK, pp. 267–288.

Cooley, R.N. (1992) The use of RFLP analysis, electrophoretic karyotyping and PCR in studies of plant pathogenic fungi. In: Stahl, U. and Tudzynski, P. (eds) *Molecular Biology of Filamentous Fungi*. VCH, Weinheim, Germany, pp. 13–26.

Cruickshank, R.H. and Wade, G.C. (1980) Detection of pectic enzymes in pectin-acrylamide gels. *Annals of Biochemistry* 107, 177–181.

Fox, G.E. and Stackebrandt, E. (1987) The application of 16S rRNA cataloguing and 5S rRNA sequencing in bacterial systematics. *Methods in Microbiology* 19, 405–458.

Fox, R.T.V. (1993) *Principles of Diagnostic Techniques in Plant Pathology*. CAB International, Wallingford, UK.

Frisvad, J.C. (1981) Physiological criteria and mycotoxin production as aids in identification of common asymmetric *Penicillia*. *Applied and Environmental Microbiology* 41, 568–579.

Frisvad, J.C. and Filtenborg, O. (1990) Secondary metabolites as consistent criteria in *Penicillium* taxonomy and a synoptic key to *Penicillium* subgenus *Penicillium*. In: Samson, R.A. and Pitt, J.I. (eds) *Modern Concepts in* Penicillium *and* Aspergillus *Classification*. Plenum Press, New York, USA, pp. 373–384.

Frisvad, J.C., Thrane, U. and Filtenborg, O. (1998) Role and use of secondary metabolites in fungal taxonomy. In: Frisvad, J.C., Bridge, P.D. and Arora, D.K. (eds) *Chemical Fungal Taxonomy*. Marcel Dekker, New York, USA, pp. 289–319.

Gardes, M., Mueller, G.M., Fortin, J.A. and Kropp, B.R. (1991) Mitochondrial DNA polymorphisms in *Laccaria bicolor, L. laccata, L. proxima*, and *L. amethystina. Mycological Research* 95, 206–216.

Geisen, R. (1998) PCR methods for the detection of mycotoxin-producing fungi. In: Bridge, P.D., Arora, D.K., Elander, R.P. and Reddy, C.A. (eds) *Applications of PCR in Mycology*. CAB International, Wallingford, UK, pp. 243–266.

Goodfellow, M. and James, A.L. (1994) Rapid enzyme tests in the characterization and identification of microorganisms. In: Hawksworth, D.L. (ed.) *The Identification and Characterization of Pest Organisms*. CAB International, Wallingford, UK, pp. 289–301.

Gutteridge, C.S., Mackey, B.M. and Norris, J.R. (1980) A pyrolysis gas–liquid chromatography study of *Clostridium botulinum* and related organisms. *Journal of Applied Bacteriology* 49, 165–174.

Hawksworth, D.L. (1974) *The Mycologists Handbook*. CAB International, Wallingford, UK.

Hillis, D.M. and Dixon, M.T. (1991) Ribosomal DNA: molecular evolution and phylogenetic inference. *Quarterly Reviews in Biology* 66, 411–453.

Holmes, B., Willcox, W.R. and Lapage, S.P. (1978) Identification of *Enterobacteriaceae* by the API20E system. *Journal of Clinical Pathology* 31, 22–30.

Hulton, C.S.J., Higgins, C.F. and Sharp, P.M. (1991) ERIC sequences: a novel family of repetitive elements in the genomes of *Escherichia coli, Salmonella typhimurium* and other enterobacteria. *Molecular Microbiology* 5, 825–834.

Humble, M.W., King, A. and Phillips, I. (1977) API ZYM: a simple and rapid system for the detection of bacterial enzymes. *Journal of Clinical Pathology* 30, 275–277.

Jacobson, D.J. and Gordon, T.R. (1990) Variability of mitochondrial DNA as an indicator of relationships between populations of *Fusarium oxysporum* f.sp. *melonis. Mycological Research* 94, 734–744.

Johnk, J.S. and Jones, R.K. (1994) Comparison of whole-cell fatty acid compositions in intraspecific groups of *Rhizoctonia solani* AG1. *Phytopathology* 84, 271–275.

Keddie, R.M. and Cure, G.L. (1978) Cell wall composition of coryneform bacteria. In: Bousfield, I.J. and Callely, A.G. (eds) *Coryneform Bacteria*. Academic Press, London, UK, pp. 47–84.

Kock, J.L.F. and Botha, A. (1998) Fatty acids in fungal taxonomy. In: Frisvad, J.C., Bridge, P.D. and Arora, D.K. (eds) *Chemical Fungal Taxonomy*. Marcel Dekker, New York, USA, pp. 219–246.

Kock, J.L.F. and van der Walt, J.P. (1986) Fatty acid composition of *Schizosaccharomyces* Lindner. *Systematic and Applied Microbiology* 8, 163–165.

Komagata, K. and Suzuki, K.I. (1987) Lipid and cell wall analyses in bacteria systematics. *Methods in Microbiology* 19, 161–208.

Kuraishi, H., Katayama-Fujimura, U., Sugiyama, J. and Yokoyama, T. (1985) Ubiquinone systems in fungi. I. Distribution of ubiquinones in the major families of Ascomycetes, Basidiomycetes and Deuteromycetes, and their taxonomic implications. *Transactions of the Mycological Society of Japan* 26, 383–395.

Lamboy, W.F. (1994) Computing genetic similarity coefficients from RAPD data: correcting for the effects of PCR artefacts caused by variation in experimental conditions. *PCR Methods and Applications* 4, 38–43.

Lodwig, E.M., Bridge, P.D., Rutherford, M.A., Kung'u, J. and Jeffries, P. (1999) Molecular differences distinguish clonal lineages within east african populations of *Fusarium oxysporum* f.sp. *cubense. Journal of Applied Microbiology* 86, 71–77.

Logan, C. and Khan, A.A. (1969) Comparative studies of *Phoma* spp. associated with potato gangrene in Northern Ireland. *Transactions of the British Mycological Society* 52, 9–17.

Lund, B.M. (1965) A comparison by the use of gel electrophoresis of soluble protein components and esterase enzymes of some Group D streptococci. *Journal of General Microbiology* 40, 413–419.

Manczinger, L. and Polner, G. (1985) Cluster analysis of carbon source utilization patterns of *Trichoderma* isolates. *Systematic and Applied Microbiology* 9, 214–217.

Marriott, A.C., Archer, S.A. and Buck, K.W. (1984) Mitochondrial DNA in *Fusarium oxysporum* is a 46.5 kilobase pair circular molecule. *Journal of General Microbiology* 130, 3001–3008.

Mateos, P., Jimenez-Zurdo, J., Chen, J., Squartini, A.S., Haack, S.K., Martinez-Molina, E., Hubbell, D. and Dazzo, F.B. (1992) Cell-associated pectinolytic and cellulolytic enzymes in *Rhizobium leguminosarum* biovar. *trifolii*. *Applied and Environmental Microbiology* 58, 1816–1822.

Meyer, W., Morawetz, R., Borner, T. and Kubicek, C.P. (1992) The use of DNA-fingerprint analysis in the classification of some species of the *Trichoderma* aggregate. *Current Genetics* 21, 27–30.

Minikin, D.E. and Goodfellow, M. (1980) Lipid composition in the classification and identification of acid-fast bacteria. In: Goodfellow, M. and Board, R.G. (eds) *Microbiological Classification and Identification*. Academic Press, London, UK, pp. 189–256.

Paterson, R.R.M. and Bridge, P.D. (1994) *Biochemical Techniques for Filamentous Fungi. IMI Technical Handbooks*, Vol. 1. CAB International, Wallingford, UK.

Paterson, R.R.M. and Buddie, A.G. (1991) Rapid determination of ubiquinone profiles in *Penicillium* by reversed phase high performance thin-layer chromatography. *Letters in Applied Microbiology* 13, 133–136.

Persson, P. and Sletten, A. (1995) Fatty acid analysis for the identification of *Erwinia carotovora* subsp. *atroseptica* and *E. carotovora* subsp. *carotovora*. *EPPO Bulletin* 25, 151–156.

Priest, F. and Austin, B. (1993) *Modern Bacterial Taxonomy*, 2nd edn. Chapman & Hall, London, UK.

Rutherford, M.A., Paterson, R.R.M., Bridge, P.D. and Brayford, D. (1993) Coumarin utilization patterns of pathogenic races of *Fusarium oxysporum* causing vascular wilt in cotton. *Journal of Phytopathology* 138, 209–216.

Rutherford, M.A., Bridge, P.D., Paterson, R.R.M. and Brayford, D. (1995) Identification techniques for special forms of *Fusarium oxysporum*. *EPPO Bulletin* 25, 137–142.

Saiki, R.K., Gelfand, D.H., Stoffel, S., Scharf, S.J., Higuchi, R., Horn, G.T., Mullis, K.B. and Erlich, H.A. (1988) Primer-directed enzymatic amplification of DNA with a thermostable DNA Polymerase. *Science* 239, 487–491.

Sambrook, J., Fritsch, E.F. and Maniatis, T. (1989) *Molecular Cloning: a Laboratory Manual*, 2nd edn. Cold Spring Harbor Laboratory Press, Cold Spring Harbor, New York, USA.

Saylor, G.S. and Layton, A.C. (1990) Environmental applications of gene probes. *Annual Review of Microbiology* 44, 625–648.

Schaad, N.W., Smith, O.P., Bonde, M.R., Peterson, G.L., Beck, R.J., Hatziloukas, E. and Panopoulos, N.J. (1994) Polymerase chain reaction methods for the detection of seedborne plant pathogens. In: Hawksworth, D.L. (ed.) *The Identification and Characterization of Pest Organisms*. CAB International, Wallingford, UK, pp. 461–471.

Schlick, A., Kuhls, K., Meyer, W., Lieckfeldt, E., Borner, T. and Messner, K. (1994) Fingerprinting reveals gamma-ray induced mutations in fungal DNA: implications for identification of patent strains of *Trichoderma harzianum*. *Current Genetics* 26, 74–78.

Seal, S.E., Jackson, L.A. and Daniels, M.J. (1992) Use of tRNA consensus primers to indicate subgroups of *Pseudomonas solanacearum* by polymerase chain reaction amplification. *Applied and Environmental Microbiology* 58, 3759–3761.

Skerman, V.B.D. (1969) *Abstracts of Microbiological Methods*. John Wiley & Sons, London, UK.

Skerritt, J.H. and Appels, R. (1995) An overview of the development and application of diagnostic methods in crop sciences. In: Skerritt, J.H. and Appels, R. (eds) *New Diagnostics in Crop Sciences*. CAB International, Wallingford, UK, pp. 1–32.

Smith, J.J., Offord, L.C., Holderness, M. and Saddler, G.S. (1995) Pulsed-field gel electro-phoresis analysis of *Pseudomonas solanacearum*. *EPPO Bulletin* 25, 163–168.

Spitzer, E.D., Lasker, B.A., Travis, S.J., Kobayashi, G.S. and Medoff, G. (1989) Use of mitochondrial and ribosomal DNA polymorphisms to classify clinical and soil isolates of *Histoplasma capsulatum*. *Infection and Immunity* 57, 1409–1412.

St Leger, R.J. (1995) The role of cuticle-degrading proteases in fungal pathogenesis in insects. *Canadian Journal of Botany* 73 (Suppl. 1), S1119–S1125.

Stead, D.E. (1995) Profiling techniques for the identification and classification of plant pathogenic bacteria. *EPPO Bulletin* 25, 143–150.

Tharagonnet, D., Sisson, P.R., Roxby, C.M., Ingham, H.R. and Selkon, J.B. (1977) The API ZYM system in the identification of Gram-negative anaerobes. *Journal of Clinical Pathology* 30, 505–509.

Thomas, V., Rutherford, M.A. and Bridge, P.D. (1994) Molecular differentiation of two races of *Fusarium oxysporum* special form *cubense*. *Letters in Applied Microbiology* 18, 193–196.

Typas, M.A., Griffen, A.M., Bainbridge, B.W. and Heale, J.B. (1992) Restriction fragment length polymorphisms in mitochondrial DNA and ribosomal RNA gene complexes as an aid to the characterization of species and sub-species populations in the genus *Verticillium*. *FEMS Microbiology Letters* 95, 157–162.

Typas, M.A., Mavridou, A. and Kouvelis, V.N. (1998) Mitochondrial DNA differences provide maximum intraspecific polymorphism in the entomopathogenic fungi *Verticillium lecanii* and *Metarhizium anisopliae*, and allow isolate detection/identification. In: Bridge, P., Couteaudier, Y. and Clarkson, J. (eds) *Molecular Variability of Fungal Pathogens*. CAB International, Wallingford, UK, pp. 227–237.

Versalovic, J., Koeuth, T. and Lupski, J.R. (1991) Distribution of repetitive DNA sequences in eubacteria and application to fingerprinting bacterial genomes. *Nucleic Acids Research* 19, 6823–6831.

Vos, P., Hogers, R., Bleeker, M., Reijans, M., van de Lee, T., Homes, M., Frijters, A., Pot, J., Peleman, J., Kuiper, M. and Zabeau, M. (1995) AFLP: a new concept for DNA finger-printing. *Nucleic Acids Research* 23, 4407–4414.

Welsh, J. and McClelland, M. (1990) Fingerprinting genomes using PCR with arbitrary primers. *Nucleic Acids Research* 18, 7213–7218.

White, T.J., Bruns, T.D., Lee, S. and Taylor, J. (1990) Amplification and direct sequencing of fungal ribosomal DNA genes for phylogenetics. In: Innis, M.A., Sninsky, D.H. and White, T.J. (eds) *PCR Protocols*. Academic Press, London, UK, pp. 315–322.

Williams, J.G.K., Kubeik, A.R., Livak, K.J., Rafalski, J.A. and Tingey, S.V. (1990) DNA polymorphisms amplified by arbitrary primers are useful as genetic markers. *Nucleic Acids Research* 18, 6531–6535.

Williams, R.A.D. and Shah, H.N. (1980) Enzyme patterns in bacterial classification and identification. In: Goodfellow, M. and Board, R.G. (eds) *Microbiological Classification and Identification*. Academic Press, London, UK, pp. 299–318.

Inoculation

J.M. Waller

CABI Bioscience UK Centre, Bakeham Lane, Egham, Surrey TW20 9TY, UK

Inoculation involves the application of inoculum (pathogen propagules) to the infection court of an intended host plant. The term may also be used more broadly to indicate the transfer of propagules to a new growth medium, thus agar plates etc. may be inoculated by transferring explants of cultures, spore suspensions, etc. to them. The choice of inoculation method depends on the following basic considerations.

Initial Considerations

Objective

Where the objective of the inoculation is to induce growth of the pathogen in host tissue, procedures that overcome any natural host resistance, such as the use of high levels of inoculum or insertion in the host tissue may be justified. It is sometimes desirable to inoculate cultures of pathogens on to their hosts in this way to restore their virulence. Where inoculation is part of a screening procedure for resistance care has to be taken that inoculum levels, inoculation techniques and host treatment do not overcome resistance that may be operable under field conditions (Russell, 1978). Inoculation may be undertaken to determine the pathogenicity of an organism or as part of Koch's postulates; in this case the relevance of the inoculation techniques to natural infection processes must be considered.

The pathogen

A knowledge of the behaviour of the pathogen, particularly stages in its life cycle that induce infection, requirements for spore germination, mode of infection, host range, disease symptoms, etc. is desirable for successful inoculation. However, the objective of the inoculation may be to determine some of these characteristics in which case careful experimentation that differentiates between known and unknown variables is required. Some idea of likely methods of inoculation can often be obtained from the taxonomic relationships of the pathogen, the origin of isolates, behaviour in the field, etc. Pathogens are variable and isolates are likely to differ in pathogenicity.

Inoculum

The different types of spores that are produced by most fungi may germinate and infect under different conditions, so it is advantageous to know the role of particular spore types in the life cycle of the pathogen. Mycelial inoculum is often appropriate for soil-inhabiting organisms. Other structures such as rhizomorphs, sclerotia, bacterial cells or virus particles may be appropriate inocula for certain diseases. Mycelial inoculum from cultures may contain toxic metabolites or be genetically or physiologically different from spore inoculum, any of which may induce an abnormal host response. Spore suspensions prepared from nutrient agars may similarly contain toxic metabolites in unnatural concentrations especially if their concentration is high. Pathogens often lose virulence and/or change genetically when cultured so that inoculum is best obtained directly from pathogens sporulating on diseased hosts, or from cultures maintained by techniques that reduce or prevent saprophytic growth (e.g. lyophilization, cryogenic storage, etc., see Chapter 37). The production of appropriate types and amounts of spores may also require the use of special culture techniques (see Chapter 38).

Quantitative aspects of inoculum also require consideration. This includes both the concentration or density of the particular inoculum (spores, bacteria, viruses) applied and the inoculum potential (Baker, 1978). Spore and bacterial concentrations can be estimated by direct counting techniques (e.g. haemocytometer), turbidimetric techniques (light absorption) or cultural techniques (dilution plating). Unnaturally high concentrations may hinder spore germination through mutual inhibition.

The host

A susceptible host is necessary for successful development of a disease after inoculation, although the detection of resistance may be the objective of the inoculation. For many pathogens to infect, the host must be at a susceptible stage in its development, or in a susceptible condition (phenotypically susceptible). Furthermore, only certain parts of the plant such as juvenile tissues, flowers or fruit may be susceptible to particular pathogens so that the inoculum must be applied to the appropriate infection court. Viruses and some other systemic pathogens may require inoculation into the subepidermal or vascular tissue of the plant. Many pathogens will only successfully infect a host to produce disease if there is some prediposing factor. It is the outcome of inoculation that is usually the relevant criterion for study and this can be assessed at the cellular, tissue or plant level depending on the objectives of the inoculation.

The environment

Environmental conditions influence plant diseases in several ways (Colhoun, 1973) and must be considered during inoculation. Firstly, preinoculation conditions may affect the susceptibility of the host. Temperature, light, nutrition and moisture may have predisposing effects (Shoeneweiss, 1975), and generally, progressive reduction of host vigour predisposes them to increasingly weaker parasites. The immediate postinoculation conditions are often very critical. Most fungal and bacterial parasites require adequate moisture for infection, often in the form of liquid water on the plant surface that must be present for several hours at a suitable temperature to allow infection. This can be achieved by placing plastic hoods over the adequately moistened plants and avoiding temperature increases through convective or radiant heating. Dew deposition can be simulated by adding warm water to cool closed containers in which plants are placed and some forms of dew chambers have been devised. Humidifiers or mist propagators can be used to maintain humid or saturated conditions in larger chambers. Prolonged periods of high humidity or particular lighting conditions may be necessary for adequate disease expression of some pathogens. Natural diurnal fluctuation of temperature and humidity and natural sunlight intensity are difficult to reproduce in greenhouses or growth chambers.

Basic Techniques

Natural exposure to inoculum

Exposing growing plants or plant parts to natural sources of inoculum, such as airborne spores, infested soil, or viruliferous insects, can be used to trap pathogenic organisms, but is subject to the variability of natural infection where escapes may confound assessments of resistance. Natural inoculum levels can be enhanced; previously infected plants can be used to 'inoculate' field plots by planting them in 'spreader' rows between or within plots of test plants, or soil-borne pathogens can be used to create 'sick plots'.

Direct application to the host surface

Propagules can be applied to leaves etc. by spraying or dusting, or by direct placement. Spraying is the most widely used technique for inoculating pathogens on to plant shoots, even for those that normally produce dry-dispersed spores. Suspensions of spores, bacterial cells, mycelial homogenates or preparations of infected host tissues to which wetting agents, nutrients, humectants or adhesives may be added can be applied by a variety of spraying mechanisms from simple atomizers to high-pressure sprayers. Controlled droplet application techniques (spinning discs) that produce uniform droplet sizes are useful for quantitative work. Micropipettes can be used to apply droplets of known size usually containing a known inoculum concentration. Attempts to standardize inoculation methods for use in fungicide or resistance screening programmes have been made by using single or multiple spraying devices automatically delivering a known volume at a set distance from the plant, which can be rotated to achieve even coverage.

Dry spores applied as a dust in a settling tower can be used in quantitative inoculation work where a critical density of spores on the leaf is required (Eyal et al., 1968). Subsequent incubation of dust-inoculated plants will require the use of a dew chamber when moisture is required for the infection process. Dry spores may also be applied to plant surfaces using cotton wool swabs, and single spores or spore clumps can be applied with a needle.

Direct application of cultures to the aerial parts of the plant can be made using discs of agar, filter paper, cellophane, plastic mesh or foam impregnated with the culture. Some mechanism to hold them in place, such as elastic or sticky tape, and to prevent desiccation may be needed. Such methods are used for stem and root pathogens, particularly where discrete local inoculation is needed.

Inoculation through soil or roots

Root pathogens may be inoculated by growing the host in soil artificially infested with inoculum. Inoculum may consist of naturally infected plant parts, such as root or leaf debris, or a culture of the pathogen on a weak natural substrate such as bran, straw or maize meal and sand or on a porous substance such as vermiculite or perlite impregnated with a nutrient solution. Cultures on nutritionally rich media are best avoided as the metabolites produced may be toxic to the roots or have a marked effect on the soil properties. Spore suspensions or dry spores, often diluted with an inert dust such as kaolin, can also be mixed with the soil. Some root pathogens require a high inoculum potential for successful infection so that the provision of inoculum in adequate quantities or with a substantial food base may be necessary. Infected wood blocks are often used as inoculum for root pathogens of woody plants. Various processes can be used to wound plant roots so that pathogen infection is facilitated. Transplanting normally has this effect; a more severe procedure is to immerse the plant roots in a spore or mycelial suspension between uprooting and transplanting.

Soil factors such as moisture, temperature and pH have a marked effect on many root diseases, so that successful inoculation of soil-borne pathogens may require a special consideration of these factors. Similarly, an important part of the soil environment is its microflora and fauna so that inoculation of plants in sterile soil may give misleading results. Nematodes interact closely with root pathogens and may need to be included if the field situation is to be represented.

Inoculation of inaccessible areas

Many pathogens require introduction to normally well-protected plant parts such as meristem tissues beneath bud scales, substomatal cavities or the interior of flowers before they can successfully infect. Others need to be placed beneath the protective epidermal tissues either through wounds or natural injection by a vector. Air-blast sprayers can be used to force inoculum suspensions into these areas or they can be injected with a fine hypodermic needle. Inoculum can also be drawn into inaccessible areas if the relevant plant parts are immersed in a spore suspension and subjected to a partial vacuum that is then suddenly released.

Inoculation requiring wounding

Infection by many pathogens is facilitated if the host surface is wounded. Minute abrasions in the cuticle caused by brushing or rubbing may allow the entry of fine particles such as viruses, or may cause the exudation of nutrients that stimulate pathogen infection. Direct entry beneath the epidermis can be achieved simply by puncturing the plant surface with a needle beneath a drop of inoculum. Entry into intercellular spaces or vascular elements is facilitated if the plant is under a slight moisture stress. Syringes equipped with a fine hypodermic needle can be used to infiltrate inoculum under pressure into intracellular spaces or vascular tissue. For injecting the vascular tissue of woody plants hollow taps can be screwed into the sapwood or held in previously drilled holes by an expandable pressure ring at the trunk surface.

Slivers of contaminated or infected wood used as inoculum can be forced into tissues; larger wounds can be plugged with agar cultures or other material or a liquid suspension can be infiltrated under pressure. Alternatively, saw blades, knives, etc. can be smeared with the inoculum, which is driven into the plant as the wound is made. Unspecialized facultative parasites can often be induced to produce lesions in this way, but care is needed in interpreting the results of such unnatural inoculation methods.

Inoculation of detached plant parts

Simple laboratory inoculations can be attempted using detached plant organs such as leaves or leaf discs maintained in a viable state by floating them on a solution of kinetin (up to 20 p.p.m.) or benzimidazole (up to 150 p.p.m.) and exposing them to light. Fruits can be similarly maintained with their peduncle immersed in the solution. The inoculation of tissue cultures can also be used to maintain cultures of obligate pathogens, but the reaction of the tissue culture may be very different from that of the intact plant (Ingram and Helgeson, 1980).

Inoculation of germinating seeds and seedlings can be readily achieved by wrapping them in a moist absorbent paper towel to which inoculum has been added.

Special Considerations for Non-fungal Pathogens

Viruses

These can be inoculated by grafting or budding infected plant parts on to healthy plants. Dodder (*Cuscuta* spp.) is also able to transmit many viruses and mollicutes from diseased to healthy plants.

Mechanical inoculation of viruses is commonly achieved by rubbing host leaves with the sap (or tissue extract) of infected plants to which a mild abrasive such as celite or fine carborundum powder has been added. Puncturing with a needle is also effective. Various associated treatments can enhance the infectivity (Yarwood, 1973).

Many viruses cannot be mechanically transmitted and a suitable vector has to be used. The vector (e.g. a hemipteran insect) is fed on a diseased plant and then allowed to feed on the intended host. Factors such as pre-acquisition fasting of the vector, acquisition time, latent period in the vector, duration of retention of infectivity and loss of infectivity on moulting vary according to the transmission characteristics (persistence) of the virus. Most soil-borne viruses are transmitted by nematodes or fungi, and successful inoculation of these may require the presence of suitable numbers and life stages of their vectors.

Bacteria

Inoculation of bacteria into plant shoots can be accomplished by infiltration with a fine-needled hypodermic syringe, air-blast atomizers, which will force inoculum into stomata or small ruptures between epidermal cells, needle puncture methods (a ring of fine needles held in a sterilizable holder can be forced through the leaf on to a pad of cotton wool held below, which is soaked with inoculum), or by spraying. High humidity is usually important in the postinoculation phase. The more severe techniques are most successful in achieving infection but may produce rather atypical symptoms. The more luxuriant or tender parts of plants that have been subjected to mild water stress immediately before inoculation are infected more readily by the less severe methods. Root inoculation by soil-borne bacteria can be achieved by drenching the soil with a bacterial suspension, but the root-dip method at transplanting, although unnaturally severe, is more successful in achieving infection.

Quarantine Precautions

The need for taking strict precautions when performing plant inoculations with cultures of pathogenic organisms must always be borne in mind. If the pathogen is not of local origin care must be taken to ensure that the local plant disease legislation requirements are met.

Many techniques have been developed for particular pathosystems and purposes. Several have been standardized for large-scale screening of pesticides or the progeny of plant breeding. References to these can be obtained by scanning the literature related to pesticide or resistance screening. Some more general techniques are indexed in the *Review of Plant Pathology* under Apparatus and Techniques. General reviews that are still relevant are found in Tuite (1969), Kiraly (1970), Kado (1972) and Waterston (1968).

References

Baker, R. (1978) Inoculum potential. In: Horsfall, J.G. and Cowling, E.B. (eds) *Plant Diseases: an Advanced Treatise*, Vol. 2. Academic Press, London, UK, pp. 137–157.

Colhoun, J. (1973) Effects of environmental factors on plant diseases. *Annual Review of Phytopathology* 11, 343–364.

Eyal, Z., Clifford, B.C. and Caldwell, R.M. (1968) A settling tower for quantitative inoculation of leaf blades of mature small grain plants with uredospores. *Phytopathology* 58, 530–531.

Ingram, D.S. and Helgeson, J. (eds) (1980) *Tissue Culture Methods for Plant Pathologists*. Blackwell, Oxford, UK, 272 pp.

Kado, C.I. (1972) Mechanical and biological inoculation principles. In: Kado, C.I. and Agrawal, H.O. (eds) *Principles and Techniques in Plant Virology*. Van Nostrand Reinhold, New York, USA, pp. 3–31.

Kiraly, Z. (ed.) (1970) *Methods in Plant Pathology*. Akademiai Kiado, Budapest, 509 pp.

Russell, G.E. (1978) *Plant Breeding for Pest and Disease Resistance*. Butterworths, London, UK, 485 pp.

Shoeneweiss, D.F. (1975) Predisposition, stress and plant disease. *Annual Review of Phytopathology* 13, 193–211.

Tuite, J. (1969) *Plant Pathological Methods: Fungi and Bacteria*. Burgess, Minneapolis, Minnesota, USA, 239 pp.

Waterston, J.M. (1968) Inoculation. *Review of Applied Mycology* 47, 217–222.

Yarwood, C.E. (1973) Quick drying versus washing in virus inoculation. *Phytopathology* 63, 72–76.

Epidemic Modelling and Disease Forecasting

M.W. Shaw

Department of Agricultural Botany, School of Plant Sciences,
The University of Reading, Reading RG6 6AS, UK

Introduction

There are at least two reasons to model an epidemic or forecast disease. First, it is essential to forecast how severe disease will be in order to judge whether it is worth making some alteration in crop management, such as sowing a more resistant cultivar or spraying with a fungicide. Secondly, in order to clarify what types of intervention are likely to be useful it is important to understand the dynamics of a disease. A forecast may be a direct numerical prediction of the amount of disease in a crop, but more commonly it is a prediction of the probability of loss in a series of categories – severe, moderate or light, for example. To make either sort of forecast it is necessary to have some kind of model of how disease will develop in the crop. This may be as simple as a curve showing average amount of disease in an average crop at a particular time of the year, or as complicated as a complete simulation of the growth of crop, the development of the pathogen population and the relation of both to the weather.

This chapter discusses practice and problems in developing models of disease in crops, and how these models may be used to produce forecasts capable of helping with decision making.

Modelling Changes in Pathogen Populations and Disease Over Time

The damage caused by disease is in part controlled by the size of the pathogen population, which is governed by birth, death, immigration and emigration.

A good general introduction to this field is Begon *et al.* (1996). It is useful to distinguish two phases in the life cycle of most pathogens: multiplication within the crop, and survival between crops and transfer to the new crop. The observational and modelling problems posed differ between the two phases.

Disease progress within crops

There has been much discussion of appropriate models to fit to disease prog-ress data within an annual crop. Campbell and Madden (1990) provide a good introduction, as does Thresh (1983). The most commonly used mathematical models are the exponential (also, confusingly, sometimes called logarithmic), the logistic and the Gompertz (Table 25.1). These may be applied in continuous time, i.e. where the population is at least potentially measured and considered at every instant. This is usually appropriate for diseases with many successive, overlapping generations during one cropping season. The models may also be applied in discrete time, suitable for applications where generations of pathogen do not overlap and measurements are made once per generation (typically annually).

Data on changes in disease over time are needed to fit the parameters of any model. The ideal data set has many points closely spaced during times when disease is changing rapidly, and enough data when disease is scarce or common to determine final asymptotes and initial amounts of disease effectively. This implies a bare minimum of nine to ten data points; if this is not possible, other approaches to data analysis should be considered. The parameters of a non-linear growth-rate model can be estimated efficiently by any of several statistical programmes. There are several problems that need to be considered. The common procedure of transforming data to a log or logit scale and then doing a conventional linear regression can introduce severe bias. This is because very small changes in disease observed, if severity or incidence is close to 0% or to 100%, cause large changes in the transformed values. In models with an upper asymptote, the estimates of the rate and asymptote parameters will depend on each other. Assuming an upper asymptote of 100% disease will bias estimates of rates downwards (Neher and Campbell, 1992) if the true asymptote is less. This could have serious effects if the rates are then used to judge when actions need to be taken in a crop. Lastly, there will often be correlations between the 'errors' on different dates, which require very specialist statistical input.

Less often considered, but equally serious, are the different error structures applying to observations at low and medium disease severities. Relative bias (i.e. the ratio between the true mean value and the mean value that would be realized in many repeat samples) and proportionate error (i.e. the standard deviation as a proportion of the mean) are much larger when disease is scarce than when it is common (see Chapter 26). The bias is almost always upwards, so small amounts of disease are overestimated. The effect is to reduce estimates of growth rate in a systematic way that is undetectable during the analysis. Although proportionate error will often be large when disease is scarce, the absolute error is usually small. Statistical software will have assumptions about the variability of observations built in which may be incorrect for disease

Table 25.1. The measured pathogen population is symbolized by y and time by t. The population that would be reached eventually is symbolized K. The rate parameter in each case is symbolized r. e is a mathematical constant, close to 2.718. Since r represents the linear rate of increase of a different function of the data for each equation, it will have a different value if each equation is fitted to the same data set. If the maximum of y in a data set is much smaller than K, the exponential and logistic estimates of r will be very similar; it will also be impossible to obtain good estimates of K. If disease is measured as a proportion of the host population or area, K is sometimes assumed to be 1. However, this assumption has in practice often proved to be false, which leads to very misleading estimates of r.

		Exponential	Logistic	Gompertz
Continuous	Function	$y = y_0 e^{rt}$	$y = \dfrac{Ky_0 e^{rt}}{1 + y_0 e^{rt}}$ or $\dfrac{y}{K-y} = \dfrac{y_0}{K-y_0}\, e^{rt}$	$y = K\left(\dfrac{y_0}{K}\right)^{e^{rt}}$
	Inverse	$\log_e y = \log_e y_0 + rt$	$\log_e\left(\dfrac{y}{K-y}\right) = \log_e\left(\dfrac{y}{K-y}\right) + rt$	$-\log_e\left(-\log_e\left(\dfrac{y}{K}\right)\right) = -\log_e\left(-\log_e\left(\dfrac{y_0}{K}\right)\right) + rt$
Discrete	Function	$y_t = \lambda^t y$	$y_t = \lambda\, y_{t-1}\left(1 - \dfrac{y_{t-1}}{K}\right)$	[a]Unlikely to be useful
	Inverse	$\log_e y = \log_e y_0 + rt$	None	None

[a]This function has rather extreme density-dependence, and careful consideration needs to be given to whether it is appropriate (May, 1980).

data; it is desirable to check these assumptions. The generalized linear model framework is a useful one, available for example in Genstat (NAG, Oxford, UK) and SAS (SAS Institute, Carey, North Carolina, USA). A good introduction to this is Crawley (1993). The method allows the error structure of the data to be specified explicitly and independently of the statistical model describing how mean disease changes with time. (Generalized linear models should not be confused with the general linear model facility in SAS.)

To choose among the various models requires consideration of both statistical measures of fit and the purpose of the model. Goodness-of-fit measures such as R^{*2} help, but it is vital to look at residual patterns, and to consider reasons for good and bad fit. Where possible there is much to be said for complementing proper statistically valid estimation of a model chosen for sound reasons with visual plots on log and logit scales, to appreciate better the patterns visible. (Such plots are not suitable for estimating parameters because the errors around the data points are distorted by the transformation, as discussed above.) It is also desirable to consider the consequences of any failure of fit for the purpose to which the model is to be put. For example, a logistic fitted to data that approximate a Gompertz will tend to underestimate epidemic progress early on, while a Gompertz fitted to a more symmetric epidemic will do the reverse. Whether this is important depends on the use to which the results will be put. The difference is likely to be slight, but, as mentioned above, fitting a function with an asymptote of 100% to data with a true asymptote that is lower may have a much more serious effect. For forecasting purposes, other uncertainties and sampling difficulties are likely to be much more important than the relatively small differences among models with broadly sigmoidal behaviour.

Interpretation of the parameters of non-linear models is not always easy, and data are often too sparse (relative to measurement error) to allow good fits. It may often be preferable to regard progress data simply as providing information about rates of disease or pathogen increase over time, rather than trying to fit a unified model over an entire season. For diseases that multiply within a crop, it is likely to be sensible to refer rates of increase to a *per capita* basis. In this case it is useful to plot the data on logarithmic scales, estimate *per capita* rates of increase from the slopes, and then look for factors determining differences in these *per capita* rates of increase. If natural logarithms (logarithms to base e) are used, the slope of a graph (i.e. increase in $\log_e$ disease/time) is the *per capita* rate of increase. This slope can then be related to biological and physical factors.

One factor causing decrease in rate of multiplication is inevitably crowding. This will slow pathogen increase close to initial infections even when the pathogen is uncommon in the crop as a whole. If pathogen dispersal distances are quite large it may be reasonable to regard all plants in a crop as equally likely to receive propagules, in which case the infection rate of propagules will just be directly proportional to the frequency of healthy plants in the crop. This is a sensible null hypothesis concerning the effects of crowding, and can be tested by plotting data transformed by the multiple infection transformation (Zadoks and Schein, 1979).

Increase at a steady exponential rate slowed only by crowding implies the logistic, symmetric pattern of disease increase (Table 25.1). Perhaps not surprisingly this pattern is rarely an exact fit to disease progress curves (Berger, 1981), but it is often a good approximation. However, it is unlikely that levels of disease where crowding becomes important are of great interest for forecasts because intervention will usually be desirable much earlier. Decisions made when disease severity has reached 30–40% are likely to be too late and the impact of disease on the crop will be catastrophic. Furthermore, large absolute errors in forecasting very severe disease are unlikely to be important in reaching management decisions. The difference between forecasting 70% crop loss and 50% crop loss is unlikely to alter decisions substantially.

A problem in applying many simple models of disease progress is that their biological justification assumes the host is not growing. When the host is growing the models may not apply; and if they do, it can only be by coincidence. In fact, relative disease severity may appear static for some parts of the season, or decline when host growth outweighs pathogen growth. To deduce how favourable weather, cultivar and other factors are for the pathogen requires that pathogen and host growth be measured separately. In turn this means that disease measurement needs to be related to host tissue turnover. For example, if disease can only be seen on the youngest three leaves, it is important none the less to keep track of absolute leaf position, so that host growth and pathogen growth can be disentangled.

Beresford and Royle (1988) introduced the concept of pathocron to describe the ratio between the leaf emergence interval and the pathogen incubation period; disease cannot be observed on leaves younger than the pathocron, and monitoring them is not only pointless, it can be actively misleading.

An important recent observation is the dramatic effect that increasing plant resistance or decreasing favourability of the environment can have on the *variability* of results. Slight differences in rate or initial pathogen abundance arising by chance may produce very large final differences (Kleczkowski *et al.*, 1996). Such fluctuations are intrinsic to the system and cannot be forecast by observing external factors. The information required for disease management in a system with behaviour like this is the size of the variation that may be expected under apparently identical conditions.

Simulations

Considerable interest has been shown in the past in building up predictions of population change under varying weather conditions by using experimentally determined relationships between the rates of individual components of the life cycle and environmental factors. For example, suppose (1) it were required to calculate changes in disease on a host that was not actively growing, and (2) a pathogen produced short-lived spores that required rain to disperse and infect. Then, given information about the relationships between temperature and latent period, temperature and spore production, rainfall and successful spore dispersal, rainfall, temperature and infection success, and initial information about the amount and age of pathogen in the crop, it should be possible to calculate population changes for a period thereafter (France, 1984). Such an approach requires a great deal of information and is usually impossible without some simplifying assumptions. The field has a considerable literature and some interesting successes in certain areas of study; however, in plant disease epidemiology it has so far led only rarely to models used successfully for active advice. Exceptions exist. One example comes from north Italy, where a simulation model of wheat brown rust has given excellent predictions and is stated to be in routine use to assess the need for fungicide spraying (Rossi *et al.*, 1997).

The biological work involved in constructing a simulation model in which all parameters are well estimated is very considerable. It is also often necessary to simplify relationships where data are impossible to gather accurately in the field. Typical areas in which dramatic simplifying assumptions are needed are: the relation between rainfall and leaf wetness (Huber and Gillespie, 1992); the types of wetness needed by a pathogen to infect or sporulate (Royle and Butler, 1986); the relation between host resistance and the population density of the pathogen growing on it; and the process of spore dispersal and deposition (Aylor, 1990). The effects of different possible methods of creating such simplified links on model output require very careful examination, by repeatedly examining the model with data sets including extreme cases, under different assumptions. If many different such simplifications can be identified, a complete testing of their consequences may be impractical. In such cases it is important to explore a subset of possible combinations in an organized way. For example, it may be appropriate to take a random sample of possible simplified models for detailed examination.

Software that makes model construction and documentation relatively easy is widely available. A typical package is Modelmaker (Cherwell Scientific, Oxford, UK), which has a useful feature for fitting a model to observed data sets, and for repeatedly running models with slight variations in the parameters or input data, so as to give some indication of the uncertainty in model results. However, it should be remembered that the single most likely source of error is in model structure – omitting relationships or important factors or including erroneous ones – about which no mechanical procedure can provide any guidance.

Transfer between crops and large-scale processes

There has been much less interest in modelling the transfer of pathogens between crops and the survival of pathogens during the off-season. This is not because the processes are unimportant in determining the effects of disease on crop yield, but because pathogens are usually rare during the off-season and information on population changes is therefore very hard to gather. Some information may often be gleaned by examining changes in disease related to different sowing dates or distances from previous crops. Little quantitative work suitable for use in management has been done that relates the popularity of a crop in an area or distance to the nearest previous crop to the need for particular control actions. An intriguing exception is the work of de Jong (1996) on the efficiency of control achieved by roguing transplanted leeks infected with *Puccinia porri* in relation to the distance between leek fields.

Forecasting and Decision Making

A forecast must be based on a factor that is usually or commonly limiting to the population processes; a model is needed to specify how the factor acts to limit the population. The variety is wide but some examples can illustrate the range. In *Lettuce mosaic potyvirus* in the central California valley, the final level of disease was simply and closely related to the proportion of infected seed (Zink *et al.*, 1956). In Florida, *Colletotrichum*-induced postbloom citrus drop is limited by, and can be forecast using, rainfall and a measurement of inoculum during flowering (Timmer and Zitko, 1996); in more humid areas, these factors are not limiting and cannot be used to guide sprays. In UK cereal growing, the risk of damage to cereals by *Mycosphaerella graminicola* is related to the risk of serious disease on the top two to three leaves of the crop. This is limited by the probability of rain moving spores to upper leaves, which is in turn limited by the period during which upper and lower leaves overlap (Lovell *et al.*, 1997). In this way the rate of stem extension during the early summer becomes the limiting factor for the epidemic, which can be used to assess risk and guide spray decisions.

Data and experiments needed

The direction of any research intended to provide advice on risk from a particular disease must depend on what is already known, and cannot be separated from the study of the population dynamics of the pathogen. Many of the data needed have already been discussed above; they can be summarized as repeated sets of information on host growth, on pathogen population or disease development within the host, on evolution within that population, and on environmental conditions. Useful information will be much easier to gather if the life cycle is already understood; however, where the life cycle is incompletely understood such data may still be useful, and also give useful insights into the life cycle. The data must come from observations, repeated in different environments and years, in which the information needed is recorded, preferably in strictly comparable conditions (see Chapter 26). Ideally factors believed to be limiting will also be manipulated in controlled experiments so as to reduce confounding with other factors. For example, if initial inoculum is believed to limit disease development, it may be possible to add extra cultured inoculum to some plots.

Between crops, it is hard to say anything general about the data required without specifying how the disease survives and transfers to the new crop. The factors likely to be involved are: the weather during the over-seasoning period; the distance travelled by the propagules produced by the over-seasoning inoculum and the area density of sources (the risk of many diseases will increase the more common a crop is, because a new crop will on average be closer to a source of inoculum) (Sun and Zeng, 1994); the survival time of perennial structures such as sclerotia; the permissiveness of intervening crops or fallows; and the weather or crop susceptibility during the period when propagules are infecting a new crop. An example of what can be done is the set of 'stem rust rules' developed to predict severe epidemics of *Puccinia graminis* on wheat in north India (Nagarajan and Singh, 1975). These are based on synoptic weather patterns likely to lead early in crop growth to the coincidence of long-distance spore movement with conditions suitable for infection.

Climate and weather monitoring

Standard climatic data come from instruments displayed in a specific way, which is designed to be as reproducible as possible from sampling location to sampling location. These data are widely available. If necessary, complete weather stations and software to handle the data can be bought ready to use. There are three distinct problems in using this kind of data. First is sparsity of sampling. We need to know rainfall on the crop of interest, not on another crop, and rainfall can vary on a very small scale. However, the concept of measured rainfall used here is still the standard one – it describes the rain captured by a rain gauge set out in a particular way. It does not tell us how wet each square centimetre of ground actually becomes, which exemplifies the second problem: that the crop is immensely heterogeneous and modifies its environment. It is unlikely to be useful to attempt to monitor on a very small scale, because standardization will be harder to achieve and estimates of averages or population distributions will be harder to make than from standardized instrument exposures (Penrose and Nicol, 1996). The third problem is that the variables controlling the life cycle of a pathogen are likely to be very indirectly linked to the standard measures. For the purposes of understanding disease progress it will be useful to know how the microclimate affects disease, for example, that free surface water is needed for infection or that gusts greater than a certain speed are needed to detach spores (Aylor, 1990; Geagea *et al.*, 1997). However, to try directly to measure aspects of the environment affecting pathogens is very difficult, both because the instrumentation is often not available or too delicate to employ continuously and routinely, and because of the sampling problems already mentioned.

The best compromise is probably to try to link the available, standardized measurements to pathogen behaviour empirically, either with or without using intermediate modelling links between the measured variables and the microclimatic variables that directly influence pathogens. For example, measurements of relative humidity and temperature differences between air and ground might be used to deduce dew deposition. Basic physics may be used to deduce drying times of droplets of different sizes in different positions in the canopy, to arrive at estimates of the proportion of leaf area experiencing certain dew periods (Zhang and Gillespie, 1990; Huber and Gillespie, 1992). The research effort remains very large. A shorter-term approach would simply be to look for relationships between dew-forming periods based on relative humidity and temperature gradients, and infection observed in inoculated plants. In either case, the data provide a measure of how favourable conditions were for pathogen infection, which may form part of a prediction of disease severity.

The data most likely to be useful are: temperature, because all the organisms involved have vital rates greatly affected by temperature; rainfall, because this affects both transfer of pathogens, loss of spores from the air and patterns of persistence of wetness/high humidity in infection courts; humidity, because this influences the humidity at the infection courts; insolation, for the same reason and because it affects host condition and water balance; and wind, because it affects pathogen transport and is an indicator of general weather systems. Leaf wetness is in principle useful but only if a standardized way of exposing the instrument can be developed (Huband and Butler, 1984; Penrose and Nicol, 1996). Generalized weather system information (e.g. anticyclonic) and wind direction can both be important, especially if transfer between crops is important, or as predictors of particular conditions such as dew or frost.

Finding correlates of loss or disease

Two approaches have been taken to search for relationships between disease and weather data, which may be characterized as the open and the focused. In an open search, the data sets are simply searched for climatic variables that appear to correlate well with disease measures, such as severity at a particular date. This search can be quite systematic, but can only look at functions of the variables that have been specified by the researcher, such as sums, averages or counts of days with particular properties. The method will tend to miss unexpected combinations of conditions, for example, sunny days with frost overnight (which could predispose to certain diseases), unless these are explicitly searched for. If a relationship is found, the challenge is then to discover the biological reason why it works. For example, Daamen and Stol (1992) have shown a strong correlation between *Septoria* severity in wheat in The Netherlands and the amount of sunshine in the previous August. It is possible that this reflects slow decay of inoculum in hot, sunny years, but corroborative data are obviously needed before the factor could safely be used in risk prediction.

This approach has led to the publication of a number of successful examples (Coakley, 1988), and has had some forceful advocates. However, the significance of correlations found is hard to judge since so many have been tested. (For example, repeated searches were made for relationships in artificial data sets in which no true relationship existed. Disease levels in 15 years were represented by 15 random numbers between 0 and 1, and weather by 15 lists of 360 completely random numbers, each representing a year's data on a single 'weather' variable. Correlations were calculated between the 'disease' and averages over subsets of all possible durations less than a year in the 'weather' data, and the best correlation found (M.W. Shaw, unpublished data). The 5% significance level for the best correlation was actually 0.87, rather than 0.44 given in tables for a single conventional correlation.) Only tests in several further years, not used in model development, can determine how good the relationships really are; many conventional tests of the robustness of a correlation are invalid because of the selection of the data. Resampling methods might be adapted to provide better guides to the probability of a particular relationship arising by chance, and should now be computationally feasible, if more demanding than pathologists have been accustomed to (Davison and Hinkley, 1997).

In the focused approach, the researcher knows the climatic variable involved and looks at this variable, possibly in combination with others or in a wide variety of possible forms, simply to see whether it does relate. This avoids the pitfalls of coming up with a random relation among variables, but at the cost of providing no guidance as to what to do next if the relationship is poor. It also depends on good knowledge of which climatic variable is limiting and approximately when in the season, which may not be easy to deduce correctly from the pathogen life cycle.

Common problems in both approaches are the small size of good comparable data sets, the strong intercorrelation there tends to be between ostensibly different locations (e.g. if England has a hot summer, all sites in England will tend to have a hot summer; if disease is low, many sites really only correspond to a single degree of freedom).

Most other forecast methods posit some model of how the pathogen population increases in the crop and use this to obtain the possibly non-linear relation between environmental factors and disease. The simplest model is one of exponential growth; the next simplest is of logistic increase. For example, in the EPIPRE system of generating spray advice for wheat (Daamen, 1991), logistic rates of increase were used to predict amounts of disease in the future based on that at present, according to the logistic equation

$$(futuredisease) = presentdisease\ e^{rt}/(1 + presentdisease\ e^{rt})$$

The rate parameter r was then modified by rules embedded in the program to allow for soil types, cultivars, etc; these rules were based on observations of rates observed in field experiments made in different locations, cultivars, etc. Similar systems are now incorporated in a number of computer-based decision support systems in many countries.

Risk versus forecast

Distinctions have sometimes been drawn between risk assessment and forecasting. The difference is largely in when the forecast is made and in the focus of the advice: a risk assessment tends to assess directly the question of whether a management action will be profitable, whereas a forecast answers an intermediate question about the severity of disease in the future. Provided the focus remains on providing effective advice there seems little point in worrying about the distinction.

Forecasting is likely to be useful in two sets of circumstances. First, when disease severity or timing varies greatly from year to year or place to place, so that it is clear that some management actions are sometimes necessary and sometimes not. Secondly, when consumer pressure, material, application or environmental costs make it essential to ensure very accurate timing of management actions. Generally speaking the actions most likely to be variable in a grower's plans are environmental modifications such as ventilation, heating or irrigation, and agrochemical applications. Forecasts therefore need to address the application, timing and size of these variables.

What sorts of forecasts are possible?

If a pathogen infects at a particular crop growth stage, then management actions and forecasts that support them should be based on the host and pathogen condition at that stage. Even advance knowledge of exactly when a growth stage will be reached may be very useful. In some cases good phenological models of plant development are available, which will give early warning of when flowering (for example) will occur (Jamieson et al., 1998; Wang and Engel, 1998). By itself, obviously, a host-based forecast does not allow actions to be omitted in any particular season, but availability of host at times when pathogen propagules are likely to be present or infective may determine whether or not a spray is needed.

Because most host–pathogen relationships are weather-sensitive, the weather can limit the severity of disease in many pathosystems, by modifying the rate of increase of the pathogen or the damage resulting from it. If the limiting weather conditions can be determined, as discussed above, it may be possible to use them to guide management actions, by two routes. First, it may be possible to produce a true forecast of disease severity based on weather factors alone. Secondly, and much more commonly in practice, knowledge of the conditions in which the pathogen has a rapid rate of multiplication can be used to construct an index or complicated function of weather variables that should be correlated to pathogen growth rate. In turn, this is often correlated to first appearance of disease or to disease severity; this should be true provided the initial amount of inoculum in the crop is not overwhelmingly more important. The value of this index can then be experimentally correlated to treatment frequency or timing, usually picking treatments by intuition and modifying them ad hoc to avoid excessive risk or excessive treatment. This has proved extremely useful in a number of intensively managed crops (Gleason et al., 1995), but is pointless and possibly counterproductive if growth-rate is not the limiting factor (Ellison et al., 1998).

In some pathosystems, inoculum is regularly a factor limiting disease severity at the time of infection of the harvested parts. If a simple means of assessing the amount of inoculum is available it can be used to guide management practice. The means of assessing inoculum must depend on the pathogen. Examples include: the indices of migratory aphid numbers and infectivity published via the Rothamsted Experiment Station in the UK (Plumb and Carter, 1991); estimates of the proportions of oilseed rape (Canola) petals infected by Sclerotinia sclerotiorum at flowering (Turkington and Morrall, 1993); the experimental systems under consideration at the time of writing involving spore traps directly linked to quantitative immunoassays for Alternaria spp. and Botrytis cinerea (Bossi and Dewey, 1992); or simply visual detection of disease after a fixed scouting effort. There are several requirements that must be fulfilled before such an inoculum-based approach will work. In particular, inoculum must be very uniformly distributed, since otherwise very intensive sampling will be needed to determine inoculum levels. This is unlikely to be economic. Modern immunological and molecular diagnostic techniques may obviate some problems in identifying the amount or type of inoculum present, but they do not overcome problems with sampling.

The discussion above should have made it clear that only in a minority of cases will any one of these approaches function effectively; several possible limiting factors may need to be considered simultaneously to produce an effective forecast or risk estimate. It is in these circumstances that simulation models are often considered. However, as discussed already, the field experimental data required to produce an adequate model rapidly become very daunting. Furthermore, if reliable relationships are hard to find between severity and the factors operating in the current season that are expected to control it, it may be that parts of the system that are actually important have been omitted. These might include weather relationships or interactions with predators or hyperparasites during the off-season, or the amount of disease in the previous season. Once again, the research time required to establish securely relationships which operate only occasionally will only be feasible for very important targets in very stable agroecosystems.

Problems with transfer to users

The history of plant pathology is littered with proposed forecasting schemes that have seen no use in practice. Several problems may make an apparently wonderful scheme fail in practice. First, the scheme may be too imprecise to be useful, or be inadequately tested. With very short research funding cycles this becomes almost inevitable: timescales devised for molecular biology or graduate training are simply inappropriate for testing forecasting schemes over many seasons. Secondly, the savings in time or money may be too small or, if management complexity becomes too great, actually negative. Thirdly, an improvement in average profitability may be bought at the cost of increased risk, in effect asking a grower to sacrifice their insurance policy. Fourthly, the informal and intuitive rules already used by growers may be close to optimal without the explicit scheme. Fifthly, there may be no suitable management action useable at the time the forecast or risk estimate can be made; this is a common situation, for example, with protectant fungicides or diseases that develop very rapidly relative to the plant and grower reaction times.

In successfully implemented schemes, the first three problems have usually been overcome by involving growers, so that the scheme evolves with understanding and advice from growers, and their support for continuing refinement. The mental model of development–testing–implementation is not appropriate here, since testing and implementation are intermingled. However, for this to be an effective development route, there has to be a good promise of savings and acceptable levels of risk right from the start. Growers interested in innovation may then become involved; as the methods are developed, more and more sceptical growers will be drawn in (Guerin and Guerin, 1994). Schemes which promise large savings only if close to perfect, and which therefore need lengthy testing, are unlikely to be successful, and run the risk of being overtaken by other technological changes.

References

Aylor, D.E. (1990) The role of wind in the dispersal of fungal pathogens. *Annual Review of Phytopathology* 28, 73–92.

Begon, M., Mortimer, M. and Thompson, D.J. (1996) *Population Ecology: a Unified Study of Animals and Plants*, 3rd edn. Blackwell, Oxford, UK, 247 pp.

Beresford, R.M. and Royle, D.J. (1988) Relationships between leaf emergence and latent period for leaf rust (*Puccinia hordei*) on spring barley, and their significance for disease monitoring. *Zeitschrift für Pflanzenkrankheiten und Pflanzenschutz* 95, 361–371.

Berger, R.D. (1981) Comparison of the Gompertz and logistic equations to describe plant disease progress. *Phytopathology* 71, 716–719.

Bossi, R. and Dewey, F.M. (1992) Development of a monoclonal antibody based immuno-detection assay for *Botrytis cinerea*. *Plant Pathology* 41, 472–482.

Campbell, C.L. and Madden, L.V. (1990) *Introduction to Plant Disease Epidemiology*. John Wiley & Sons, New York, USA, 532 pp.

Coakley, S.M. (1988) Variation in climate and prediction of disease in plants. *Annual Review of Phytopathology* 26, 163–181.

Crawley, M.J. (1993) *GLIM for Ecologists*. Blackwell, Oxford, UK, 379 pp.

Daamen, R.A. (1991) Experiences with the cereal pest and disease management system EPIPRE in the Netherlands. *Danish Journal of Plant and Soil Science* 85, 77–87.

Daamen, R.A. and Stol, W. (1992) Surveys of cereal diseases and pests in the Netherlands. 5. Occurrence of *Septoria* spp. in winter wheat. *Netherlands Journal of Plant Pathology* 98, 369–376.

Davison, A.C. and Hinkley, D.V. (1997) *Bootstrap Methods and Their Application*. Cambridge University Press, Cambridge, UK, 257 pp.

Ellison, P., Ash, G. and McDonald, C. (1998) An expert system for the management of *Botrytis cinerea* in Australian vineyards 2. Validation. *Agricultural Systems* 56, 209–224.

France, J. (1984) *Mathematical Models in Agriculture: a Quantitative Approach to Problems in Agriculture and Related Sciences*. Butterworths, London, UK, 335 pp.

Geagea, L., Huber, L. and Sache, I. (1997) Removal of urediniospores of brown (*Puccinia recondita* f.sp. *tritici*) and yellow (*P. striiformis*) rusts of wheat from infected leaves submitted to a mechanical stress. *European Journal of Plant Pathology* 103, 785–793.

Gleason, M.L., Macnab, A.A., Pitblado, R.E., Ricker, M.D., East, D.A. and Latin, R.X. (1995) Disease warning systems for processing tomatoes in eastern north America: are we there yet? *Plant Disease* 79, 113–121.

Guerin, L.J. and Guerin, T.F. (1994) Constraints to the adoption of innovations in agricultural research and environmental management: a review. *Australian Journal of Experimental Agriculture* 34, 549–571.

Huband, N.D.S. and Butler, D.R. (1984) A comparison of wetness sensors for use with computer or microprocessor systems designed for disease forecasting. *British Crop Protection Conference: Pests and Diseases 1984*. BCPC Publications, Croydon, UK, pp. 633–638.

Huber, L. and Gillespie, T.J. (1992) Modeling leaf wetness in relation to plant disease epidemiology. *Annual Review of Phytopathology* 30, 553–579.

Jamieson, P.D., Brooking, I.R., Semenov, M.A. and Porter, J.R. (1998) Making sense of wheat development: a critique of methodology. *Field Crops Research* 55, 117–127.

de Jong, P.D. (1996) A model to study the effect of certification of planting material on the occurrence of leek rust. *European Journal of Plant Pathology* 102, 293–295.

Kleczkowski, A., Bailey, D.J. and Gilligan, C.A. (1996) Dynamically generated variability in plant–pathogen systems with biological control. *Proceedings of the Royal Society B* 263, 777–783.

Lovell, D.J., Parker, S.R., Hunter, T., Royle, D.J. and Coker, R.R. (1997) Influence of crop growth and structure on the risk of epidemics by *Mycosphaerella graminicola* (*Septoria tritici*) in winter wheat. *Plant Pathology* 46, 126–138.

May, R.M. (1980) Non-linear phenomena in ecology and epidemiology. *Annals of the New York Academy of Science* 357, 267–281.

Nagarajan, S. and Singh, H. (1975) Indian stem rust rules: an epidemiogical concept on the spread of wheat stem rust. *Plant Disease Reporter* 59, 670–672.

Neher, D.A. and Campbell, C.L. (1992) Underestimation of disease progress rates with the logistic, monomolecular and gompertz models when maximum disease intensity is less than 100 percent. *Phytopathology* 82, 811–815.

Penrose, L.J. and Nicol, H.I. (1996) Aspects of microclimate variation within apple tree canopies and between sites in relation to potential venturia inaequalis infection. *New Zealand Journal of Crop and Horticultural Science* 24, 259–266.

Plumb, R.T. and Carter, N. (1991) The use and validation of the infectivity index as a method of forecasting the need to control barley yellow dwarf virus in autumn sown crops in the United Kingdom. *Acta Phytopathologica et Entomologica Hungarica* 26, 59–62.

Rossi, V., Racca, P., Giosuè, S., Pancaldi, D. and Alberti, I. (1997) A simulation model for the development of brown rust epidemics in winter wheat. *European Journal of Plant Pathology* 103, 453–465.

Royle, D.J. and Butler, D.R. (1986) Epidemiological significance of liquid water in crop canopies and its role in disease forecasting. In: Ayres, P.G. and Boddy, L. (eds) *Water, Fungi and Plants*. Cambridge University Press, Cambridge, UK, pp. 139–156.

Sun, P. and Zeng, S.M. (1994) Modeling the interregional disease spread. *Zeitschrift für Pflanzenkrankheiten und Pflanzenschütz* 101, 545–549.

Thresh, J.M. (1983) Progress curves of plant virus disease. *Advances in Applied Biology* 8, 1–85.

Timmer, L.W. and Zitko, S.E. (1996) Evaluation of a model for prediction of postbloom fruit drop of citrus. *Plant Disease* 80, 380–383.

Turkington, T.K. and Morrall, R.A.A. (1993) Use of petal infestation to forecast sclerotinia stem rot of canola: the influence of inoculum variation over the flowering period and canopy density. *Phytopathology* 83, 682–689.

Wang, E.L. and Engel, T. (1998) Simulation of phenological development of wheat crops. *Agricultural Systems* 58, 1–24.

Zadoks, J.C. and Schein, R.D. (1979) *Epidemiology and Plant Disease Management*. Oxford University Press, New York, 427 pp.

Zhang, Y. and Gillespie, T.J. (1990) Estimating maximum droplet wetness duration on crops from nearby weather station data. *Agricultural and Forest Meteorology* 51, 145–158.

Zink, F.W., Grogan, R.G. and Welch, J.E. (1956) The effect of the percentage of seed transmission upon subsequent spread of lettuce mosaic virus. *Phytopathology* 46, 662–664.

Design of Experiments

J. Riley

IACR-Rothamsted, Harpenden, Hertfordshire AL5 2JQ, UK

Introduction

In order to determine the effect of any intervention on plant or insect material a study must be designed in such a way that the estimate of the effect of the intervention is not obscured by any other effect. Suppose a new fungicide is sprayed on to one coffee bush and a second is sprayed with a well-known fungicide. A comparison of disease incidence from the two bushes will indicate the efficacy of the new fungicide compared to the old. But if the bushes had different levels of disease incidence before they were sprayed, the estimate of the treatment comparison would not be reliable. Both bushes should be as similar as possible at the start, in their genotype, age, nutrient availability and levels of disease and pest infestation.

This is a very simple example, but one that underlies the efficiency of all experimental design. Complex theory of experimental design has developed to support many different agricultural and environmental studies. To assist in the understanding of the most important principles of this theory, some fundamental ideas are presented here.

Statistical Requirements

To achieve appropriate and accurate recommendations from experiments good design needs to be employed, taking account of the ideas of *replication*, *randomization* and *independence*. When treatments are applied to experimental units, such as coffee trees, plots of potatoes, or individually potted sunflower plants, adequate *replication* of each differentially treated

©CAB *International* 2002. *Plant Pathologist's Pocketbook*
(eds J.M. Waller, J.M. Lenné and S.J. Waller)

experimental unit is essential to estimate the treatment effects. Variability that may exist between the experimental units prior to treatment application can also be estimated if enough experimental units are available. The relative sizes of the 'between treatment' variability and the existing 'between unit' variability indicates the effectiveness of the treatments. The optimum number of replicate experimental units depends on the degree of precision required in the estimation process and the expected levels of variability.

To prevent bias in the results *randomization* of treatment application to experimental units is important. If all the trees in one corner of an experiment receive the same treatment, and this corner benefits from residual fertilizer from previous studies, these trees will be advantaged and the effect of that treatment will appear to be much larger than the effect of the other treatments in the less advantaged area. It is necessary to ensure all treatments are equally likely to be in that corner so that they all have the same chance of success. Randomization of treatments to trees achieves this equitable allocation.

To satisfy the requirements of many commonly used statistical tests, it is desirable that data measured on each experimental unit are *independent* of each other. Thus the treated trees should not influence each other. If sufficient replication has been obtained and if treatments have been allocated randomly to trees, the likelihood is that independence will be achieved.

In many pathological trials, measurements are available for samples, or subunits, taken from within each experimental unit. These provide treatment replication that is neither randomized nor independent. The samples in any one experimental unit are all treated the same and are interdependent in the same environment. For example, three branches measured on one tree are likely to give more similar results than three branches taken from another tree regardless of the treatments each tree receives. Thus *independence* between branches, or subunits, does not exist. The variability between these non-independent branches within the trees must be estimated separately from the independent variability between the experimental units, here the whole trees. The incorrect use in statistical tests of non-randomized, and thus interdependent, measures in place of true, randomized replication has been called *pseudoreplication* by Hurlbert (1984). He gives an extensive description of appropriate and inappropriate design structures and demonstrates the extent of misusage of pseudoreplication in published ecological studies.

Specific Designs

Particular designs have been developed which are applicable to many standard experimental studies and which, if used correctly, are robust to the influence of sources of variability external to the study. These include completely randomized designs, randomized complete block designs, Latin squares, split plot designs and incomplete block designs. These experimental design structures are distinct from treatment design structures, such as new treatments versus controls, treatments involving increased dosages and factorial treatment combinations. Such treatment design structures can be incorporated in any of the above experimental design structures.

Experimental design structures

Where the environment in which the experiment is to be done is known to be homogeneous, replicated plots of the chosen different treatments can be placed randomly. This is known as a completely randomized design. Estimates of the variation between the treatments and the average variation between the units within the treatments can then be determined using analysis of variance techniques. Estimates of the treatment effects and their standard errors can be determined from the model underlying the analysis of variance. All good statistical software packages will provide this information. An example of a completely randomized design is given in Fig. 26.1.

When the environment is not homogeneous, unless an extra design feature is incorporated, the estimates of the treatment effects will be biased. Such a design feature is commonly called blocking and works in the following way: if the experimental area is largely dry but a small area is near a stream and is very moist, then some plots may need to be located in the moist area, in order to achieve the size of plot and desired level of replication. To ensure that no one treatment is wholly, and unfairly, located in the moist area, an equal number of plots of each treatment should be placed in that area. An equal number of plots of each treatment will be placed in the dry area. The plots in the moist area are said to comprise a block. The plots in the dry area are said to comprise another block. Using analysis of variance techniques, the variability between the different blocks can be determined, as can the variability between the treatments and the variability within the treatments. This is an example of a randomized complete block design, here with two blocks. Note the following important features.

- Blocks do not need to be laid out in the field in a square or rectangular shape. Blocking is not so much a physical structure as a statistical device to provide more efficient estimation of treatment effects. Blocking ensures that all treatments are equally represented in each of the different environments. Within a set of plots represented by any one block, the environment should be homogeneous.
- Heterogeneity – and thus variability – from one block to another is to be expected as it is designed into the experiment. A significant block effect in an analysis of variance is desirable. An insignificant block effect should be

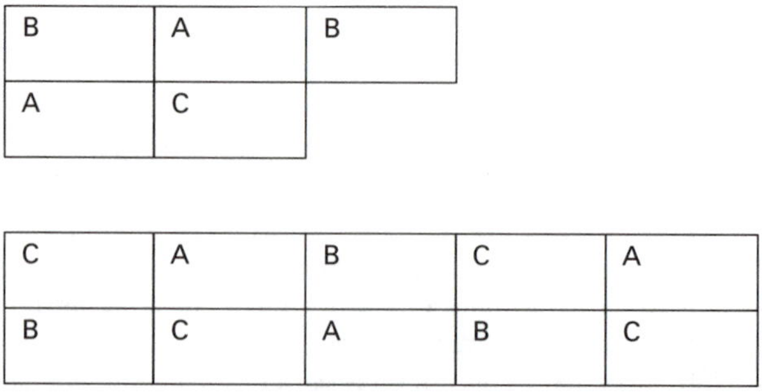

Fig. 26.1. A completely randomized design for three treatments A, B, C with five replications of each. The plots are located to ensure homogeneity of environment across all plots.

an indication to explore the data or the chosen model more closely: the blocks may appear ineffective, but they may in fact have been located wrongly in the field.

- Replicates and blocks are not interchangeable terms although blocks usually correspond to sets of replicated treatments. Within any one block there can be any number of replicates of each treatment.

An example of a randomized complete block design with two blocks of three treatments, A, B, C, each replicated twice in a block is given in Fig. 26.2. In a randomized complete block design, a single blocking structure is employed. Sometimes, the nature of the environmental variability demands that more than one blocking structure be used. In a Latin square design two blocking structures are used, to account for variability in two directions. The first blocking structure may relate for example to variation in the field; the second, usually physically at right angles to the gradient of field variation, may relate to distance from the spread of pest or disease infestation. An example of a Latin square arrangement for five treatments A, B, C, D, E is given in Fig. 26.3.

In other designs, superblocks are used, each one corresponding to a complete copy of the experiment and each copy containing blocks appropriate to control the local variation. These superblocks are often called replicates. A further set of useful design structures that accommodate small-scale variation in the field are incomplete block designs, incorporating balanced designs, lattices and resolvable designs. Detail of these designs and others, together with appropriate analyses, can be found in Cochran and Cox (1957). Williams and Matheson (1994) provide an up-to-date summary of such designs and methods of computer-based mixed-model statistical analysis. Bateman *et al.* (1994) and Bauske *et al.* (1994) provide examples to compare their efficiencies.

Other designs exist for different situations. One design structure useful for examining the spread of diseases from a defined source are systematic designs. These should be used carefully, for, although necessary in such situations, they must not be laid down to coincide with trends in fertility in the experimental area otherwise the treatments in question will not be unbiased. An example of the use of such a design is given in Jenkyn and Bainbridge (1973). Nelson

Block 1

A	B	C
C	A	B
PATH	PATH	PATH
B	C	A
A	B	C

Block 2

Fig. 26.2. A randomized complete block design, with two replicates of each treatment in each block. The blocks separate parts of the field dissected by a path.

A	B	C	D	E
B	C	D	E	A
C	A	E	B	D
D	E	A	C	B
E	D	B	A	C

Fig. 26.3. A Latin square arrangement for five treatments with columns and rows representing two sets of variability.

and Hartman (1987) provide an extensive summary of designs and field plot technique for biological and cultural treatments.

Treatment design structures

The purpose of an experiment will dictate the choice of treatments whose effects are to be estimated. This is not a trivial issue; treatments are often expensive, some require detailed monitoring, an understanding is needed of how treatments may interact with each other or with residual treatments in the soil and, most importantly, an appreciation of the treatment design structure is necessary to choose the most appropriate analysis of the ultimately collected data. Some common treatment structures include new treatments versus controls, treatments involving increased dosages of a substance and factorial treatment combinations.

When a set of new treatments is to be compared with a control treatment, it is assumed that much is already known about the control, but little about the new treatments. The treatment structure then leads us naturally to think of comparing any new treatment with the control or the average of some of the new treatments with the control. The experimental design is not complex; replicates of all of the treatments should be randomized to the experimental units. If environmental variation exists in the experimental area, then one or more blocking structures should be used to account for this, and the treatments allocated in a balanced way within the blocking structure.

Sometimes the control treatment is given greater replication; the decision to do this must be taken with great care. They provide extra degrees of freedom for error which may be useful in reducing the standard error of a treatment versus control comparison, and provide extra plots for afterthoughts during the experiment. A guideline, but not a hard and fast rule, to the number of replicates to choose for the control is the square root of the total number of treatments. However, if space or time are short, interest may lie primarily in estimation of the new treatment effects, and the number of control plots may be sacrificed. An excellent summary of the need for extra replication of nil plots is given by Dyke (1988).

Choice of treatments that involve increased dosages of a substance, such as levels A, B and C of a spray, assumes that you are interested in comparing the efficiencies of A, B and C. Plots with zero level of spray are necessary here so that the response curve, representing mean yields plotted against dosage, can be displayed. The design of the experiment is non-complex. The necessary number of replicates of each treatment are randomized in the experimental area and blocking structures are used to account for any unwanted variability already in existence. The analysis of the data should not consist of the calculation of comparisons between each spray level and the control. Rather it should consist of the precise estimation of the shape of the response curve, and the determination of the optimum response and the corresponding dosage. For a clear treatise on the subject see Perry (1986).

Sometimes more than one type of treatment is included in an experiment. Instead of including one set of varieties or one set of doses of fungicides, both could be included in combination. To ensure that every variety and every dose is fairly represented, the varieties and doses are combined in factorial structure. For two varieties V1 and V2 and three doses of fungicide F20, F40 and F60, the six factorial treatment combinations are V1F20, V1F40, V1F60 and V2F20, V2F40, V2F60. Thus each of the varieties will be tested with each dose of fungicide. In addition the degree to which the response of variety 1 to the increased doses of fungicide differs from the response of variety 2 to the increased doses can be estimated. This is commonly called the interaction of the treatments. For factorial designs and layouts for confounded factorial designs, see Cochran and Cox (1957) and John and Quenouille (1977).

The design of such factorial experiments is no more complex than experiments with only one treatment set. The six treatment combinations above are considered as the new set of treatments and can be located in a completely randomized design, a randomized complete block design or a Latin square design in the same way as an experiment with a single set of treatments. Considerations for the replication levels, blocking structure and layout and randomization are as specified above. Figure 26.4 shows the six treatment combinations of variety and fungicide randomized in four randomized blocks laid out in the field to allow for variation in the field.

A special design also exists for factorial treatment structures. This is a split plot design. Simply, this involves a completely randomized design, a randomized block design or a Latin square design where the first set of treatments is randomized to the plots, allowing for the appropriate blocking structure. Each plot is then subdivided (split) to form subplots, the number being equal to the number of treatments in the second set. The second set of treatments is then allocated randomly to each set of subplots in each plot. This design is very effective when practical necessity demands that the first set of treatments be applied to the larger plots. Thus they may be sprays which are more easily allocated to large plots or they may be irrigation schemes whose pipework is expensive or difficult to construct. Only under such constraints should they be used. They are not as efficient as factorial designs using completely randomized, randomized block or Latin square structures, as the treatment combinations are not randomized within the experimental area. The efficiency of the estimates of the first treatment set on the larger plots will be less than that of the estimates of the second treatment set. Cochran and Cox (1957) give a

Block 1	V2F40	V2F60	V1F20	V2F20	V1F60	V1F40
Block 2	V1F20	V2F20	V1F40	V2F60	V2F40	V1F60
Block 3	V1F60	V1F40	V2F60	V2F40	V1F20	V2F20

V2F20	V1F40	V2F60
V1F60	V2F40	V1F20

Block 4

Fig. 26.4.   A factorial treatment structure in a randomized block design.

clear description of the design and analysis of a split plot design and the calculation of appropriate standard errors for estimates of the treatment means.

Underlying Models

In any simple comparative experiment variability can be subdivided into that which is of primary interest, the treatment variability, and that which is typically termed 'residual variation'. This residual variation may result from several sources. Each unit of measurement can be described by a model containing terms representing the average measurement for all the units plus the extra effects caused by these two sources of variation. In pathology experiments, if a single measurement is taken on each experimental unit, there will be three terms in the following model, Model A:

individual unit = average for all units + effect of treatment + residual effect
measurement

where residual effect represents the variability between units within the treatments and which may result from locational features such as different soil types, surrounding crops or foliage, variable soil structures or differential plot management.

When the unit of measurement is one of a number of samples from a treated unit, the model will have four terms as in Model B:

sample   = average for   + effect of   + variability   + variability between
measurement   all samples   treatment   between units   samples within
                                              within treatments   units

where variability between samples within units may result from selection of unequal or non-identical samples or poor data recording or imprecise measurement equipment.

It is important to determine precisely the relative sizes of these sources of variability. The treatment variability is likely to be of most interest to the experimenter. However, if the other sources are not identified nor their effects isolated and quantified, the treatment effects may be obscured and recommendations based on the trial results will not be reliable.

When blocking structures are included in an experimental design, a term to account for the variation accounted for by the blocks must be included in the model. Thus for measurements collected from plots in a randomized complete block design, the underlying model, Model C, would be:

individual plot = average for + effect of block + effect of treatment + residual effect
measurement   all plots

For Latin square designs, two terms would be included to account for the two blocking structures, commonly called rows and columns. For split plot designs in a randomized block structure, the model would be as Model B. The difference would be in the analysis, where the treatment effects would be subdivided to account for the effects of each treatment set and their interactions with each other.

When the unit of measurement is one of a number of samples taken from a treated unit, and the experiment is subject to one or more blocking structures, the model is Model D:

sample   = average + effect + effect of + variability + variability
measurement   for all   of   treatment   between   between
               samples   block   units within   samples
                                                treatments   within units

It is possible to use analysis of variance to subdivide the variability in a set of measurements according to the appropriate model components. For Model A the analysis of variance table would appear as:

Source of Variation	Degrees of freedom	Sums of squares (ss)	Mean square (ms)
Treatment effect	$t-1$	Treatments ss	Treatment ms
Residual (between units)	$rt-t$	Residual ss	Residual ms
Total minus average	$rt-1$	Total ss	

where t = the number of treatments and r = the number of replicate units of each treatment.

An F test of differences between treatment effects can be calculated by dividing the treatment ms by the residual ms, and comparing the result with tabulated values of the F distribution with $t-1$ and $rt-t$ degrees of freedom. For Model B the analysis of variance table would appear as:

Source of Variation	Degrees of freedom	Sums of squares (ss)	Mean square (ms)
Treatment effect	$t-1$	Treatments ss	Treatment ms
Residual (between units)	$rt-t$	Between units ss	Between units ms
Residual (between samples within units)	$rtf-rt$	Between samples ss	Between samples ms
Total minus average	$rtf-1$	Total ss	

where f = the number of samples in each unit and rtf is the total number of samples in the whole experiment.

To test for differences between treatment effects, an F test may be done, as described above, using the treatment mean square and the residual (between units) mean square on $t - 1$ and $rt - t$ degrees of freedom. The residual mean square representing variability between samples within units should not be used in the F test. Nor should a combination of the residual between-units variability and the residual between samples within units variability be used. The treatments have been applied at the unit level and therefore their effects should be tested at the unit level.

The relative sizes of the residual (between-unit) variation and the residual (between samples within units) variation indicate which source of residual variation is most prevalent. An F test between the two using $rt - t$ and $rtf - rt$ degrees of freedom can be done if necessary. Typically the two differ. Information on their relative sizes can be used to design further trials more efficiently by minimizing the effects of the greater residual source.

Analysis of variance structures for Models C and D are:

Model C

Source of variation	Degrees of freedom	Sums of squares (ss)	Mean square (ms)
Block effect	$b - 1$	Blocks ss	Blocks ms
Treatment effect	$t - 1$	Treatments ss	Treatment ms
Residual (between units)	$rt - t$	Residual ss	Residual ms
Total minus average	$rtb - t - b + 1$	Total ss	

where t = the number of treatments, b = the number of blocks, r = the number of replicate units of each treatment and rtb = the total number of units in the whole experiment.

Model D

Source of variation	Degrees of freedom	Sums of squares (ss)	Mean square (ms)
Block effect	$b - 1$	Blocks ss	Blocks ms
Treatment effect	$t - 1$	Treatments ss	Treatment ms
Residual (between units)	$rtb - t - b + 1$	Between units ss	Between units ms
Residual (between samples within units)	$rtb(f - 1)$	Between samples ss	Between samples ms
Total minus average	$rtbf - 1$	Total ss	

where b = the number of blocks, t = the number of treatments, r = the number of replicates of units, f = the number of samples per unit and $rtbf$ = the total number of samples in the whole experiment.

In Model D as in Model B, tests of significance of the treatment effects should be done by comparing the treatment ms with the residual (between units) ms on $t - 1$ and $rtb - t - b + 1$ degrees of freedom.

Other Special Statistical Features

Whenever a model is fitted to data from a field experiment, each measured unit has a corresponding residual term representing variation that cannot be ascribed to known sources. The set of residuals from all the measured units consists of a set of numbers, some positive and some negative and which sum to zero. The square of each of these values, when added together, forms the residual mean square for each appropriate analysis when all known sources of variation have been accounted for. This set of residuals is invaluable to check for influential data values, appropriateness of design structures and the fit of the model. To check design structures, the plotting of these residuals in field layout can be informative. The residuals in the following field layout show that the first plot in Block 3 has an unusually low measurement and should be checked.

Block 1	−0.4	−0.2	0.1	0.3	0.2
Block 2	−0.3	−0.2	0.0	0.1	0.4
Block 3	−0.8	−0.1	0.0	0.5	0.4

They also show that the negative and positive residuals are not evenly distributed across the experimental area. This indicates that the blocking structures may have been laid out inappropriately. A more suitable division of the plots into blocks would have been at right angles to the direction of those chosen. One block should have been located within the area with negative residuals and another within the area with positive residuals.

Another use of residuals is to adjust plot measurements for the effects of immediately adjacent plots. Thus in variety trials, some varieties are larger than others and have a competitive advantage. In spraying trials, nil treatments or very low dosages of insecticide are often applied as well as very strong doses. Those plots with nil or very low doses may then harbour greater infestation and may influence the immediately adjacent plots to a greater or lesser extent depending on the strength of the treatments on these plots. The different treatments will not then be equally, or fairly, represented and their estimates will be biased.

There are two ways to handle this. The first is to design fairness into the experiment by using a design balanced for nearest neighbours. The second is to adjust for unfairness at the analysis stage by using a nearest neighbour analysis. A nearest neighbour design is one where each treatment has every other pair of treatments as its neighbours. Dyke and Shelley (1976) described the design of a fungicide trial on barley that was balanced for neighbours. The layout of the four treatments A, B, C, D was:

A, BCAD, ADBC, DBAC, ACBD, BDAC, DCAB, ABDC, BCDA, DCBA, B

Each foursome is a block, each treatment is neighboured by every other possible pair and the extra plots, A and B, at each end of the line are to provide the necessary neighbour influence. Since then much research into neighbour-balanced designs has been undertaken by statisticians and lists of appropriate designs are available (see, for example, Azais *et al.*, 1993; David *et al.*, 1996).

To adjust for influential neighbour effects at the analysis stage, the residual values are important. An early method (Papadakis, 1937) used for each plot the average of the residuals on the neighbouring plots as a covariate in the analysis of variance. This approach has been developed and explored to incorporate the creation of covariates from different functions of sets of residuals (Wilkinson *et al.*, 1983; Besag and Kempton, 1986). Jenkyn and Dyke (1985) and Hide *et al.* (1995) give examples relevant to plant pathology.

Data Quality and Disease Scoring

The importance of data quality has been acknowledged for many years. The impact of inaccurate measurement on compiled national statistics can severely distort the representativeness of crop production figures (FAO, 1982). Poor data quality can render an analysis nonsensical. Riley *et al.* (1983) illustrated how data recorded to increasing degrees of imprecision corresponded to increasingly meaningless results in analysis of variance tables, and Tricker (1990) demonstrated the effects of rounding of data on the robustness of a range of other popular statistical tests.

In plant pathology trials, disease scoring is important and, unless done well and at least consistently over the whole experiment, it can provide several distorted results. Fielding (1992) showed that marked differences exist between the ability of different people to score damage in leaves and that training is essential to establish standards against which damaged leaves could be assessed. Dyke (1988) demonstrated the effect on statistical calculations of a gross error of measurement and encouraged the regular scoring of plots throughout an experiment. He recommended marking scores of disease, soil quality and irregularity of growth in whole numbers on a sketch of the field layout with the treatment allocations not marked. Variation in disease within a plot should lead to the plot being scored an average of the internal scores, but clear notes should be kept of its appearance. The ultimate scores should then be used at the end of the experiment to determine whether unusual data values are linked to treatments and should be retained, or whether they are gross errors and should be replaced with missing values. For an example of scoring in disease trials see Juliatti and Maluf (1995) (see also Chapter 4).

Another useful technique to understand the presence of disease and its spread in time is to relate disease scores to climate data. This is particularly important in long-term experiments (Cross and Berrie, 1995; Kocks and Zadoks, 1996). The simplest way to examine changing patterns is to plot both disease incidence and weather variables on the same graph. Correlation of disease incidence with weather measurements in the period immediately preceding the appearance of the disease will indicate whether there is influence and this can be explored further.

Acknowledgements

IACR-Rothamsted is grant aided by the Biotechnology and Biological Sciences Research Council of the UK.

References

Azais, J.M., Bailey, R.A. and Monod, H. (1993) A catalogue of efficient neighbour designs with border plots. *Biometrics* 49, 1252–1261.

Bateman, G.L., Hornby, D., Payne, R.W. and Nicholls, P.H. (1994) Evaluation of fungicides applied to soil to control naturally-occurring take-all using a balanced-incomplete-block design and very small plots. *Annals of Applied Biology* 124, 241–251.

Bauske, E.M., Hewings, A.D., Kolb, F.L. and Carmer, S.G. (1994) Variability in enzyme-linked immunosorbent assays and control of experimental error by use of experimental designs. *Plant Disease* 78, 1206–1210.

Besag, J.E. and Kempton, R.A. (1986) Analysis of field experiments using spatial statistics. *Biometrics* 42, 231–251.

Cochran, W.G. and Cox, G.M. (1957) *Experimental Designs*. John Wiley & Sons, New York, USA, 611 pp.

Cross, J.V. and Berrie, A.M. (1995) Field experimentation on pests and diseases of apples and pears. *Aspects of Applied Biology* 43, 229–239.

David, O., Kempton, R.A. and Nevison, J.M. (1996) Designs for controlling interplot competition. *Journal of Agricultural Science, Cambridge* 127, 285–288.

Dyke, G.V.D. (1988) *Comparative Experiments with Field Crops*, 2nd edn. Griffin, London, UK, 262 pp.

Dyke, G.V.D. and Shelley, C.F. (1976) Serial designs balanced for effect of neighbours on both sides. *Journal of Agricultural Science, Cambridge* 87, 303–305.

FAO (1982) *Estimation of Crop Areas and Yields in Agricultural Statistics*. Food and Agriculture Organization, Rome, Italy, 104 pp. + annexes.

Fielding, W.J. (1992) Damage assessment by eye: some Caribbean observations. *Field Crops Research* 30, 183–186.

Hide, G.A., Welham, S.J., Read, P.J. and Ainsley, A.E. (1995) Influence of planting seed tubers with gangrene (*Phoma foveata*) and of neighbouring healthy, diseased and missing plants on the yield and size of potatoes. *Journal of Agricultural Science, Cambridge* 125, 51–60.

Hurlbert, S.H. (1984) Pseudoreplication and the design of ecological field experiments. *Ecological Monographs* 54, 187–211.

Jenkyn, J.F. and Bainbridge, A. (1973) Problems of designing field experiments with air-dispersed pathogens. *Rothamsted Experimental Station Annual Report for 1972*, Part 1, 133 pp.

Jenkyn, J.F. and Dyke, G.V. (1985) Interference between plots with plant pathogens. *Aspects of Applied Biology* 10, 75–85.

John, J.A. and Quenouille, M.H. (1977) *Experiments: Design and Analysis*. Charles Griffin, London, UK, 289 pp.

Juliatti, F.C. and Maluf, W.R. (1995) Inheritance of resistance to an isolate of Tospovirus (TSWV). *Fitopatologia Brasileira* 20, 39–47.

Kocks, C.G. and Zadoks, J.C. (1996) Cabbage refuse piles as sources of inoculum for black rot epidemics. *Plant Disease* 80, 789–792.

Nelson, L.A. and Hartman, J.R. (1987) Plant-diseases; epidemiology. Techniques, experimental design, field-experimentation in plant-pathology. In: *Biological and Cultural Tests for Control of Plant Diseases,* 1987, No. 2. American Phytopathological Society, St Paul, Minnesota, USA, 80 pp.

Papadakis, J.S. (1937) Methode statistique pour des experiences sur champ. *Bulletin de l'Institut pour l'Amelioration des Plantes a Salonique* 23.

Perry, J.N. (1986) Multiple comparison procedures: a dissenting view. *Journal of Economic Entomology* 79, 1149–1155.

Riley, J., Bekele, I. and Shrewsbury, B. (1983) How an analysis of variance is affected by the degree of precision of the data. *BIAS* 10, 18–43.

Tricker, A.R. (1990) The effect of rounding on the significance level of certain normal test statistics. *Journal of Applied Statistics* 17, 31–38.

Wilkinson, G.N., Eckert, S.R., Hancock, T.W. and Mayo, O. (1983) Nearest neighbour analysis of field experiments (with discussion). *Journal of the Royal Statistical Society, B* 45, 151–211.

Williams, E.R. and Matheson, A.C. (1994) *Experimental Design and Analysis for Use in Tree Improvement*. CSIRO, Melbourne, Australia, 169 pp.

Plant Health and Quarantine

J.M. Waller

CABI Bioscience UK Centre, Bakeham Lane, Egham, Surrey TW20 9TY, UK

Plant health and quarantine regulations are concerned with the principle of pathogen exclusion and sometimes with eradication. Movement of crop species throughout the world as agriculture has developed, has resulted in many crops being grown in regions away from their centres of origin or diversity and where their coevolved pests and pathogens are absent. This has also sometimes resulted in the development of 'new-encounter' pathogens that can then pose a threat to crops grown in their native regions. On a local scale, it may be necessary to prevent newly arrived pathogens from spreading within countries or regions and perhaps to attempt their eradication. On a smaller scale, it is desirable to prevent diseases from entering the crop in the first place and this can be done through the use of certified disease-free seed material. General accounts of aspects of plant health and quarantine are in Hewitt and Chiarappa (1977), Ebbels and King (1979) and Kahn (1989).

In its strict sense, plant quarantine refers to the holding of plants in isolation until they are believed to be healthy. Common usage has, however, broadened the meaning of the term, and plant quarantine is here taken to include other aspects of plant health concerning the regulation of the movement of living plants, living plant parts or plant products between politically defined territories or ecologically distinct parts of them. The terms intermediate quarantine and post-entry quarantine are used, respectively, to denote the detention of plants in isolation for inspection during transit or after arrival at their final destination.

Most countries are aware of the desirability of delaying for as long as possible the arrival of exotic pathogens and take action to prevent their spread by introducing legislation and setting up organizations to prevent their entry. Plant quarantine legislation varies from country to country but is now largely harmonized throughout most geographic regions. In most cases it restricts or prohibits the importation of the pests or pathogens themselves, plants on which they might be living, soil that might be infested, foodstuffs that might carry them, and packing materials, particularly those of plant origin. Good legislation is as brief and clear as possible, at the same time being easy to interpret, gives adequate protection without interfering more than is essential with trade, and contains only restrictions which are scientifically justifiable. The General Agreement on Tariffs and Trade (GATT) is explicit on the need for an importing country to base import regulations on a sound scientific basis, which requires adequate pest risk assessment (PRA). This must be based on biological and ecological knowledge of the pest organism[1] and of conditions relevant to its establishment as a pest organism in the importing country (Kranz, 1998).

When plants are imported there are certain principles that, if followed, ensure that as few risks as possible are taken.

1. Import from a country where, for the crop in question, pathogens that may be particularly damaging are absent. Guidance on the presence or absence of pathogens can be obtained from country and regional disease lists (Chapter 28), from the CAB *International Distribution Maps of Plant Diseases* and *Crop Protection Compendium*, and from the appropriate regional or national plant protection organizations. The pest situation must be similarly investigated. Caution is required in applying this principle as lack of a record of a pathogen does not necessarily mean that it is absent but may indicate that insufficient surveys have been done. The definition of a serious pest or pathogen does not always rest with its specific identity. Many species may occur as forms that differ in virulence or host range. In these cases, care is needed to distinguish between these variants, which are often identical morphologically.

2. Import from a country with an efficient plant quarantine service, so that inspection and treatment of planting material before despatch will be thorough, thus reducing the likelihood of contaminated plants being received.

3. Obtain planting material from the safest known source within the selected country.

4. Obtain an official certificate of freedom from pests and diseases from the exporting country. The value of this will vary with the stature of the issuing plant quarantine service but the receipt of such a certificate should never be relied upon as a complete guarantee that the consignment is healthy. Treatment of the material in the country of origin may be done; this should be noted on the certificate.

5. Import the smallest possible amount of planting material; the smaller the amount the less the chance of its carrying infection, and inspection as well as postentry quarantine, if necessary, will be simplified.

6. Inspect material carefully on arrival and treat (dust, spray, fumigate, heat treat) as necessary.

7. Import the safest type of planting material, e.g. seeds are usually safer than vegetative material, unrooted cuttings than rooted. The use of axenic cultures of meristem tip tissues (micropropagation) or plants derived from other forms

---

[1] A plant pest is defined by the International Plant Protection Convention as 'any species, strain or biotype of plant, animal or pathogenic agent injurious to plants or plant products'.

of tissue culture for the international exchange of germplasm material has outstanding advantages as such tissues can be expected to be free from latent infections by viruses, phytoplasmas, etc. as well as other pathogens that are more readily detectable by visual means.

8. If other precautions are not thought to be adequate, the consignment for import should be subject to intermediate or postentry quarantine (see below). Such quarantine must be carried out at a properly equipped station with suitably trained staff, otherwise it is worse than useless, as it gives a false feeling of security.

It is also important to ensure that plant material arrives in good condition and for this proper packing is essential.

Legislation

Exporters of plant material should be familiar with the plant import regulations of the recipient country in order to ensure that the consignment is covered by the necessary documents. An international phytosanitory certificate of freedom from pests and diseases is usually needed and this may require special declarations for particular plants. Knowledge of the regulations ensures that only suitable and acceptable material is sent and these can be obtained from the relevant official regional or national plant protection organizations. The main source of such information on international plant quarantine is the Secretariat of the International Plant Protection Convention (IPPC) based at the Food and Agriculture Organization of the United Nations, and can be accessed via the FAO web site www.fao.org. This provides, *inter alia*, details of regional and national plant protection organizations. Information is also available in the *FAO Plant Protection Bulletin* in which summaries of new plant quarantine laws are frequently published. Regional organizations also publish information.

Inspection, Treatment and Certification

Inspection of plant material is an important part of plant quarantine procedure, and may be done both in the exporting country, before issue of a health certificate, and after arrival, as a double check and to detect any pests or diseases that may have become evident during transit. The value of inspection is dependent on the skill and experience of the inspecting staff, who must be familiar with the appearance of planting material of all types, when healthy, and be able to recognize the presence of pests and diseases. The harmonization of International Standards for Phytosanitary Measures (ISPM) is currently being developed by the IPPC Secretariat.

Various publications to guide inspectors have been prepared. Data sheets on individual organisms of quarantine importance have been issued by the European and Mediterranean Plant Protection Organization (Smith *et al.*, 1997). Some countries have prepared Plant Quarantine Manuals (e.g. Singh, 1979; Griffin, 1997). The CAB *International* set of description sheets of pest organisms and associated maps are useful sources of information and the CAB *International Crop Protection Compendium* provides an interactive and easily searchable reference work of particular value to all concerned with plant health. The FAO provides an official plant and pest information service (GPPIS) online, which is used by the IPPC Secretariat. FAO/IPGR also publish guidelines for the safe movement of plant germplasm and this series now covers many crop species.

Treatments used in plant quarantine are generally similar to those used in general agricultural practice for pest and disease control, and include dusting, spraying, and dip treatment with pesticides, subjecting material to heat or low temperature, and fumigation. To be useful in quarantine, treatments must, however, be completely effective; levels of partial control acceptable for field crops are not good enough. Irradiation is now being investigated as a method for eradicating pest organisms from agricultural materials (Sommer and Mitchell, 1986; Hallman, 1998)

Post-entry and Intermediate Quarantine

Despite every precaution of inspection, certification and treatment, it is not always possible to guarantee that a consignment is completely free from pathogens. In doubtful cases it is advisable to subject plants to a period of growth in isolation, under strict supervision, a procedure normally done in the importing country (post-entry quarantine). This situation arises particularly in relation to the importation of plant germplasm needed for crop improvement when the relevant *FAO/IPGR Guidelines for the Safe Movement of Germplasm* should be consulted. Often this requires containment in specially constructed closed quarantine facilities comprising growth rooms or glasshouses with controlled access, ventilation, etc. and the services of relevant scientists. Summaries of post-entry quarantine techniques and the design of quarantine plant houses are given by Smee and Setchell (1974) and Smee (1975).

When direct importation of plants to a country's own quarantine station is considered too dangerous, quarantine during transit from the country of origin (intermediate quarantine) may be required. This is often done, for example, for rubber, banana and cacao. The requirements of an intermediate station are similar to those for a post-entry station. Intermediate quarantine inspection must always be followed by post-entry quarantine after arrival of the consignment at its final destination.

During post-entry or intermediate quarantine, plants must be kept under close supervision, so that any pest or disease that appears may be immediately detected and grown under optimum conditions, so that symptoms are not masked by physiological disturbances. Various tests may have to be done to check whether plants are free from infection. Virus diseases are among the most frequently encountered and are the most difficult to guard against so indexing by grafting on to indicator plants, immunological or DNA testing is often undertaken (Barker and Torrance, 1997).

Exotic pest organisms and materials other than imported plant germplasm that might harbour them may be need to be received or kept at certain localities within countries, e.g. at research institutes. Regulations governing the receipt, use and containment of such material form part of most countries' plant health regulations.

International Aspects

International action in plant protection, and particularly in plant quarantine, was encouraged by the FAO with the establishment in 1951 of the International Plant Protection Convention. The text of the convention was revised in 1979 and again in 1997; 111 countries are now signatories of the convention. The aim of the convention is to promote and harmonize as far as possible plant quarantine legislation, regulations and associated practices. The 1997 revised text emphasizes cooperation and exchange of information toward the objective of global harmonization. It includes an agreement with the World Trade Organization on the application of sanitary and phytosanitary measures (WTO-SPS agreement) and this has led to the establishment of International Standards for Phytosanitary Measures (ISPMs). Signatories to the convention are obliged to provide information on the pests present in their countries and to provide official contact points to facilitate information exchange.

Current ISPMs include:

- Principles of plant quarantine related to international trade.
- Guidelines for pest risk analysis.
- Code of conduct for the import and release of exotic biological control agents.
- Requirements for the establishment of pest-free areas.
- Glossary of phytosanitary terms.
- Guidelines for surveillance.
- Export certification system.
- Determination of pest status in an area.
- Guidelines for pest-eradication programmes.
- Requirements for establishment of pest-free places of production and pest-free production site exchange.

Regional agreements and organizations have been created to safeguard the interests of groups of neighbouring countries with similar plant protection problems. This may be associated with closer trade and political linkages (Ebbels, 1993). Regional action is particularly desirable to prevent a pathogen or pest absent from a whole area from being introduced into any part of the area, as its entry into one territory would immediately endanger neighbouring countries. For example, *Microcyclus ulei,* the cause of South American leaf blight of rubber, is absent from South-East Asia; its establishment in any one country there would endanger all others in the vicinity.

Regional plant quarantine organizations are:

- The European and Mediterranean Plant Protection Organization (EPPO), established in 1951 (Mathys, 1977), covering 43 countries.
- The Inter-African Phytosanitary Council (IAPSC) (Addoh, 1977), covering 52 countries.

- Organismo Internacional Regional de Sanidad Agropecuaria (OIRSA), covering eight territories in Central America (Berg, 1977).
- Asia and Pacific Plant Protection Commission (APPPC), covering 24 countries, established in 1956 (Reddy, 1977).
- The Caribbean Plant Protection Commission (CPPC), established in 1967 (Berg, 1977), having 23 member countries including South and Central American countries bordering the Caribbean.
- The North American Plant Protection Organization (NAPPO) having as members Canada, USA and Mexico.
- Communidad Andina, established in 1969, covering five Andean countries.
- Comite Regional de Sanidad Vegetal para el Conosur, established in 1980, covering five southern cone countries.

Other Plant Health Measures under Legislative Control

Containment and eradication

This can be attempted during the early stages of introduction of an exotic pest organism and there are many examples (see Ebbels and King, 1979). Containment of a new disease outbreak involves restricting the movement into and out of an affected area combined with various treatments to prevent carriage of spores or other propagules by vehicles etc. The decision to adopt these measures depends as much on available resources as on the biological properties of the pest organism and the terrain in which it has been located. Eradication of a disease is usually difficult once it has entered field crops, as spread has usually occurred by the time the disease has been recognized and latent infections are not detected. Both containment and eradication require skilled resources to be deployed in the field and careful and rapid enforcement if they are to be successful.

Disease-free planting material

Seed certification schemes attempt to provide growers with planting material free from pest organisms and true to variety type and of good quality (see also Chapter 29). Planting material of vegetatively propagated crops is particularly prone to transmitting virus diseases and requires close examination during propagation and bulking for distribution. The certification process and the marketing of certified material often forms part of a country's plant health legislation and requires both the technical facilities to raise and propagate disease-free material and an efficient inspection and monitoring service.

References and Further Reading

Addoh, P.G. (1977) The International Plant Protection Convention: Africa. *FAO Plant Protection Bulletin* 25, 164–166.

Barker, H. and Torrance, L. (1997) Importance of biotechnology for germplasm health and quarantine. In: Callow, J.A., Ford-Lloyd, B.V. and Newbury, H.J. (eds) *Biotechnology and Plant Genetic Resources: Conservation and Use.* Biotechnology in Agriculture Series No. 19, CAB International, Wallingford, UK, pp. 235–254.

Berg, G.H. (1977) The International Plant Protection Convention: Central America and the Caribbean. *FAO Plant Protection Bulletin* 25, 160–163.

Chock, A.K. (1977) Government consultation on the International Plant Protection Convention. *FAO Plant Protection Bulletin* 25, 15–25.

Chock, A.K. and Mulders, J.M. (1978) Digest of plant quarantine regulations. *FAO Plant Protection Bulletin* 26, 1–4.

bbels, D.L. (ed.) (1993) *Plant Health and the European Single Market.* Monograph no. 54, British Crop Protection Council, Farnham Royal, UK, 416 pp.

Ebbels, D.L. and King, J.E. (eds) (1979) *Plant Health. The Scientific Basis for Administrative Control of Plant Parasites.* Blackwell, Oxford, UK, 322 pp.

FAO/IPGRI Technical Guidelines for the Safe Movement of Germplasm (Guidelines for many crops published in the series.) International Plant Genetic Resources Institute (IPGRI), Rome, Italy.

van der Graaff, N.A.(1993) International plant protection; the role of FAO. *Netherlands Journal of Plant Pathology* 99 (Suppl. 3), 93–102.

Griffin, R. (1997) Inspection methodology for plant quarantine. *Arab Journal of Plant Protection* 15, 140–143.

Hallman, G.J. (1998) Ionizing radiation quarantine treatments. *Anais da Sociedade Entomologica do Brasil* 27(3), 313–323.

Hewitt, W.B. and Chiarappa, L. (eds) (1977) *Plant Health and Quarantine in International Transfer of Genetic Resources.* CRC Press, Cleveland, Ohio, 346 pp.

Kahn, R.P. (ed.) (1989) *Plant Protection and Quarantine.* Vol. 1. *Biological Concepts,* 226 pp.; Vol. 2. *Selected Pests and Pathogens of Quarantine Significance,* 265 pp.; Vol. 3. *Selected Crops,* 215 pp. CRC Press, Boca Raton, Florida.

Kahn, R.P. (1990) Plant quarantine as a disease control strategy in tropical countries. In: Raychaudhuri, S.P. and Verma, J.P. (eds) *Review of Tropical Plant Pathology.* Vol. 6. *Techniques and Plant Quarantine.* Today & Tomorrow's Printers and Publishers, New Delhi, India, pp. 151–180.

Kahn, R.P. (1991) Exclusion as a plant disease control strategy. *Annual Review of Phytopathology* 29, 219–246.

Kranz, J. (1998) Assessment of disease risks for crops and sites. In: Martin, K., Muther, J. and Auffarth, A. (eds) *Agroecology, Plant Protection and the Human Environment: Views and Concepts.* PLITS 16, 87–99.

Mathys, G. (1977) European and Mediterranean Plant Protection Organisation. *FAO Plant Protection Bulletin* 25, 152–156.

Mathys, G. and Baker, E.A. (1980) An appraisal of the effectiveness of quarantines. *Annual Review of Phytopathology* 18, 85–101.

McGee, D.C. (ed.) (1997) *Plant Pathogens and the World-wide Movement of Seeds.* APS Press, St Paul, Minnesota, USA, 109 pp.

Reddy, D.B. (1977) The International Plant Protection Convention: Plant Protection Committee for the South East Asia and Pacific Region. *FAO Plant Protection Bulletin* 25, 157–159.

Singh, K.G. (1979) *Plant Quarantine Manual.* Malaysian Department of Agriculture, Kuala Lumpur, 208 pp.

Smee, I. (1975) The post-entry quarantine of imported plant material in Australia. *PANS* 21, 168–174.

Smee, I. and Setchell, P.J. (1974) *Design of Plant Houses for Australian Quarantine Purposes.* Australian Department of Health, Canberra, Australia, 10 pp.

Smith, I.M. and Charles, L.M.F. (1999) *Distribution Maps of Quarantine Pests for Europe.* CAB International, Wallingford, UK, 768 pp.

Smith, I.M., McNamara, D.G., Scott, P.R. and Holderness, M. (eds) (1997) *Quarantine Pests for Europe, edn 22. Data Sheets on Quarantine Pests for the European Union and for the European and Mediterranean Plant Protection Organization.* CAB International, Wallingford, UK, 1425 pp.

Sommer, N.F. and Mitchell, F.G. (1986) Gamma irradiation – a quarantine treatment for fresh fruits and vegetables. *Horticultural Science* 21, 356–360.

Regional and Country Lists of Plant Diseases

Revised by J.M. Waller

CABI Bioscience UK Centre, Bakeham Lane, Egham, Surrey TW20 9TY, UK

Up-to-date published lists of plant diseases in particular countries are not readily available in the more accessible literature. However, countries are obliged to maintain records of these under the International Plant Protection Convention and they may often be obtainable as unpublished reports from the official National Plant Protection Office, addresses of which can be accessed via the IPPC area of the FAO web site (see Chapter 27).

Regional Lists

Africa

Anon. (1979) A bibliography of lists of plant diseases and fungi 1. Africa. *Review of Plant Pathology* 58, 305–308.

Brown, A.C.P. (1969) *A Distribution List of the More Important Pathogens of Economic Plants with Particular Reference to Africa*. Commonwealth Mycological Institute, Kew, UK, 95 pp.

Caribbean

Anon. (1970) *The Major Pests and Diseases of Economic Crops in the Caribbean*. FAO Caribbean Plant Protection Commission, Port of Spain, Trinidad, 25 pp.

Central America

McGuire, J.U. and Crandall, B.S. (1967) *Survey of Insect Pests and Plant Diseases of Selected Food Crops of Mexico, Central America and Panama.* USDA, International Agricultural Development Service, 157 pp.

Europe

Oudemans, C.A.J.A. (1919–1924) *Enumeratio Systematica Fungorum,* 5 vols. M. Nijhoff, The Hague.

South America

Viegas, A.P. (1961) *Indice de fungos da America do Sul.* (Index of the fungi of South America.) Instituto Agronomico Campinas, Sao Paulo, Brazil, 921 pp.

South Pacific

Commonwealth Mycological Institute (1975) A bibliography of lists of plant diseases and fungi V. Australasia and Oceania. *Review of Plant Pathology* 54, 963–966.

Dingley, J.M., Fullerton, R.A. and McKenzie, E.H.C. (1981) *Survey of Agricultural Pests and Diseases.* Technical Report Vol. 2. *Records of Fungi, Bacteria, Algae, and Angiosperms Pathogenic on Plants in Cook Islands, Fiji, Kiribati, Mue, Tonga, Tuvalu, and Western Samoa.* South Pacific Bureau for Economic Co-operation, UN Development Programme, FAO, v + 485 pp.

Firman, I.D. (1978) Plant pathology in the region served by the South Pacific Commission. *Review of Plant Pathology* 57, 85–90.

Firman, I.D. (1978) Bibliography of plant pathology and mycology in the area of the South Pacific Commission 1820–1976. *Technical Paper, South Pacific Commission* No. 176, 78 pp.

Country Lists

Afghanistan

Roivainen, O. (1984) *A List of Crop Pests and Diseases Occurring in Afghanistan.* Food and Agriculture Organization, Rome, Italy, 17 pp.

Angola

Serafim, F.J.D. and Serafim, M.C. (1968) Lista des doencas de culturas de Angola. (List of diseases of crops of Angola.) *Serie Tecnica Instituto de Investiga, cdo de Angola* No. 2, 22 pp.

Antigua

Walker, P.T., Waller, J.M. and Evans, A.A.F. (1974) *The Insect Pests, Plant Diseases and Nematodes of Crops in Antigua*. Ministry of Overseas Development, London, 40 pp.

Argentina

Marchionatto, J.B. (1951) *Los hongos parasitos de las plantas*. (Fungi parasitic on plants.) Buenos Aires, Acme Agency, 118 pp.

Australia

Morschel, J.R. (1975) *Plant Diseases Recorded in Australia and Overseas. Part 1. Vegetable Crops*. Australian Department of Health, Canberra, 67 pp.

New South Wales

Anon. (1976) *45th Annual Plant Disease Survey for the 12 Months Period Ending 30th June 1975*. New South Wales Department of Agriculture, Rydalmere, 30 pp.
See also other papers in this series.
Anon. (1977) *Plant Disease Survey 1975–1976*. New South Wales Department of Agriculture, Rydalmere, 52 pp.

Northern Territory

Pitkethley, R.N. (1970) A preliminary list of plant diseases in the Northern Territory. *Technical Bulletin, Primary Industries Branch, Northern Territory Administration, Darwin* No. 2, 30 pp.

Queensland

Simmonds, J.H. (1966) *Host Index of Plant Diseases in Queensland*. Queensland Department of Primary Industries, Brisbane, Australia, 111 pp.

South Australia

Warcup, J.H. and Talbot, P.H.B. (1981) *Host–pathogen Index of Plant Diseases in South Australia*. Waite Agricultural Research Institute, Adelaide, Australia, 114 pp.

Tasmania

Sampson, P.J. and Walker, J. (1982) *An Annotated List of Plant Diseases in Tasmania*. Department of Agriculture, Hobart, 121 pp.

Victoria

Fisher, E.E. and Freeman, H. (1959–1961) Plant diseases recorded in Victoria. Part 1. Sections 1–4. *Journal of the Department of Agriculture, Victoria* 57, 679–683; 58, 404–406; 59, 61–64; 114–117.

Freeman, H. (1964) Plant diseases recorded in Victoria. Part 2. Section 1. *Journal of the Department of Agriculture, Victoria* 62, 235–237.

Western Australia

MacNish, G.C. (1967) Supplementary list of diseases recorded on various hosts in Western Australia. *Bulletin, Western Australia Department of Agriculture* No. 3481, 8 pp.
See also previous lists by various authors in the *Journal of Agriculture Western Australia* 1959–1964.

Austria

Wodicka, B. (1980) Bericht uber das Auftreten wichtiger Krankheiten und Schadlinge an Kulturpflanzen in Osterreich im Jahre 1978. (Report on the occurrence of important diseases and pests on cultivated plants in Austria in 1978.) *Pflanzenschutzberichte* 46, 61–72.
See also other reports in this series.

Azores

Dennis, R.W.G., Reid, D.A. and Spooner, B. (1977) The fungi of the Azores. *Kew Bulletin* 32, 85–136.

Bangladesh

Ishaque, M. and Talukdar, M.J. (1967) Survey of fungal flora of East Pakistan. *Agriculture Pakistan* 18, 17–26.

Barbados

Norse, D. (1974) Plant diseases in Barbados. *Phytopathological Papers* No. 18, 38 pp.

Belize

Anon. (1977?) *Annual Report of the Department of Agriculture, Belize, for the year 1973.* Department of Agriculture, Belmopan, Belize, 23 pp. Appendix lists main diseases.

Bermuda

Waterston, J.M. (1947) The fungi of Bermuda. *Bulletin, Department of Agriculture Bermuda* No. 23, 305 pp.

Bolivia

Alandia, S. and Bell, F.H. (1957) Diseases of warm climate crops in Bolivia. *FAO Plant Protection Bulletin* 5, 172–173.

Bell, F.H. and Alandia, S. (1957) Diseases of temperate climate crops in Bolivia. *Plant Disease Reporter* 41, 646–649.

Brazil

Galli, F., Tukeshi, H., Torres de Carvalho, P. de C., Balmer, E., Kimati, H., Nogueira Cordoso, C.O. and Lima Salgado, C. (1968) *Manual defitopatoiogia; doencas das plantas e seu controle.* (Manual of phytopathology; plant diseases and their control.) Ceres, Sao Paulo, 640 pp.

Brunei

Peregrine, W.T.H. and Kassim bin Ahmad (1982) Brunei: a first annotated list of plant diseases and associated organisms. *Phytopathological Papers* No. 27, 87 pp.

Bulgaria

Khristov, A. (1972) *Opredelitel na bolestite po rasteniyata.* (A guide to plant diseases.) 2nd edn. Zemizdat, Sofia, 498 pp.

Burma

Thaung, M.M. (1970) New records of plant diseases in Burma. *PANS 16*, 638–640.

Tun, S. (1970) Plant protection in Burma. *Beitrage zur Tropischen und Subtropischen Landwirtschaft und Veterindrmedezin* 8, 287–294.

Canada

Conners, I.L. (1967) An annotated index of plant diseases in Canada and fungi recorded on plants in Alaska, Canada and Greenland. *Publications, Canada Department of Agriculture* No. 1251, 381 pp.

Canary Islands

Gjaerum, H.B. (1976) A review of the fungal flora of the Canary Islands. In: *Biogeography and Ecology in the Canary Islands.* Junk, The Hague, pp. 287–296.

Chile

Mujica, F.R. and Oehrens, E.B. (1967) Addenda a flora fungosa chilena 2. (Addenda to Chilean mycoflora 2.) *Boletin Tecnico Departmento de Investigacion Agricola, Chile* No. 27, 81 pp.

Mujica, F.R. and Vergara, C. (1945) *Flora Fungosa Chilena*. (Chilean fungus flora.) Ministerio de Agricultural, Santiago, 199 pp.

Mujica, F.R. and Vergara, C. (1961) Addenda a flora fungosa chilena 1. (Addenda to Chilean mycoflora 1.) *Boletin Tecnico Departmento de Investigacion Agricola, Chile* No. 6, 60 pp.

China

Hanson, H.C. (1963) *Diseases and Pests of Economic Plants of Central and South China, Hong Kong and Taiwan (Formosa)*. American Institute of Crop Ecology, Washington, USA, 184 pp.

Tai, F.L. (1979) *Sylloge fungorum Sinicorum*. Science Press, Academia Sinica, Beijing, 1527 pp.

Colombia

Orjuela, N.J. (1965) Indice de enfermedades de plantas cultivadas en Colombia. (List of diseases of plants cultivated in Columbia.) *Boletin Tecnico Instituto Colombiano Agropecuario* No. 11, 66 pp.

Cuba

Rosenada, M.F. (1973) Catalogo de enfermedades de plantas Cubanas. (Catalogue of Cuban plant diseases.) *Serie Agricola, Academia de Ciencias de Cuba, Instituto de Investigaciones Tropicales* No. 27, 78 pp.

Cyprus

Zyngas, J. Ph. (1973) *The Cyprus Fungi*. Department of Agriculture, Nicosia, Cyprus, 56 pp.

Czechoslovakia

Anon. (1959–1962) *Zemedelska fytopatologie*. (Agricultural phytopathology.) Czechoslovak Academy of Agricultural Sciences, Prague, 4 vols. 704; 776; 714; 1086 pp.

Denmark

Anon. (1980) *Plantesygdomme i Danmark 1979. 96 Arsoversigt*. (Plant diseases in Denmark in 1979, 96th annual survey.) State Plant Pathology Institute, Lyngby, 51 pp.

See also other reports in this series.

Dominica

Critchett, C. (1974) *A Final Report on Recommendations in Plant Disease and Pest Control for St. Vincent and Dominica 1970/72*. Ministry of Overseas Development, London, UK, 96 + 9 pp.

Dominican Republic

Ciferri, R. (1961) Mycoflora Domingensis integrate. *Quaderno Istituto Botanico, Universita Pavia* No. 19, 539 pp.

Ecuador

Molestina, E. (1942) Indice preliminar de las principales enfermedades y plagas de la agriculture en el Ecuador. (Preliminary index of the principal diseases and pests of agriculture in Ecuador.) *Boletin del Departamento de Agricultura, Ecuador* No. 15, 25 pp.

Egypt

El-Helaly, A.F., Ibrahim, I.A., Assawah, M.W., Elarosi, H.M., Abou-el Dahab, M.K., Michail, S.H. and Abd-el-Rahim, M.A. (1963) General survey of plant diseases and pathogenic organisms in the U.A.R. (Egypt) until 1962. *Research Bulletin, Faculty of Agriculture, Alexandria University* No. 8, 107 pp.

El Salvador

Crandall, B.S., Abrego, L. and Patino, B. (1951) A check list of diseases of economic plants of El Salvador, Central America. *Plant Disease Reporter* 35, 545–554.

Estonia

Jarva, L. and Parmasto, E. (1980) Eesti seente koondnimestik. (A list of Estonian fungi.) Academy of Science, Tartu, 331 pp.

Ethiopia

Stewart, R.B. and Yirgou, D. (1967) Index of plant diseases in Ethiopia. *Bulletin, Experiment Station, College of Agriculture, Haile Selassie University* No. 30, 95 pp.

Faeroes

Moller, F.H. (1945, 1958) *Fungi of the Faeroes. Part I. Basidiomycetes.* 295 pp. *Part II. Myxamycetes, Archimycetes, Phycomycetes, Ascomycetes and Fungi Imperfecti.* Ejnar Munksgaard, Copenhagen, 286 pp.

Fiji

Firman, I.D. (1972) A list of fungi and plant parasitic bacteria, viruses and nematodes in Fiji. *Phytopathological Papers* No. 15, 36 pp.

Graham, K.M. (1971) *Plant Diseases in Fiji.* HMSO, London, UK, xxvii + 251 pp.

France

Viennot-Bourgin, C. (1956) Mildious, oidiums, caries, charbons, rouilles des plantes de France. (Mildews, oidiums, bunts, smuts, rusts of the plants of France.) *Encyclopedie Mycologique* No. 26, 317 pp.; No. 27, 98 pp.

French Polynesia

Reddy, D.B. (1973) Preliminary list of pests and disease of principal crops in French Polynesia. *Technical Document, FAO Plant Protection Committee for the South East Asia and Pacific Region* No. 89, 5 pp.

Gabon

Barat, H. (1966) *Notes de Phytopathologie Gabonnaise (Mission du 20 au 26 juillet 1966).* (Notes on plant diseases in Gabon (Survey of 20–26 July 1966).) IRAT, 15 pp.

Germany (East)

Ramson, A., Erfurth, P., Herold, H. and Sachs, E. (1981) Das Auftreten der wichtigsten Schaderreger in der Pflanzenproduktion der Deutschen Demokratischen Republik im Jahre 1980 mit Schlussfolgerungen fur die weitere Arbeit im Pflanzenschutz. (The occurrence of the most important damage causing agents in the plant production of the German Democratic Republic in 1980 with conclusions as to further work in plant protection.) *Nachrichtenblattfur den Pflanzenschutz in der DDR* 35, 85–101.

See also other reports in this series.

Ghana

Leather, R.I. (1959) Diseases of economic plants in Ghana other than cacao. *Bulletin, Ministry of Food and Agriculture, Ghana* No. I, vii + 40 pp.

Piening, L.J. (1962) A check list of fungi recorded from Ghana. *Bulletin, Ministry of Agriculture, Ghana* No. 2, vii + 130 pp.

Greece

Pantidou, M.E. (1973) *Fungus-host index for Greece*. Benaki Phytopathological Institute, Athens, Greece, 382 pp.

Guatemala

Schieber, E. and Sanchez, A. (1968) Indice preliminar de las enfermedades de las plantas en Guatemala. (Preliminary list of diseases of plants in Guatemala.) *Boletin Tecnico Ministerio de Agricultura, Guatemala* No. 25, 55 pp.

Guinea

Kranz, J. (1963, 1965) Fungi collected in the Republic of Guinea. I. Collections from the rain forest. II. Collections from the Kindia area in 1962. III. Collections from the Kindia area in 1963/64, and host index. *Sydowia* 17, 132–138; 174–185; 19, 92–107.

Guyana

Bisessar, S. (1962) *A Revised List of Diseases of Economic Plants in British Guiana, 1962.* Ministry of Agriculture, Georgetown, 24 pp.

Haiti

Benjamin, C.R. and Slot, A. (1969) Fungi of Haiti. *Sydowia* 23, 125–163.

Hawaii

McCain, A.H. and Trujillo, E.E. (1967) Plant diseases previously unreported from Hawaii. *Plant Disease Reporter* 51, 1051–1053.

Parris, G.K. (1940) A check list of fungi, bacteria, nematodes, and viruses occurring in Hawaii, and their hosts. *Plant Disease Reporter Supplement* No. 121, 91 pp.

Raabe, R.D. (1966) Check list of plant diseases previously unreported in Hawaii. *Plant Disease Reporter* 50, 411–414.

Hong Kong

Leather, R.I. and Hor, M.N. (1969) A preliminary list of plant diseases in Hong Kong. *Agricultural Bulletin, Hong Kong* No. 2, 64 pp.

Hungary

Voros, L. and Leranth, J. (1974) Review of the mycoflora of Hungary. Part XII. Deuteromycetes; Moniliales and Myceliales. *Acta Phytopathologica Academiae Scientiarum Hungaricae* 9, 333–361.
See also other papers in this series published in the same journal since 1967.

Iceland

Jørstad, I. (1952) The Uredinales of Iceland. *Skrifter utgitt av det Norske Videnskapsakademi i Oslo* No. 2, 87 pp.
Jørstad, I. (1963) Icelandic parasitic fungi apart from Uredinales. *Skrifter utgitt av det Norske Videnskapsakademi i Oslo* No. 10, 72 pp.

India

Bilgrami, K.S., Jamaluddin and Rizwi, M.A. (1979) *Fungi of India*. Part I *List and References*, 467 pp. Part 2 *Host Index and Addenda*. Today and Tomorrow's Printers and Publishers, New Dehli, India, 268 pp.
Butler, E.J. and Bisby, G.R. (Revised by Vasudeva, R.S.) (1960) *The Fungi of India*. Indian Council of Agricultural Research, New Delhi, India, 552 pp.
Chakravarti, B.P., Hegde, S.V., Ahmed, S.R., Gupta, D.K. and Rangarajan, M. (1973) *Phytopathogenic Bacteria of India and Bibliography (1892–1972)*. University of Udaipur, Udaipur, 77 pp.
Raychaudhuri, S.P. and Nariani, T.K. (1977) *Virus and Mycoplasma Diseases of Plants in India*. Oxford and IBH Publishing Co, New Delhi, India, vi + 102 pp.

Indonesia

Semangun, H. (1992) *Host Index of Plant Diseases in Indonesia*. Gadjah Mada University Press, Yogyakarta.
Triharso, Kaselan, J. and Christanti (1975) List of diseases of important economic crop plants already reported in Indonesia. *Bulletin, Fakultas Pertanian Universitas Cadjah Mada, Yogyakarta* No. 14, 60 pp.

Iran

Ershad, D. (1977) Fungi of Iran. *Publication, Department of Emtany, Iran* No. 10, 277 pp.

Iraq

Mathur, R.S. (1968) *The Fungi and Plant Diseases of Iraq*. Kanpur, India, 90 pp.

Israel

Kenneth, R., Barkai-Golan, R., Chorin, M., Dishon, I., Katan, Y., Netzer, D., Palti, J. and Volcani, Z. (1970) *A Revised Checklist of Fungal and Bacterial Diseases of Vegetable Crops in Israel.* Volcani Institute of Agricultural Research, Bet Dagan, Israel, 39 pp.
Kenneth, R., Palti, J., Frank, Z.R., Anikster, Y. and Cohn, R. (1975) A revised checklist of fungal and bacterial diseases of field and forage crops in Israel. *Special Publication, Volcani Center* No. 36, 70 pp.
Palti, J., Solel, Z., Sztejnberg, A., Arenstein, Z., Bar-Joseph, M., Barkai-Golan, R., Cohn, R., Fleisher, Z., Golan, Y., Orion, D., Pappo, S., Pinkas, Y., Sciffmann-Nadel, M., Shabi, E. and Tanne, E. (1977) A check list of diseases of fruit crops in Israel. *Phytoparasitica, Supplement* No. 1, 53 pp.

Italy

Ciferri, R. and Camara, C. (1962–1963) Tentativo di elencazione dei funghi italiani. 1. Erisifali. 2. Uredinali. 3. Peronosporales. (An attempt at listing Italian fungi. 1. Erysiphales. 2. Uredinales. 3. Peronosporales.) *Quaderni Laboratorio Crittogamico, Istituto Botanico della Universita, Pavia* Nos 21, 46 pp.; 23, 98 pp.; 30, 39 pp.

Ivory Coast

Resplandy, R., Chevaugeon, J., Delassus, M. and Luc, M. (1954) Premiere liste annotee de champignons parasites de plantes cultivees en Côte d'Ivoire. (First annotated list of parasitic fungi of cultivated plants in the Ivory Coast.) *Annales des Epiphyties* 5, 1–61.

Jamaica

Naylor, A.G. (1974) *Diseases of Plants in Jamaica.* Ministry of Agriculture, Kingston, Jamaica, 129 pp.

Japan

Anon. (1966) *List of Important Diseases and Pests of Economic Plants in Japan.* Nihon Tokushu Noyaku Seizo K.K., Tokyo, 591 pp.

Kampuchea

Soonthronpoot, P. (1969) List of plant diseases in Cambodia. *Technical Document, FAO Plant Protection Committee for the South East Asia and Pacific Region* No. 70, 23 pp.

Kenya

Kung'u, J.N. and Boa, E. (1997) *Kenya Checklist of Fungi and Bacteria on Plants and Other Substrates.* CABI Bioscience, Egham, UK, 96 pp.
Ondieki, J.J. (1973) Host lists of Kenya fungi and bacteria. (A ten-year supplement 1961–1970). *East African Agricultural and Forestry Journal* 38 (Special Issue), 31 pp.

Korea, North

Lim, H.W. (1963) (A list of parasitic fungi of the main economic plants.) *Sengmulkhak* 2, 3744.

Korea, South

Anon. (1972) *A List of Plant Diseases, Insect Pests, and Weeds in Korea.* Korean Society of Plant Protection, Seoul, 424 pp.

Laos

Manser, P.D. (1970) Provisional list of diseases on various crops in Laos. *Technical Document, FAO Plant Protection Committee for the South East Asia and Pacific Region* No. 73, 10 pp.

Lebanon

Khatib, H., Davet, P. and Sardy, G. (1969) *Liste des Maladies des Plantes au Liban 1959–1969.* (List of plant diseases in Lebanon 1959–1969.) Institute of Agricultural Research, Fanar, Lebanon, 44 pp.

Libya

Kranz, J. (1965) A list of plant pathogenic and other fungi of Cyrenaica (Libya). *Phytopathological Papers* No. 6, 24 pp.
Pucci, E. (1965) Lista preliminare delle malattie delle piante osservate in Tripolitania del 1959 al 1964. Sintomi, danni, lotta. (Preliminary list of plant diseases observed in Tripolitania from 1959 to 1964. Symptoms, damage, control.) *Rivista di Agricultura Subtropicale e Tropicale* 49, 337–375.

Madeira

Gjaerum, H.B. (1970) Fungi from the Canary Islands and Madeira. *Cuadernos de Botanica Canaria* 9, 3–7.

Malagasy Republic

Dadant, R., Rasolofo, Mme. and Baudin, P. (1960) *Liste des maladies des plantes cultivees a Madagascar.* (List of the diseases of cultivated plants in Madagascar.) Institut de la Recherche Agronomique, Madagascar, 94 pp.

Malawi

Peregrine, W.T.H. and Siddiqi, M.A. (1972) A revised and annotated list of plant diseases in Malawi. *Phytopathological Papers* No. 15, 36 pp.

Malaysia

Singh, K.G. (1980) A check list of host and disease in Malaysia. *Bulletin, Ministry of Agriculture Malaysia* No. 154, 280 pp.

Sabah

Liu, P.S.W. (1977) A supplement to a host list of plant diseases in Sabah, Malaysia. *Phytopathological Papers* No. 21, 50 pp.
Williams, T.H. and Liu, P.S.W. (1976) A host list of plant diseases in Sabah, Malaysia. *Phytopathological Papers* No. 19, 67 pp.

Sarawak

Keuh, T.-K. (1976) New plant disease records for Sarawak for 1973 and 1974. *Sarawak Museum Journal* 24, 217–225.
Turner, G.J. (1971) Fungi and plant disease in Sarawak. *Phytopathological Papers* No. 13, 55 pp.

West Malaysia

Singh, K.G. (1973) A checklist of host and disease in Peninsular Malaysia. *Bulletin, Division of Agriculture Malaysia* No. 132, 189 pp.

Malta

Wheeler, B.E.J. (1957) *A Plant Disease Survey of Malta.* Department of Information, Malta, 30 pp.

Mariana Islands

Adair, C.N. (1971) A review of fungi reported from the Mariana Islands. *Micronesica* 7, 79–83.

Mauritius

Orieux, L. and Felix, S. (1968) List of plant diseases in Mauritius. *Phytopathological Papers* No. 7, 48 pp.

Mexico

Rodriguez, S.H. (1972) Enfermedades parasitarias de los cultivos agricolas en Mexico. (Parasitic diseases of agricultural crops in Mexico.) *Folleto Miscelaneo INIA* No. 23, 58 pp.

Morocco

Rieuf, P. (1969–1971) Parasites et saprophytes des plantes au Maroc. (Parasites and saprophytes of plants in Morocco.) *Cahiers de la Recherche Agronomique* 27, 1–178; 28, 179–357; 29, 362–463; 30, 469–570.

Mozambique

de Canalho, T. and Mendes, O. (1958) *Doencas de plantas em Mocambique.* (Plant diseases in Mozambique.) Direccao de Agricultura e Florestas, Lourenco Marques, 84 pp.

Nepal

Khadka, B.B. and Shah, S.M. (1967) Preliminary list of plant diseases recorded in Nepal. *Nepal Journal of Agriculture* 2, 47–76.
Khadka, B.B., Shah, S.M. and Lawat, K. (1968) Plant diseases in Nepal (a supplementary list). *Technical Document, FAO Plant Protection Committee for the South East Asia and Pacific Region* No. 66, 12 pp.
Manandhr, K.L. and Shah, S.M. (1975) List of plant diseases in Nepal (second supplement). *Technical Document, FAO Plant Protection Committee for the South East Asia and Pacific Region* No. 97, 11 pp.

Netherlands

Anon. (1961) Overzicht van de belangrijkste ziekten en plagen van landbouwgewassen en hun bestrijding. (Survey of the most important diseases and pests of agricultural crops and their control.) *Verslagen Plantenziektenkundige Dienst Wageningen* No. 92, 160 pp.

New Caledonia

Huguenin, B. (1966) Micromycetes de Nouvelle-Caledonie. (Microfungi of New Caledonia.) *Cahiers ORSTOM, Biologie* 1, 61–83.

New Zealand

Pennycook, S.R. (1989) *Plant Diseases Recorded in New Zealand*. Vols 1, 2 and 3. Plant Diseases Division, DSIR, Auckland, New Zealand, 958 pp.

Nicaragua

Litzenberger, S.C. and Stevenson, J.A. (1957) A preliminary list of Nicaraguan plant diseases. *Plant Disease Reporter Supplement* 243, 19 pp.

Niger

Kranz, J. and Hammat, H. (1979) Niger. New records of phytopathogens on cultivated plants. *FAO Plant Protection Bulletin* 27, 97–99.

Nigeria

Bailey, A.G. (1966) A check-list of plant diseases in Nigeria. *Memorandum, Department of Agricultural Research Ibadan* No. 96, 33 pp.

Norway

Ramsfjell, T. and Fjelddalen, J. (1962) *Sjuidommer og skadedyr pd jordDraksvekster*. (Diseases and pests on agricultural crops.) Bondenes Forlag, Oslo, 196 + xv pp.

Oman

Waller, J.M. and Bridge, J. (1978) Plant diseases and nematodes in the Sultanate of Oman. *PANS* 24, 313–326.

Pakistan

Ghafoor, A. and Khan, S.A.J. (1976) *List of Diseases of Economic Plants in Pakistan*. Ministry of Food, Agriculture and Underdeveloped Areas, Karachi, 85 pp.
Mirza, J.H. and Qureshi, M.S.A. (1978) *Fungi of Pakistan*. University of Agriculture, Faisalabad, iv + 311 pp.

Panama

Rios, E.A.E. (1982) *Catalogo de enfermedades de las plantas en la Republica de Panama*. (Catalogue of diseases of plants in the Republic of Panama.) Published by the author, 123 pp.

Papua New Guinea

Shaw, D. (1963) Plant pathogens and other micro-organisms in Papua and New Guinea. Annotated list of references to plant pathogens and miscellaneous fungi in West New Guinea. *Research Bulletin, Department of Agriculture Papua and New Guinea* No. I, 82 pp.

Paraguay

Spegazzini, C. (1922) Fungi paraguayenses. *Anales del Museo Nacional de Historia Natural de Buenos Aires* 31, 355–450.

Peru

Bazan de Segura, C. (1959) *Principales enfermedades de los plantas en el Peru.* (Principal plant diseases in Peru.) Sociedad Nacional Agraria, Peru, 70 pp.

Philippines

Tangonan, N.G. and Quebral, F.C. (1992) *Host Index of Plant Diseases in the Philippines*, 2nd edn. Department of Science and Technology (DOST), Metro Manila, 273 pp.

Poland

Malachowska, D., Babilas, W., Grajek, E., Kagan, F., Lewartowski, R., Piekarczyk, K., Walczak, F. and Romankow-Zmudowska, A. (1977) Wystepowanie wazniejszych chorób i szkodników roslin w Polsce w roku 1976 oraz prognozy ich pojawu i szkodliwosci w roku 1977. (The occurrence of the main diseases and pests of plants in Poland in 1976 and the forecast of their appearance and harmfulness in 1977.) *Transactions of the 17th Scientific Conference, Institute of Plant Protection, Poznan, Poland*, 427–459.
See also previous papers in this series.

Portugal

Dias, M.R. de S. and Lucas, M.T. (1981) Fungi Lusitaniae. XXVII. (Portuguese fungi, XXVII.) *Agronomia Lusitana* 40, 135–144.
See also previous papers in the series.

Puerto Rico

Stevenson, J.A. (1975) The fungi of Puerto Rico and the American Virgin Islands. *Contributions of the Reid Herbarium* No. 23, 743 pp.

Romania

Bontea, V. (1953) *Ciaperci parazite si saprofite din Republica Populara Romdna*. (Parasitic and saprophytic fungi in the Romanian People's Republic.) Editura Academiei Republicu Populare Romane, Bucharest, 639 pp.

Hulea, A. and Savescu, A. (eds) (1977) *Rasptndirea bolilor si daundtorilor plantelor cultivate in Romania in perioada 1961–1971*. (Distribution of plant diseases and pests in Romania during 1961–1971.) Centrul de Material Didactic si Propaganda Agricola, Bucharest, 190 pp.

St Vincent

Critchett, C. (1974) *A Final Report on Recommendations in Plant Disease and Pest Control for St Vincent and Dominica 1970/72*. Ministry of Overseas Development, London, 96 + 9 pp.

Samoa

Johnston, A. (1963) Host list of plant diseases in Samoa. *Technical Document, FAO Plant Protection Committee for the South East Asia and Pacific Region* No. 35, 5 pp.

Samoa, American

Firman, I.D. (1975) Plant diseases in the area of the South Pacific Commission. 2. American Samoa. *Information Document South Pacific Commission* No. 38, 7 + 3 pp.

Samoa, Western

Reddy, D.B. (1970) A preliminary list of pests and diseases of plants in Western Samoa. *Technical Document, FAO Plant Protection Committee for the South East Asia and Pacific Region* No. 77, 15 pp.

Saudi Arabia

Natour, R.M. (1970) A survey of plant diseases in Saudi Arabia. *Beitrage zur Tropischen und Subtropischen Landwirtschaft und Tropenveterindrmedezin* 8, 65–70.

Senegal

Bouhot, D. and Mallamaire, A. (1965) *Les principales maladies des plantes cultivees au Senegal*. (The principal diseases of cultivated plants in Senegal.) 2 vols. 292 pp., 160 pp. Dakar, Senegal.

Seychelles

Mathias, P.L. (1971) Pest and disease problems in the Seychelles. *Proceedings of the 6th British Insecticide and Fungicide Conference 1971*, pp. 835–841.

Sierra Leone

Deighton, F.C. (1956) *Diseases of Cultivated and Other Economic Plants in Sierra Leone.* Government Printing Department, Sierra Leone, 76 pp.

Singapore

Tan, L.K. and Lim, G. (1970) Fungal diseases of vegetables and ornamentals in Singapore. *Revue de Mycologie* 35, 47–51.

Solomon Islands

Brown, J.F. (1973) Microorganisms associated with plant diseases in the British Solomon Islands Protectorate. *Information Document, South Pacific Commission* No. 32, 27 pp.

Somalia

Golato, C. (1967) *Malattie delle piante coltivate in Somalia.* (Diseases of cultivated plants in Somalia.) Istituto Agronomico per l'Oltremare, Firenze, Italy, 147 pp.

South Africa

Gorter, G.J.M.A. (1977) Index of plant pathogens and the diseases they cause in cultivated plants in South Africa. *Science Bulletin, Department of Agricultural Technical Services* No. 392, 177 pp.

Gorter, G.J.M.A. (1981) Index of plant pathogens, II, and the diseases they cause in wild growing plants in South Africa. *Science Bulletin, Department of Agriculture and Fisheries* No. 398, 84 pp.

Gorter, G.J.M.A. (1982) A newly revised guide to South African literature on plant diseases. *Technical Communication, Department of Agriculture, Republic of South Africa* No. 179, 65 pp.

Sri Lanka

Abeygunawardena, D.V.W. (1969) *Diseases of Cultivated Plants. Their Diagnosis and Treatment in Ceylon.* Apothecaries' Co. Ltd, Colombo, 289 pp.

Sudan

Tarr, S.A.J. (1955) *The Fungi and Plant Diseases of the Sudan*. Commonwealth Mycological Institute, Kew, UK, 127 pp.

Tarr, S.A.J. (1963) A supplementary list of Sudan fungi and plant diseases. *Mycological Papers* No. 85, 31 pp.

Surinam

Del Prado, F.A. (1966) Overzicht van de belangrijkste ziekten en plagen van landbouwgewassen. (Survey of the most important diseases and pests of crop plants.) *Mededelingen Landbouwproefstation in Suriname* No. 37, 40 pp.

Syria

Mulder, D. (1958) Plant diseases of economic importance in the Northern Region, United Arab Republic. *FAO Plant Protection Bulletin* 7, 1–5.

Taiwan

Anon. (1979) *List of Plant Diseases in Taiwan*. Taiwan Plant Protection Society, Taichung, 404 pp.

Tanzania

Ebbels, D.L. and Allen, D.J. (1978) A supplementary and annotated list of plant diseases, pathogens and associated fungi in Tanzania. *Phytopathological Papers* No. 22, 89 pp.

Riley, E.A. (1960) A revised list of plant diseases in Tanganyika Territory. *Mycological Papers* No. 75, 42 pp.

Thailand

Giatgong, P. (1980) *Host Index of Plant Diseases in Thailand*, 2nd edn. Ministry of Agriculture and Cooperatives, Bangkok, 118 pp.

Togo

Niemann, E., Lare, M., Tchinde, J. and Zakari, I. (1972) Beitrag zur Kenntnis der Pflanzenkrankheiten und -schadlinge Togos. (Contribution to the knowledge of plant diseases and pests of Togo.) *Zeitschrift fur Pflanzenkraniheiten und Pflanzenschutz* 79, 595–619.

Tonga

Daft, G.C. (1973) A supplementary list of diseases in Tonga. *Technical Document, FAO Plant Protection Committee for the South East Asia and Pacific Region* No. 88, 3 pp.

Johnston, A. (1965) Plant diseases in Tonga. *Technical Document, FAO Plant Protection Committee for the South East Asia and Pacific Region* No. 44, 4 pp.

Trinidad and Tobago

Martyn, E.B. (1951) Diseases of economic plants in Trinidad and Tobago. *Journal of the Agricultural Society of Trinidad and Tobago* 51, 475–506.

Tristan da Cunha

Wace, N.M. and Dickson, J.H. (1965) Biology of the Tristan da Cunha Islands. II. The terrestrial botany of the Tristan da Cunha Islands. *Philosophical Transactions of the Royal Society, B* 249, 273–360.

Turkey

Karel, G. (1958) *A Preliminary List of Plant Diseases in Turkey.* Ministry of Agriculture, Ankara, 44 pp.

Kurçman, S. (1977) Determination of virus diseases on cultural plants in Turkey. *Journal of Turkish Phytopathology* 6, 27–48.

Uganda

Emechebe, A.M. (1975) *Some Aspects of Crop Diseases in Uganda.* Makerere University, Kampala, 43 pp.

United Kingdom

Baker, J.J. (1972) Report on diseases of cultivated plants in England and Wales for the years 1957–1968. *Technical Bulletin, Ministry of Agriculture, Fisheries and Food* No. 25, 322 pp.

Foister, C.E. (1961) The economic plant diseases of Scotland: a survey and check list covering the years 1924–1957. *Technical Bulletin, Department of Agriculture and Fisheries for Scotland* No. 1, iv + 209 pp.

Moore, W.C. (1959) *British Parasitic Fungi.* Cambridge University Press, Cambridge, UK, xvi + 430 pp.

Uruguay

Koch de Brotos, L. and Boasso, C. (1955) Lista de enfermedades de los vegetales en el Uruguay. (List of plant diseases in Uruguay.) *Publicacion, Ministerio de Ganaderia y Agricultura Montevuleo* No. 106, 65 pp.

USA

Anon. (1960) Index of plant diseases in the United States. *Agriculture Handbook, United States Department of Agriculture* No. 165, iv + 531 pp.

USSR

Dement'eva, M.I. (1970) *Fitopatologiya.* (Phytopathology.) Izdatel'stvo Kolos, Moscow, 464 pp.
Khokhryakova, M.K. (ed.) (1971–1978) *Ukazatel' vozbuditelel bolezue'sel' skokhokhozyal-stvennykh rastenii.* (Index of pathogens of diseases of agricultural plants.) Parts 1–5. Institute of Plant Protection, Leningrad.

Vanuatu

Johnston, A. (1963) Host list of plant diseases in the New Hebrides. *Technical Document, FAO Plant Protection Committee for the South East Asia and Pacific Region* No. 27, 9 pp.

Venezuela

Diaz Polanco, C. and Salas de Diaz, G. (1973) Nuova lista de patogenos en las plantas cultivadas en Venezuela. (New list of pathogens on cultivated plants in Venezuela.) *Boletin Especial Sociedad Venezolana de Fitopatologia* No. 2, 61 pp.

Vietnam

My, H.-T. (1966) *A Preliminary List of Plant Diseases in South Vietnam.* Directorate of Research, Ho Chi Minh City, 63 pp.

Virgin Islands, American

Stevenson, J.A. (1975) The fungi of Puerto Rico and the American Virgin Islands. *Contributions of the Reid Herbarium* No. 23, 743 pp.

Yemen Arab Republic

Kamal, M. and Agbari, A.A. (1980) Revised host list of plant diseases recorded in Yemen Arab Republic. *Tropical Pest Management* 26, 188–193.

Yugoslavia

Buturovic, D. (1972) Lista uzrocnika oboijenja bilja u Bosni i Hercegovini utvrtenih do 1971 godine. (A list of causal agents of diseases of plants in Bosnia and Hercegovina registered up to 1971.) *Zbornik Radova, Sarajevo,* 167–188.

Janezic, F. (1957) Indeks rastlinskih bolezni v Sloveniji. (Index of plant diseases in Slovenia.) *Zbornik za Kmetijstvo in Gozdarstvo* 3, 39–86.

Zaire

Buyckx, E.J.E. (ed.) (1962) Precis des maladies et des insectes nuisibles rencontres sur les plantes cultivees au Congo, au Rwanda et au Burundi. (Compendium of diseases and noxious insects found on cultivated plants in the Congo, Rwanda and Burundi.) *Publications de l'Institut National pour l'Etude Agronomique du Congo Belge,* Hors Serie, 708 pp.

Zambia

Angus, A. (1962–1966) *Annotated List of Plant Pests and Diseases in Zambia (Northern Rhodesia) Recorded at the Plant Pathology Laboratory, Mount Makulu Research Station.* Parts 1–7, supplement, 384 pp.

Zimbabwe

Rothwell, A. (1975) A revised list of plant diseases in Rhodesia – Additions, 1966–72. *Kirkia* 10, 295–307.

Whiteside, J.O. (1966) A revised list of plant diseases in Rhodesia. *Kirkia* 5, 87–196.

Seed Health

29

Revised by J.M. Waller

CABI Bioscience UK Centre, Bakeham Lane, Egham, Surrey TW20 9TY, UK

Introduction

The health of seed is of fundamental importance to the production of healthy crops. Seed, soil and climate are the three external factors necessary for agriculture to commence. Some crops such as tuberous plants and high value perennials are propagated by vegetative means, but it is estimated that 90% of food crops produced in the world are reared from true seed. A wide range of microorganisms are carried on or in seed and most are components of the normal crop-associated biodiversity including both harmless saprobes and damaging pathogens. A wide variety of pathogens are seed-borne; some may be opportunistic necrotrophs only harmful to damaged seeds or seedlings, others may be directly parasitic on the seed itself, and others such as smuts and viruses may be specialized biotrophs that cause systemic infections of the growing plant. Seed-borne pathogens have been extensively catalogued by Richardson (1990). The epidemiology of seed-borne diseases is particularly significant for both crop health and the success of the pathogen. Pathogens carried and dispersed with the propagules of their hosts have a unique advantage as an early parasitic association is almost assured but a plant emerging from a diseased seed is almost doomed from the start of its life. Seed-borne inoculum can initiate epidemics and introduce new diseases or pathotypes to new areas and is therefore of quarantine significance (see Chapter 27).

For vegetatively propagated crops, many of the pathogens associated with the growing crop can be directly transferred via the propagule as there is usually no barrier between mother plant and offset. Systemic infections of viruses and bacteria, soil-borne fungi and nematodes may all be efficiently dispersed on tubers, rooted cuttings, budwood, etc. True seed, however, generally poses fewer dangers of transmitting systemic infections, but many pathogens are specialized in the seed-borne mode of transmission; this synopsis will concentrate on pathogens transmitted through true seed. The biology and control of seed-borne diseases is the subject of several specialized texts (Neergaard, 1977; Agarwal and Sinclair, 1988; Maude, 1996).

Modes of Seed Transmission

Pathogens are often transmitted as contaminants of seed either adhering to seed coats or in crop debris that accompanies seed. Resting fungal structures such as sclerotia (*Claviceps* spp.) or spores (*Tilletia* spp.) are examples of specialized pathogens transmitted in this fashion but spores or other dormant structures of many foliage pathogens inevitably contaminate seeds during harvesting and not all are removed during seed-cleaning processes.

Direct infection of the seed coat is a common route for transmission of many seed-borne pathogens. These often spread to the developing seed directly from diseased foliage and a large number of foliage pathogens are transmitted in this way (*Septoria*, *Phoma*, etc. and some bacteria). Infection usually occurs at the flowering stage subsequently spreading to the outermost tissues of the seed.

Other pathogens may infect the deeper tissues of the seed including the embryo. These include many systemic pathogens such as viruses, some bacteria and fungi. Often these gain entry through direct systemic infection from the mother plant; some may be pollen transmitted whereas others infect through the flower. Many are specialized seed-borne pathogens with this route as their main method of survival and transmission. Seed transmission of fungal pathogens of the vascular system is uncommon but can occur under certain circumstances.

Disease Development

The risk of seed-borne pathogen inoculum giving rise to disease problems in growing crops depends on many factors from the initiation of infection in the developing plant to epidemic development in the growing crop and depends on a range of environmental factors affecting these processes. Although seed-borne pathogens can provide the initial inoculum for the development of epidemics, many survive and are dispersed by other means. The large number of necrotrophs that occur on or in seed often also survive on crop debris in the soil or on alternate hosts. Therefore, the exact role that seed-borne inoculum has in epidemic development of such diseases is not always clear. Furthermore, the pathogenic inoculum carried on seed may not always give rise to infected plants or the number of infected seeds in a seed lot may be insufficient to pose a real threat to the crop as a whole. Therefore, both quantitative and qualitative aspects of seed health are important and methods to assess both of these need to be take into consideration (Colhoun, 1983).

Seed Health Testing

Methods of testing seeds for the presence of pathogens are of four main types. These are: (i) direct examination; (ii) examination after incubation, either for the presence of pathogens or symptoms; (iii) isolation and culturing of pathogens on agar plates etc.; and (iv) biochemical or immunological techniques to detect the presence of particular pathogens. Many of these techniques have been refined for particular crops or pathogens and approved internationally by the International Seed Testing Association (ISTA) as standard procedures.

Some pathogens can be detected by more than one method, but the results usually differ quantitatively. The method to be preferred will usually depend on the purpose for which the test is undertaken; for example, when tests are being carried out for quarantine purposes the most sensitive test would normally be used. Other factors to consider are the relative cost, time taken to obtain results, and reliability of the test result in indicating the amount of disease likely to be transmitted. The presence of an organism associated with seed is not necessarily an indication of pathogenicity, and even if the organism is a known pathogen of the crop it may not necessarily be transmitted by the seed to cause disease in the crop derived from that seed.

Direct examination

Direct examination methods are the least useful for obtaining accurate quantitative information about infection levels, but they are of use in preliminary screening to detect heavily infected seed stocks. It is usually only the more grossly diseased seeds that are visible and in samples with such seeds the presence of additional symptomless seeds can be expected. Negative results from such tests cannot be used as an indication of freedom from disease.

The morphology of the seed may be changed, so knowledge of the appearance of healthy seed is necessary, and also of obvious pathogen structures such as fungal sclerotia and nematode galls. Seeds may be discoloured or misshapen and some infections produce characteristic staining such as a pink colour of cereal seeds indicating *Fusarium* spp. The usefulness of colour differences is limited, however, since they may be inconsistent, non-specific, or cultivar related. In white-seeded cultivars of *Phaseolus* the brown lesions caused by *Colletotrichum lindemuthianum* allow affected seeds to be removed by electronic sorting equipment. The presence of the pathogen can sometimes be directly observed as spore masses, bacterial exudate or fungal fruiting bodies such as dark stroma or pycnidia but these are more readily detected by examination under a stereoscopic microscope. Seed samples may also be contaminated with fungal bodies, such as sclerotia of ergot (*Claviceps* and *Typhula* in samples of grass and clover seed), or seeds may be 'blind' and contain nematodes (*Anguina* ear cockles in wheat) or fungal spores (*Tilletia*).

An extension of the direct examination method is to soak seeds to allow spores or nematodes to float off, to be detected either in washing fluid or collected on a filter. Such methods are, however, difficult to use to obtain quantitative results although they may be used to estimate viabilty of inoculum by for example vital staining using tetrazolium. A further extension of the direct examination method is to process seeds to examine those parts in which the infection is to be found. The best known is the embryo test to detect *Ustilago nuda* in wheat and barley (Rennie and Seaton, 1975). Seeds are soaked in 5% NaOH at 20–22°C for 22 h. The embryos separate and are collected by washing through sieves of mesh sizes 3.5, 2.0 and 1.0 mm. The embryos are then dehydrated in 95% ethyl alcohol for 2 min, separated from chaff by differential flotation in lactophenol/water and then cleared by heating, in a fume cupboard, in water-free lactophenol to boiling point. Embryos can then be examined in lactophenol for the presence of characteristic mycelium in the scutellum. A similar method can be used for *Sclerospora graminicola* in *Pennisetum* (Shetty *et al.*, 1978) and *U. nuda* in wheat, with the use of trypan blue to aid in observation of the mycelium.

Incubation methods

There are many variations of the methods under this heading, but basically they rely on the ability of pathogens, especially fungi, either to produce identifiable structures on the seed when incubated on moist blotters in a humid atmosphere, or to produce symptoms or other evidence of infection.

Seeds are normally spaced on blotters. Several layers of filter paper may be used, but thicker blotting paper has better moisture-retaining properties. They are incubated in plastic containers which allow the passage of near ultra violet light (NUV, 320–380 nm) known to enhance the sporulation of many fungi. The usual routine is 12 h NUV and 12 h darkness. Seeds of temperate crops are usually incubated at 14–20°C, but those of tropical crops are often incubated at 28°C. As with other methods, the reliability of the results tends to depend more on the ability of the observer to recognize pathogens or their effects than on the provision of precisely controlled conditions of incubation.

Seed is not usually pretreated to eliminate superficial or saprobic organisms, but seeds of rice and other tropical crops, which are often heavily contaminated, may be treated with a solution of sodium hypochlorite, containing the equivalent of 1% chlorine, for 10 min before incubation. Incubation is usually for 7 days, but information may be obtained in as little as 2 days with some fungi. For small hyaline fungi the use of dark blotters provides a background against which they are more easily seen.

A low-power (×10–50) stereoscopic microscope, preferably with a moving stage, is necessary for examination. Fluorescent or optic fibre lighting systems allow good lighting to be obtained without too much drying out of the material being examined. With 100 seeds on a 20 × 10 cm blotter infected seed can be marked as observed, colour coded for different organisms, with ink pens or water colour pencils, and totalled at the end of the examination. For further study, isolations can be easily made by transferring spores to agar or microscope slides either with a needle or fine watchmaker forceps.

During incubation germination usually begins and symptoms may be observed such as pycnidia on the seed coat and dark brown or black streaks, especially on the coleoptile. Fruiting bodies and lesions often appear towards the end of the incubation period of small seeds when the nutrients have been exhausted. Other effects may also be noted as substances produced by diseased seeds seep out on to the blotter causing visible stains.

Germination may be unnecessary, and the development of shoots and roots may be inhibited to allow easier observation. A 200 p.p.m. solution of the sodium salt of 2,4-D (2,4-dichlorophenoxyacetic acid) in water prevents germination without affecting fungal development. Incubating seeds can also be frozen at −18 to 20°C for 24 h after a few days and then re-incubated for 5–7 days. The deep freezing kills the seed and the subsequent fungal development is often more abundant, but whether the method is more useful than normal blotter or agar methods depends on the particular host–pathogen combination.

The roll-towel or paper-dolly method of germination testing used in many seed-testing stations, especially for cereals and legumes, also allows observations of symptoms to be made. Seeds are placed on a sheet of damp paper towelling covered by another layer, the lower 5 cm turned over, and the whole rolled up and secured by elastic bands. The rolls are incubated under humid conditions, standing vertically in baskets to provide a space saving and inexpensive test. Characteristic lesions or fungal structures may develop on seedling shoots or the pathogen may spread to the paper towel where it may sporulate or produce other characteristic structures.

Growing-on tests are an extension of the incubation methods. They have the disadvantage of taking more time, space and materials than blotter methods, and being more subject to environmental conditions. They can be used simply as a means to gain time to allow symptom expression, e.g. for *lettuce mosaic virus*, or in combination with a stress test (the Hiltner method) where seed is deep-sown in a heavy or water-retentive substrate, e.g. sand, china clay granules, or brickstone. Under these conditions pathogens may be detected that do not appear in an ordinary germination test.

Isolation and culturing methods

These are probably the most reliable and widely developed methods for detecting the majority of seed-borne fungi, and the requisite facilities for medium preparation and aseptic techniques should be available to any worker in seed pathology. Most methods depend on incubation of seeds on agar plates to enable the pathogen to grow out from the seed and sporulate. The expense of Petri dishes and media is offset by the speed with which a batch of plates can be recorded using macroscopic features of colonies with little need by experienced observers for microscopic observation.

Seeds are generally pretreated with a mild disinfectant before plating on agar to prevent overgrowth of the plates by saprobic fungi. The usual pretreatment is 10 min in a 1% solution of chlorine in water (10% dilution of a standard stock solution of sodium hypochlorite or proprietary chlorine-based dairy disinfectant), but the duration is not critical for most seeds. Dilute solutions of calcium hypochlorite or mercuric chloride can also be used. Except with the mercury pretreatment, washing is not necessary and seeds can be transferred directly after blotting off excess liquid before plating.

Potato dextrose agar is the most widely used medium but encourages rapid growth of saprobes and is not conducive to the sporulation of many fungi; the addition of a small quantity of streptomycin or similar antibiotic will limit the growth of bacteria. Growth media with a lower sugar content are preferable (see Chapter 38). Water agar has the advantage that it provides no external nutrients and tends to favour growth of those fungi able to use nutrients present in the seed. Characteristic structures produced by the pathogen in the agar can also be readily observed. Plates are normally incubated for 7 days at temperatures of 18–22°C, or 25–28°C for tropical crop seeds. Large seeds (peas, beans and maize) are plated five per 8.5 cm Petri dish, smaller seeds ten to a dish.

Many seed health tests are carried out to detect specific pathogens, and selective media have been developed to allow recognition of particular pathogens without interference or contamination by other organisms. Agar plating methods are unsuitable for pathogens that are affected by competing or antagonistic organisms on the seed, for example *Pyricularia oryzae* is readily overgrown on agar by *Alternaria padwickii*. They are also less useful for the detection of bacteria. These and some other fungi present in small quantities may be extracted from seed by a range of washing and soaking techniques and the pathogens detected by plating out the resulting liquid. This may also be applied to ground-up seed to detect internal inoculum.

Indirect cultural methods include inoculating seed extracts into indicator plants. Host leaves can be used as an enrichment medium to be inoculated with extracts from seeds under test; bacterial pathogens of *Phaseolus* beans can be detected by soaking seeds in water for a few hours and then inoculating healthy host seedlings with the soak water.

Immunodiagnostic and nucleic acid methods

The advantages of these methods are that they allow accurate identification of often minute amounts of seed inoculum of specific pathogens. Serological techniques have been used for some time and are more applicable to viruses and bacteria, but techniques applicable to fungi are developing rapidly (see Chapter 22). The ELISA technique is very sensitive and can detect infections in single seeds among thousands and colorimetric techniques enable quantitative assays. It is now widely used as a standard technique for detecting most seed-borne viruses. There are many different forms of ELISA, some of which have been developed for seed testing (Lange, 1986; Ball and Reeves, 1992). Nucleic acid methods are now being rapidly developed but the methods are still expensive and require further development before they can be applied to routine seed health screening. Nucleic acid probes are more applicable to identification of organisms once they have been isolated from seeds and not to routine seed health testing. The PCR technique has the potential for wider application for the detection of seed-borne inoculum in seed extracts. (See Chapter 23 and Reeves, 1995.)

Production of Healthy Seeds

The production of healthy seed crops is a major prerequisite for production of healthy seed and a knowledge of the epidemiology of the pathogen and especially of how it infects seed is necessary. Seed crop inspection and roguing out of diseased plants is often required as part of the certification process for healthy seed (essential for vegetatively propagated crops). Appropriate harvesting and seed-cleaning techniques can reduce the extent of contaminating pathogens and badly diseased seeds. There is a range of methods for treating seed to eradicate pathogens and some of the first disease control measures were used on seed-borne pathogens of cereals. These involved soaking seeds in brine and other substances; seed soaking in selected pesticides is still used today. Most countries have certification schemes supported by legislative measures to enable farmers to select healthy seed and these are coordinated through the ISTA. These make use of recommended 'tolerance levels' in seed of particular seed-borne pathogens. For quarantine purposes tolerances are generally very strict.

Seed treatment practices to control seed-borne diseases involve the use of chemical, physical and biological methods. A wide range of fungicides especially those with systemic or eradicant properties have been applied to seeds. Seed application of fungicides has many advantages; very small amounts of active ingredient are required and these can be precisely targeted compared to that required and achieved by field application to growing crops. Not only can seed-borne pathogens be controlled but protection can also be given to young plants against many pathogens, both soil borne and airborne which attack them in early stages of development. Systemic compounds applied to seed have been used specifically for the control of airborne pathogens such as mildews. Some pathogens have developed resistance to seed-applied fungicides especially some systemic compounds but some *Pyrenophora* spp. have developed resistance to organomercurial compounds used on oats. Compounds are often approved for use on seeds by national regulatory bodies taking into account efficacy and environmental considerations. Application of pesticides to seed can be undertaken on a small scale by farmers using dusts or slurries, but the most effective way is to use specialized machinery in a commercial operation and can involve application of 'cocktails' encapsulated around the seed in a pellet. There is a wide range of modern techniques and Jeffs (1986) has provided a review of this subject (see also Chapters 32 and 33).

Physical control methods include the use of heat and hot water treatment to eradicate deep-seated infections from seed. A temperature of about 50°C for 20–40 min is commonly used. The success of the methods depends on even heat distribution throughout the seed lot and requires careful monitoring to ensure this without exposing some seeds to excess time–temperature regimes that will lead to loss of viability.

316 J.M. Waller

Biological control of seed-borne fungi has been developed recently mainly using bacterial antagonists such as *Bacillus subtilis* and *Pseudomonas cepacia* and some products are available commercially, mainly against soil-borne seedling diseases (see also Chapter 34).

References and Further Reading

Agarwal, V.K. and Sinclair, J.B. (1988) *Principles of Seed Pathology*. Vol. I, 176 pp. and Vol. II, 166 pp. CRC Press, Boca Raton.
Anon. (1981) *Handbook on Seed Health Testing*. Section 2: Working Sheets. International Seed Testing Association, Zurich.
Ball, S. and Reeves, J. (1992) Application of rapid techniques to seed health testing – prospects and potential. In: Duncan, J.M. and Torrance, L. (eds) *Techniques for the Rapid Detection of Plant Pathogens*. Blackwell Scientific Publications, Oxford, UK, pp. 193–207.
Brlansky, R.H. and Derrick, K.S. (1979) Detection of seed-borne plant viruses using serologically specific electron microscopy. *Phytopathology* 69, 96–100.
Coulhoun, J. (1983) Measurement of inoculum per seed and its relationship to disease expression. *Seed Science and Technology* 11, 665–671.
Cruz, M. and Fernandez, A.I. (1979) La immunofluorescencia en el diagnostico de semillas y hojas de plantas enfermas con bacterial. *Ciencias Sanidad Vegetal* No. 17, 9 pp.
Diekmann, M. (1993) *Seed-borne Diseases in Seed Production*. International Center for Agricultural Research in the Dry Areas (ICARDA), Aleppo, Syria, 81 pp.
Hutchins, J.D. and Reeves, J.C. (eds) (1997) *Seed Health Testing*. CAB International, Wallingford, UK, 263 pp.
International Seed Testing Association (1993) International rules for seed testing, 1993. *Seed Science and Technology* 21, Supplement, 296 pp.
Jafarpour, B., Shepherd, R.J. and Grogan, R.G. (1979) Serologic detection of bean common mosaic virus and lettuce mosaic virus in-seed. *Phytopathology* 69, 1125–1129.
Jeffs, K.A. (ed.) 1986) *Seed Treatment*. BCPC Publications, Thornton Heath, UK.
Kritzman, G. and Netzer, D. (1978) A selective medium for isolation and identification of *Botrytis* spp. from soil and onion seed. *Phytoparasitica* 6, 3–7.
Kushi, K.K. and Khare, M.N. (1979) Comparative efficacy of five methods to detect *Macrophomina phaseolina* associated with *Sesamum* seeds. *Indian Phytopathology* 31, 258–259.
Lange, L. (1986) The practical application of new developments in test procedures for the detection of viruses in seed. *Developments in Applied Biology* 1, 269–281.
Lange, L. and Heide, M. (1986) Dot immuno binding (DIB) for detection of virus in seed. *Canadian Journal of Plant Pathology* 8, 373–379.
Langerak, C.J., Franken, A.A.J.M. and Martin, T. (1994) Diagnostic methods for the detection of plant pathogens in vegetable seeds. In: *Seed Treatment: Progress and Prospects: Proceedings of a Symposium held at the University of Kent, Canterbury, 5–7 January 1994*. BCPC Monograph No. 57 British Crop Protection Council, Farnham, UK, pp. 169–178.
Lima, J.A.A. and Purcifull, D.E. (1980) Immunochemical and microscopical techniques for detecting blackeye cowpea mosaic and soybean mosaic viruses in hypocotyls of germinated seeds. *Phytopathology* 70, 142–147.
Lister, R.M. (1978) Application of the enzyme-linked immunosorbent assay for detecting viruses in soybean seed and plants. *Phytopathology* 68, 1393–1400.
Mathur, S.B. and Cunfer, B.M. (eds) (1993) *Seed-borne Diseases and Seed Health Testing of Wheat*. Jordbrugsforlaget, Frederiksberg, Denmark, 168 pp.
Maude, R.B. (1996) *Seedborne Diseases and Their Control*, CAB International, Wallingford, UK, 280 pp.

McGee, D.C. (1988) *Maize Diseases: a Reference Source for Seed Technologists.* APS Press, St Paul, Minnesota, USA, 149 pp.

McGee, D.C. (1992) *Soybean Diseases: a Reference Source for Seed Technologists.* APS Press, St Paul, Minnesota, USA, 151 pp.

McGee, D.C. (1995) Epidemiological approach to disease management through seed technology. *Annual Review of Phytopathology* 33, 445–466.

Mew, T.W. and Misra, J.K. (eds) (1994) *A Manual of Rice Seed Health Testing.* International Rice Research Institute (IRRI), Manila, Philippines, 113 pp.

Neergaard, P. (1977) *Seed Pathology,* 2 vols. Macmillan, London, UK, 1187 pp.

Ralph, W. (1977) Problems in testing and control of seed-borne bacterial pathogens: a critical evaluation. *Seed Science and Technology* 5, 735–752.

Reeves, J.C. (1995) Nucleic acid techniques in testing for seed borne diseases. In: Skerritt, J.H. and Appels, R. (eds) *New Diagnostics in Crop Science.* CAB International, Wallingford, UK, pp. 127–149.

Rennie, W.J. and Seaton, R.D. (1975) Loose smut of barley. The embryo test as a means of assessing loose smut infection in seed stocks. *Seed Science and Technology* 3, 697–709.

Richardson, M.J. (1990) *An Annotated List of Seed-borne Diseases.* 4th edn. International Seed Testing Association, Zurich, Switzerland.

Russell, T.S. (1988) Inoculum thresholds of seedborne pathogens. Some aspects of sampling and statistics in seed health testing and the establishment of threshold levels. *Phytopathology* 78(6), 880–881.

Schaad, N.W. and Donaldson, R.C. (1980) Comparison of two methods for detection of *Xanthomonas campestris* in infected Crucifer seeds. *Seed Science and Technology* 8, 383–391.

Sharon, E., Okon, Y., Bashan, Y. and Henis, Y. (1981) Leaf enrichment: a method for detecting small numbers of phytopathogenic bacteria in seeds and symptomless leaves of vegetables. *Phytoparasitica* 9, 250.

Shetty, H.S., Mathur, S.B. and Neergaard, P. (1980) *Sclerospora graminicola* in pearl millet seeds and its transmission. *Transactions of the British Mycological Society* 74, 127–134.

Warham, E.J., Butler, L.D. and Sutton, B.C. (1976) *Seed testing of maize and wheat; a laboratory guide.* CIMMYT, El Batan, 84 pp.

Cultural Control

A.J. Termorshuizen

Biological Farming Systems, Wageningen University,
Marijkeweg 22, 6709 PG Wageningen, The Netherlands

Introduction

Cultural control consists of all methods a farmer can apply to control plant diseases excluding chemical, resistance breeding and biological control methods. Quarantine measures, which are taken to prevent the introduction of a pathogen, are treated in Chapter 27.

Characteristic to cultural practices is that their effects on the agroecosystem are complex, usually exhibiting direct effects on the pathogen and indirect effects via the host and the soil ecosystem (Cook and Baker, 1983). Well-known examples are the multiple effects of fertilization and irrigation. If the indirect effects on the pathogen prevail, the effect of cultural practices vary according to their effects on the host and the environment. Therefore, effects of cultural measures on the pathogen can be difficult to predict. A well-documented example of unexpected effects of cultural control is soil inundation that dramatically reduced populations of *Fusarium oxysporum* f.sp. *cubense* (the causal agent of banana wilt) (Stover *et al.*, 1953). However, the small amount of inoculum of this fungus that survived the inundation recolonized the soil quickly to levels that were much higher than before inundation (Stover, 1979) because competitors had also been decimated.

The wide definition of cultural control adopted here allows for multiple types of classification. Cultural control methods can be related to the timing of application (before, during, or after crop growth), to the type of effect on the pathogen (prevention of pathogen establishment, reduction of initial population, or reduction in the build-up of the pathogen population), or to the area of application (plant, field, farm, village, province, nation). Here cultural control methods are divided into:

©CAB *International* 2002. *Plant Pathologist's Pocketbook*
(eds J.M. Waller, J.M. Lenné and S.J. Waller)

1.   Cultural practices that are inevitably part of cultivating the crop, such as fertilization, where the phytopathologist is interested in the influence on disease development of, for example, timing, amount and type of fertilization;
2.   Cultural practices that have the foremost or single goal of preventing, suppressing, or killing pathogens, such as seed or soil heat treatment.

Due to the multitude of cultural control methods available it is impossible to present an exhaustive appraisal here. Rather, an overview is given of the variety of cultural control methods and the inherent complexity of their effects.

Cultural Practices That Influence Plant Diseases

Several cultural practices, such as nitrogen fertilization or sowing time, affect some diseases in different ways depending on the form of their application. For example, ammonium fertilization can have an opposite effect on disease progression to nitrate fertilization, and early sowing can produce an opposite effect to late sowing. The optimal application of cultural practices to combat diseases is therefore achieved when it is known which diseases are likely to develop in a certain field, either from experience or by pathogen detection assays. The effects of cultural practices on diseases and on soil-borne pathogens have recently been reviewed (Rush *et al.*, 1997; Abawi and Widmer, 2000).

Crop rotation

Crop rotation is one of the most effective strategies to control soil-borne pathogens. It is one of the oldest cultural practices that maintains soil fertility and soil structure and that controls serious weed problems.

A major phytopathological goal of crop rotation is to prevent the build-up of persistent structures of pathogens such as chlamydospores and sclerotia. Thus, many dicotyledonous crops show a decline after repeated cultivation for two or more consecutive years (e.g. Oyarzun *et al.*, 1993). However, monocotyledonous crops can often be grown in continuous culture for years without any substantial increase in diseases or decrease in yield. Rotation to suppress disease is primarily achieved by growing non-hosts. Generally, a one-to-four rotation precludes the build-up of plurivorous soil-borne pathogens such as *Rhizoctonia solani*, *Sclerotinia sclerotiorum* and *Verticillium dahliae*. Hosts can also be used in a rotation, provided that the pathogens do not cause great damage and that the crop is harvested before the pathogen has formed its survival structures. For example, in The Netherlands sugarbeet does not contribute to the build-up of microsclerotia of *V. dahliae* in soil, because it is harvested well before the pathogen forms its survival structures.

The positive effect of wide crop rotations is counterbalanced by the desire of the farmer to grow the most valuable crop as frequently as possible, which will then usually entail the use of pesticides. To reduce the usage of pesticides some countries place legal requirements on farmers to practice rotation. For example, to prevent the build-up of large populations of potato cyst nematodes, it is not permitted to grow potatoes every year in The Netherlands.

A decrease of disease after years of continuous cropping occurs if populations of antagonists of the pathogen develop. Such a decline has been described for take-all (*Gaeumannomyces graminis*) in wheat and is known as take-all decline. In the first years, the pathogen causes increasing damage but in the following years, these effects gradually decline. Decline phenomena may be sensible starting points for searching for effective antagonists of pathogens, or for components of disease suppression.

Mixed cropping and intercropping

Cropping mixed varieties in a field (mixed cropping) or different crops in alternating rows (intercropping) are old and widely used cropping systems. In fact, increasing the genetic diversity in the field can have the same effect on soil microbiota. In mixed cropping, diseases develop more slowly due to interception of the pathogen (Akanda and Mundt, 1996) or its vector. Besides the effect on disease suppression, intercropping uses resources more efficiently and soils are generally less prone to erosion.

Selection of cultivar to be grown

The most effective way to limit the risks of a plant disease is to choose cultivars that exhibit the highest resistance or tolerance levels. Organic farmers depend primarily on selecting appropriate cultivars and on applying wide rotations. However, cultivars that often have characteristics that are desired by consumers or the processing industry are often highly susceptible to several pathogens, e.g. potato cultivars Russet Burbank and Bintje. If alternative control measures are absent, farmers shift to growing lower-value but more tolerant cultivars, such as in sugarbeet fields that are found to be infested with *Polymyxa betae*.

Tillage

Conventional tillage carries with it the incorporation of organic matter into soil. Organic matter that is incorporated into soil is colonized by saprotrophic fungi, and plant pathogens play a role if they have competitive saprotrophic ability to colonize the organic matter, or if they are already present on the organic matter. If the organic matter has previously been infected by airborne plant pathogens, then incorporation into soil usually affects these pathogens negatively. This is because many airborne pathogens have low competitive saprotrophic abilities and low survival capabilities. Certain soil-borne pathogens such as *R. solani* can be strongly stimulated by amendments of fresh organic matter (see sections on fertilization and organic amendments). For competition-sensitive soil-borne pathogens such as *Pythium* spp. it is optimal to incorporate organic matter after it has been colonized by saprotrophic fungi. Deep-incorporation of organic matter that has been infected by soil-borne pathogens has been advocated but this exercise is expensive and the soil structure and organic matter content may be affected negatively. In general, tillage activities need to be balanced for effects on soil structure, erosion and fertility and phytopathological considerations.

Planting time

Planting time influences the time the pathogen has access to susceptible plant tissue, which is most relevant to pathogens causing seedling diseases. Sowing during cool and damp weather promotes growth of *Pythium* spp. and low temperatures prolong the susceptibility of the seedlings. On the other hand, early sowing of sugarbeets is recommended to reduce damage caused by *Beet necrotic yellow vein virus* (BNYVV), which is vectored by *P. betae*, which has its optimum at 25°C. Plant stage, and consequently sowing time, strongly influences tolerance of wheat against several snow moulds. If wheat is sown in late summer the larger plants are apparently tolerant against leaf mould, and if wheat is sown in October the very young plants are thought to escape infection (Cook and Baker, 1983).

Fertilization

Inorganic soil amendments are primarily used to fertilize a crop but they should be used concurrently to influence plant disease development (Marschner, 1995). This can be achieved through changes in the resistance or tolerance of the host plant, in the pathogen or in the microbial community influencing the pathogen. Changes in the resistance or tolerance of the host plant can for example occur by changes in cell wall composition that affect pathogen infection or by changes in the concentration or rate of production of phytoalexins. Apparent resistance can be achieved when rates of host development and pathogen development are asynchronized so that the host is resistant at the time of highest activity of the pathogen. Like most cultural practices, effects of mineral nutrition on plant diseases are usually complex and are therefore difficult to generalize. An example of an indirect effect of fertilization is by alteration of the humidity of the crop stand, which may enhance the development of fungi such as *S. sclerotiorum*.

In general, fertilization to manage diseases is optimal when it is also optimal for plant growth (Marschner, 1995). The influence of mineral nutrition on plant resistance is relatively small in highly susceptible or highly resistant cultivars but can be substantial in moderately susceptible or partially resistant cultivars. Information from the literature as to what levels of fertilization are optimal to manage diseases are generally confounded by lack of information on the concentration of the major nutrients in soil and plant tissue.

Effects of mineral nutrition can be unexpectedly low if the mineral accumulation site in the plant is different from the infection site of the pathogen. For example, silicon supply increases the resistance to multiple fungal diseases in grass species, such as the blast fungus (*Pyricularia oryzae*) in rice. However, silicon is translocated mainly to the mature leaves, whereas only the young rice leaves are infected by the blast fungus. Nevertheless, silicon supply is sufficiently effective to eliminate enhanced susceptibility of rice to the blast fungus caused by application of high levels of nitrogen.

Liming is an effective method to reduce club root caused by *Plasmodiophora brassicae*, but under these circumstances common potato scab caused by *Streptomyces scabies* is increased.

Application of nitrogen generally increases the susceptibility to obligate parasites and decreases it to facultative parasites. The increased susceptibility to obligate parasites at nitrogen levels that are optimal for plant growth can be due to decreased phytoalexin biosynthesis and a higher proportion of non-mature plant tissue. Nitrogen-induced plant growth also reduces the silicon content in the leaves, which can increase the susceptibility to some diseases. The form of nitrogen fertilization can have large effects on disease incidence (Henis and Katan, 1975). This can be related to preferences of the plant for one form of nitrogen but it can also result from toxic effects on the pathogens by formation of ammonia from ammonium (e.g. several nematodes and *Sclerotium rolfsii*) or by formation of nitrite from urea (e.g. *F. oxysporum* f.sp. *cubense*).

In contrast to nitrogen, application of potassium to potassium-deficient soils elicits increased resistance responses to both facultative and obligate pathogens. The explanation is that at potassium-deficient levels, the formation of high-molecular compounds that are related to several resistance mechanisms, such as proteins, starch and cellulose, is impaired.

Calcium generally enhances resistance against diseases, including post-harvest diseases, by regulating the stability of biomembranes and cell walls. When the calcium content is low, exudation into the apoplast of sugars and other simple compounds that can be used by parasites is enhanced. Many bacteria and fungi invade plant tissue by producing extracellular pectolytic enzymes. The activity of these enzymes is reduced if the middle lamella contains high concentrations of calcium polygalacturonates.

Micronutrients can have various effects, which have been reviewed by Engelhard (1989). The most well-known include the role of boron, manganese and copper in the production of the plant defence-related phytoalexins and lignin.

Irrigation

The effects of irrigation on plant diseases can be divided into effects on transportation of inoculum and effects on growth of the pathogen in soil. High water levels impose oxygen stress on the plants, which may increase their susceptibility to several root and foot rot pathogens. On the other hand, while under normal conditions sorghum is resistant against *Macrophomina phaseolina*, the causal agent of stalk rot, under dry conditions sorghum is highly susceptible, especially at the flowering stage. Irrigation can dramatically influence spread of the disease if spores are transported by streaming water. This danger is also evident in ebb-and-flow systems applied in greenhouse potting cultures. In this and other systems where water is being reused, sterilization of recirculation water, e.g. by ozone or heat treatment, is necessary.

The moisture content of soil greatly influences the dynamics of soil-borne pathogens; *Pythium* and soil-borne *Phytophthora* species are well-known examples of pathogens that flourish in wet soil conditions. Naturally, the more the pathogen is favoured and the more the plant is stressed, the more root rot, foot rot and damping-off diseases develop. High-precision irrigation using drippers close to the stem base increases variability of soil moisture content and thus may increase the resistance of the soil ecosystem against growth of plant pathogens.

Cultural Control of Plant Diseases

Since most cultural control measures are quite expensive, knowledge of the origin of the disease is essential. For example, if a soil-borne disease is primarily spread by irrigation, there is no point in disinfecting seed or solarizing soil. Local measures to remove inoculum are more successful for soil-borne diseases than for airborne diseases.

Prevention

Prevention starts with the use of uninfected, preferably certified, planting material. The most generally applied cultural practice to prevent disease is sanitation. Farmers must make sure that they do not transfer contaminated materials (equipment, soil, etc.) from field to field or from season to season. Machinery should not damage harvest products. Insects vectoring viruses can be kept out of greenhouses by covering the windows with specially designed nets.

Destruction or removal of inoculum

Soil heat treatment and soil inundation aim to kill pathogens directly. Lethal doses of heat are known for a range of plant pathogens under moist conditions (Bollen, 1993). With soil solarization, pathogens are killed by heating up the soil by covering irrigated soil with transparent plastic sheeting. However, sufficiently high temperatures can only be reached in mediterranean or warmer climates. In cooler regions soils can be treated with steam, although this method is quite expensive and is therefore limited to high-value crops in greenhouses. Soil inundation has been shown to be effective for the control of several pathogens on soils with a suitable hydraulic conductivity. Soil heat treatment and soil inundation also affect many other organisms and care should be taken that the soil is not recolonized by the inoculum remaining in the deeper soil layers.

Removal of infected plants and termination of cropping for a number of years is especially important for new or rare diseases. In The Netherlands this has been or is advocated for *Erwinia amylovora* (causing fire blight in fruit trees), *Ralstonia solanacearum* (brown rot in potato), *Ophiostoma ulmi* (Dutch elm disease) and *Synchytrium endobioticum* (potato wart disease).

Suppression of inoculum dispersal

Splash-dispersed pathogens such as *Phytophthora cactorum* and *Phytophthora fragariae* in strawberries can be suppressed by mulching the soil with straw. Installation of vertical barriers in soil has been advocated to prevent the spread by rhizomorphs of *Armillaria* spp. on ornamental trees. Mixed cropping, inter-cropping (see above) or cropping smaller areas, slow down the build-up of airborne diseases. It is also advisable not to grow a crop next to a field that had the same crop the previous year, since inoculum may have been formed on the stubbles that remained on the field.

Suppression of build-up of new inoculum

Build-up of new inoculum after the winter or dry season is always related to the presence of food substrate. Therefore, suppression of build-up of new inoculum should focus on restricting the availability of the adequate food substrate, or, if this is unavoidable, by making sure that the food substrate is colonized by strong competitors. Ringing of trees one or two years prior to felling prevents the saprophytic colonization of the remaining root system by *Armillaria* spp. and the subsequent colonization of the surrounding soil by massive amounts of rhizomorphs. Epidemics of *Phytophthora infestans* can be retarded significantly by carefully covering potato refuse heaps and removal of potato volunteers. Inappropriate management of organic matter may increase inoculum of soil-borne pathogens considerably. For example, use of non-mature compost can lead to a great increase of *Rhizoctonia solani* in soil.

Organic amendments

Certain organic amendments may have a profound effect on plant pathogens. With nitrogen-rich amendments the effect is often based on the formation of ammonia, which is toxic to many fungi. However, in many cases the effects are indirect through stimulation of the resident microflora or through stimulation of the resistance of the host. The more specific the amendment the more likely it is that its effect will be large and dependent on the environment. However, the price and transport costs of these amendments, such as blood meal, is usually high in contrast to less specific amendments such as compost. Much effort is being put into defining compost characteristics that correlate with disease suppression. In some cases the standing crop can be used as organic amendment. In a recent study it appeared that broccoli cultivated into soil greatly diminished inoculum of *Verticillium dahliae* (Subbarao and Hubbard, 1996).

New Developments

A key feature in developing new cultural methods is the communication between the scientist, who relies on the wishes and experiences of the farmer, and the farmer, who relies on the ingenuity of the scientist.

Catch crops

Catch crops are crops that are used to reduce the inocula of nematodes, e.g. *Tagetes* spp. produce toxins that reduce levels of *Pratylenchus* spp. Susceptible crops can also be used to hatch resting structures of cyst nematodes, after which the crops are harvested early, before completion of the life cycle of the nematodes. There has as yet been no success in using catch crops to reduce fungal soil infestations.

Slow sand filtration

Sand filtration has been used within living memory to clean drinking water, and it is being used increasingly to disinfect water that is recirculated in greenhouses (Wohanka, 1993). *Pythium* spp. and *Phytophthora* spp. can be eliminated successfully, but not *F. oxysporum* (van Kuik, 1994; Runia *et al.*, 1997). Percolation velocity, temperature and sand grain size appear to be among the most critical variables influencing inactivation. Slow sand filtration can be especially useful for greenhouse farmers who cannot invest in more expensive control measures such as heat treatment and ozonation of recirculation water.

Biological soil disinfestation

Biological soil disinfestation was developed recently by Blok *et al.* (2000). It is a non-chemical control method aimed at imposing a general soil anaerobiosis by incorporating green plant material into moist soil and covering the soil with thick ensilage plastic for at least 6 weeks. This resulted in a reduction in inoculum density by at least 95% of a number of soil-borne pathogens including *F. oxysporum* f.sp. *asparagi*, *V. dahliae*, *Pratylenchus penetrans*, *Meloidogyne chitwoodi* and *Globodera pallida*. The method is still applicable only for high-value crops because the plastic is expensive and the land needs to stay fallow for some time in late summer. However, in climatic zones where (sub)lethal temperatures are reached the incubation time may be much shorter. More effective organic amendments may be identified in the future.

Conclusions

Naturally the attitude of the farmer to cultural practices should be to optimize the cultivation with all measures that reduce pathogen establishment and development. Although cultural practices mostly relate to sustainable agriculture, several direct control methods definitely do not (e.g. the high energy demands of soil steaming and the need of plastic in solarization and biological soil disinfestation). For economic reasons, rotation in western countries has narrowed considerably over the last decades, making the use of unsustainable control measures necessary. It is important to realize that for many crops measures are available to permit sustainable agriculture, but that farmers are forced to reach production levels that make many of the environmentally benign cultural practices inapplicable for economic reasons. Therefore, it is likely that the increasing legal restriction of the application of chemical crop protection will stimulate the implementation of existing and the development of new cultural measures.

References

Abawi, G.S. and Widmer, T.L. (2000) Impact of soil health management practices on soilborne pathogens, nematodes and root diseases of vegetable crops. *Applied Soil Ecology* 15, 37–47.

Akanda, S.I. and Mundt, C.C. (1996) Effects of two-component wheat cultivar mixtures on stripe rust severity. *Phytopathology* 86, 347–353.

Blok, W.J., Lamers, J.G., Termorshuizen, A.J. and Bollen, G.J. (2000) Control of soilborne plant pathogens by incorporating fresh organic amendments followed by tarping. *Phytopathology* 90, 253–259.

Bollen, G.J. (1993) Factors involved in inactivation of plant pathogens during composting of crop residues. In: Hoitink, H.A.J. and Keener, H.M. (eds) *Science and Engineering of Composting*. Renaissance Publishers, Worthington, Ohio, pp. 301–318.

Cook, R.J. and Baker, K.F. (1983) Agricultural practices and biological control. In: *The Nature and Practice of Biological Control of Plant Pathogens*. APS Press, St Paul, Minnesota, USA.

Engelhard, A.W. (1989) *Management of Diseases with Macro- and Microelements*. APS Press, St Paul, Minnesota, USA.

Henis, Y. and Katan, J. (1975) Effect of inorganic amendments and soil reactions on soil-borne plant diseases. In: Bruehl, G. (ed.) *Biology and Control of Soil-borne Plant Pathogens*. American Phytopathological Society, St Paul, Minnesota, USA, pp. 100–106.

van Kuik, A.J. (1994) Eliminating *Phytophthora cinnamomi* in a recirculated irrigation system by slow sand filtration. *University of Gent, Medical Faculty Landbouww* 59(3a), 1059–1063.

Marschner, H. (1995) *Mineral Nutrition of Higher Plants*, 2nd edn. Academic Press, London, UK.

Oyarzun, P., Gerlagh, M. and Hoogland, A.E. (1993) Relation between cropping frequency of peas and other legumes and foot and root rot in peas. *Netherlands Journal of Plant Pathology* 99, 35–44.

Palti, J. (1981) *Cultural Practices and Infectious Crop Diseases*. Springer-Verlag, Berlin, Germany.

Runia, W.T., Michielsen, J.M.P.G., van Kuik, A.J. and van Os, E.A. (1997) Elimination of root-infecting pathogens in recirculation water by slow sand filtration. In: *Proceedings of the 9th International Congress of Soilless Culture, St. Helier, Jersey*. International Society of Soilless Culture, pp. 395–407.

Rush, C.M., Piccinni, G. and Harveson, R.M. (1997) Agronomic measures. In: Rechcigl, N.A. and Rechcigl, J.E. (eds) *Environmentally Safe Approaches to Crop Disease Control*. CRC Press, Boca Raton, Florida, pp. 243–282.

Stover, R.H. (1979) Flooding of soil for disease control. In: Mulder, D. (ed.) *Soil Disinfestation*. Elsevier, Amsterdam, pp. 19–28.

Stover, R.H., Thornton, N.C. and Dunlap, V.C. (1953) Flood fallowing for eradication of *Fusarium oxysporum* f.sp. *cubense*. I. Effect of flooding on fungal flora of clay loam soils in Viva Valley, Honduras. *Soil Science* 76, 225–238.

Subbarao, K.V. and Hubbard, J.C. (1996) Interactive effects of broccoli residue and temperature on *Verticillium dahliae* microsclerotia in soil and on wilt in cauliflower. *Phytopathology* 86, 1303–1310.

Wohanka, W. (1993) Slow sand filtration and UV radiation; low-cost techniques for disinfection of recirculating nutrient solution of surface water. In: *Proceedings of the 8th International Congress of Soilless Culture*. International Society of Soilless Culture, pp. 497–511.

Disease Resistance

J.M. Waller[1] and J.M. Lenné[2]

[1]CABI Bioscience UK Centre, Bakeham Lane, Egham, Surrey TW20 9TY, UK; and [2]ICRISAT, Patancheru, Andhra Pradesh 502 324, India

Introduction

Host plant resistance is the main method of control for most diseases. Resistance to disease is a property common to all plants and particular species are susceptible to very few of the many pathogenic organisms that are known to exist; most plants are non-hosts to most pathogens. Although there are many necrotrophic pathogens and opportunistic invaders of weakened plant tissues that have very wide host ranges, biotrophic pathogens tend to have very restricted host ranges. Enhancement of disease resistance in plants is a major component in the control of most diseases. This can be achieved both through growing plants in conditions that enhance resistance mechanisms already operational in the plant, or by incorporating genetically controlled resistance factors into the plant by breeding and selection and recently by genetic engineering. The study of the many aspects of disease resistance in plants is a major task of plant pathology and incorporates molecular biology, biochemistry, physiology, genetics, epidemiology and population dynamics. It continues to be a subject of much debate and research that cannot be completely covered here. This synopsis will outline the major characteristics of disease resistance in plants and provide a bibliography for further reading and access to the extensive literature on this subject.

Resistance Mechanisms

Natural barriers to infection exist in all plants. Mechanical barriers include the cuticle, epidermal tissues, cork layers in bark and on healed wounds and thickened cell walls in many tissues. Other resistance mechanisms may involve morphological features that protect susceptible parts of the plant such as flowers or buds from for example smut diseases. Damage to any of these barriers allows entry of many groups of pathogens including some viruses transmitted by mechanical means, bacteria and many fungal pathogens. Cultural practices that avoid damage to crops and promote rapid healing of wounds facilitate the operation of these defence mechanisms.

Although many of these barriers are preformed, they may be reinforced following infection and others produced. These include thickening and suberization or lignification of cell walls adjacent to fungal hyphae and production of tyloses in xylem tissues. Chemical barriers include the gums, tannins and other substances present in the outer tissues of some plant organs. For example, catechol in red onion scales imparts resistance to smudge (*Colletotrichum circinans*) and saponins, such as avenacin in oats, are implicated in resistance (Osbourn, 1996). Toxins produced by some pathogens (Graniti *et al.*, 1989) are closely involved with the pathogenic process and resistance to these is often associated with the plant's ability to neutralize the toxin.

Many different biochemical resistance reactions are produced as a response to infection and these have been extensively studied often as part of understanding the process of pathogenesis (Bailey and Deveral, 1983; Fritig and Legrand, 1993). When tissue is damaged, enzyme systems are stimulated and typically protective polyphenolic substances are produced. These effects are triggered by exogenous substances such as pectolytic and cellulytic enzymes and toxins produced by the pathogen as part of the process of pathogenicity. These substances are commonly produced by necrotrophic fungi whereas biotrophic fungi do little initial damage as they penetrate plant tissues. Phytoalexins are toxic biochemicals produced by living cells as a response to infection or damage and have been widely implicated as mechanisms of disease resistance (Kuc, 1994). The type of phytoalexin produced is dependent on the type of plant. Pterocarpans, such as pisatin from pea and phaseolin from beans, are characteristic of legumes, whereas sesquiterpenes, such as rishitin from potato and capsidiol from capsicum, are characteristic of solanaceous crops. Generally, pathogens of particular crops are more tolerant of the phytoalexins produced by them than are non-pathogens. Most of the above resistance mechanisms are non-specific in that they generally operate against all potential pathogens; however, specialized, usually biotrophic, pathogens can avoid them and more specific mechanisms then come into play.

The hypersensitive reaction of plant cells is a defence response especially effective against biotrophic pathogens that cause little apparent damage and can thus escape the defence reactions referred to above. Cells invaded by a pathogen die and the biotrophic pathogen cannot survive. The hypersensitive reaction is largely responsible for the specificity of the host–parasite relationships of many biotrophic pathogens, and the ability of the pathogen to avoid it enables a 'compatible' parasitic relationship to proceed. This is mediated through interactions between virulence factors in the pathogen and corresponding resistance genes in the host involving biochemical signalling systems that are the subject of much current research. Virulence in some cases is sometimes associated with the lack of avirulence genes that produce elicitors recognized by the host cell receptors that stimulate the hypersensitive response. Genetic changes in the pathogen can overcome this type of host resistance resulting in the emergence of new, virulent pathogen races. The formation of both phytoalexins and pathogenesis-related proteins are factors associated with the hypersensitive response.

Systemic acquired resistance can occur in susceptible plants in response to localized infections (Stitcher *et al.*, 1997). This resistance is apparently non-specific being effective against a wide range of pathogens but declines with time; it is associated with raised levels of pathogenesis-related proteins, peroxidase, salicylic acid and other compounds. It is dependent on the translocation of some, as yet undetermined, signal within the plant.

Factors Influencing Disease Resistance

Environmental predisposition to disease is widely recognized. Mechanical or physiological damage caused by abiotic agencies is a major factor in this (see also Chapters 17 and 18). Similarly, damage caused by other biotic agencies, such as rodents, insects or other pathogens, can allow entry of pathogens or hamper the operation of active defence mechanisms. Field-grown plants are attacked by a range of pest and disease organisms often simultaneously and these may have negative or positive (synergistic) effects on diseases. The synergistic effects of nematodes and root-infecting pathogens are well recognized (Chapter 13). Drought stress predisposes many plants to root and stem rots caused by a range of fungal pathogens, e.g. *Macrophomina phaseolina*, *Fusarium* and *Phytophthora* spp. (Schoenewiess, 1986). Mineral nutrition is also important for the optimal performance of resistance mechanisms, whereas excessive nitrogen can lead to increased susceptibility to foliage pathogens; low levels of other elements or general imbalances also reduce resistance (Chapter 30). Many of these environmental factors operate through effects on the physiological processes of the host but endogenous physiological factors also affect resistance. The most obvious of these is age; many pathogens tend to infect plants at particular stages in their growth (seedlings, fruits, etc.) and older foliage is more susceptible to necrotrophic pathogens. Physiological stress imposed by fruiting is a common factor that increases susceptibility, for example coffee plants are more susceptible to rust (*Hemileia vatatrix*) when heavily bearing. Many of these effects are associated with a decrease in available sugars.

Genetic factors

Most non-pathogen-specific resistance is polygenic or at least depends on the interaction of a variety of intrinsic factors under genetic control. Although often influenced by environmental factors, this type of resistance is not readily overcome by genetic changes in pathogens. By contrast, single or a few genes in the host usually control pathogen-specific resistance. This major gene resistance may be overcome by genetic changes in the pathogen, a series of virulence genes in the pathogen interacting with corresponding resistance genes in the host giving a differential host–pathogen interaction. The gene-for-gene mechanism was first elucidated by Flor (1946) and forms the basis of much of the effort on breeding crops for disease resistance. Whether or not resistance can be overcome by genetic changes in the pathogen and the related epidemiological effects is the basis for vertical and horizontal resistance (Vanderplank, 1963, 1984; Robinson, 1987). The 'selection pressure' imposed by host resistance genes has major effects on the pathogen population (Wolfe and Caten, 1987) and the resultant effect on the pathogen is critical to the durability of resistance. The genetics of resistance in plants has been widely studied because of both its significance in plant breeding and the challenge of understanding the molecular biological mechanisms involved (Day and Jellis, 1987; Simmons, 1991; Crute *et al.*, 1997).

Effects on disease epidemiology

Single or major gene resistance to specific pathogens often imparts immunity or at least a high degree of resistance and is effective at the initial stages of infection, but once overcome by virulence changes in the pathogen, it often has no effect on subsequent disease progress. These are characteristics of vertical resistance. This type of resistance is easy to assess, can be manipulated by conventional plant breeding techniques and forms the basis of much of the resistance deployed in crops especially to biotrophic pathogens. However, it is often not durable as selection pressure imposed on the pathogen population results in the emergence of new virulent races able to overcome it, especially among airborne pathogens capable of rapid rates of population development. Resistance that operates after infection restricts the pathogen's subsequent development in the host plant. This can be manifested in longer incubation and latent phases and suppressed or restricted sporulation, all of which delay or prevent the epidemic progress of the pathogen and its deleterious effects on the host plant populations. This type of resistance is generally non-specific and generally not overcome by genetic changes in pathogens although it is often only partial and influenced by extrinsic factors. Stable, non-specific resistance that reduces the rate of epidemic progression is characteristic of horizontal resistance. Plants possess a range of different types of resistance whether classified by mechanisms, genetics or effects and they are often interdependent.

Tolerance

The effect that diseases have on plants depends on how the disease affects the host physiology and is expressed in the symptoms of the disease and in an agricultural context on the qualitative or quantitative aspects of the yield that the plant produces. The two are not always related, as some diseases producing spectacular symptoms on older leaves may have little effect on yield, whereas others that produce few obvious symptoms (some viruses and root pathogens) may have marked effects on yield. Plants are more tolerant of diseases if they show less severe disease symptoms and/or suffer less yield decline than those that are less tolerant. The distinction between tolerance and resistance is arguable. There are many factors involved in this process, but plants stressed through environmental or physiological factors are less tolerant of the damage caused by diseases than are those growing under ideal conditions. This may be due to predisposition to disease and suppression of resistance mechanisms or it may be that the plant is less able to compensate for physiological damage caused by the pathogen.

Breeding and Selection for Resistance

Disease resistance is only one of several factors incorporated into plant breeding programmes, the main thrust of which is to achieve a suitable combination of many desirable properties. Typically, sources of resistance need to be identified, usually from wild relatives or 'land races' of crop species, which can be incorporated in a crossing programme involving pedigree breeding and back-crossing during which individual progeny are carefully selected for incorporation at each stage. Mass selection from interbreeding populations may also be used where characters are under polygenic control. More recently, exotic sources of resistance may be incorporated by direct genetic modification using genetic engineering techniques without recourse to traditional breeding techniques. Resistance in tobacco to *Tobacco mosaic virus* was obtained by incorporating the gene for the virus coat protein into the host; this was the first example of a genetically engineered crop. Readers should refer to texts on plant breeding and genetic engineering for more information on these topics (see references and further reading below).

For selection of resistance at all stages in the breeding programme knowledge of pathogen behaviour is necessary so that screening can be effective and realistic (i.e. is a true reflection of the natural process). Quick, easily standardized and readable screening tests tend to select for those resistance mechanisms that operate at the initial infection stage and provide qualitative data. Both for this reason and because of the practical limitations to breeding and resistance screening, major gene resistance has been favoured. Following advances in molecular biology it is now possible to identify genetic sequences associated with resistance or other characters and to detect these in breeding progeny through the use of DNA probes or PCR; this marker-assisted selection considerably shortens the time required for progeny screening (Nelson and Leung, 1994). Resistance that affects the later stages of disease development and has impacts on epidemic progress is more difficult both to assess and to manipulate in breeding programmes. Nevertheless, this can be ascertained during natural disease screening under field conditions, which has to be the ultimate test of new crop varieties. Field testing under conditions that are often location specific is a critical stage in the assessment of the effectiveness of disease resistance (and of other characteristics) (Buddenhagen, 1993). The ultimate users of the products of plant breeding are farmers and recently greater farmer participation in the crop improvement process has been promoted especially in developing countries (Witcombe *et al.*, 1996).

Durability of resistance (Lamberti *et al.*, 1983; Johnson, 1984; Jacobs and Parlevliet, 1993) is a major challenge in resistance breeding particularly for perennial crops where change of cultivar to deploy new effective resistance against changing pathogen virulence is not a practical option within the time frame of most agricultural enterprises. There are many examples of single gene resistance that have proved relatively durable especially against slow-moving soil-borne pathogens, but generally polygenic quantitative resistance or horizontal resistance is the most durable (Simmons, 1991). Various breeding strategies are utilized to improve durability and interacting additive or epistatic effects between genes have been implicated. Gene pyramiding aims to incorporate a range of different resistance factors into plants making it more difficult for simple genetic changes in the pathogen to be effective against them. Other factors can assist durability, especially the way in which the resistance is deployed within the cropping system and those which hamper the development and spread of virulence within the pathogen population. These include the use of multilines and variety mixtures with different resistance genes, cultural measures to disrupt pathogen life cycles and other disease control measures. Although apparently durable resistance may not be complete, it can still form an important component of integrated disease management strategies.

International Perspective

Crop improvement through plant breeding is an international activity because most crops are grown in regions remote from their centres of diversity where genetic characteristics useful to plant breeding may be found (Lenné and Wood, 1991). A network of International Agricultural Research Centres (IARCs) established under the auspices of The Consultative Group for International Agricultural Research (CGIAR) maintain germplasm banks of the world's major staple crop species and undertake extensive plant breeding programmes in partnership with national agricultural research institutions to improve crop species. This includes the need to improve resistance to diseases and IARCs monitor the reaction of crop germplasm across a series of multilocational disease-screening nurseries.

References and Further Reading

Allen, D.J., Lenné, J.M. and Waller, J.M. (1999) Pathogen biodiversity; its nature, characterisation and consequences. In: Wood, D. and Lenné, J.M. (eds) *Agrobiodiversity: Characterisation, Utilization and Management*. CAB International, Wallingford, UK, pp. 123–153.

Bailey, J.A. and Deveral, B.J. (eds) (1983) *The Dynamics of Host Defence*. Academic Press, Sydney, Australia.

Brown, A.H.D., Clegg, M.T., Kahler, A.L. and Weir, B.S. (eds) (1990) *Plant Population Genetics, Breeding, and Genetic Resources*. Sinauer Associates, Sunderland, Massachusetts, USA.

Buddenhagen, I.W. (1983) Breeding strategies for stress and disease resistance in developing countries. *Annual Review of Phytopathology* 21, 385–409.

Chet, I. (ed.) (1993) *Biotechnology and Plant Disease Control*. Wiley-Liss, New York, USA, 373 pp.

Crute, I.R., Holub, E.B. and Burdon, J.J. (eds) (1997) *The Gene-for-Gene Relationship in Plant Parasite Interactions*. CAB International, Wallingford, UK.

Day, P.R. and Jellis G.J. (eds) (1987) *Genetics and Plant Pathogenesis*. Blackwell, Oxford, UK, 352 pp.

van den Elzen, P.J.M., Bevan, M.W., Harrison, B.D. and Leaver, C.J. (1994) *The Production and Uses of Genetically Transformed Plants*. Chapman & Hall, London, UK.

Flor, H.H. (1946) Genetics of pathogenicity in *Melampsora lini*. *Journal of Agricultural Research* 73, 335–357.

Fritig, B. and Legrand, M. (eds) (1993) *Mechanisms of Plant Defence Response*. Kluwer Academic, Dordrecht, The Netherlands, 401 pp.

Gatehouse, A.M.R., Hilder, V.A. and Boulton, D. (eds) (1992) *Plant Genetic Manipulation for Crop Protection*. CAB International, Wallingford, UK, 266 pp.

Graniti, A., Durbin, R.D. and Ballio, H. (eds) (1989) *Phytotoxins and Plant Pathogenesis*. Springer-Verlag, Berlin, Germany, 508 pp.

Hammerschmidt, R. and Kuc, J. (eds) (1995) *Induced Resistance to Disease in Plants*. Kluwer, Dordrecht, The Netherlands.

Heitefuss, R. and Williams, P.H. (eds) (1976) *Physiological Plant Pathology*. Springer-Verlag, Berlin, Germany, 890 pp.

Jacobs, Th. and Parlevliet J.E. (eds) (1993) *Durability of Plant Resistance*. Kluwer Academic, Dordrecht, The Netherlands, 375 pp.

Johnson, R. (1984) Analysis of durable resistance. *Annual Review of Phytopathology* 22, 309–378.

Johnson, R. and Jellis, G.J. (eds) (1992) *Breeding for Disease Resistance*. Kluwer Academic Publishers, Dordrecht, The Netherlands, 205 pp.

Johnson, R. and Knott, D.R. (1992) Specificity in gene-for gene interactions between plants and pathogens. *Plant Pathology* 41, 1–4.

Kuc, J. (1994) Phytoalexins, stress metabolism and disease resistance in plants. *Annual Review of Phytopathology* 33, 275–297.

Lamberti, F., Waller, J.M. and van der Graff, N. (1983) *Durable Resistance in Crops*. Plenum, New York, USA.

Lenné, J.M. and Wood, D. (1991) Plant disease and the use of wild germplasm. *Annual Review of Phytopathology* 29, 35–63.

Nelson, R.J. and Leung, H. (1994) The use of molecular markers to characterize pathogen populations and sources of disease resistance. In: Teng, P.S., Heong, K.L. and Moody, K. (eds) *Rice Pest Science and Management*. International Rice Research Institute (IRRI), Manila, Philippines, pp. 173–192.

Nicholson, R.L. and Hammerschmidt, R.E. (1992) Phenolic compounds and their role in disease resistance. *Annual Review of Phytopathology* 30, 369–389.

Osbourn, A. (1996) Pre-formed anti-microbial compounds and plant defence against fungal attack. *Plant Cell* 8, 1821–1831.

Robinson, R.A. (1987) *Host Management in Crop Pathosystems*. Macmillan, New York, USA, 263R pp.

Russell, G.E. (1978) *Plant Breeding for Pest and Disease Resistance*. Butterworths, London, UK, 485 pp.

Schoenewiess, D.F. (1986) Water stress predisposition to disease – an overview. In: Ayres, P.G. and Boddy, L. (eds) *Water, Fungi and Plants*. British Mycological Society Symposium 11. Cambridge University Press, Cambridge, UK, pp. 157–174.

Sidhu, G.S. (1987) Host–parasite genetics. *Plant Breeding Reviews* 5, 393–433.

Simmons, N.W. (1991) Genetics of horizontal resistance to disease of crops. *Biological Reviews* 66, 189–241.

Stitcher, L., Mauch-Mani, B. and Metrauxm, J.P. (1997) Systemic acquired resistance. *Annual Review of Phytopathology* 35, 235–270.

Swords, K.M.M., Liang, J.H., Shah, D.M., Liang, J.H. and Setlow, J.K. (1997) Novel approaches to engineering disease resistance in crops. In: *Genetic Engineering: Principles and Methods*, Vol. 19. Plenum, New York, USA, pp. 1–13.

Vanderplank, J.E. (1963) *Plant Diseases: Epidemics and Control*. Academic Press, New York, USA, 349 pp.

Vanderplank, J.E. (1982) *Host–Pathogen Interactions in Plant Disease*. Academic Press, New York, USA, 207 pp.

Vanderplank, J.E. (1984) *Disease Resistance in Plants*, 2nd edn. Academic Press, Orlando, Florida, USA, 194 pp.

Witcombe, J.R., Joshi, A., Joshi, K.D. and Sthapit, B.R. (1996) Farmer participatory crop improvement. 1. Varietal selection and breeding methods and their impact on bio-diversity. *Experimental Agriculture* 32, 445–460.

Wolfe, M.S. and Caten, C.E. (1987) *Populations of Plant Pathogens: Their Dynamics and Genetics*. Blackwell, Oxford, UK, 280 pp.

Fungicides

D. Hollomon

Long Ashton Research Station, University of Bristol, Long Ashton, Bristol BS18 9AF, UK

Types of Fungicides

Chemicals used to control plant diseases fall into three groups according to their biological mode of action.

Sterilants and fumigants

These have a wide range of activity, but are generally not applied to growing plants. They eradicate pathogens (and pests) and effectively sterilize soil, buildings and produce entering the store. They are multisite inhibitors.

Protectants

When applied to either seed or growing crops, these establish a protective chemical barrier covering all host surfaces in order to prevent infection. Most protectants are not systemic and require repeated applications to ensure coverage of new foliage. Recently introduced protectant fungicides, such as strobilurins and quinolines, move as vapour within the crop and accumulate in the wax layer of new leaves as they emerge. In terms of the amount used worldwide protectant fungicides are the largest.

©CAB *International* 2002. *Plant Pathologist's Pocketbook*
(eds J.M. Waller, J.M. Lenné and S.J. Waller)

Eradicants

Applied to either seed or growing crops, these are absorbed and move upwards in the xylem to systemically protect the plant. Pathogen growth is arrested, but seldom are members of this group fungicidal. On woody crops they tend to be used as protectants. Usually they act at one biochemical target site.

Formulation

Commercially available fungicides consist of a mixture of the active ingredient (a.i.) and other substances, including inert diluents, wetting agents, stickers and emulsifiers. These aid distribution of the active ingredient over crop surfaces and retention on seed. Formulations often contain mixtures of different active ingredients (especially mixtures of protectant and systemic fungicides) to broaden the range of diseases controlled and as part of anti-resistance strategies. Different formulations incorporating different amounts of the same active ingredient are used for distinct purposes. Particular methods of application may require the use of particular formulations.

Granular formulations are increasingly common because of the improved safety of handling the concentrate. The active ingredient is usually incorporated at a rate of 30–80%, and together with an inert diluent, such as kaolin, and a wetting agent, they are manufactured as a prill. In earlier formulations the kaolin was finely milled and used as a wettable powder. In both cases the active ingredient is finally applied as a suspension in water.

Emulsifiable concentrates are liquid formulations in a solvent that is miscible with water. Solvents such as xylene were commonly used but have been replaced with less hazardous materials.

Dusts are seldom used today. Other formulations may be used for special purposes, including liquid seed dressings, paints, gels and oil formulations for low volume applications.

Nomenclature

The names by which fungicides are known can be rather confusing, as there are usually several names referring to one substance. First, the chemical names describing the structure and composition of the chemicals may be straightforward for fairly simple compounds, but with complex organic compounds different chemical names may be used to describe the same compound according to usages in different countries; the usage in the USA differs somewhat from that used in Britain and Europe. Secondly, the common name may be identical to the chemical name with simple compounds, or it may be an abbreviated or simplified derivative of the chemical name when this is complex. Generally, most chemicals are best referred to by their common names, which are approved by the International Standards Organization and national groups such as the British Standards Institution and the American National Standards Institute. Thirdly, the trade names under which different formulations of the same compound are marketed by different chemical companies vary widely. Trade names are a particular source of confusion, as they may become generally, but wrongly, accepted as common names where one company has a marketing monopoly of a particular chemical in one area. However, all marketed fungicides should state clearly the common (standard) name of the fungicide (a.i.) and the amount of it contained in the formulated product.

An example:
Chemical name: Methyl 1*H*-benzimidazol-2-ylcarbamate
Common name: Carbendazim, or methyl benzimidazole carbamate (MBC)
Formulation: Bavistin (BASF)

Usage

Successful control of plant diseases depends on correct usage of fungicides. Diseases must be accurately diagnosed. Table 32.1 shows the usage of some fungicide groups against different diseases worldwide. Because of environmental side effects chemicals should be used carefully, taking account of the following points.

- Always use the correct fungicide for the disease requiring control.
- Follow label recommendations. Use recommended application rates, which have been derived by manufacturers through a careful series of field experiments to give the best overall control.
- Apply at the optimum time and observe the correct interval between application and harvest.
- Use the correct method of application for the product taking adequate precautions when handling chemicals. Avoid spillage and contamination of skin and clothes; never smoke, eat or drink while handling chemicals; wear appropriate protective clothing. Dispose of empty containers, excess chemical and washings from machinery in a safe way. Some containers are now made from biodegradable plastics. Some safety precautions are printed on the label; further details can be found in the up-to-date edition of *The Safe Use of Poisonous Chemicals on the Farm* published by HMSO for the UK Ministry of Agriculture, Fisheries and Food, and in the British Crop Protection Council book *Using Pesticides* (Anon, 1996).

Sterilants and Fumigants

Chemicals used for soil sterilization include formaldehyde, chloropicrin and methyl bromide. All are extremely toxic and must be handled with extreme care. In many countries these compounds are covered by some form of poisons legislation and may not be available for sale. Application must be by a licensed operator and involves injection under polythene sheeting. There is worldwide agreement to phase out methyl bromide early in the 21st century, although it is expected that it will be retained for specific uses where alternatives are not available. Other sterilants, such as dazomet, and

Table 32.1. List of fungicides and antibiotics. This list is not exhaustive but covers the main chemical groups. For a complete list of currently available products, their full chemical names, formulations and approved uses consult *The Pesticide Manual* (Tomlin, 2000) or local pesticide guides.

Fungicide group	Mode of action	Main targets for disease control
Main groups		
2-Aminopyrimidines	Adenosine deaminase	Powdery mildews
Anilinopyrimidines	Methionine biosynthesis	Cereal eyespot, *botrytis*, apple scab
Azoles (DMIs)	Sterol biosynthesis	All pathogen groups except oomycetes
Benzimidazoles	Anti-tubulin	All pathogen groups except oomycetes
Carboxanilides	Mitochondrial respiration (complex II)	Basidiomycetes
Copper salts	Multisite	Oomycetes
Dithiocarbamates	Multisite	All pathogen groups
Guanidines	Multisite	Cereal seed-borne diseases, apple scab
Morpholines (including spiroketalamines)	Sterol biosynthesis	Powdery mildews, rusts and banana Sigatoka disease
Phenylamides	RNA synthesis	Oomycetes
Phenylpyrolles	Osmoregulation	*Botrytis*, Fusarium blights
Phosphorothiolates	Fatty acid synthesis	Rice blast
Quinolines	Not known	Powdery mildews
Strobilurins (also called QOI inhibitors)	Mitochondrial respiration (complex III)	All pathogen groups
Tin salts	Multisite	Potato blight
Other chemistry		
Acibenzolar-*S*-methyl	Activate host resistance	All pathogen groups including viruses and bacteria
Chloroneb	Multisite	Seed-borne pathogens
Chlorothalonil	Multisite. Sulphydryl reagent	All pathogen groups
Cymoxanil	Not known	Potato blight, downy mildews
Dichlofluanid	Multisite	*Botrytis* and other protected crop diseases
Diethofencarb	Anti-tubulin	*Botrytis*
Dimethomorph	Anti-tubulin	Potato blight, downy mildews
Fluazinam	Mitochondrial respiration probably complex I	Potato blight
Phosphonates	Not known	Some oomycetes
Probenazole	Not known but probably activates host resistance	Rice blast
Pyrimidine carbinols	Sterol biosynthesis	All pathogen groups except oomycetes
Sulphur	Multisite	Powdery mildews
Tricyclazole	Melanin biosynthesis	Rice blast

Continued

Table 32.1. *Continued*

Antibiotics	Mode of action	Diseases
Blasticidin-S	Protein synthesis	Rice blast
Kasugamycin	Protein synthesis inhibitor	Rice blast
Streptomycin	Protein synthesis	Bacterial diseases including fire blight
Validomycin	Carbohydrate metabolism	Rice sheath blight

metam-sodium are available as granules that release toxic methyl isothio-cyanate when in contact with moist soil. Toxic vapour from all these sterilants may take several days to be released from soil, and it is usual to sow a rapidly germinating plant species as a test, such as cress, before planting. All these chemicals are general toxicants, which eradicate pathogens, nematodes, soil insects and troublesome weeds.

Protectant Fungicides: Non-systemic

Widespread use of fungicides began in the 19th century, especially on grapes for controlling downy and powdery midews. These were inorganic compounds including formulations of sulphur, copper, tin, mercury and arsenic. Only the first three are in use today, often as colloidal salts to improve rain fastness. Copper fungicides in particular have a broad spectrum of activity including bacterial activity, which most other fungicides lack, and their cheapness is a valuable asset. Copper salts are also used as wood preservatives.

By the 1930s the first modern organic fungicides had been introduced. These dithiocarbamates are now used as salts with various metals such as zinc, sodium and manganese. One of the first to be developed, thiram, is widely used as a seed treatment today, and the mixed zinc and manganese salt of ethylene-bis-dithiocarbamate (Mancozeb) is a useful partner in antiresistance strategies. A number of other organic compounds have been developed as protectant fungicides; perhaps the most important is the pthalonitrile, chlorothalonil.

The mode of action of these protectant fungicides is not fully understood, but they clearly inhibit many biochemical steps either as chelating or sulphydryl reagents. They are also toxic to plant cells and consequently it is important that they remain on the leaf surface. Because they are not systemic repeated applications are needed to protect rapidly growing crops.

Protectant Fungicides: Systemic

Recently, protectant activity has been coupled with movement within the crop either as a vapour, or in the xylem transport sytem of the plant. Strobilurins (including oxazolidinediones), quinolines and anilinopyrimidines all inhibit the infection process. Some, such as the strobilurin, azoxystrobin, have limited eradicant activity whereas the quinoline, quinoxyfen, which is only active against powdery mildews, is just a protectant. Although their mode of action is not always known, unlike other protectant fungicides they are likely to be site-specific inhibitors. The strobilurins all inhibit electron transport in mitochondria, interacting with the cytochrome bc_1 complex but not necessarily all at the same site in the target protein. All can give long-lived protection, including to leaves not exposed when sprayed, and require application prior to any disease being observed. Consequently, they have the potential to change treatment schedules, with the emphasis shifting to applications made early in crop growth. Equally important is the ability of strobilurins to control oomycetes as well as ascomycetes, basidiomycetes and fungi imperfecti, allowing diseases such as grape powdery mildew (*Uncinula necator*) and downy mildew (*Plasmopara viticola*) to be controlled with a single treatment.

Eradicant Fungicides

Since the late 1960s there has been a substantial development of this type of systemic fungicide, and they are now at the core of many fungicide-use programmes. All move upward in the plant's transpiration stream. Many are based on MBC and related compounds, which all interfere with tubulin function. Examples of the benzimidazoles include benomyl, carbendazim, thiophanate and thiabendazole, and these were the first truly systemic fungicides to be used on a wide scale. By far the largest group of eradicant systemic fungicides are the triazoles and related heterocyclic nitrogen compounds, which all inhibit sterol biosynthesis at the 14α demethylation step (14DM). Some can be used as seed treatments but in general they have quite harsh effects on germination of many crop plants. More than 25 sterol 14α demethylation inhibitors (DMIs) are used commercially in agriculture and horticulture, and an equivalent number are used in medicine. Like the benzimidazoles they control a wide range of diseases, but not oomycetes. Other steps in sterol biosynthesis have been exploited as targets for morpholine and spiroketalamine fungicides, which have a narrower disease control spectrum, but which have particularly useful and rapid action against most powdery mildews. Although oomycetes are not, it seems, fungi but algae, several very active systemic compounds are available to control these destructive diseases, including phenylamides, dimethomorph, fluazinam, cymoxanil and phosetyl-Al. Systemic fungicides may be used against one disease even though they may control several others. These are listed in Table 32.1 but more detailed information may be obtained from *The UK Pesticide Guide* (Whitehead, 2000) or *The Pesticide Manual* (Tomlin, 2000).

Antibiotics

Antibiotics usually act systemically and some are used commercially as fungicides and bactericides. Griseofulvin was the first, and perhaps the most significant, since it demonstrated that systemicity was possible, and so encouraged the search for the systemic fungicides we have today. Blasticidin and kasugamycin were used against rice blast but they have largely been replaced by more active compounds not prone to resistance. Cycloheximide, streptomycin and tetracycline have all been used on high-value cash crops, but phytotoxicity is often a problem.

Systemic Acquired Resistance

Even when infection occurs, most plants attempt to prevent it by marshalling a number of barriers in the region of penetration. These include inhibitory compounds, carbohydrates to encase infection hyphae, and localized and necrotic cell death. Successful defence depends on rapid respones to signals received as pathogens attempt to penetrate host cells, and these signals can be in some way systemic. Consequently, inoculation with one pathogen can generate immunity in other parts of the plant not only to that pathogen, but non-specifically to viruses, bacteria and fungi. As more fundamental information has emerged about this process of systemic acquired resistance (SAR), a concept of disease control has developed based on enhancing this resistance response. Some 20 years ago, the highly successful rice blast fungicide probenazole was introduced, which may act in this way, since it has no intrinsic toxicity against the rice blast fungus. The SAR concept of disease control has recently been given impetus by the introduction of acibenzolar-S-methyl, which has no direct fungicidal activity, yet controls a wide range of fungal pathogens in crops treated earlier with the chemical. It has been registered for use in at least one country, but as a micronutrient rather than a fungicide (see Chapter 31).

Resistance

Resistance can have a major impact on fungicide efficacy but although it has been identified in over 300 fungal pathogens, the vast majority have been generated in the laboratory. Perhaps no more than 30 fungi have become resistant in practice. The risk of resistance in practice depends on both the chemical class and the pathogen (Fig. 32.1; Brent and Hollomon, 1998). Risk is particularly high where intensive use is made of site-specific persistent fungicides to control pathogens that reproduce profusely and rapidly. Of equal interest are fungicides where the risk is low, such as multisite inhibitors, morpholines and fungicides that activate host resistance and do not directly affect pathogens. Compared with insecticide resistance there are fewer problems associated with fungicide resistance, partly because there is greater chemical diversity available, which can be incorporated into successful anti-resistance strategies.

Selection pressures can be reduced further by confining the use of effective fungicides to a period when they are particularly valuable. Cultural methods that prevent carry-over of resistant strains between crops and seasons are also useful. Resistance is largely managed through use of fungicide mixtures where partners have different modes of action, and ideally one partner should be a multisite inhibitor. Alternating fungicides with different modes of action is

Chemical group	Basic fungicide risk	Low (1)	Medium (2)	High (3)
Benzimidazoles, Dicarboximides, Phenylamides	High (3)	3	6	9
Carboxanilides, DMIs, Phosphorothiolates, Anilinopyrimidines, Phenylpyrroles, Strobilurins	Medium (2)	2	4	6
Coppers, Dithiocarbamates, Melanin inhibitors, Phthalimides, Sulphur, SAR-inducers	Low (1)	1	2	3
Basic fungicide risk / Basic disease risk		Seed-borne (e.g. *Pyrenophora ustilago*) Soil-borne (e.g. *Phytopthora*) Cereal eyespot Cereal rust Rice sheath blight	Barley *Rhynchosporium* Wheat septoria	Apple scab Banana Sigatoka Cereal powdery mildew Grape *Botrytis* Potato blight Citrus *Penicillium* Rice blast

Fig. 32.1. Resistance risk. This diagram shows the interaction between chemical group and disease and includes the best estimates based on currently available knowledge. Combined risk: 1 = low, 2–6 = medium, 9 = high. From Brent and Hollomon (1998).

also possible but this exposes the 'at risk' fungicide during the period it is used alone (see Brent, 1995).

Fungicide Use and Diagnostics

To meet increasing costs and environmental pressures fungicides need to be used only when necessary and at optimum timing. Many diseases are not easy to identify from symptoms and fungicides often need to be applied pre-symptomatically. Accurate diagnostic tools would be useful, and immuno-based ELISA assays have been available for some time to identify a few cereal and other diseases. More recently, DNA-based assays exploiting the PCR have been developed, which can give quantitative measures of disease. So far these assays have had limited use in decision making but much ongoing research is beginning to assemble the data needed to do this. By using DNA techniques borrowed from medical diagnostics it is possible to identify resistance, at least when it is target-site based, and this should avoid waste of fungicides against diseases where they are not effective.

References and Further Reading

Anon. (1996) *Using Pesticides: a Complete Guide to Safe, Effective Spraying*. British Crop
    Protection Council, Farnham, UK.
Brent, K.J. (1995) *Fungicide Resistance in Crop Pathogens: How Can It Be Managed?* FRAC
    Monograph 1, GIFAP, Brussels.
Brent, K.J. and Hollomon, D.W. (1998) *Fungicide Resistance: the Assessment of Risk*. FRAC
    Monograph 2, GCPF, Brussels.
Hewitt, H.G. (1998) *Fungicides in Crop Protection*. CAB International, Wallingford, UK.
Lyr, H. (1995) *Modern Selective Fungicides: Properties, Applications, Mechanisms of Action*.
    Fisher Verlag, Jena, Germany.
Tomlin, C.D.S. (2000) *The Pesticide Manual*, 12th edn. British Crop Protection Council,
    Farnham, UK.
Whitehead, R. (2000) *The UK Pesticide Guide 1998*. CAB International, Wallingford, UK,
    601 pp.

The Application of Chemicals for Plant Disease Control

G.A. Matthews

*International Pesticide Application Research Centre,
Imperial College at Silwood Park, Ascot, Berkshire SL5 7PY, UK*

Introduction

Application of sulphur dust on vines was recorded over 2000 years ago, but it was the discovery of Bordeaux mixture at the end of the 19th century that led to the development of modern pesticide application techniques. Lodeman (1896) described the earliest knapsack sprayers, the horse-drawn forerunner of tractor equipment and even the use of a rotary brush to apply fungicides for blight control on potatoes. This chapter describes the latest types of fungicide application.

Plant pathogens are not readily visible to the naked eye, so preventative fungicides need to be applied before disease symptoms appear. Accurate timing of an application, for example in relation to the susceptible stages of the crop, or periods when the pathogen is particularly active (e.g. spore dispersal), and weather conditions (especially wetness) favouring the pathogen are most important, but well-timed treatments can fail if the fungicide is not effectively deposited or is not sufficiently persistent on the surface infected with the pathogen. Crop growth, erosion of fungicides by rainfall and other factors determine the duration of protection and when repeat applications are needed. Other epidemiological factors such as the rate of disease progress and the latent period must also be considered when formulating a rational spraying schedule for disease control. Traditionally, many fungicides have been applied diluted in very large volumes of water so that all the plant surface is wetted. This high volume (HV) treatment wastes chemical as most drips from foliage and, in many cases, an infection on the undersurface of leaves is not controlled, unless the fungicide has a translaminar or vapour action.

Ideally seed of a cultivar with disease resistance is sown, but as this is not always possible, various methods of applying fungicides are required:

- Seed treatment
- Soil treatment
- Foliar treatment
    Dusts
    Sprays
    Mists and fogs
- Stem treatment

Seed Treatment

Soil- and seed-borne pathogens infecting young seedlings can be controlled by seed treatment if sufficient toxicant can be applied evenly to the surface of each seed, without causing phytotoxicity. The aim is to penetrate surface irregularities and kill the spores and mycelia of pathogenic fungi before sowing. Pelleted seed, when small seeds are coated with other inert materials, facilitates incorporation of a fungicide with the seed. Sulphuric acid used to delint cotton seeds will also control pathogens on the outer surface of the seeds.

As dust formulations can cause an inhalation hazard, the trend has been to use liquid or suspensions of fungicides, often in the form of slurries, which generally give better cover and retention on the seed than dry materials. A colour dye or bitter-tasting substance may be used with the pesticide to show that the seed has been treated and discourage birds and other organisms from ingesting it. A sticker may also be added to ensure the coating is not lost during sowing. Seed treatment is normally done by the seed supplier and should always be applied with machinery specially designed for the purpose. On a small scale, a 'Rotastat' seed is treated by operating a foot-pedalled bicycle chain to rotate the seed container while pesticide is applied with a spinning disc driven by a battery-operated motor.

Fungicides with a systemic action can control deep-seated or internal infections, such as those of the cereal loose smuts, whereas previously hot water treatment was used. Some systemic fungicides are also applied to cereals as seed treatments to control certain airborne diseases, such as mildews, as they are absorbed by the seedling and translocated to the shoot.

Fungicidal soaks are used for certain seeds, but the seeds have to be dried after treatment. Thiram-soaked seed has been widely used to control certain seed-borne diseases of vegetables, notably leaf spot of celery (*Septoria apiicola*), and more recently dry application of carbendazim-generating fungicide⁻ ˙ ᵖroved very effective against onion neck rot (*Botrytis allii*). white rot (*Sclerotium cepivorum*) is achieved using fungicide-ⁱed.

ₐnting material such as sugarcane or pineapple setts may gicide suspensions before planting to protect them from . Root dips may also be used at transplanting to control ιbroot (*Plasmodiophora brassicae*).

345

Soil Treatment

Apart from fungi that exist in soil or plant debris as mycelium or spores, bacterial and virus pathogens, various pests and nematodes are also present in most soils. Crop rotation is an important way of reducing the levels of pathogen, but soil treatment before planting may still be required. Soil solarization is recommended where high temperatures can be achieved by covering the soil with a plastic sheet for a sufficiently long period. Chemical treatment using a fumigant is expensive and is used mainly on high-value crops such as strawberries on relatively small areas. Some of the most widely used fumigants such as methyl bromide are being phased out as the gases are no longer environmentally acceptable. Heat treatment using steam will destroy most of these organisms, but this is also economic only for high-value fruit or horticultural crops, especially for seed beds or propagation soil in nurseries.

Heat treatment

Steam is usually applied either over the surface or by injection into the soil. Surface treatment is cheaper and involves blowing the steam under a heavy plastic sheet fixed down firmly at its edges. Steam trapped under the sheet penetrates the soil. A temperature of 65–75°C for at least 10 min reached at depths of 15–20 cm after 6–8 h is usually effective. Steam is injected into the soil through variously designed perforated metal pipes or drainage tiles, the latter being permanently situated in the soil. Whichever injection method is used, it is usual to treat the soil for approximately 30 min but because of the small area treated at any one time the sheet method is quicker although generally less efficient. Toxicities caused by nitrite, ammonium or manganese ions, which sometimes occur when soil is heated at 100°C with steam, can be avoided by mixing the steam with an equal volume of air, which reduces the temperature to about 83°C. Electrical heating may be used to sterilize very small quantities of soil.

Chemical treatment

Methyl bromide[1] usually mixed with chloropicrin has been the most popular chemical soil sterilant as the fumigant will penetrate soil efficiently. Before application, the soil must be finely cultivated to a depth of at least 40 cm, the moisture content must be approximately 70% of the soil's field capacity and the soil temperature must be above 10°C. Methyl bromide should be applied by trained operatives as a gas under prelaid polythene sheeting (150 gauge) or preferably impermeable multilayer sheeting, which is kept in place for

---

[1] Use of methyl bromide is due to be phased out in EU countries by 2005 and its use in developing countries restricted from 2015.

48–96 h. In some situations a roll of polythene is laid over the soil as methyl bromide is injected behind a tractor-mounted applicator. Other fumigants are chloropicrin, dazomet and metam-sodium. A solid, such as dazomet, may be scattered on to the soil and incorporated by cultivation. After a treatment period, which varies with the fumigant used, the soil must be ventilated and all traces of the fumigant dispersed before cropping. Before using treated soil a test with cress seeds will indicate whether phytotoxicity is likely to occur.

Soil treatment to control nematodes is often now performed with a granular nematicide. Hand-operated to tractor-mounted equipment is available, the latter often fitted to seeders so that the granules are placed to the side and below the seed to protect the developing roots. Some systemic fungicides may be applied as granules to the soil, for example to control coffee leaf rust.

Although largely superseded by seed treatment, certain fungicides that are not phytotoxic may also be applied as a drench to soil at sowing or postplanting to control root diseases, usually of horticultural crops. Systemic fungicides such as metalaxyl and furalaxyl, effective against oomycetes, and carbendazim-generating fungicides, effective against powdery mildews and some other leaf diseases, can be applied to soil for translocation to the aerial parts of plants.

Fungicides are also applied to compost to protect seedlings and cuttings from soil-borne pathogens. They may be applied as dusts to the soil surface and raked in, or incorporated into peat or compost blocks, or applied as postplanting drenches of dilute suspensions. Low concentrations can be used in nutrient solutions for crops grown by the nutrient film technique.

Foliar Treatment

Dusts

Dusting is rarely used nowadays as it is much less efficient than spraying, especially in terms of persistence. However, sulphur dust in particular is still occasionally used on some crops such as grapes. Although a major advantage of dusting is that it obviates the need to transport water, the relative cost of active ingredient is very much higher due to the cost of transporting a high proportion of inert filler.

Sprays

Spraying machinery

The majority of sprayers have a tank, pump and one or more hydraulic nozzles with associated valves and hoses. They vary in size from small hand-held to very large self-propelled vehicles. Some fungicides are applied from aircraft. The nozzles break up ('atomize') the liquid into a large number of small droplets and effectively distribute the fungicide. Apart from hydraulic nozzles, twin-fluid, vortical and rotary nozzles may be used on certain types of equipment. One or more fans may be used to produce an air flow to assist the delivery of the spray from the nozzle to the foliage.

Nozzles

There are several types of hydraulic nozzles, but the most commonly used for fungicide application are the cone and standard fan nozzles. In cone nozzles the spray liquid is forced through tangential slots, so that it rotates as it emerges through a circular orifice and forms a conical diverging spray. The standard fan nozzle, with an elliptical-shaped orifice, produces droplets by the break-up of a flat fan-shaped sheet of liquid formed by the angular impingement of two opposing jets within the nozzle. Both types produce droplets with a range of sizes, but the average size decreases with increasing pressure and/or decreasing orifice size and flow rate. Fungicide sprays are normally applied at 3 bar pressure. Cone nozzles are preferred where the target is a complex of broadleaves, for example when treating potato and coffee crops, whereas the fan nozzles are suited to cereal crops, such as wheat. Each manufacturer uses a code to identify the size of the nozzle, but a standard code has been introduced, which indicates the type of nozzle, spray angle, output (litres per minute) and recommended pressure; thus F110/1.2/3 will be a 110° fan nozzle emitting 1.2 l min^{-1} at 3 bar pressure. To achieve adequate coverage of foliage, nozzles should be selected to give a fine or medium spray quality. In some circumstances a very fine spray or mist is needed (see next section).

Nozzles are commonly made from a hard-wearing plastic as the cost of the injection moulding process is less than the manufacture of metal or ceramic nozzles. The latter are also often mounted in plastic so that with an international colour-coding system, the output of a nozzle tip can be easily seen. Where water is of poor quality and contains abrasive particles of sand, the ceramic tips are better. To reduce the risk of nozzle blockage during a spray a filter should be fitted to each nozzle. In most situations a 50 mesh filter is adequate.

Several other hydraulic nozzle types are now available. These include the deflector, pre-orifice and air-induction types. All these produce a larger average droplet spectrum, with fewer droplets below 100 μm thus reducing the risk of spray drifting downwind, especially where an 'unsprayed buffer zone' is required. Although these are suitable for herbicide application, in general coarser sprays from these nozzles are less satisfactory for fungicide treatments unless they are being used to apply a soil drench.

A high-velocity airstream (80–1000 m s^{-1}), often produced by a centrifugal fan, may be used to atomize the spray liquid. Generally, higher velocity air and lower liquid feed rates produce smaller drops, so it is important to avoid applying too high a flow rate with this equipment. Like hydraulic nozzles, the droplet spectrum from an airshear nozzle is often wide. A narrower droplet spectrum is achieved using a rotary atomizer. Liquid fed on to the rotating disc or wire cage is thrown from its edge by centrifugal force. Increasing the rotational speed decreases the droplet size provided the flow rate is kept low. With controlled droplet application (CDA) the droplet size is selected to suit the target; thus for fungicide sprays droplets of 70–150 µm volume median diameter (VMD) would be suitable. Rotary nozzles were initially used primarily to apply ultralow volumes of oil-based formulations at less than 5 l ha^{-1}, but special formulations are more expensive, so the trend has been to use water-based formulations at very low volume (5–30 l ha^{-1}). An adjuvant may be added to reduce the evaporation of water from in-flight droplets, improve rain-fastness of deposits or increase uptake of systemic fungicides into plants.

Droplet size

Droplet size is measured in micrometres (µm) diameter and the simplest, though not the most precise, indication of the performance of the atomizing device is the VMD, half the volume of liquid being in the form of droplets smaller than this diameter. Sprays may be defined generally as:

Fog:	VMD < 50 µm and > 10% by volume smaller than 30 µm
Mist:	VMD 50–100 µm and < 10% by volume smaller than 30 µm
	(Fogs and mists are collectively referred to as very fine sprays)
Fine spray:	VMD 100–200 µm
Medium spray:	VMD 200–400 µm
Coarse spray:	VMD > 400 µm

A system of reference nozzles to define the boundaries of each spray quality has now been established so that spray quality of nozzles can be measured with a range of different equipment. This system allows the whole droplet spectrum to be considered, and combined with wind-tunnel studies assesses the drift index, i.e. the potential for downwind spray drift.

Pumps

Many manually operated knapsack or mechanized sprayers are fitted with either a diaphragm or piston pump. Associated with these two types of pump, the sprayer has an air chamber to even out the variations of pressure obtained by the pulsating pump. Alternatively on some smaller tractor-mounted sprayers a roller-vane pump may be used. On aircraft, a centrifugal pump is preferred. Downstream of all pumps there should be a pressure regulating or control flow valve to ensure that spray is delivered at a constant pressure. Pump output on tractor-mounted and other large sprayers should be sufficient for the total number of nozzles (with highest recommended output for the crops to be treated) on a boom plus a surplus to feed back into the tank and provide agitation. The input side of the pump must be protected by a large capacity filter that is cleaned after each spray.

Air assistance

When treating tree crops, projection of spray into the tree canopy is needed, so orchard spraying equipment is usually fitted with an axial fan or similar air delivery system. To maintain the droplets in the airstream without fall-out on the soil, droplet size should be < 150 µm VMD. It is important for the foliage to move to allow better penetration into the canopy and enhance deposition of the smaller droplets, so air turbulence is better than a high velocity narrow airjet. Air volume displaced needs to be sufficient to displace air within the crop canopy.

Spraying techniques

The application of HV sprays, applying $> 700\,1\,ha^{-1}$ is still widely practised. The perception is that a large proportion of the plant surface is covered with the spray liquid, which also flows into crevices inaccessible to the direct spray. However, any surfactant in the formulation is so diluted that the fungicide is mainly deposited at the edge of the leaves where the surplus liquid drips off, the remainder of the leaf surface often having a very patchy deposit. The large volume is often applied by hand-directed nozzle lances supplied through flexible hoses, attached to a vehicle-mounted machine. The operator is invariably heavily contaminated with the spray. This method is expensive in terms of labour and is not practical in certain tropical situations where water is scarce, so alternative lower volume application techniques are now being introduced. Some growers add the pesticide to irrigation water, an application technique known as chemigation.

High volumes are also applied through a series of nozzles arranged on horizontal booms for ground crops or on vertical masts for tree crops, although by careful selection of the orifice size, coverage can be improved with less wasted on the ground.

Low volume spraying, applying smaller drops of more concentrated fungicides in less water ($70–400\,1\,ha^{-1}$ on ground crops, $70–600\,1\,ha^{-1}$ on trees, depending on the amount of foliage and height of the crop) is more economical than the HV technique. Deposition of lower volumes is improved where air assistance is used. The spray deposit covers a smaller proportion of the plant surface as a pattern of discrete droplets, but uptake of systemic fungicides through the cuticle or spread of chemical over the leaf surface enables control to be obtained. Thus, the mode of action and formulation of the fungicide play important roles in the effectiveness of low volume application.

Small areas are treated with knapsack motorized mist blowers, especially for tropical tree crops. A recent innovation is the mounting of a fan behind the operator with a spinning disc to provide a turbulent air flow to treat coffee and similar low bushes. On larger areas tractor-mounted or trailed equipment is required. Sprayers with an axial fan are widely used, but adaptations to duct air closer to the upper part of trees or using cross-flow fans are also suitable. On some small tree or bush crops planted in single lines, a tunnel sprayer encloses the spray during treatment to minimize downwind drift and allows some of the spray to be recirculated.

Very low volume (VLV) spraying ($10–50\,1\,ha^{-1}$), preferably with CDA equipment, is now being used for crop disease control, as it confers operational and economic advantages of increased work rates over conventional methods, especially in situations where water is scarce. The technique has the advantage that treatments can be made rapidly when conditions favour the pathogen, thus allowing better timing of a fungicide application. In some cases it is effective when applying substantially less pesticide than in conventional spraying. As there are fewer droplets per unit area of leaf, effectiveness is even more dependent than in the low volume technique on redistribution of the active fungicide from the initial deposits to uncovered areas.

Whichever spraying technique is adopted, the application of a specified amount of fungicide on a given area does not guarantee successful disease control. The operation must ensure optimum deposition of the fungicide on the target surfaces and this will depend on the knowledge and skill of the operator as well as on the suitability of the equipment. Most systemic fungicides move in the transpiration stream of the plant and diffuse between cells, but few are actively translocated in the phloem. Therefore, when applying these chemicals, even coverage of plants is still required.

Spray distribution on plant surfaces is easily seen by the use of a fluorescent dye, which can be suspended in the spray liquid. Samples of leaves or other parts of plants can be examined, when dry, in a darkened room under ultraviolet light (wavelength, 300–400 nm). The fluorescent tracer technique is best as the actual plant surface is examined, but an alternative method is to attach small pieces of water-sensitive paper to plants, preferably on both the under and upper surfaces of leaves.

Mists and fogs

With the higher costs of labour associated with high volume sprays, crops in glass and plastic houses are often now protected by space treatments with thermal or cold fogs. Thermal fogs can be generated either by pulse jets or via an engine exhaust. Pulse jet machines are more satisfactory for applying fungicides. Few specially formulated pesticides are available for fog generators, but most fungicides can be used in these machines if they are mixed with a suitable carrier. Thermal fogging equipment must always be attended with the operator wearing full protection to prevent inhaling the very small particles produced.

Cold fogging equipment uses a vortex of air to shear the spray liquid into small droplets. These are often electrically operated and can be set up in a glasshouse with a timer so that applications are made in the evening when no persons are in the glasshouse. The air flow should be initiated before the fungicide is applied and for a period afterwards to ensure adequate distribution of the fog. When equipment is static the deposits tend to be higher close to the nozzle.

Fogs should be uniformly directed over the top of the crop and never directly at the foliage; they should not be applied in the early morning, in bright sunlight, at temperature extremes, or when crops are wet or under moisture stress. Fogs deposit about 95% of the fungicide on the upper surfaces of leaves, but with translaminar effects can give adequate disease control comparable to HV sprays and are much quicker to apply. The dosage is usually less than that recommended for HV sprays as there is less lost to the ground, but less persistence is expected with a space treatment.

In a few cases thermal fogging has been used in plantations preferably with a closed canopy, for example to control some rubber leaf diseases. In these situations there must be a temperature inversion and no wind, otherwise the fog will be dispersed too quickly and little fungicide will be deposited on the trees.

Fogging does present an inhalation hazard so an alternative is to apply a mist. Motorized knapsack mistblowers with airshear nozzles and spinning disc nozzles mounted in front of a small fan are used for low and very low volume application in glasshouses, the latter being suitable for localized applications. Some fungicides (e.g. dicloran for *Botrytis* control) can be applied as smokes, which are released when the device is ignited. Even distribution of the smoke is required. Tecnazene for *Botrytis* control and sulphur for powdery mildew control can be vaporized in electrical thermostatically controlled thermal vaporizers.

Stem Treatment

Some fungicides have been applied as a paste to the surface of cuts to prevent pathogen infection, especially after pruning. An antagonistic fungus has been applied to the cutting blade of secateurs to limit invasion by pathogens such as silver leaf on pears. On some high-value trees a fungicide has been applied to the tree trunk by drilling a small hole at an angle downwards and injecting the fungicide into the hole. The hole is then covered with a fungicidal paste. In an experimental technique on cocoa, copper fungicide was impregnated into an absorbent pad attached to the trunk so that the chemical was washed over pods by rain to control *Phytophthora* black pod disease.

Selected Bibliography

Jeffs, K.A. (ed.) (1978) *Seed Treatment. CIPAC Monograph*, no. 2. Collaborative International Pesticide Analytical Council, Harpenden.

Lodeman, E.G. (1896) *The Spraying of Plants*. Macmillan, London, UK.

Matthews, G.A. (1999) *The Application of Pesticides to Crops*. IC Press, London, UK.

Matthews, G.A. (2000) *Pesticide Application Methods*, 3rd edn. Blackwell Science, Oxford, UK.

Matthews, G.A. and Thornhill, E.W. (1994) Pesticide application equipment for use in agriculture. *FAO Agricultural Services Bulletin Vols I and II*.

Morgan, W.M. (1979) Application of iprodione by thermal fogging for the control of grey mould of tomato. *Annals of Applied Biology* 93, 21–29.

Southcombe, E.S.E., Miller, P.C.H., Ganzelmeier, H., Miralles, A. and Hewitt, A.J. (1997) The international (BCPC) spray classification system including a drift potential factor. *Proceedings of the Brighton Crop Protection Conference*, Brighton, pp. 371–380.

Sreenivasan, T.N., Pettitt, T.R. and Rudgard, S.A. (1990) An alternative method of applying copper to control cocoa pod disease. *Proceedings of the Brighton Crop Protection Conference*, Brighton, pp. 583–588.

Biological Control of Fungal Plant Pathogens

S.S. Navi and R. Bandyopadhyay

ICRISAT, Patancheru, Andhra Pradesh 502 324, India

In the past century, science, medicine and engineering have invented, and begun to develop, amazing technologies to benefit all aspects of our life and to address the challenges with which we are faced. The exploitation of existing and modified biological systems to combat and control both animal and plant diseases is one such technology. Biological control or biocontrol represents an attractive alternative for the future because of the many concerns about pesticide use.

Many fungi successfully compete with, antagonize or actively parasitize other fungi and these interactions, which influence the balance of the natural mycoflora of the soil, rhizosphere and phyllosphere, have been exploited by humans to control pathogenic fungi. The traditional method of covering wounds caused by removal of diseased fruit tree branches with clay and horse manure was instinctively applying the principle of microbial competition by using a mixture rich in microorganisms, which held the pathogens at bay. Since the early 1950s plant pathologists have realized the potential of fungi for biocontrol of plant diseases either alone or in conjunction with chemical fungicides and in the last decade several lines of research have passed from the experimental stage through field trials to commercial usage in the agricultural and horticultural industries.

Control methods are based on foliar spray, soil treatment, seed dressing, root protection of trees, stump treatment, treatment of pruning wounds and direct spraying of suspensions of natural hyperparasites on to fungal lesions.

©CAB *International 2002. Plant Pathologist's Pocketbook*
(eds J.M. Waller, J.M. Lenné and S.J. Waller)

Foliar Sprays

Several commercial formulations can be sprayed for biocontrol (Table 34.1). Rose powdery mildew, caused by *Sphaerotheca pannosa* var. *rosae*, has been controlled in commercial greenhouses under experimental conditions using the yeast-like fungus *Sporothrix flocculosa*. The control was as effective as fungicide application (Bélanger *et al.*, 1994). *Gliocladium roseum*, a filamentous fungus, is an effective and versatile antagonist against grey mould disease of strawberry caused by *Botryotinia fuckeliana* (*Botrytis cinerea*). The isolates of *G. roseum* (10^6 conidia ml^{-1}) consistently suppressed *B. fuckeliana* both under greenhouse and field conditions. The performance of the antagonist was equal to or better than that of other leading antagonists and the fungicide (Sutton *et al.*, 1997). Similarly, the isolates of *G. roseum* performed better than other antagonists and captan when tested against *B. fuckeliana* in raspberry field plots (Sutton *et al.*, 1997).

Soil Treatment

Some success has been achieved by adding spore suspensions of antagonists to unsterilized soils where, however, they run the risk of declining in numbers as a result of competition with the resident microbial flora. Non-aflatoxigenic strains of *Aspergillus flavus* have been used to competitively exclude aflatoxin-producing strains of *A. flavus* to reduce aflatoxin contamination in several crops such as maize, groundnut and cottonseed. Significant reduction in aflatoxin contamination in corn has been reported using a non-aflatoxigenic strain of *A. flavus* (NRRL 21882) and *Aspergillus parasiticus* (NRRL 21369). Rice grains colonized with the non-aflatoxigenic strains were applied on rows with a Gandy applicator pulled behind a tractor when maize plants were 30–60 cm high (Dorner *et al.*, 1999). Preparation of inexpensive carriers of non-aflatoxigenic strains is detailed in Bock and Cotty (1999).

Seed Dressing

Dressing of muskmelon and watermelon seeds with *Kalisena* SD and *Kalisena* SL (bioformulations of *Aspergillus niger* strain AN 27) was effective in controlling Fusarium wilt (*Fusarium oxysporum melonis*) thereby improving yield and quality of the fruits in resource-poor farmers' fields in India (Sen, 2000). In addition, these formulations were also effective against *Rhizoctonia solani* and *Pythium* spp. Seedling diseases of barley and wheat, caused by *Fusarium culmorum* and *Bipolaris sorokiniana*, have been controlled by treating seeds with *G. roseum* and *Idriella* (*Microdochium*) *bolleyi* under field conditions. The efficacy of the biocontrol agents was as good as fungicidal seed treatment (Knudsen *et al.*, 1995).

Table 34.1. Some commercially available biocontrol products used against plant diseases caused by fungi.

Biofungicide	Biocontrol organism	Target pathogen/disease	Crop	Formulation	Application method	Manufacturer/distributor
AQ10	*Ampelomyces quisqualis* isolate M-10	Powdery mildew	Apples, cucurbits, grapes, ornamentals, strawberries and tomatoes	Water-dispersible granule	Spray	Ecogen, Inc., 2005 Cabot Blvd. West, Langhorne, PA 19074; Phone: 1-215-757-1590; Fax: 1-215-752-2461; or PO Box 4309, Jerusalem, Israel; Phone: 972-2-733212; Fax: 972-2-733265
Bio-Fungus (formerly Anti-Fungus)	*Trichoderma* spp.	*Sclerotinia*, *Phytophthora*, *Rhizoctonia solani*, *Pythium* spp., *Fusarium*, *Verticillium*	Flowers, strawberries, trees and vegetables	Granular, wettable powder, sticks and crumbles	Applied after fumigation; incorporated in soil; sprayed or injected	De Ceuster, Meststoffen N.V. or DCM, Forstsesteenweg, 30, B-2860 St Katelijne-Waver, Belgium; Phone: 32-15-31-22-57; Fax: 32-15-31-36-15; Email: dcm@dcmpronatura.com
Aspire	*Candida oleophila* I-182	*Botrytis* spp., *Penicillium* spp.	Citrus, pome fruit	Wettable power	Postharvest application to fruit as drench, drip or spray	Ecogen, Inc.,
Binab T	*Trichoderma harzianum* (ATCC 20476) and *Trichoderma polysporum* (ATCC 20475)	Pathogenic fungi that cause wilt, take-all, root rot, and internal decay of wood products and decay in tree wounds	Flowers, fruit, ornamentals, turf and vegetables	Wettable powder and pellets	Spray, mixing with potting substrate, mixing with water and painting on tree wounds, inserting pellets in holes drilled in wood	Bio-Innovation AB, Bredholmen, Box 56, S-545 02, Algaras, Sweden; Phone: 46-506-42005; Fax: 46-506-42072; or Henry Doubleday Research, Association Sales, Ltd, Ryton on Dunsmore, Coventry CV8 3LG, UK
Biofox C	*Fusarium oxysporum* (non-pathogenic)	*Fusarium oxysporum*, *Fusarium moniliforme*	Basil, carnation, cyclamen, tomato	Dust or alginate granule	Seed treatment or soil incorporation	S.I.A.P.A., Via Vitorio Veneto 1 Galliera, 40010, Bologna, Italy; Phone: 39-051-815508; Fax: 39-051-812069

Product	Organism	Target pathogen	Crop	Formulation	Application	Manufacturer/contact
Contans	*Coniothyrium minitans*	*Sclerotinia sclerotiorum* and *Sclerotinia minor*	Canola, sunflower, groundnut, soybean and vegetables (lettuce, bean, tomato)	Water-dispersible granule	Spray	Prophyta Biologischer Pflanzenschutz GmbH, Inselstrasse 12, D-23999 Malchow/Poel, Germany; Phone: 49-38425-230; Fax: 49-38425-2323; Web site: www.prophyta.com
Fusaclean	*Fusarium oxysporum* (non-pathogenic)	*Fusarium oxysporum*	Asparagus, basil, carnation, cyclamen, gerbera, tomato	Spores, microgranule	In drip to rock wool; incorporate in potting mix; in row	Natural Plant Protection, Route d'Artix B.P. 80,64150, Nogueres, France; Phone: 33-559-84-10-45; Fax: 33-559-84-89-55 (Registration pending)
Koni	*Coniothyrium minitans*	*Sclerotinia sclerotiorum* and *S. minor*	Cucumber, lettuce, capsicum, tomato and ornamental flowers in greenhouse production	Granule	Granules incorporated into soil or soil-less mix	BIOVED, Ltd, Ady Endre u. 10, 2310 Szigetszentmiklos, Hungary; Phone: 36-24-441-554; Email: boh8457@helka.iif.hu
Paecil (also known as Bioact)	*Paecilomyces lilacinus*	Various nematode species	Banana, tomatoes, sugarcane, pineapple, citrus, wheat, potatoes and others	Dry spore concentrate	Seedling or soil drench	Technological Innovation Corporation Pty Ltd, Innovation House, 124 Gymnasium Dr, Macquarie University, Sydney, NSW 2109, Australia; Phone: 61-2-9850-8216; Fax: 61-2-9884-7290; Web site: www.ticorp.com.au

Continued

Table 34.1. *Continued*

Biofungicide	Biocontrol organism	Target pathogen/disease	Crop	Formulation	Application method	Manufacturer/distributor
Polyversum (formerly Polygandron)	*Pythium oligandrum*	*Pythium* spp., *Fusarium* spp., *Botrytis* spp., *Phytophthora* spp., *Aphanomyces* spp., *Alternaria* spp., *Tilletia caries*, *Pseudocercosporella herpotrichoides*, *Gaeumannomyces graminis*, *Rhizoctonia solani*, *Sclerotium cepivorum*	Vegetables (tomatoes, potatoes, pepper, cucumbers, *Brassicaceae* vegetables), fruits (grapes, strawberries, citrus), legumes, cereals, canola, forest nurseries and ornamental plants	Wettable powder	Root and stem drench, spray	Biopreparaty Ltd, Tylisovska 1, 160 00, Prague 6, Czech Republic; Phone: (4202)-311-42-98; Fax: (4202)-3332-12-17; Email: biopreparaty@mbox.vol.cz
Primastop	*Gliocladium catenulatum*	*Pythium* spp., *Rhizoctonia solani*, *Botrytis* spp., *Didymella* spp.	Greenhouse crops	Wettable powder	Drench and incorporation	Kemira Agro Oy, Porkkalankatu 3, PO Box 330, 00101 Helsinki, Finland; Phone: 358-0-13-211; Fax: 358-0-694-1375. US Distributor: Ag Bio Development Inc., 9915 Raleigh St, Westminster, CO 80030; Phone: 303-469-9221; Fax: 303-469-9598
Protus WG	*Talaromyces flavus*, isolate V117b	*Verticillium dahliae*, *Verticillium albo-atrum*, and *Rhizoctonia solani*	Tomato, cucumber, strawberry, rape oilseed	Water dispersible powder containing ascospores	Soil or seed treatment, soil drench, root dip application	Prophyta Biologischer Pflanzenschutz GmbH, Inselstrasse 12, D-23999 Malchow/Poel, Germany; Phone: 49-38425-230; Fax: 49-38425-2323; Web site: www.prophyta.com

Product	Organism	Target pathogens	Host	Formulation	Application	Supplier
Root Pro	*Trichoderma harzianum*	*Rhizoctonia solani, Pythium* spp., *Fusarium* spp., and *Sclerotium rolfsii*	All susceptible flower and vegetable plants	Fungal spores mixed with peat and other organic material	Agent is mixed into growing media at time of seeding or transplanting	Mycontrol Ltd., Alon Hagalil M.P. Nazereth Ellit 17920, Israel; Phone/Fax: 972-4-9861827; Email: mycontro@netvision.net.il
RootShield T-22G, T-22 Planter Box (also sold as Bio-Trek)	*Trichoderma harzianum* Rifai strain KRL-AG2 (T-22)	*Pythium* spp., *Rhizoctonia solani, Fusarium* spp., and *Sclerotinia homeocarpa*	Trees, shrubs, transplants, all ornamentals, tomato, cucumber, bean, cabbage, maize, cotton, groundnut, potato, sorghum, soybean, sugarbeet, turf and greenhouse ornamentals	Granules, dry or wettable powder	Granules added in-furrow with granular applicator, mixed with soil or potting medium, broadcast application to turf. Powder mixed with seeds in planter box, or in commercial seed treatment slurry or with water and added as soil drench	Bioworks, Inc. (formerly TGT, Inc.), 122 North Genesee St, Geneva, NY 14456, USA; Phone: 1-315-781-1703; Fax: 1-315-781-1793
Rotstop, P.g. suspension	*Phlebia gigantea*	*Heterobasidion annosum*	Trees	Spores in inert powder	Spray, chain saw oil	Kemira Agro Oy, Porkkalankatu 3, PO Box 330, 00101 Helsinki, Finland; Phone: 358-0-13-211; Fax: 358-0-694-1375
SoilGard (formerly GlioGard)	*Gliocladium virens* GL-21	Damping-off and root rot pathogens especially *Rhizoctonia solani* and *Pythium* spp.	Ornamental and food crop plants grown in greenhouses, nurseries, homes, and interior scapes	Granules	Granules are incorporated in soil or soil-less growing media prior to seeding	Thermo Trilogy, 9145 Guilford Road, Suite 175, Columbia, MD 21046, USA; Phone: 1-301-604-7340; Fax: 1-301-604-7015
Supresivit	*Trichoderma harzianum*	Various fungi				Borregaard BioPlant, Helsingforsgade 27B, DK-8200 and Arhus N, Denmark

Continued

Table 34.1. *Continued*

Biofungicide	Biocontrol organism	Target pathogen/ disease	Crop	Formulation	Application method	Manufacturer/distributor
Trieco	*Trichoderma viride*	*Rhizoctonia* spp., *Pythium* spp., *Fusarium* spp., root rot, seedling rot, collar rot, red rot, damping-off, Fusarium wilt	Mustard, *masoor*, oilseeds, soybean, cotton, chillies, chickpeas, green *arhar/dal*, tobacco, cardamom, turmeric, *moong, udad, chawli,* tea, coffee, rubber, ginger, tomato, sugarcane, citrus, grapes, sunflower, cereals, vegetables and others	Powder	Dry or wet seed, tuber, or set dressing or soil drench, spread/ broadcast over field	Ecosense Labs (I) Pvt. Ltd, 54 Yogendra Bhavan, J.B. Nagar, Andheri (E), Mumbai – 400 059, India; Phone: 834-9136/830-0967; Fax: (91-22) 822-8016; Email: ecosense.mamoo@gems.vsnl.net.in
Trichodex	*Trichoderma harzianum*	Primarily *Botrytis cinerea*, also *Collectotrichum* spp., *Fulvia fulva, Monilinia laxa, Plasmopara viticola, Pseudoperonospora cubensis, Rhizopus stolonifer, Sclerotinia sclerotiorum*	Cucumber, grape, nectarine, soybean, strawberry, sunflower, tomato	Wettable powder	Spray	Makhteshim Chemical Works, Ltd, PO Box 60, Beer Sheva, Israel; Phone: (Main Office) 972-3-5179351; (Factory) 972-7-6296615; (US Office): 551 5th Ave Suite 1100, New York, NY 10176, USA; Phone: 1-212-661-9800

Trichopel, Trichoject, Trichodowels, Trichoseal	*Trichoderma harzianum* and *T. viride*	*Armillaria, Botryosphaeria, Chondrosternum, Fusarium, Nectria, Phytophthora, Pythium, Rhizoctonia*		Agrimm Technologies, Ltd, PO Box 13-245, Christchurch, New Zealand; Phone: 64-13-366-8671; Fax: 64-13-365-1859
Trichoderma 2000 (formerly 'TY')	*Trichoderma* sp.	*Rhizoctonia solani, Sclerotium rolfsii, Pythium* spp., *Fusarium* spp.	Nursery and field crops	Incorporated into soil or potting medium	Mycontrol, Ltd, Alon Hagalil M.P. Nazereth Eit 17920, Israel; Phone/Fax: 972-4-9861827; Web site: mycontro@netvision.net.il

Source: Summarized from www.barc.usda.gov/psi/bpdl/bpdlprod/bioprod.html

Mention of a trademark or proprietary product does not constitute a guarantee or warranty of the product by the authors, editors or CAB *International*, and does not imply its approval to the exclusion of other products that may also be suitable. This list is provided for information purposes only.

Root Protection

Root and stem damage caused by species of *Pythium*, *Fusarium*, *Botrytis*, *Phytophthora*, *Aphanomyces*, *Alternaria*, *Rhizoctonia* and *Sclerotium* occurring on vegetables, fruits, legumes, cereals and ornamental plants, as well as *Tilletia caries* and *Gaeumannomyces graminis* on cereals, can be controlled using Polyversum, which is a commercially available biofungicide formulation (Table 34.1).

Stumps and larger roots of trees killed by *Armillaria mellea* provide a source of infection to other trees. An early integrated method of chemical and biological control was developed in citrus groves in California (Bliss, 1951). The infected stumps were removed mechanically and the soil treated with carbon disulphide, after which *Trichoderma viride*, present in the surrounding soil, colonized the root fragments replacing the *Armillaria*.

In Britain stumps of broad-leaved trees are a major source of *A. mellea* infection. Treatment of the cut surfaces with a 40% solution of ammonium sulphamate followed by a basidiospore suspension of *Coriolus versicolor* gave four-year protection of birch stumps by introduction of a successful competitor to the *Armillaria*, which is harmless to the living tree (Rishbeth, 1976).

Heterobasidion annosum is the cause of butt-rot and results in considerable losses in pine and other conifer plantations. When thinning plantations, application of *Phlebia gigantea*, 10^4 conidia per 16 cm diameter stump, introduces a fungus, which competes with the *H. annosum* present in the infected roots. The *Phlebia* itself forms sporophores during the year following its application to the stumps and these provide long-term protection by furnishing a further source of spores. This method is not suitable when felling whole plantations as *P. gigantea* cannot compete with well-established infection by *H. annosum* that persists in the roots and infects later plantings (Rishbeth, 1963, 1976).

Pruning Wound Protection

Many pathogens invade through pruning cuts and fungicidal paints do not give lasting protection as cracking of the paint layer after application allows access to the spores of pathogens; such paints may even encourage invasion by suppressing the natural microbial flora, which in an undisturbed state provides competition to potential pathogens. *Cladosporium cladosporioides*, for instance, protects pruning cuts against infection by *Nectria galligena* (European canker of apple trees) providing better, longer-lasting and less-expensive protection than that of fungicides. *Trichoderma* is now extensively used to protect pruning wounds (see section on control of silver leaf). It is essential that a spore suspension of the protecting fungus should be applied to the wound immediately and for this to be achieved modified pruning shears have been developed, which are connected to a reservoir holding a suspension of spores, which delivers spores directly to the cut surface via grooves in the cutting blades (Corke, 1980).

Mercier and Wilson (1994) stated that the presence of naturally occurring microflora of apple wounds did not interfere with the biocontrol of storage rot of apple (*B. cinerea*) by *Candida oleophila* and in some cases was even more beneficial. Grey mould rot (*B. cinerea*) of apple was successfully controlled under storage conditions using *C. oleophila* (Mercier and Wilson, 1995). Colonization of apple wounds by natural microflora, including pathogens such as *B. cinerea*, was reduced using *C. oleophila*. However, several commercial formulations like 'Binab T' developed from *Trichoderma harzianum* (ATCC 20476) and *Trichoderma polysporum* (ATCC 20475) are available to protect pathogenic fungi that cause wilt, take-all, root rot, internal decay of wood products and decay in tree wounds, and postharvest diseases of various fruits under storage conditions. Such formulations can be applied in the form of spray, mixing with potting substrate, mixing with water and painting on tree wounds, inserting pellets in holes drilled in wood, drench or drip (Table 34.1).

Preinoculation

Preinoculation as described above has been used in Australia using large numbers of spores of *Fusarium lateritium* against infection by *Eutypa armeniacae*, the cause of dieback of apricot trees. *F. lateritium* is tolerant of a concentration of benomyl 10 times higher than is the pathogen and inoculation with a suspension of 10^4 spores ml^{-1} of a weak solution of benomyl raised the protective effect from 74% with spores alone to 98%, and this protection was long term. *Endothia parasitica* damages chestnut trees in Europe and America. Successful protection has been achieved by inoculation with hypovirulent strains unable to produce symptoms. By far the widest commercially used fungus for protection and cure of fungal infections, however, is *Trichoderma*. For example, Phytophthora foot rot of *Gerbera* (*Phytophthora cryptogea*) has been effectively controlled using *T. viride* at 300–600 g m^{-3} applied to peat (Orlikowski, 1995).

Control of Silver Leaf Disease

As a result of cooperative work between scientists at Long Ashton Research Station and pilot plant tests in Sweden, *Trichoderma* is now extensively used in commercial orchards against silver leaf disease of apple, pear, plum and cherry caused by *Chondrostereum purpureum*. A formulation of spores of the American Type Culture Collection cultures ATCC 20475 and 20476 has been developed in Sweden. This is marketed both as a wettable powder for use after dilution (10 g l^{-1} water) in wound protection, and as pellets for insertion in holes drilled in the trunks of infected trees. Such materials are subject to patent laws, which vary between countries; they are explained and the background of commercialization is outlined by Ricard (1981) and Templeton *et al.* (1980).

Shoot Treatment

The phyllosphere is the habitat of many saprophytic microorganisms, which not only compete with the spores of pathogenic fungi on the surface of the leaf but also sometimes activate the plant's defensive mechanisms in advance of attack by a pathogen. There are also many fungi that live as hyperparasites on the lesions of such fungi as rusts and assessment of these hyperparasites as components of an integrated control programme has given some promising results. Reports of the use of *Tuberculina maxima* grown in culture and then sprayed on to lesions of *Cronartium ribicola* on *Pinus monticola* in the USA, of *Scytalidium uredinicola* used against *Endocronartium harknessii* (western gall rust) on *Pinus banksiana* and *Pinus contorta* in Canada and of *Verticillium lecanii* against *Uromyces dianthi* on carnations in glasshouses in the UK are encouraging.

Cross-protection

In this system, an avirulent or hypovirulent strain of a pathogenic fungus is allowed to invade the host. The host is not damaged, but by a variety of mechanisms not yet entirely understood, the avirulent strain may act directly against the pathogen already in the plant, or it may induce the plant to inhibit an invading pathogen (Cook and Baker, 1983). For example, avirulent strains of *F. oxysporum* have been used to prevent various wilt diseases of vegetables, potatoes and ornamentals caused by virulent isolates of *F. oxysporum*.

Bacteria

The effects of plant growth-promoting rhizobacteria are due at least in part to the natural biological control of root pathogens and similar effects have been shown to occur on plant shoots. Fluorescent pseudomonads, *Streptomyces* spp. and *Bacillus subtilis* are known bacterial antagonists. Kodiak (Gustafson Inc., Dallas, Texas, USA) is a commercial preparation of *B. subtilis* for soil application. Mechanisms of antagonism include the production of siderophores and antibiotics. Weller (1988) gives more information on the use of bacteria for biological control of soil-borne pathogens.

References and Further Reading

Baker, R. and Dickman, M.B. (1992) Biological control with fungi. In: Metting. F.B. Jr (ed.) *Soil Microbial Ecology: Applications in Agricultural and Environmental Management*. Marcel Dekker, New York, USA, pp. 275–305.

Bélanger, R.R., Labbé, C. and Jarvis, W.R. (1994) Commercial-scale control of rose powdery mildew with fungal antagonist. *Plant Disease* 78, 420–424.

Bliss, D.E. (1951) The destruction of *Armillaria mellea* in citrus soils. *Phytopathology* 41, 665–683.

Bock, C.H. and Cotty, P.J. (1999) Wheat seed colonized with atoxigenic *Aspergillus flavus*: characterization and production of a biopesticide for aflatoxin control. *Biocontrol Science and Technology* 9, 529–543.

Boland, G.J. and Kuykendall, L.D. (eds) (1997) *Plant–Microbe Interactions and Biological Control*. Marcel Dekker, New York, USA, 442 pp.

Campbell, R. (ed.) (1989) *Biological Control of Plant Pathogens*. Cambridge University Press, Cambridge, UK, 218 pp.

Cook, R.J. and Baker, K.F. (1983) *The Nature and Practice of Biological Control of Plant Pathogens*. American Phytopathological Society, St Paul, Minnesota, USA, 589 pp.

Corke, A.T.K. (1980) Biological control of tree diseases. *Long Ashton Research Station Report 1979*, 190–198.

Dorner, J.W., Cole, R.J. and Wicklow, D.T. (1999) Aflatoxin reduction in corn through field application of competitive fungi. *Journal of Food Protection* 62, 650–656.

Hornby, D. (ed.) (1990) *Biological Control of Soil-borne Plant Pathogens*. CAB International, Wallingford, UK, 479 pp.

Katan, J. (1991) Interactions of roots with soil-borne pathogens. In: Waisel, Y., Eshel, A. and Kafkafi, U. (eds) *Plant Roots: the Hidden Half*. Marcel Dekker, New York, USA, pp. 823–836.

Kloepper, J.W. (1992) Plant growth-promoting rhizobacteria as biological control agents. In: Metting. F.B. Jr (ed.) *Soil Microbial Ecology: Applications in Agricultural and Environmental Management*. Marcel Dekker, New York, USA, pp. 255–274.

Knudsen, I.M.B., Hockenhull, J. and Jensen, D.P. (1995) Biocontrol of seedling diseases of barley and wheat caused by *Fusarium culmorum* and *Bipolaris sorokiniana*: effects of selected fungal antagonists on growth and yield components. *Plant Pathology* 44, 467–477.

Mercier, J. and Wilson, C.L. (1994) Colonization of apple wounds by naturally occurring microflora and introduced *Candida oleophila* and their effect on infection by *Botrytis cinerea* during storage. *Biological Control* 4, 138–144.

Mercier, J. and Wilson, C.L. (1995) Effect of wound moisture on the biocontrol by *Candida oleophila* of gray mold rot (*Botrytis cinerea*) of apple. *Postharvest Biology and Technology* 6, 9–15.

Orlikowski, L.B. (1995) Studies on the biological control of *Phytophthora cryptogea* Pethybr. Et Laff. II. Effectiveness of *Trichoderma* and *Gliocladium* spp. in the control of Phytophthora foot rot of gerbera. *Journal of Phytopathology* 143, 341–343.

Ricard, J.L. (1981) Commercialisation of *Trichoderma* based mycofungicide: some problems and solutions. *Biocontrol News and Information* 2, 95–98.

Rishbeth, J. (1963) Stump protection against *Fomes annosus*. III. Inoculation with *Peniophora gigantea*. *Annals of Applied Biology* 52, 63–77.

Rishbeth, J. (1976) Chemical treatment and inoculation of hardwood stumps for control of *Armillaria mellea*. *Annals of Applied Biology* 82, 57–70.

Sen, B. (2000) Biological control: a success story. *Indian Phytopathology* 53, 243–249.

Stoll, G. (1988) *Natural Crop Protection Based on Local Farm Resources in the Tropics and Subtropics*, 3rd edn. Tropical Agroecology No. 1. Margraf Publishers Scientific Books, Weikersheim, Germany, 188 pp.

Sutton, J.C., Gang Peng, D.W.L., Yu, H., Zhang, P. and Valdebenito-Sanhueza, R.M. (1997) *Gliocladium roseum* a versatile adversary of *Botrytis cinerea* in crops. *Plant Disease* 81, 316–328.

Templeton, C.E., Smith, R.J. and Klomparens, W. (1980) Commercialisation of fungi and bacteria for biological control. *Biocontrol News and Information* 1, 291–294.

Weller, D.M. (1988) Biological control of soil-borne plant pathogens in the rhizosphere with bacteria. *Annual Review of Phytopathology* 26, 379–407.

Plant Pathogens for Biological Control of Weeds

H.C. Evans

CABI Bioscience UK Centre, Silwood Park, Ascot, Berkshire SL5 7PY, UK

Introduction

A comprehensive account of the use of plant pathogens for biological control of weeds was included in the Second Edition of the *Plant Pathologist's Pocket Book* (Hasan, 1983). The present chapter summarizes the developments over the intervening 2 decades and attempts to assess the potential of plant pathogens for the control of weeds. For a more comprehensive coverage of the subject, the following reviews should be consulted: Cullen and Hasan (1988); McRae (1988); Wapshere *et al.* (1989); Hasan and Ayres (1990); Charudattan (1991); Morris (1991); TeBeest (1991, 1993); Watson (1991); Mortensen (1997) and McFadyen (1998).

Biological control of weeds using plant pathogens is still a relatively new strategy in weed management. Two seemingly distinct approaches have been adopted: classical biological control of exotic or alien weeds, which involves the introduction of coevolved, highly specific pathogens, or natural enemies in general, from the centre of origin of the target weed species; and, inundative control, in which pathogens are mass-produced and applied as formulated products (bioherbicides). However, recent developments have demonstrated that these strategies need not necessarily be mutually exclusive.

For classical biological control, the concepts are simple and follow a logical approach, initially developed and refined by entomologists over the last century. The premise is based on the theory that plants, once freed of their natural enemy complex, become ecologically fitter and, therefore, more competitive than those subject to natural control. If climatic factors are favourable, then there are few barriers to regulate growth and this may result in population explosions with the subsequent development of weed invasions. Thus, alien plant species are invariably introduced, either deliberately or

©CAB *International* 2002. *Plant Pathologist's Pocketbook* (eds J.M. Waller, J.M. Lenné and S.J. Waller)

accidentally, into a new geographic area without their coevolved natural enemies, or at best with only a partial complement. Classical biological control aims to redress this imbalance by searching in the centre of origin of the weed target, identifying the plant pathogens present, selecting and screening those that appear to impact most on that target, and, finally, introducing and releasing them in the exotic area. The single, most important criterion is that the selected pathogen agents must be highly specific to the target weed.

The inundative approach involves the mass production, formulation and application of a product (bioherbicide), based on a pathogen that may or may not be specific to the weed target. Typically, the latter is an indigenous or naturalized plant species that has escaped from its natural enemy complex because of agricultural practices. For example, the weed pathogens within an agricultural ecosystem may be constrained by various factors, such as poor interseasonal survival or limited dispersal capacity within a managed cropping system, whereby their populations fail to build up sufficiently rapidly to impact significantly on their weed hosts. Invariably, this approach necessitates a much higher level of technology than for classical biological control. If the product is to be marketed commercially then this has to be registered and subject to many of the safety standards demanded of more traditional chemical herbicides. Thus, in sharp contrast to classical biological control, considerable investment may be needed and a commercial strategy adopted in order to recuperate the investment and thence to generate profits for the investors. These two approaches are now analysed and compared using past and present examples.

Classical Biological Control

The deliberate introduction of any alien organism into an ecosystem involves a certain degree of unpredictability, and hence risk. Therefore, any classical biological control programme has to be subjected to a rigorous, scientifically based risk assessment before an exotic agent can be released. In the case of plant pathogens for the biological control of weeds, these risks have often been considered to be unacceptable, based not on scientific evidence but on emotive and historical associations with crop disease epiphytotics. Perhaps this is not surprising given the fact that many of the major food and commodity crops have been decimated in the past by coevolved pathogens that have caught up with their hosts in exotic situations often with catastrophic socioeconomic consequences (Large, 1940). Indirectly, however, such infamous examples from agriculture demonstrate just how effective plant pathogens can be, spreading rapidly and efficiently within and between host populations. Ironically, the very fact that the rigid host specificity has been maintained provides circumstantial evidence to prove that similar coevolved pathogens of weeds are inherently safe and will not expand their host ranges when released into new ecosystems. The stability of natural enemies has been emphasized by Marohasy (1996) who critically analysed the case histories of insect biocontrol agents implicated in host 'shifts'. From this study, it was concluded that, although more than 600 agents have been transferred between geographic regions for the biological control of weeds, there have been surprisingly few documented cases of expansion of host range, all of which proved to be the result of predictable behavioural responses. Indeed, one important advantage of biocontrol over chemical control is now considered to be its evolutionary stability: in a coevolved association, the natural enemy adapts to genetic changes in the host, but is genetically stable outside of this association. Thus, in contrast to chemical herbicides, which have been described as 'evolutionarily evanescent' due to problems of weed resistance (Holt and Hochberg, 1997), host natural enemy interactions are probably permanent and, therefore, sustainable.

Host specificity tests provide the information on which to base the risk assessment and, thereby, play the central role in any classical biological control project. The protocol followed for screening of pathogens has been adapted from the centrifugal, phylogenetic method devised by Wapshere (1974) to evaluate insect biological control agents. This involves greenhouse screening of the selected biocontrol agent against a range of plant species, starting with those most closely related to the target weed and progressing to more and more distantly related taxa. As an in-built safeguard, additional test species representing regionally important crop and amenity plants are also included in the final test list. In fact, the initial protocol proved to be so stringent that modifications in the testing sequence were proposed in order to reduce the chances of rejecting potentially useful agents (Wapshere, 1989). However, as pointed out by Evans and Ellison (1990) and Evans (1995a, 2000), the safety requirements for plant pathogens are consistently more demanding than those for insect agents with significantly more test species being required to be screened.

Ironically, some of the pioneering and still most successful weed pathology programmes were completed before this protocol was fully in place. Hasan (1983) discussed the example of the rust *Puccinia chondrillina*, with a Mediterranean centre of origin, which was released in Australia in the early 1970s for control of skeleton weed (*Chondrilla juncea*). Within a relatively short period of time, weed infestations had been reduced by 99% to a density approaching that recorded in Europe (Cullen and Hasan, 1988), with an estimated annual saving of over US$12 million, resulting from increased crop yield and reduced herbicide usage (Mortensen, 1986). The cost:benefit ratio has been put at 112:1 compared to that for chemical herbicides which averages 4–5:1 (Tisdell, 1990). Control has proven to be sustainable and the pathogen so specific that previously unknown resistant biotypes were identified within the susceptible population. Using isoenzyme techniques for biotype–pathotype matching, additional rust strains have since been introduced (Chaboudez *et al.*, 1992; see Table 35.1). Another early example involved mistflower (*Ageratina riparia*) and a white smut (*Entyloma ageratinae*), which was introduced from Jamaica into Hawaii after screening against a relatively short and disparate test list, and, perhaps more contentiously, with an incomplete knowledge of its taxonomy and true origin (Barreto and Evans, 1988). Such basic information is now an essential prerequisite before host range screening can even be considered. It should be added that spectacular control of the weed was achieved in Hawaii and, moreover, the pathogen has proved to be highly specific (Trujillo, 1985). However, despite these early successes, only a small number of pathogens have been deliberately released as classical biological control agents compared to insect agents. This is partly due to 'pathophobia', often apparent in even the

Table 35.1. Fungal pathogens introduced as classical biological control agents of alien weeds.

Fungal pathogen	Alien weed	Country of origin	Country (date) of introduction	Control status
Maravalia cryptostegiae (Uredinales: Chaconiaceae)	*Cryptostegia grandiflora* (Asclepiadaceae)	Madagascar	Australia (1994)	Spreading rapidly; aided by dispersal from helicopters
Phaeoramularia eupatorii-odorati[a] (Hyphomycetes)	*Ageratina adenophora* (Asteraceae)	Mexico	South Africa (1987)	Limited impact
Entyloma ageratinae (Ustilaginales: Tilletiaceae)	*Ageratina riparia* (Asteraceae)	Jamaica[b]	Hawaii, USA (1975); South Africa (1989); New Zealand (1998)	Substantial to complete control; Established and spreading; severe defoliation; Recent introduction; established and spreading
Puccinia carduorum (Uredinales: Pucciniaceae)	*Carduus nutans* (Asteraceae)	Turkey	USA (1987)	Decreases seed production and accelerates senescence; overall impact unknown
Puccinia cardui-pycnocephali	*Carduus pycnocephalus* (Asteraceae)	Italy	Australia (1993)	Impact unknown; established and spreading
Puccinia chondrillina	*Chondrilla juncea* (Asteraceae)	Italy	Australia (1971, 1980, 1982, 1996)[c]; USA (1976)[d]; Argentina (1982)	Extremely effective; high level of control of major weed biotype; Highly effective in California; Low impact
Puccinia abrupta var. *partheniicola*	*Parthenium hysterophorus* (Asteraceae)	Mexico	Australia (1991)	Established, but impact not significant
Puccinia melampodii		Mexico	Australia (1999)	Established and spreading
Puccinia xanthii	*Xanthium strumarium* (Asteraceae)	USA[e]	Australia (1974)	Spectacular results in Queensland; no longer a problematic weed
Uromyces heliotropii (Uredinales: Pucciniaceae)	*Heliotropium europaeum* (Boraginaceae)	Turkey	Australia (1991)	Spreading sporadically; impact unknown
Uromyces galegae	*Galega officinalis* (Fabaceae)	France	Chile (1973)	Established; impact unknown
Colletotrichum gloeosporioides f.sp. *clidemiae*	*Clidemia hirta* (Melastomataceae)	Panama	Hawaii, USA (1986)	Established; die-back reported

Continued

Table 35.1. *Continued*

Fungal pathogen	Alien weed	Country of origin	Country (date) of introduction	Control status
Uromycladium tepperianum (Uredinales: Pileolariaceae)	*Acacia saligna* (Mimosaceae)	Australia	South Africa (1987)	Highly effective; galling affects seed production and causes tree death; 80–100% reduction reported
Diabole cubensis (Uredinales: Ravenelliaceae)	*Mimosa pigra* (Mimosaceae)	Mexico	Australia (1996)	Recent establishment
Sphaerulina mimosae-pigrae (Dothideales: Mycosphaerellaceae) (Anamorph = *Phloeospora mimosae-pigrae*)		Mexico	Australia (1994)	Established and spreading[f]
Septoria passiflorae (Coelomycetes)	*Passiflora tripartita* (Passifloraceae)	Colombia	Hawaii, USA (1996)	Established and spreading
Cercospora rodmanii (Coelomycetes)	*Eichhornia crassipes* (Pontederiaceae)	USA	South Africa (1991)	Established; still being assessed
Phoma clematidina (Coelomycetes)	*Clematis vitalba* (Ranunculaceae)	USA	New Zealand (1996)	Establishment not yet confirmed
Phragmidium violaceum (Uredinales: Phragmidiaceae)	*Rubus constrictus* *Rubus ulmifolius* (Rosaceae)	Germany	Chile (1973)	Weed density decreasing; premature defoliation retards stem lignification predisposing plants to frost damage and invasion by secondary pathogens
Phragmidium violaceum	*Rubus fruticosus*	Europe[g] France	Australia (1984) Australia (1991)	Causing population decline Specific strain, accelerating decline in weed populations

Pathogen	Weed (family)	Origin	Release location (year)	Status
Septoria sp.[h] (Coelomycetes)	Lantana camara (Verbenaceae)	Ecuador	Hawaii, USA (1997)	Status unknown
Puccinia evadens	Baccharis halmifolia (Asteraceae)	USA	Australia (1998)	Established
Septoria myricae[h,i]	Myrica faya (Myricaceae)	USA (North Carolina)	Hawaii, USA (1997)	Status unknown
C. gloeosporioides f.sp. miconiae[h]	Miconia calvescens (Melastomataceae)	Brazil	Hawaii, USA (1997)	Established; low initial impact but now spreading

[a]Despite a recent redescription of this species (Morgan-Jones, 1997), its identification remains in doubt due to confusion over host records; originally, accidentally introduced into Australia (1954) on an insect biocontrol agent and, subsequently, into New Zealand (1962); also documented from China (1982).

[b]Both weed and pathogen naturalized in Jamaica; pathogen later discovered in Mexico, the true centre of origin (Barreto and Evans, 1988).

[c]Additional strains (pathotypes) introduced from Italy and Turkey against remaining weed biotypes.

[d]Reported in Canada in 1992, natural spread from USA.

[e]This is a well-documented example of a highly successful 'accidental' introduction; there is a suspicion that illegal releases were made to circumvent the official, regulatory channels, viewed by some farmers as being too slow and cumbersome.

[f]Also being mass-produced in liquid culture and sprayed onto weed infestations; one of the few examples of an augmentative or inundative approach being adopted for classical biological control agents.

[g]Illegal introduction from unknown European source; new strain from France released in 1991 after prolonged host specificity–virulence screening of a complex of European strains. Rust detected in New Zealand in 1984, thought to be the result of natural spread.

[h]Data from *International Bioherbicide Group News* 6 (2), 13–14 (1997).

[i]Original host of this pathogen was the North American species *Myrica cerifera*; *M. faya*, the target weed, originates from several Atlantic islands (Madeiras, Azores).

most informed regulatory bodies, as well as to poorly defined sources of funding which, typically, has to be generated from the public rather than the private sector. Julien and Griffiths (1998) listed most of those pathogens that have been introduced so far (see Table 35.1). All are fungal species, since it is highly unlikely that either bacteria or viruses will ever be considered as classical agents by biocontrol practitioners and quarantine authorities, not only because of the inherent 'pathophobia' linked with these microorganisms, but also because of the requirement for vectors or supplementary dissemination. Table 35.1 clearly shows that the early introductions were based exclusively on biotrophs, particularly rust fungi, because of their proven host specificity and efficient dispersal mechanisms. However, a recent trend has been to select necrotrophs, notably coelomycete genera such as *Colletotrichum* and *Septoria*, for those weed targets where suitable biotrophs have not been found, or simply have not coevolved with these hosts. It is probable that, because of the relatively inefficient modes of dispersal of these slime-spored fungi, augmentation and inundative application will be necessary to move the pathogens between, and even within, weed populations (Evans, 1995b, 2000).

An additional constraint to the exploitation of the classical approach concerns the potential threat posed by exotic pathogens, not to non-target plants, but, ironically, to the weed populations themselves, in those situations where the target plant may have some perceived economic benefits. Such conflicts of interest, highlighted by the different names for the European plant *Echium plantagineum* in Australia (Paterson's curse or Salvation Jane) depending on the region and the agroecosystem involved, resulted in the Biological Control Act of 1984 which provided, for the first time, a legal basis for the release of exotic biocontrol agents (Cullen and Delfosse, 1985).

Despite the fact that classical biological control is coming under increased scrutiny (Howarth, 1991; Simberloff and Stiling, 1996), the success rate has been high and there appears to be considerable potential in the use of fungal pathogens as classical agents, particularly since it is becoming recognized as the most environmentally and economically acceptable, long-term strategy for managing alien, invasive weeds (Morris, 1991; Watson, 1991; Evans, 1995b; Barreto and Evans, 1997; McFadyen, 1998).

Inundative Biological Control (Bioherbicides)

In the early 1970s, at about the same time that plant pathogens were first being considered for classical biological control of exotic weeds, attention was also being focused on the exploitation of indigenous fungi for management of endemic weeds, particularly in the USA (Templeton *et al.*, 1979). The strategy employed was simple: to bulk-up the fungal pathogen, typically an easily cultured, rapidly sporulating necrotroph, and apply it at an early stage in the weed-growing season in order to simulate natural epiphytotics that normally occur too late in the season to prevent crop loss. In effect, this early and massive application of inoculum overcomes the constraints in the life cycles of many of the microbial agents (typically coelomycete fungi and bacteria) that were being considered as exploitable for bioherbicide development at that time, especially those with poor intercrop survival, slow inoculum build-up and inefficient dispersal. In practice, the technology, as well as the regulatory processes and hence the economics, have proven to be more complex than envisaged initially. Thus, despite the early optimism, progress over the past two decades has been slow. The constraints in the development of bioherbicides have been analysed (Auld and Morin, 1995; Greaves, 1996; Mortensen, 1997) and the major ones are now summarized using case studies.

Templeton (1992) detailed the steps taken to develop and promote public acceptance of Collego®, one of the first registered bioherbicides, for control of northern jointvetch (*Aeschynomene virginica, Leguminosae*) in the southern USA, based on *Colletotrichum gloeosporioides* f.sp. *aeschynomene*. During this pioneering study, close collaboration was maintained with the US Environmental Protection Agency (EPA), who much later published guidelines for the registration of bioherbicides and biorationals in general. These were less complicated than those previously required for the registration of Collego®, a process which took 13 years, but Templeton (1992) has since argued that these should be further simplified, especially those relating to toxicology and environmental fate, in order to reduce the already considerable registration costs and thereby promote interest in and development of bioherbicides. He used the evidence gained over the 20 years of field experience with several commercial products, as well as with over 100 experimental products, to demonstrate that bioherbicides are environmentally safe, genetically stable and have high public acceptance. Undoubtedly, however, the economics of registration, coupled with inadequate or non-commercial market sizes, in addition to the technological constraints, especially relating to formulation chemistry, are the main reasons why so many promising mycoherbicides have failed to reach the commercial phase.

According to Cook *et al.* (1996), only three bioherbicides were registered for control of weeds in the USA, as of 1995: Collego®; DeVine®, based on fresh rather than formulated preparations of *Phytophthora palmivora* for use against stranglervine (*Morrenia odorata, Cucurbitaceae*); and Dr Biosedge®, based on a strain of *Puccinia canaliculata* specific to and virulent against yellow nutsedge (*Cyperus esculentus, Cyperaceae*). The latter product incorporates a unique formulation of rust spores and a low-dose chemical herbicide. However, this has failed, so far, to reach the market, purportedly due to problems with the mass production of rust inoculum, whereas the other two products have been commercialized for over 15 years. Nevertheless, despite their technological success and high farmer acceptance, they have not been highly profitable, and have thus failed to attract the agrochemical companies, mainly because of the limited market size. Significantly, and unlike classical biological control agents, both pathogens were not specific to the target weed (or weed genus). The '*forma specialis*' of *C. gloeosporioides* was found subsequently to infect crop species within other genera of the *Leguminosae*, and the *P. palmivora* strain was known to attack cucurbitaceous crops, such as melon. However, since the product was developed exclusively for use in orchard crops, it was assessed to be safe if not applied within a specified distance (200 m) of annual crops, based on epidemiological data. Similarly, Collego® was targeted for use only in rice ecosystems, far removed from susceptible leguminous crops. A similar extension of host range was also shown by *C. gloeosporioides* f.sp.

malvae, registered as BioMal™ in Canada for control of round-leaved mallow (*Malva pusilla*, *Malvaceae*). In this case, registration was granted only after it had been demonstrated that the risk to susceptible crops, such as safflower, was acceptable (Greaves, 1996; Mortensen, 1997).

The bioherbicide potential of other, plurivorous, virulent plant pathogens is now being actively investigated. A recent example concerns using the silverleaf fungus, *Chondrostereum purpureum*, for control of black cherry (*Prunus serotina*, *Rosaceae*), an invasive North American species that is a serious threat to conifer plantations, as well as to native woodlands in The Netherlands (de Jong *et al.*, 1990). Silverleaf is, of course, a well-known problem in commercial stone fruit orchards, which, until recently, was a notifiable disease in the UK. Conceptual and simulation models were developed, based on epidemiological, macrometeorological and air pollution data, in order to predict the risks posed to non-target hosts by artificially increased populations of the pathogen. The components of the risk analysis that were identified during this study are outlined in Fig. 35.1. Basically, it was shown that the risk to crop plants is high up to 500 m from the treated area but negligible at 5000 m. Subsequently, the Dutch Plant Protection Service confirmed that the use of *C. purpureum* for weed control was acceptable, except in those situations where the site to be treated was within 500 m of commercial stone fruit plantations, since, after this distance, the background contamination was as high as that from the added inoculum. The bioherbicide, Biochon®, is currently being marketed by Koppert Biological Systems as an environmentally friendly solution to undesirable tree regrowth. The use of this pathogen for management of weedy, endemic, deciduous trees in conifer plantations and amenity areas is also being evaluated in Canada (Prasad, 1994), and a commercial product (Ecoclear™) has recently been registered in North America for use in forestry and public rights-of-way (Shamoun and Hintz, 1998).

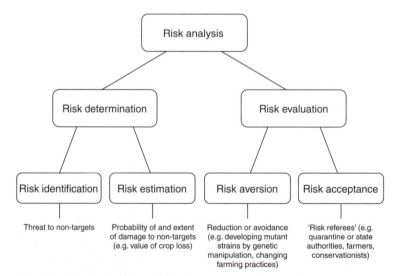

Fig. 35.1.   Components of a risk analysis (adapted from de Jong *et al.*, 1990).

The development and registration of another wood rot fungus, *Cylindro-basidium laeve*, for control of alien wattle trees (*Acacia* spp., *Mimosaceae*) has also been reported in South Africa (Morris *et al.*, 1998). Production of the fungal inoculant, known as Stumpout®, is at the cottage-industry level, employing a relatively simple formulation technology. Such examples may point the way towards how best to exploit plant pathogens for economic, safe and sustainable weed control, not just in resource-poor countries, but also in 'high-tech' countries like the UK where invasive woody weeds, such as rhododendron, are difficult or expensive to control by conventional means.

In complete contrast, advanced technology and large companies are currently involved in the development of bioherbicides in Japan, not only in crop protection but also in the highly lucrative leisure industry. The most troublesome weed in golf courses is annual bluegrass (*Poa annua*) and chemical herbicides are either non-selective or now considered to be environmentally undesirable. A highly specific, bacterial endophyte, *Xanthomonas campestris* pv. *poae*, was registered in 1997 under the name Camperico®, and constitutes the first bacterial herbicide to reach the commercial market (Imaizumi and Fujimori, 1998). The product is applied after mowing and invades via the cut surfaces. Golf consortia in both Japan and the USA are willing to pay astronomical amounts to protect the greens (S. Imaizumi, personal communication) and, thus, this has all the ingredients of a commercially successful venture.

Formulation

There is no doubt that formulation has played a key role in the marketing of bioherbicides, such as Camperico®, in order to overcome problems with storage, establishment and efficacy in the field. Essentially, formulation is mixing the active ingredient, in this case the biological propagule, with a carrier or solvent and other adjuvants in order to develop a product that can be stored, for at least one year, before being effectively applied to the target weed with safe and consistent results. The latter has been one of the major stumbling blocks in promoting bioherbicides, and one of the main aims of formulation, which is involving more and more advanced chemistry frequently borrowed from the food industry, has been to overcome these inconsistencies. This can be done by reducing dew requirements, stimulating spore germination, improving virulence, increasing spreadability and stickability, and protecting against sunlight and desiccation. The technology involved in the formulation of bioherbicides has been reviewed comprehensively by TeBeest *et al.* (1992), Green *et al.* (1997) and Greaves *et al.* (1998).

Summary

Plant pathogens are potentially valuable additions to the arsenal of weapons for use against weeds. Classical biological control offers a cheap, pollution-free and sustainable method for the management of invasive weeds, not only in natural ecosystems but also in those situations where, because of the vast areas involved, conventional forms of control are either uneconomic, difficult to implement or environmentally undesirable, or a combination of these. Notable, even spectacular successes have been achieved with the relatively few exotic pathogens (*c.* 20) released so far, bringing significant financial and environmental returns. The role of bio- or microbial herbicides in agriculture, however, is still problematic and insignificant. Nevertheless, because of pressures to reduce the reliance on chemical herbicides, which can only increase, bioherbicides could make a significant contribution to weed control in the future, once the well-documented constraints have been overcome, particularly through improved formulation and marketing.

The over-riding concern, to both regulatory authorities and to the public in general, in using plant pathogens for weed control is their potential threat to non-targets. This is especially true for classical biological control where alien pathogens are introduced into new ecosystems. However, all exotic introductions are based on a risk and threat analysis crucial to which is centrifugal, phylogenetic, host-range screening, which has proven to be an extremely useful and reliable tool in risk assessment (Evans, 2000).

Far from relying on rigid host specificity, which is axiomatic with employing coevolved pathogens for classical biological control, bioherbicides are now being developed using highly virulent, plurivorous crop pathogens. Here, the risk analysis is dependent on interpreting epidemiological rather than host range data. A code of conduct has recently been prepared to ensure and promote the safe use of biocontrol agents, including plant pathogens (FAO, 1996). This code should be adhered to for the introduction of exotic plant pathogens, as well as for risk assessments of indigenous pathogens, so that microbial biodiversity can be successfully exploited as an environmentally effective tool for management of weeds.

References

Auld, B.A. and Morin, L. (1995) Constraints in the development of bioherbicides. *Weed Technology* 9, 638–652.

Barreto, R.W. and Evans, H.C. (1988) Taxonomy of a fungus introduced into Hawaii for biological control of *Ageratina riparia* (Eupatoriae: Compositae), with observations on related weed pathogens. *Transactions of the British Mycological Society* 91, 81–97.

Barreto, R.W. and Evans, H.C. (1997) Role of fungal biocontrol of weeds in ecosystem sustainability. In: Palm, M.E. and Chapela, I.H. (eds) *Mycology in Sustainable Development*. Parkway Publ. Inc., Boone, North Carolina, USA, pp. 183–210.

Chaboudez, P., Hasan, S. and Espiau, C. (1992) Exploiting the clonal variability of *Chondrilla juncea* for use in Australia. In: *Proceedings, First International Weed Congress*. Weed Society of Victoria, Melbourne, Australia, pp. 118–121.

Charudattan, R. (1991) The mycoherbicide approach with plant pathogens. In: TeBeest, D.O. (ed.) *Microbial Control of Weeds*. Chapman & Hall, New York, London, pp. 24–57.

Cook, R.J., Bruckart, W.L., Coulson, J.R., Goettel, M.S., Humber, R.A., Lumsden, R.D., Maddox, J.V., McManus, M.L., Moose, L., Meyer, S.F., Quimby, P.C., Stack, J.P. and Vaughn, J.L. (1996) Safety of microorganisms intended for pest and plant disease control: a framework for scientific evaluation. *Biological Control* 7, 333–351.

Cullen, J.M. and Delfosse, E.S. (1985) *Echium plantagineum*: catalyst for conflict and change in Australia. In: Delfosse, E.S. (ed.) *Proceedings, Sixth International Symposium on Biological Control of Weeds*. Agriculture Canada, Ottawa, Canada, pp. 249–292.

Cullen, J.M. and Hasan, S. (1988) Pathogens for the control of weeds. *Philosophical Transactions of the Royal Society of London* 318, 213–224.

Evans, H.C. (1995a) Pathogen–weed relationships: the practice and problems of host range screening. In: Delfosse, E.S. and Scott, P.R. (eds) *Proceedings, Eighth International Symposium on Biological Control of Weeds*. DSIR/CSIRO, Melbourne, Australia, pp. 539–551.

Evans, H.C. (1995b) Fungi as biocontrol agents of weeds: a tropical perspective. *Canadian Journal of Botany* 73, 58–64.

Evans, H.C. (2000) Evaluating plant pathogens for biological control of weeds: an alternative view of pest risk assessment. *Australasian Plant Pathology* 29, 1–14.

Evans, H.C. and Ellison, C.A. (1990) Classical biological control of weeds with microorganisms: past, present, prospects. *Aspects of Applied Biology* 24, 39–49.

FAO (1996) *International Standards for Phytosanitary Measures*. Secretariat of the International Plant Protection Convention, Rome, Italy.

Greaves, M.P. (1996) Microbial herbicides – factors in development. In: Copping, L.G. (ed.) *Crop Protection Agents from Nature*. Royal Society of Chemistry, Cambridge, UK, pp. 444–467.

Greaves, M.P., Auld, B.A. and Holloway, P.J. (1998) Formulation of microbial herbicides. In: Burges, H.D. (ed.) *Formulation of Biopesticides, Beneficial Microorganisms, Nematodes and Seed Treatments*. Kluwer Academic, London and New York, pp. 203–233.

Green, S., Stewart-Wade, S.M., Boland, G.J., Teshler, M.P. and Liu, S.H. (1997) Formulating microorganisms for biological control of weeds. In: Boland, G.J. and Kuykendall, L.D. (eds) *Plant–Microbe Interactions and Biological Control*. Marcel Dekker, New York, USA, pp. 249–281.

Hasan, S. (1983) Plant pathogens and biological control of weeds. In: Johnston, A. and Booth, C. (eds) *Plant Pathologist's Pocketbook*, 2nd edn. Commonwealth Agricultural Bureau, Farnham Royal, UK, pp. 269–274.

Hasan, S. and Ayres, P.E. (1990) The control of weeds through fungi: principles and prospects. *New Phytologist* 115, 201–222.

Holt, R.D. and Hochberg, M.E. (1997) When is biological control evolutionarily stable (or is it)? *Ecology* 78, 1673–1683.

Howarth, F.G. (1991) Environmental impacts of classical biological control. *Annual Review of Entomology* 36, 485–509.

Imaizumi, S. and Fujimori, T. (1998) The bacterium *Xanthomonas campestris* pv. *poae* to control *Poa annua*. *Abstracts of Fourth International Bioherbicide Workshop*, University of Strathclyde, p. 13.

de Jong, M.D., Scheepens, P.C. and Zadoks, J.C. (1990) Risk analysis for biological control: a Dutch case study in biocontrol of *Prunus serotina* by the fungus *Chondrostereum purpureum*. *Plant Disease* 174, 189–194.

Julien, M.H. and Griffiths, M.W. (1998) *Biological Control of Weeds. A World Catalogue of Agents and their Target Weeds*, 4th edn. CAB International, Wallingford, UK, 223 pp.

Large, E.C. (1940) *The Advance of the Fungi*. J. Cape, London, UK, 488 pp.

Marohasy, J. (1996) Host shifts in biological weed control: real problems, semantic difficulties or poor science? *International Journal of Pest Management* 42, 71–75.

McFadyen, R.E.C. (1998) Biological control of weeds. *Annual Review of Entomology* 43, 369–393.

McRae, C.F. (1988) Classical and inundative approaches to biological weed control compared. *Plant Protection Quarterly* 3, 124–127.

Morgan-Jones, G. (1997) Notes on Hyphomycetes. *Mycotaxon* 61, 363–373.

Morris, M.J. (1991) The use of plant pathogens for biological weed control in South Africa. *Agriculture, Ecosystems and Environment* 37, 239–255.

Morris, M.J., Wood, A. and Den Breeyen, A. (1998) Development and registration of a fungal inoculant to prevent regrowth of cut wattle tree stumps in South Africa. *Abstracts of Fourth International Bioherbicide Workshop*, University of Strathclyde, UK, p. 15.

Mortensen, K. (1986) Biological control of weeds with plant pathogens. *Canadian Journal of Plant Pathology* 8, 229–231.

Mortensen, K. (1997) Biological control of weeds using microorganisms. In: Boland, G.J. and Kuykendall, L.D. (eds) *Plant–Microbe Interactions and Biological Control*. Marcel Dekker, New York, USA, pp. 223–248.

Prasad, R. (1994) Influence of several pesticides and adjuvants on *Chondrostereum purpureum* – a bioherbicide agent for control of forest weeds. *Weed Technology* 8, 445–449.

Shamoun, S.F. and Hintz, W.E. (1998) Development of *Chondrostereum purpureum* as a biological control agent for red alder in utility rights-of-way. In: Wagner, R.G. and Thompson, D.G. (eds) *Third International Conference on Forest Vegetation Management*. Ontario Ministry of Natural Resources, Ontario, Canada, pp. 308–310.

Simberloff, D. and Stiling, P. (1996) How risky is biological control? *Ecology* 77, 1965–1974.

TeBeest, D.O. (1991) *Microbial Control of Weeds*. Chapman & Hall, New York and London, 284 pp.

TeBeest, D.O. (1993) Biological control of weeds with fungal plant pathogens. In: Jones D.G. (ed.) *Exploitation of Micro-organisms*. Chapman & Hall, London, UK, pp. 1–17.

TeBeest, D.O., Yang, X.B. and Cisar, C.R. (1992) The status of biological control of weeds with fungal pathogens. *Annual Review of Phytopathology* 30, 637–657.

Templeton, G.E. (1992) Regulatory encouragement of biological weed control with plant pathogens. In: Charudattan, R. and Browning, H.W. (eds) *Regulations and Guidelines: Critical Issues in Biological Control*. Institute of Food and Agricultural Sciences, University of Florida, Gainesville, USA, pp. 61–63.

Templeton, G.E., TeBeest, D.O. and Smith, R.J. (1979) Biological weed control with mycoherbicides. *Annual Review of Phytopathology* 17, 301–310.

Tisdell, C.A. (1990) Economic impact of biological control of weeds and insects. In: Mackauer, M., Ehler, L.E. and Roland, J. (eds) *Critical Issues in Biological Control*. Intercept Press, Andover, UK, pp. 301–316.

Trujillo, E.E. (1985) Biological control of hamakua pa-makani with *Cercosporella* sp. in Hawaii. In: Delfosse, E.S. (ed.) *Proceedings, Sixth International Symposium on Biological Control of Weeds*. Agriculture Canada, Ottawa, Canada, pp. 661–671.

Wapshere, A.J. (1974) A strategy for evaluating the safety of organisms for biological weed control. *Annals of Applied Biology* 77, 201–211.

Wapshere, A.J. (1989) A testing sequence for reducing rejection of potential biological control agents for weeds. *Annals of Applied Biology* 114, 515–526.

Wapshere, A.J., Delfosse, E.S. and Cullen, J.M. (1989) Recent advances in biological control of weeds. *Crop Protection* 8, 227–250.

Watson, A.K. (1991) The classical approach with plant pathogens. In: TeBeest, D.O. (ed.) *Microbial Control of Weeds*. Chapman & Hall, New York, pp. 3–23.

Safety in the Laboratory

B.J. Ritchie

CABI Bioscience UK Centre, Bakeham Lane, Egham, Surrey TW20 9TY, UK

General

A microbiology laboratory can be a hazardous environment if basic safety precautions, good practice and common sense are not applied when engaged in any kind of laboratory work. A clean and properly maintained laboratory shows a professional attitude towards the work in progress and to the well-being of colleagues. Laboratory benches and other surfaces should be kept clean and tidy, as should fridges, incubators and culture rooms. It is preferable to clean up after each stage of an operation, as overcrowding leads to accidents. Personnel and equipment are valuable; ensure neither are put at risk. Familiarize yourself with any new equipment or chemical before use; do not wait until a problem arises before seeking advice. Smoking, drinking and eating must not be permitted in laboratories, where there is a high risk of inhaling aerosols of fungus spores or ingesting chemicals. Protective clothing should be worn when and where appropriate; if you are not sure what is required for a task, seek advice and instruction from a capable or qualified colleague. Always wear laboratory coats when engaged in laboratory work and launder them regularly; a dirty laboratory coat is unsightly and a health hazard. Remember – always wash your hands before leaving the laboratory. Do not take risks with your own health and safety or that of others around you.

Equipment

Equipment should only be used when training in its correct use has been given. Expensive microscopes are easily damaged if used incorrectly or stored in poor environmental conditions. Autoclaves and other pressure vessels can explode or cause severe scalding if used inappropriately by untrained personnel. Many modern techniques require sophisticated and expensive apparatus; if available, make use of the training offered by suppliers of such equipment. Apparatus not in use should be returned to its proper storage place in a clean and working condition. Never use equipment where the safety guards or covers have been removed. Do not attempt repairs on electrical equipment while it is still connected to the electricity supply.

Chemicals

Chemicals should be stored in a manner that ensures their purity and the safety of laboratory personnel and the environment. Safety data sheets are supplied with chemicals; ensure all users of each substance are aware of any potential dangers. Maintain an inventory of chemicals and reagents; this ensures that expensive materials are not bought unnecessarily because staff are unaware of what has already been purchased and is available for use. All chemicals should have the laboratory receipt date written on the label; this enables a check to be maintained on 'shelf-life'. Laboratory reagents and chemicals should be placed on the appropriate storage shelves immediately after use with their labels to the front. Vessels containing unlabelled and unidentifiable liquids or solids must not be left to others for disposal. Flammable liquids must be stored in a metal cabinet that has a containment floor; this allows spills to be contained and in case of fire, will prevent flammable liquids fuelling and spreading a fire. Keep only small amounts, no more than 250 ml, of such liquids on the bench for everyday use. Chemicals with noxious fumes require storage in a cabinet that is vented outside the building. Work with such chemicals should only be carried out in a fume hood; if a fume hood is not part of the laboratory equipment consideration must be given as to whether the task should be performed at all!

Many reagents and stains have been formulated and used over a long period of time. Within the last few years laboratory workers have become much more aware of the hazards of using some of these time-honoured recipes. An example is lactophenol; this laboratory reagent has been used (with or without a stain) by generations of mycologists, pathologists and nematologists. It has now been realized that phenol is a suspected carcinogen and prolonged contact over many years is hazardous to health. Some of the techniques where lactophenol was formerly used have been modified to eliminate phenol or, where no appropriate substitute has been devised, to limit its use.

Culture Media and Stains

Media and stain recipes can contain hazardous substances, e.g. antibiotics and fungicides. Care must be taken when preparing substances for incorporation in media; follow all safety precautions and if required use fume hoods to avoid contaminating the laboratory, and protective equipment to avoid inhaling or ingesting potentially hazardous materials. If a recipe instructs incorporation of ingredients in a specific order, do so to avoid unstable reactions occurring. Stains prepared in the laboratory should be dated (this enables a check to be maintained where limited 'shelf-life' is important) and labelled with appropriate warning signs to remind users of potential hazards.

Plant Material

Plant material must be examined as soon as possible after receipt, especially if isolations are necessary. Material that is retained on a temporary basis should be stored in trays or paper bags and not left loose on benches. Remember to keep plant material, a potential source of mite infestation, separate from axenic cultures. Dispose of old and unwanted plant material on a regular basis. To ensure exotic plant pathogens are not released into a new environment, it is essential that plant material of a non-indigenous nature be destroyed either by autoclaving or thorough incineration when it is no longer required for examination.

Quarantine Facilities

Quarantine laboratories or glasshouses require strict rules for operation. Only personnel with requisite training should be allowed to use the facility. Sterilizing equipment and chemicals must be available within a quarantine facility. These provisions enable destruction of potentially hazardous microorganisms, pests and exotic plant species and thus prevent them escaping into a new environment.

Containment Facilities

Culture work can be carried out in cabinets. The two most commonly used are laminar flow benches and microbiology safety cabinets. Laminar flow or 'clean air' benches provide a gentle flow of filtered air into a partly enclosed chamber or hood and must only be used for tasks that will not create aerosols of microorganisms, e.g. plate-pouring, tissue culture, subculture of non-sporulating microorganisms. Class I or class II microbiological safety cabinets can be used for subculturing sporulating microorganisms or when engaged in techniques that generate aerosols. If these are not available, an area of bench can be partly screened off using a framework covered with polythene and the area kept very clean by swabbing with industrial alcohol, 70% ethanol or 4% sodium hypochlorite.

No type of cabinet or hood should be used as a substitute for good aseptic practice; by maintaining clean work areas and cultivating good technique it is possible to work on the laboratory bench. Those who complain of contamination should examine their working practices very carefully and endeavour to improve.

Inappropriate use of cabinets can lead to laboratory contamination and health problems amongst staff. It is amazing that in an age of information overload, there are still those who have not grasped the correct usage of the various cabinet types available.

Plant Pathogenic Fungal and Bacterial Cultures

Plant pathogenic fungi and bacteria would not normally pose a threat to humans; however, it must be remembered that many fungi are opportunistic and widespread saprophytes, and their full potential to effect human health is rarely known. Fungi and other microorganisms are able to enter the body through the mouth, lungs, broken or unbroken skin and the conjunctiva. In a laboratory environment the hazards increase; microorganisms are grown in vast quantities and subculturing or other manipulation increases the risk of infection if good techniques and standards of hygiene are not maintained. Those most at risk are personnel on chemotherapy or long-term antibiotic treatment or those with an immune deficiency disorder. Plant inoculation techniques can directly introduce abnormal quantities of microorganisms directly into the human body. A slip with a scalpel or inoculating needle can miss the plant tissue completely and the pathologist can be inoculated instead! Spray inoculation, if carried out carelessly, can cause aerosols containing high concentrations of propagules to drift over the operator allowing potentially harmful spores to enter the body via the lungs or conjunctiva.

Risks increase when harvesting from bulk cultures or large quantities of spores. Fungi especially can cause allergic reactions and mycotoxicoses (poisoning). Avoid direct contact with the fungi and any materials and containers in which they have been grown or been in contact with. Good technique and standards of hygiene are essential.

Contamination of Cultures

Contamination of cultures can be a problem if laboratories in which fungal cultures are prepared or stored are not kept clean and dust-free. All sources of debris on which fungi might sporulate should be removed regularly. Adequate ventilation is desirable in warm countries but this brings with it an increased risk of contamination by airborne spores.

Mites can cause major problems by feeding on fungal cultures and causing contamination and will rapidly move from plate to plate in an incubator and can cause a major outbreak in 2 or 3 days if undetected. Mites can be seen in fungal cultures as white objects, just detectable to the naked eye. Ragged colony margins or growth of contaminant fungi or bacteria forming trails may denote their presence. They thrive in warm temperatures and high humidity and survive on organic detritus. General hygiene and prevention are better than having to control an outbreak. Examine all new material when it enters the laboratory; if possible have separate areas or rooms for clean and dirty material. Plant material should not be kept in the same incubator as cultures. Old surplus plant material and cultures should be disposed of quickly and safely by autoclaving or incinerating.

Culturing, Preservation and Maintenance of Fungi

D. Smith

CABI Bioscience UK Centre, Bakeham Lane, Egham, Surrey TW20 9TY, UK

General Principles

Collectively fungi are ubiquitous and grow in a wide variety of environments utilizing a vast array of substrates both natural and artificial. Some are host specific or have growth requirements that prevent them being grown in culture. Generally, fungi grow best on media that are formulated from the natural materials from which they were isolated. Optimal growth conditions are required but the avoidance of selection of variants from within the population, strain deterioration and contamination are essential when maintaining cultures in the longer term.

There are several methods of maintaining a culture collection of fungi and all aim to retain them in the condition in which they were at the time of isolation. The methods adopted depend on the number of cultures to be kept and the time and facilities available. The methods described here are used at CABI Bioscience UK Centre and provide information on the culturing, preservation and maintenance of fungi.

Growth

The growth requirements for fungi may vary from strain to strain, although cultures of the same species and genera tend to grow best on similar media. However, the majority of fungi can be maintained on a relatively small range of media although some fungi deteriorate when kept on the same medium for prolonged periods, so different media should be alternated from time to time.

©CAB *International 2002. Plant Pathologist's Pocketbook*
(eds J.M. Waller, J.M. Lenné and S.J. Waller)

Experience at CABI *Bioscience* is that cultures grow more satisfactorily on media freshly prepared in the laboratory, especially natural media such as vegetable decoctions, which are usually easy and relatively cheap to prepare with limited facilities. Small quantities can be sterilized using a domestic pressure cooker and, if necessary, the pH can be adjusted using drops of hydrochloric acid or potassium hydroxide and measured using pH papers. However, proprietary media are often useful and can be very important in replicating work of others. Some media for special purposes such as assay work will require very careful preparation. See Chapter 38 for further information on media and their preparation.

A wide range of media is used and personal preference for growth on particular media develops through experience. The standardization of media formulae is necessary for most work (Smith and Onions, 1994) as small variations can affect colony morphology and colour, whether particular structures are formed and may affect the retention of properties. Many fungi thrive on potato dextrose agar (PDA), but this can be too rich, encouraging the growth of mycelium with ultimate loss of sporulation, so a period on potato carrot agar (PCA), a starvation medium, may encourage sporulation. All sorts of vegetable decoctions are possible and apart from the advantages of standardization it is reasonable to use what is readily available, e.g. yam media might be preferable to potato media in the tropics. The introduction of pieces of tissue, such as rice, grains, leaves, wheat straw or dung, often produces good sporulation. Plant pathogenic organisms are sometimes difficult to culture and may respond to added sterile host tissue or media produced from host plant extracts.

Certain fungi will only grow or sporulate in the presence of another organism, often another fungus. If pure cultures are required, it is sometimes satisfactory to grow the commensal and kill it off by boiling in a water bath, and re-inoculate the agar with the principal.

Temperature

The majority of filamentous fungi are mesophilic, growing at temperatures within the range of 10–35°C, with optimum temperatures between 15 and 30°C. Some may also be thermotolerant whereas others (e.g. a wide range of *Fusarium* and *Penicillium* species) are psychrotolerant, but they can still grow within the range of 20–25°C. A small number (e.g. *Chaetomium thermophilum*, *Penicillium dupontii*, *Thermoascus aurantiacus*) are thermophilic and will grow and sporulate at 45°C or higher, but fail to grow below 20°C. A few fungi (e.g. *Hypocrea psychrophila*) are psychrophilic and are unable to grow above 20°C. Generally fungi are grown in the laboratory between the temperatures of 20 and 25°C.

Light

Many species grow well in the dark, but others prefer daylight and some sporulate better under near ultraviolet light (see below). Most leaf- and stem-inhabiting fungi are light sensitive and require light stimulation for sporulation. As far as possible similar conditions to those found in nature should be reproduced, e.g. fungi growing on exposed stems and leaves prefer direct sunlight. At CABI *Bioscience* most cultures are grown in glass-fronted incubators to allow in the light, or in illuminated incubators. Some fungi are diurnal and require the transition from periods of light to dark to sporulate.

Fungi that require near ultraviolet light (wavelength 300–380 nm) for sporulation must be grown in plastic Petri dishes or plastic universal bottles for 3–4 days before irradiation. Glass is not suitable, as it is often opaque to ultraviolet light. Nutritionally weak media such as PCA are more suitable for inducing sporulation as rich media promotes vegetative growth. At CABI *Bioscience* three 1.22 m fluorescent tubes (a near ultraviolet light tube, Phillips TL 40 W/08, between two cool white tubes, Phillips MCFE 40 W/33) are placed 130 mm apart. A time switch gives a 12 h on/off cycle. The cultures are supported on a shelf 320 mm below the light source and are illuminated until sporulation is induced (Smith and Onions, 1994).

Aeration

Nearly all fungi are aerobic and, when grown in tubes or bottles, obtain sufficient oxygen through cotton wool plugs or loose bottle caps (which should not be screwed down tightly during the growth of cultures). A few aquatic *Hyphomycetes* require additional aeration, by bubbling sterile air through liquid culture media for example, to enable normal growth and sporulation to occur.

pH

Filamentous fungi vary in pH requirements. Most common fungi grow well over the range pH 3–7, although some can grow at pH 2 and below (e.g. *Moniliella acetoabutans*, *Aspergillus niger*, *Penicillium funiculosum*).

Water activity

All organisms need water for growth, but the amount required varies widely. Although the majority of filamentous fungi require high levels of available water, a few are able to grow at low water activity (e.g. *Eurotium* species, *Xeromyces bisporus*).

Culture of *Basidiomycetes*

Spore germination

Basidiospores from fresh sporophores will often germinate on 5% malt extract agar at 25°C. Spores from dry untreated herbarium material can often be revived by incubating in a darkened, saturated atmosphere at 27°C, e.g. by leaving overnight in a 'moist chamber' Petri dish. The spores are then agitated in sterile water and plated out (Watling, 1963).

Sporophore production

1.   Pack a boiling tube to within 2.5 cm of the plug with sawdust containing 5% maize meal and 2 g bone meal. Moisten and autoclave for 20 min at 121°C. Inoculate the sawdust and incubate until a sporophore stipe grows up to the mouth of the tube.

   From a wood block about 7.5 cm³, cut out a shallow disk of the same diameter as the mouth of the tube and autoclave at 121°C for 45 min. Remove the plug from the tube and insert the mouth of the tube into the hole in the wood block. Cover with a cold sterilized polythene bag (turned inside out) and secure with a rubber band.

2.   Inoculate sawdust mixture as above. When the sporophore forms round the plug, suspend the tube horizontally over a water bath with cotton wool round the neck acting as a wick to maintain high humidity.

3.   A pad of shredded block paper pulp, 10 cm in depth, previously soaked in water and partially squeezed dry, is placed into a small milk bottle or similar container and covered with Lange's nutrient agar containing 0.5% malt extract. The agar should be 1–2 cm in depth at the bottom of the bottle with a thin coating left over the pulp.

Also see Badcock (1943), Tamblyn and Da Costa (1958) and Watling (1963).

Culture Mites: Cleaning and Mite Prevention

Mites (mostly *Tyroglyphus* and *Tarsonemus* spp.) are a very common cause of problems with fungal cultures most of which are susceptible to infestation. These occur naturally in soil and on organic material and can therefore enter the laboratory in a variety of ways. Mites damage cultures by eating them (a heavy infestation can completely strip the colonies from an agar plate) and by contamination through spores and bacteria carried on and in their bodies from external sources. The mites commonly found associated with fungal cultures are about 0.25 mm in length, just visible as tiny white dots to the naked eye, so infestation can easily go undetected. Given favourable conditions of high humidity and temperature they breed rapidly and spread quickly. Many cultures can be infested before they are noticed. Infested cultures have a deteriorated look and this is often the first indication of their presence.

General hygiene and preventative precautions should be used starting with the examination of all material entering the laboratory; a separate room for checking and processing dirty material is desirable. The sealing of incoming cultures, storage in a refrigerator or some form of screening and quarantine system can be helpful, as it is possible for cultures with only a light infestation at the time of receipt to develop a heavy infestation later.

Hygiene

All work surfaces must be kept clean and cultures protected from aerial and dust contamination, for example by storage on protected shelves. The work benches and cupboards should be regularly washed with an acaricide, especially as soon as infestation is suspected. At CABI *Bioscience*, surfaces are washed down with Actellic 25EC, left for sufficient time for it to have an effect (at least 15 min, often left overnight) and then cleaned off with alcohol. Actellic is not noticeably fungicidal but is of moderate toxicity and irritating to skin therefore all contact must be avoided. In the concentration at which it is used (3% v/v: 30 ml of stock to 1 l distilled water) it is much less toxic. Plastic gloves and a vapour filter mask should be worn when handling it. As mites appear to become resistant to some chemicals the acaricide should be changed from time to time. Mite-infested cultures should be removed immediately and if possible sterilized, and all cultures in the immediate area should be checked and isolated from the rest. Carry out mite cleaning operations at the end of the working day wherever possible.

Fumigation

This method is used as a last resort or when moving into new premises and it should be carried out by a licensed specialized company. Where large numbers of important cultures or specimens are involved these items can be fumigated off site in specialized equipment. Unaffected cultures should be removed or protected as fumigation generally involves the use of chemicals that are toxic to fungi.

Mechanical and chemical barriers

The culture bottles, tubes or plates can be stood on a platform surrounded by water or oil, or on a surface inside a barrier of petroleum jelly or other sticky material to prevent access by crawling mites. At CABI *Bioscience* universal bottles and cotton wool-plugged tubes are sealed by the cigarette paper method. The pores of the paper allow free passage of air but prevent the passage of mites. Care must be taken to see that a good seal is made, and that the paper is not damaged through handling.

1.   Cut cigarette papers in half and sterilize in an oven at 180°C for 3 h.

2. Stick a cigarette paper on to the mouth of the universal bottle or tube after the plug has been pushed down using copper sulphate gelatin glue (20 g gelatin is dissolved in 100 ml water and then 2 g copper sulphate is added).

3. Burn off the excess cigarette paper up to the outer edge of the tube or bottle.

Sealing of culture containers such as Petri dishes and universal bottles with sticky tape such as 'Sellotape' or 'Scotch' tape may reduce penetration but will not act as a complete barrier. Mites eventually find their way through cracks and wrinkles.

Protected storage

The various methods of long-term storage of cultures used in culture collections prevent infestation and spread of mites. Mites generally cannot penetrate cultures preserved under mineral oil or stored in silica gel in sealed tubes or in bottles with the caps screwed down tight. Freeze-dried ampoules, being completely sealed, are impermeable to mites and they cannot penetrate ampoules stored at the ultra-low temperatures of liquid nitrogen.

For cultures in regular use, cold storage at 4–8°C definitely reduces the spread of mites but does not kill them. Storage of infested cultures in the deep freeze (< –20°C) for at least 3 days gives better control. The cultures usually remain viable, whereas the mites are usually killed. The fungus will have to be re-isolated from the original culture as the contaminants introduced by the mites will eventually grow.

Culturing in the Tropics

Contamination of culture media during isolation, sub-culturing, etc. is more of a hazard in the tropics than in temperate areas partly due to more dust and spores in the air and partly due to greater air movement from open windows, fans, air conditioners, etc. This difficulty can largely be overcome by working in a small room set aside for culturing where dust and draughts are kept to a minimum. Ideally the room should be easily surface sterilized, for example using steam or UV light before and after use. Before use the bench and floor should be washed down with disinfectant and about 15 min before operations are begun the air in the room may be sprayed with a mild disinfectant containing thymol to drop any floating spores.

Small inoculating hoods or chambers may be constructed to provide an enclosed area for transferring cultures. Such structures can be sterilized before and after use with disinfectants or irradiated. They are not usually necessary when tube cultures are used but are helpful when dealing with Petri dishes.

Suitable conditions for making contaminant-free primary isolations from diseased plant material can be improvised in an ordinary room using damp muslin or cheesecloth draped over windows and other openings. The first consideration is to avoid draughts.

Purification of Fungus Cultures

Streaking

1.   This, in its simplest form, consists of taking a small portion of fungus on a culture loop and streaking it over an agar surface in a Petri dish. As the streak progresses the spores become more and more separated till finally individual colonies, arising from few or single spores, are obtained.
2.   A suspension of spores is made in sterile water and this is streaked out on agar. By this means there is a better chance of obtaining single-spore colonies.

Dilution plates

Dilutions in sterile water

An inoculum of spores is placed in a tube of sterile water, say 10 ml. A clean sterile pipette is used to transfer an aliquot, e.g. 1 ml, of this to a tube with 9 ml of sterile water. A fresh sterile pipette is used to mix this and transfer 1 ml of this to another tube containing 9 ml of sterile water. This is continued for as many dilutions as are required. The concentration of the original inoculum can be counted by using a haemocytometer or similar device and the dilutions required to give adequate separation of spores, e.g. 5–20 per ml, made. The above are decimal dilutions, but the quantities taken over are a matter of preference. In the final dilution 1 ml of spore suspension is added to 15–20 ml of melted agar cooled to about 40°C and poured into a sterile Petri dish.

Tube dilution

Five tubes of suitable agar medium are taken, melted and allowed to cool to 45°C. One tube is taken and inoculated with a small amount of spores, soil or other material, rolled between the hands, and a plate poured. The empty tube is refilled from another tube of molten medium, rolled between the hands, and a second plate poured. This latter procedure is repeated successively with the remaining three tubes of medium.

Single spore isolation

The various methods of isolation of single microorganisms can be divided into semi-mechanical and mechanical. Filtered agar is recommended when microscopic examination for spores is involved.

Semi-mechanical methods

A rather sparse spore suspension is made on a sterile slide in sterile water by dipping a flamed, moistened loop into a sporulating culture. This spore suspension is streaked along a marked line on a very thin plate of tap-water agar and incubated overnight at about 24°C. The plate is then scanned with a stereoscopic microscope along the line and suitable germinating spores selected. Using a Borrowdale needle for preference, a cut about 2 mm square is made in the agar around a selected spore. This square and the area immediately around it are examined under the low power of a compound microscope to ensure that only one spore is involved. The agar block and germinating spore are then transferred by means of the sterile needle to an agar plate or tube.

Mechanical methods

Various cutters have been devised that can be fixed to the nosepiece of a microscope. One method (Onions *et al.*, 1981) is to use a dummy microscope objective in which the front lens is replaced by a sharp-edged metal tube, about 5 mm long and 1.5 mm diameter. This should be fairly accurately centred with the low-power objective. After locating and centring a spore under the low power, the dummy objective is then swung into place above it. The agar is cut by lowering the cutter until it reaches the bottom of the dish. Upon raising the cutter, the cylindrical block is left behind and lifted out with a needle. Another method involves drawing up a spore suspension in warm nutrient agar into fine glass capillary tubes, which can be examined microscopically and broken up into pieces each containing one spore. The pieces can then be surface sterilized and placed on an agar plate. Elaborate mechanical dissecting microscopes are available for more critical work (El-Badry, 1963).

Maintenance of a Culture Collection

The primary objective of keeping a fungus resource collection is to maintain strains in a viable state without morphological, physiological or genetic change. An important task of a collection is its custodial duty to ensure the microbial gene pool is available for research and development. Ideally complete viability and stability should be achieved, especially for important research and industrial isolates. These are of increasing importance for biotechnology and bioengineering. However, even modest teaching or research collections may have to consider additional factors such as simplicity, availability and cost.

Preservation techniques range from continuous growth through methods that reduce rates of metabolism to the ideal situation where metabolism is halted. Methods that allow growth and reproduction may allow the organism to change and adapt to laboratory conditions. Many fungi produce resistant or dormant structures that enable them to survive adverse conditions and these can be stored in the laboratory to retain viability of the organism. Healthy cultures grown under optimum conditions help to ensure this. In addition, a collection is only as good as the monitoring system to which it is subjected and collections should have ready access from the outset to specialists who know the morphological and physiological characters of the relevant fungi.

Guidelines to setting up a collection

Guidelines are outlined in the World Federation for Culture Collections (WFCC) booklet *Guidelines for the Establishment and Operation of Collections of Microorganisms* (Hawksworth *et al.*, 1990) and are discussed in principle here. It is essential that the aims and procedures of a collection are clear and well planned. Further information can be found in Hawksworth and Kirsop (1988) and Hawksworth (1985).

1. The objective of the collection must be clear, e.g. to provide a reference collection for an identification service, to support a particular research project, or as a reference collection of plant pathogens. This will enable a programme of accessioning to be established.

2. The range of organisms required must be identified. One collection will never be able to hold representatives of the 90,000 species so far described, let alone isolates of different specialized strains, from different sources, etc. What is already available can be determined through the world network of service collections; the World Data Centre for Microorganisms (WDCM), National Institute of Genetics, Japan (Sugawara *et al.*, 1993), is a good starting point to find this information.

3. The equipment and personnel available to run a well-maintained collection will be a major constraint on its eventual size.

4. Funding must be secure for the long term to ensure the continuity of the collection. Careful evaluation of this and point **3** above will help establish the number of strains that can be collected and maintained. Overstretching of resources can lead to a poor quality collection.

5. No matter what preservation and data storage facilities are used, a maintenance programme must be set up and this will be as time-consuming as collecting the strains.

6. Quality control procedures are needed to ensure viability, purity, and the identity of the strain preserved.

7. Information on strains must be verified.

A collection's aims and procedures

A collection should provide standard reference strains that must perform in specific ways. Procedures must be in place to ensure this.

1. A good collection requires good laboratory practice (GLP) and sound quality control (Smith, 1996). The WFCC guidelines (Hawksworth *et al.*, 1990) cover some of the necessary aspects. GLP guidelines developed by regulatory agencies have been discussed with reference to collections of microbial and cell cultures (Stevenson and Jong, 1992). Good techniques and procedures, management system and quality assurance carried out by qualified staff are needed (Smith *et al.*, 2001).

2. Adequate documentation must be maintained to enable each step of a procedure to be traced. This should include data on its isolation, source, dates and number of subculturings, etc.

3. Checks to confirm identities and properties should be recorded, and those staff who carried them out noted (Gams *et al.*, 1988). Creating an initial

preserved stock or seed bank that is not used for general supply but from which all subsequent re-preservations can be made avoids sequential culturing of the organism.

4. Collections must keep abreast of advancements made in technology and information management; these should be tried and proven before they are introduced into normal practice.

Acquisition of strains

Strains should be obtained within National Legislation governing access to genetic resources and within the spirit of the Convention on Biological Diversity (Davison *et al.*, 1998). Acceptance of a strain requires careful evaluation. Duplication should be avoided, new accessions should have unique properties of potential long-term interest and facilities should be available to handle them.

Accession

A unique accession number should be allocated to the strain, which should never be reassigned if the organism is later discarded. The viability, purity, identity, growth requirements and methods of maintenance and/or preservation of the strain must be determined and the information recorded against the accession number. These records can be kept by several means but should enable cross-referencing between the collection number and culture name. Initial hard copy data can be placed on a computerized database.

Preservation and Storage

There are many methods available for the preservation and storage of fungi. Continuous growth techniques involve frequent transfer from depleted medium to fresh medium providing optimum growth conditions, and there are also methods that delay the need to subculture. The latter involve storage on growth medium in the refrigerator, freezer ($-10°C$ to $-20°C$), under a layer of oil or in water. Drying, usually of the resting stage such as spores or sclerotia, can be achieved by air drying, in or above silica gel, in soil, and by freeze-drying. Metabolism can be suspended by reducing the water availability in the cells by dehydration or freezing so that there is no medium for life processes to function. The temperature of the frozen material must be reduced below $-70°C$ to achieve suspension of metabolism but to achieve conditions where no physical or chemical reaction can occur requires storage below $-139°C$ (Morris, 1981).

Wherever possible, an original culture should be preserved without subculturing and a *seed stock* should be stored separately from the *distribution stock*. It is also advisable to keep a *duplicate collection* in another secure building or site as a reserve.

After preservation, the viability, spore germination, purity and identity should be rechecked and compared with data recorded before preservation before the culture is made available outside the collection. The organism may be sent to the donor for confirmation of properties. Cultures from single spore isolations can be prepared to give a better chance of their being pure.

Preservation Methods

Growth on agar slants

Cultures are normally grown in test tubes or bottles on a nutrient medium, usually an agar gel with added nutrients. Many fungi can be maintained in this way for years but it is dependent on transfer from well-developed parts of the culture, taking care to ensure that contaminants or genetic variants do not replace the original strain. Most fungi can be grown on PCA or malt agar (MA) but some have specified growth requirements. Some fungi may need subculturing every 2–4 weeks, the majority every 2–4 months, though others may survive for 12 months without transfer.

The main disadvantages of frequent transfer are the danger of variation and loss of physiological or morphological characteristics, possible contamination by airborne spores or mites and the need for constant checking to ensure the culture remains true to type. The main advantages of frequent transfer are that collections can be kept viable for many years, the method is cheap and requires no specialized equipment and retrieval is easy.

Cultures may be stored at room temperature in a cupboard to protect the cultures from dust. However, they may dry out particularly rapidly in tropical climates and must be transferred to fresh media at least every 6 months. Table 37.1 compiles data from several sources giving temperature of storage and the period before transfer. The shelf-life given is conservative to ensure that the organisms remain viable; the majority of strains may survive longer periods if not allowed to dry out.

Storage at 4–7°C in a refrigerator or cold room can extend the transfer interval to 4–6 months from the average period of 2–4 months. Although Chu (1970) maintained some forest tree pathogens for 1 year at 5°C most organisms were best transferred after much shorter periods. Storage in a deep freeze (−7 to −24°C) will allow many fungi to survive 4–5 years between transfers, though freezing damage may occur in some.

Under mineral oil

Covering cultures on agar slants (30° to the horizontal) in 30 ml universal bottles with mineral oil prevents dehydration and slows down the metabolic activity and growth through reduced oxygen tension. This method is extensively used. Mature healthy cultures are covered by 10 mm of sterile mineral oil (liquid paraffin or medicinal paraffin specific gravity 0.830–0.890 sterilized by autoclaving twice at 121°C for 15 min). If the oil is deeper than 10 mm

Table 37.1. Shelf-life of some plant pathogenic fungi stored on agar at room temperature and in the refrigerator.

Fungus	Storage temperature (°C) and conditions	Period before transfer (years unless stated)	Reference
Allomyces	16	2–3 months	von Arx and Schipper (1978)
Alternaria bassicola	5	1	Kilpatrick (1976)
Armillaria	4–7	1	CABI *Bioscience*
Aspergillus	4	2	CABI *Bioscience*
	20–25	2–6 months	CABI *Bioscience*
Corticium	16	2–3 months	von Arx and Schipper (1978)
Forest pathogens	5	1	Chu (1970)
Mycena	16	2–3 months	von Arx and Schipper (1978)
Phytophthora	16	2–3 months	von Arx and Schipper (1978)
	4	1	Dick (1965)
Pythium	16	2–3 months	von Arx and Schipper (1978)
	4	1	Dick (1965)
Rhizopus and other low-temperature-sensitive fungi	16 (60% RH)	0.5	von Arx and Schipper (1978)
Zygomycota	4–7	1	CABI *Bioscience*
	20–25	1–2 months	CABI *Bioscience*

RH, relative humidity.

the fungus may not receive sufficient oxygen and may die, whereas if the depth is less, exposed mycelium or agar on the sides of the container may allow moisture to evaporate and the culture to dry out. At CABI *Bioscience* the universal bottles are stored with their caps loose in racks in a temperature-controlled (15–18°C) room.

Retrieval from oil is by removal of a small amount of the colony on a mounted needle, draining away as much oil as possible then streaking on to a suitable agar medium. Growth rate can often remain restricted due to adhering oil, so subculturing is necessary by re-isolating from the edge of the colony and transferring to fresh media. Inoculating an agar slope centrally sometimes gives better results as excess oil can drain down the slope allowing the fungus to grow more typically towards the top.

A wide range of fungi survive this method and some have remained viable for 40 years at CABI *Bioscience* (Table 37.2). However, many cultures deteriorate under mineral oil and must be transferred regularly at two-yearly intervals. Organisms that are sensitive to other techniques can be stored successfully in oil.

The disadvantages of oil storage are danger of contamination by airborne spores and retarded growth on retrieval. Continuous growth under adverse conditions could lead to selection of off-types. The advantages of oil storage are the long viabilities of some specimens, the survival of species that will not survive other methods of maintenance, it is generally inexpensive to operate and set up and there are no problems with mites. Preservation under oil is recommended for storage of fungi in laboratories with limited resources and facilities.

Table 37.2. Shelf-life of some plant pathogenic fungi stored on agar under mineral oil.

Fungus	Period before transfer (years)	Reference
Alternaria[a]	0.5	Sherf (1943)
Aspergillus (four strains)	20–32	Smith and Onions (1983)
Basidiomycetes (60 genera)	10	Kobayashi (1984)
Beauveria (three strains)	12	Little and Gordon (1967)
Botryosphaeria (two strains)	31	Smith and Onions (1983)
Botryosphaeria ribis	6	Smith *et al.* (1970)
Ceratocystis (three strains)	32	Smith and Onions (1983)
Corticium (three strains)	20–32	Smith and Onions (1983)
Drechslera portulacae	32	Smith and Onions (1983)
Fusarium[a]	0.5	Sherf (1943)
Fusarium oxysporum	1	Brezhneva and Khokhryakov (1971)
Fusarium (three strains)	12	Little and Gordon (1967)
Nectria pityrodes	32	Smith and Onions (1983)
Phytophthora (two strains)	32	Smith and Onions (1983)
Podospora fimbriata (two strains)	10–20	Smith and Onions (1983)
Pythium	3	Onions (1977)
Rhizoctonia (two strains)	32	Smith and Onions (1983)
Sclerotium coffeicola	32	Smith and Onions (1983)
Setosphaeria rostrata	32	Smith and Onions (1983)
Thielaviopsis basicola	12	Smith and Onions (1983)
Trichoderma harzianum	27	Perrin (1979)
Ustilago scitaminea	32	Smith and Onions (1983)
Verticillium dahliae	1	Brezhneva and Khokhryakov (1971)
Verticillium theobromae	32	Smith and Onions (1983)
Wood-inhabiting Basidiomycetes	27	Perrin (1979)
Various	15	Gutter and Bakai-Golan (1967)

[a]Retained pathogenicity.

Water storage

The method used at CABI *Bioscience* is as follows:

1. Agar blocks (6 mm³) are cut from the growing edge of a fungal colony.
2. The blocks are placed in sterile distilled water in McCartney bottles and the lids are tightly screwed down; they are stored at 20–25°C.
3. Retrieval is by removal of a block and placing, mycelium down, on a suitable growth medium.

Any growth during storage in water can be reduced if the spores or hyphae are removed from the surface of agar media and no medium is transferred.

Storage periods of 2–3 years have been obtained with species of *Phytophthora* and *Pythium* before any loss of viability was noted (Onions and Smith, 1984). There was some deterioration in pathogenicity but the majority were able to infect their host. Viability deteriorated rapidly after 2 years storage and 42% (21/50) of the isolates were dead at 5 years. Many workers have reported long-term viability of a wide range of plant pathogens for up to 10 years using this method (Figueiredo and Pimentel, 1975; Boeswinkel, 1976).

Summary of maintenance by growth techniques

For small collections, or for relatively short-term preservation, subculturing is an effective method of culture maintenance but contamination is a particular hazard. All the techniques permitting growth during preservation may allow loss of properties due to adaption to synthetic media or selection of variants. Storage of fungi in the refrigerator at 4–7°C slows down the rate of metabolism and increases the period between transfers to fresh media, almost doubling the shelf-life, but some fungi are sensitive to storage at these temperatures. One of the most important disadvantages of the mineral oil storage technique is that some fungi may adapt to it, or spontaneous mutants may grow better in the conditions provided. Some cultures grow better after recovery from oil than after maintenance by more frequent transfer techniques.

Drying and freeze-drying techniques

Most fungal spores have a lower water content than vegetative hyphae and are able to withstand desiccation. The removal of water reduces metabolism of the cell. It can be carried out by many techniques. Air drying is achieved by passing dry air over the culture or spores speeding up drying by evaporation. Drying can also be achieved by placing cells in or above an absorbant such as soil, silica gel, or other desiccant. A third way is by drying under vacuum from the frozen state or freeze-drying. Stability and long storage periods are the main advantages of freeze-drying though the expense of the modern and quite complex machinery can be a deterrent.

Drying

SILICA GEL STORAGE.   This method has proved to be very successful at CABI *Bioscience* where sporulating fungi have been stored for 7–18 years in silica gel with good revival and stability (Table 37.3).
    The following method is used at CABI *Bioscience*.

1.   One-third fill glass universal bottles with medium grain plain 6–22 mesh non-indicating silica gel and sterilize by dry heat (180°C for 3 h).
2.   Place bottles in a tray of water and freeze in a deep freeze (nominal –20°C).
3.   repare spore suspensions in cooled 5% (w/v) skimmed milk.
4.   Add the suspension to the cool gel to three-quarters wet it (approximately 1 ml) and agitate to distribute the spore suspension throughout.
5.   Store the bottles with the caps loose for 10–14 days at 25°C until the silica gel crystals dry and readily separate.
6.   Screw the caps down and store the bottles at 4°C (though storage between 20 and 25°C is quite satisfactory) in air-tight containers over indicator silica gel to absorb moisture.
7.   Retrieve the strains by scattering a few crystals on to a suitable medium.

    A wide range of sporulating fungi survive this method (Onions, 1977; Smith and Onions, 1983) but thin-walled spores, those with appendages and

Table 37.3.   Shelf-life of fungi stored on anhydrous silica gel.

Fungus	Period before transfer (years)	Reference
Arthroderma (19 strains)	3–5	Gentles and Scott (1979)
Ascomycota (53 strains)	8–11	Smith and Onions (1983)
Aspergillus	0.7	Grivell and Jackson (1969)
Epidermophyton floccosum (seven strains)	4	Gentles and Scott (1979)
Hendersonula toruloidea (five strains)	3.5	Gentles and Scott (1979)
Hymenomycetes (10 strains)	8–11	Smith and Onions (1983)
Mitotic fungi[a] (222 strains)	8–11	Smith and Onions (1983)
Nanizzia (10 strains)	4–5	Gentles and Scott (1979)
Zygomycota (20 strains)	8–11	Smith and Onions (1983)

[a]Fungi not linked to a perfect state.

mycelial cultures tend not to survive. Healthy sporulating cultures survive best. The technique is a medium-term storage method that can be used when freeze-drying is not available, though the range of fungi surviving and longevities are not as good.

Advantages of silica gel storage are that it is cheap and simple, it produces very stable cultures, there are no mite problems, and repeated inocula can be obtained from one bottle, though it is recommended that a stock bottle is kept in case contamination occurs during retrieval. Disadvantages are that it is limited to sporulating fungi, being unsuitable for *Oomycota*, mycelial fungi or fungi with delicate or complex spores, and there is the possibility of introducing contaminants by repeated retrievals.

Soil storage

At CABI *Bioscience* the method involves inoculation of double autoclaved soil (121°C for 15 min) with 1 ml of spore suspension in sterile distilled water and incubating at 20–25°C for 5–10 days depending on the rate of growth of the fungus being stored. This initial growth period allows the fungus to use the available moisture and gradually to become dormant. The bottles are then stored in a refrigerator (4–7°C). This method of storage is very successful with *Fusarium* species (Booth, 1971) and cereal pathogens (Shearer *et al.*, 1974; Reinecke and Fokkema, 1979). The initial growth may allow off-types to develop but despite this, soil storage should be used in preference to oil storage for the preservation of *Fusarium* species and other fungi that show variation under oil.

Advantages of storage in soil are: general stability of cultures; good survival; mite problems are unlikely; repeated inocula can be obtained from the same sample (although it is advisable to keep a stock culture for use only if contamination or other problems occur); and the method is inexpensive and easy. Disadvantages are that some variation may occur, fungi that cannot withstand desiccation do not survive well and there is potential for contamination when samples are retrieved.

Freeze-drying

This technique enables the retention of the shape, structure and activity of preserved biological materials and prevents the structural damage associated with most other drying techniques. There is a vast array of freeze-drying equipment on the market, ranging from laboratory bench models through pilot-scale plant to huge industrial installations. It is not essential to use expensive equipment to carry out freeze-drying successfully. The basic requirements for a system are described by Smith and Onions (1994).

Centrifugal freeze-drying, which relies on evaporative cooling, can be used successfully for the storage of many sporulating fungi (Smith, 1983), but using a shelf freeze-drier enables the optimization of cooling rate to suit the organism being freeze-dried. The heat sealing of glass ampoules or vials is preferred to butyl rubber bungs in glass vials as these may allow deterioration through leak over long-term storage. The technique is a most convenient and successful method of preserving sporulating fungi (Table 37.4) preventing contamination but retaining stability during storage, and is also ideal for the distribution of the organisms.

At CABI *Bioscience* both centrifugal freeze-drying and shelf freeze-drying are used. The latter technique can be adjusted to give cooling and warming rates suited to the individual fungus and is a one-stage process (Smith and Kolkowski, 1996). Some results are shown in Table 37.4. In general it is only the spores (e.g. conidia, ascospores and basidiospores) or resistant structures that tend to survive but Tan *et al.* (1991) successfully used trehalose as a protectant during drying and preserved hyphal colonies of some species; organisms containing melanin showed higher survival. General surveys of the process have been carried out by Jong (1978), Rowe and Snowman (1978), Alexander *et al.* (1980) and Smith (1983).

FACTORS AFFECTING SURVIVAL IN FREEZE-DRYING. The most critical points of the freeze-drying process are the selection of a suitable suspending medium, using optimal cooling rates, maintaining the frozen state during drying, retaining a residual water content of between 1 and 2% after drying and the avoidance of rehydration and contact with oxygen during processing and storage.

Suspending medium. Skimmed milk and inositol have been used successfully for many years at CABI *Bioscience* but successful use of other chemicals has been reported.

Cooling rate. Slow cooling followed by a slow drying process works best for the majority of fungi (Smith, 1986). A cooling rate of 1°C min^{-1} has proved best for the cryopreservation of fungi (Hwang, 1966).

Monitoring the frozen state. To achieve high survival levels, it is important that the temperature of the frozen material is kept below −15°C until the water content is reduced to < 5% (Smith, 1986).

Residual moisture. The residual moisture of the freeze-dried material must be prevented from falling below 1% (Smith, 1986) as the removal of structural

Table 37.4. Shelf-life of some plant pathogenic fungi stored after freeze-drying.

Fungus	Method	Inoculum processed	Storage temp. (°C)	Period before transfer (years)	Reference
Ascomycota (92 strains)	Prefrozen	Spores	4	2–19	Butterfield et al. (1974)
Aspergillus flavus	Vacuum cooling	Conidia	22–32	15	Rhoades (1970)
Aspergillus flavus	Vacuum cooling[a]	Conidia	6–7	20	Rhoades (1970)
Aspergillus fumigatus	Vacuum cooling	Conidia	6–7	20	Rhoades (1970)
Aspergillus (257 strains)	Prefrozen	Spores	4–8	2–19	Ellis and Roberson (1968)
Mitotic fungi[b] (> 400 strains)	Prefrozen	Spores	4	2–19	Butterfield et al. (1974)
Gliocladium (11 strains)	Prefrozen	Spores	4–8	2–19	Ellis and Roberson (1968)
Melampsora medusae	Prefrozen	Uredospores	1–2	5	Shain (1979)
Puccinia hordei	Vacuum cooling	Uredospores	2	1.25	Clifford (1973)
Puccinia[c]	Prefreeze	Uredospores	[d]	1	Sharp and Smith (1952)
Pyrenochaeta	Vacuum cooling	Mycelium	10	3	Last et al. (1969)
Rhizopus (three strains)	Prefreeze	Spores	4	1–19	Butterfield et al. (1974)

[a]Retained mutant characteristics; [b]fungi not linked to a perfect state; [c]retained pathogencity; [d]not given.

water may cause irreparable damage. If insufficient water is removed initial viability and stability may be good but rapid deterioration in storage occurs.

Storage condition. Once dried, ampoules can be stored at room temperature. Storage within the temperature range of –20°C to –70°C may reduce the rate of any deterioration but appears unwarranted. Heat-sealed ampoules or vials with reduced pressure or back filled with an inert gas will prevent contact with oxygen or water, which will cause deterioration.

Rehydration. This is an important factor affecting retrieval. Rehydration for 24 h in 0.1% peptone has proved to be a useful technique for reviving sensitive fungi, although 30 min rehydration in liquid medium is usually adequate.

Some rare problems with freeze-drying can occur such as the loss of morphological and physiological characteristics. Sometimes low survival rates suggest the selection of resistant cells, and genetic damage has been observed

(Heckly, 1978) although it is difficult to differentiate between this and selection of spontaneous mutants by freeze-drying. This should be avoided with careful control during freezing and prevention of overdrying.

Advantages of freeze-drying are that total sealing of the specimen protects it from contamination and infestation, it has a very long shelf-life, many isolates can be retained in a very stable condition and ampoules are easily stored and distributed. Disadvantages are that some isolates fail to survive the process, there is a low percentage viability and genetic damage may occur, therefore it is not usually suitable for starter cultures, and the process is complex and may be expensive, at least requiring a vacuum system.

Cryopreservation

Metabolism in biological material is suspended when all internal water is frozen. Although little metabolic activity takes place below –70°C, recrystalliz- ation of ice can occur at temperatures above –139°C (Morris, 1981) that can cause damage during storage. Consequently, the storage of microorgansims at the ultra-low temperature of –190°C to –196°C in or above liquid nitrogen is the best preservation method currently available (Smith, 1998). No morphological or physiological change has been observed in the 7354 isolates representing 3000 species belonging to over 695 genera stored at CABI *Bioscience* (Smith and Onions, 1994). The shelf-life is often considered as infinity: work with animal cells has estimated 32,000 years before a lethal background dosage of radiation is reached (Ashwood-Smith and Grant, 1976). Freezer storage at –80°C has been successfully used for the shorter term giving *c.* 15 years storage (Yukio, 1987; Ito, 1991).

Method

1. Prepare suspensions of fungus cells in sterile 10% (v/v) glycerol and dispense in 0.5 ml aliquots into 2 ml borosilicate glass ampoules or 2 ml polypropylene cryotubes labelled with the strain number using a permanent ink marker.
2. Place the heat-sealed glass ampoules in an erythrocin B dye bath to check for leakage. A period of at least 1 h is allowed to elapse to allow the cells to equilibrate in the glycerol.
3. Cool the ampoules or cryotubes at a suitable rate in a suitable programma- ble cooler (e.g. KRYO 10/16 series II, Planer Products Ltd). A rate of –1°C min^{-1} will allow most fungi to survive but will not give optimum recovery for all. The rate is controlled over the critical period from +5°C down to a temperature of –50°C. The initial cooling rate to 5°C is not critical and this is normally at –10°C min^{-1}.
4. When the frozen suspensions reach –50°C transfer into a 320 l liquid nitrogen storage vessel where cooling down to the final storage temperature is completed either in the liquid or vapour phase. Storage in cryotubes in a drawer rack system allows easy retrieval of such organisms.

For recovery and checking, thaw one ampoule or cryotube after four days by placing it in a circulatory water bath at +37°C or into the chamber of the cooler using a suitable thawing programme. Remove the ampoules when the last ice has melted. Do not allow the suspensions to reach the temperature of the water bath or a high chamber temperature in the cooler. Open the ampoule by scoring the preconstriction with a glass file and snap it open in a microbiological safety cabinet. Alternatively unscrew the cryotube lid and in both cases streak the contents on to a suitable growth medium.

Fungi difficult or impossible to grow in culture can be stored by this method. Rust and smuts (Kilpatrick *et al.*, 1971; Prescott and Kernkamp, 1971) and *Sclerospora* species (Gale *et al.*, 1975; Long *et al.*, 1978) have been successfully preserved by liquid nitrogen storage techniques.

Observations with a the cryogenic light microscope and subsequent viability and stability tests have enabled optimum preservation protocols to be established for some fungi (Coulson *et al.*, 1986; Smith *et al.*, 1986; Morris *et al.*, 1988; Smith and Thomas, 1998). No one cryopreservation protocol will give optimum recovery of every fungus and can be quite different for strains of the same species.

Advantages of liquid nitrogen storage are that the fungi are free of contamination in sealed ampoules, the majority of both sporulating and non-sporulating fungi survive well, and shelf-life is considered to be limitless if storage temperature is kept below –139°C. Disadvantages are that the apparatus is costly. Supply of liquid can prove expensive and if it fails the whole collection can be lost; this is a major drawback for using the technique in warmer climates and developing countries. The storage vessels must be kept in a well-ventilated room, as the constant evaporation of the nitrogen gas could displace the air and suffocate workers. The double-jacketed, vacuum-sealed storage vessels have been known to corrode and rupture. Such a failure may result in the loss of all the strains stored in it.

Selection of Preservation Techniques

The choice of method of preservation depends on the numbers and range of fungi to be preserved and the facilities available. The cost of materials and labour involved and the desired level of stability and longevity required must also be taken into consideration. Although the response of fungi is often strain specific it is often possible to predict how certain groups of fungi might be best preserved.

Oomycota

These fungi are best stored in liquid nitrogen using a cooling rate of *c.* –10°C min^{-1}, although some strains do not survive the freezing stages. Storage under mineral oil may be satisfactory for periods of up to six months and cultures can be kept viable in water storage and transferred every 2 years. These techniques may not retain particular properties of the fungi, but can be used if stability is not a priority or as a back-up to liquid nitrogen storage. Drying techniques cannot be used successfully for this group.

Zygomycota

Liquid nitrogen storage is recommended for the preservation of fungi of the Zygomycota such as *Mucor, Rhizopus* and similar genera. Most isolates can be successfully freeze-dried and remain viable for many years.

Ascomycota

The majority of Ascomycetes that can be grown in culture can be freeze-dried or cryopreserved in liquid nitrogen but those that do not sporulate well in culture survive dehydration techniques poorly. However, most of these have survived long-term storage in liquid nitrogen. Fewer species survive in silica gel although healthy heavily sporulating strains generally survive well for periods of over 8 years.

Basidiomycota

Except for yeasts, these generally grow only as mycelium in culture and can only be preserved by serial transfer on agar with or without oil, or stored in liquid nitrogen, which is the best technique for Basidiomycetes. Those fungi producing thick-walled hyphae can be freeze-dried but their viabilities are usually low. However, basidiospores harvested from fungi growing in their natural environment can usually be freeze-dried, and will survive other preservation techniques better than the mycelium does. Mycelium of wood-inhabiting Basidiomycetes can be grown and maintained on wood chips in cryopreservation below −139°C.

Rusts

Collections can be maintained in good condition in liquid nitrogen on the host or as harvested spores. *Ustilaginales* produce very disappointing cultures and survive best in liquid nitrogen though it is possible to keep them by other means. If the spores can be harvested successfully some survive freeze-drying quite well.

Mitotic fungi

Conidial fungi are relatively easy to preserve, freeze-drying being the most widely used technique but liquid nitrogen storage is the most successful technique for this group. Soil storage has been used extensively for *Fusarium* strains, some retaining viability for up to 20 years. Silica gel can also be used for the preservation of many conidial fungi, although it is less successful for Hyphomycetes.

Yeasts

The single-celled vegetative yeasts survive freeze-drying well. Most species survive silica gel storage but recovery is usually very low. Liquid nitrogen is the best preservation technique. Those that produce ascospores behave similarly to the filamentous Ascomycetes. Basidiomycete yeasts in general survive cryo-preservation best but can be freeze-dried. Silica gel storage is less successful.

Herbarium Management

A well-kept herbarium provides reference material against which fungal identities or taxonomic status can be checked or revised as well as providing records of host range and geographic distribution. Many fungi still remain to be formally described and illustrated. The taxonomic status of some of those already described may change and the re-examination of herbarium material in the light of modern developments and knowledge facilitates the correct naming of fungi. It is also important that the same fungus should be known by the same name by pathologists in different countries. However, most published records of fungal pathogens from most countries are unsupported by specimens or cultures and many common pathogens are poorly represented in large herbaria and culture collections.

Every central plant pathology laboratory should maintain a reference collection of dried plant disease specimens and dried cultures to support country lists of plant diseases. These are of value in matching new isolates or samples with authenticated specimens or as evidence for the record of disease that may become of importance in relation to plant quarantine legislation etc. The following notes merely draw attention to some general principles of herbarium organization.

A collection of herbarium specimens and dried and living cultures is basically a store of information, in many ways resembling a library. Two essential requirements are that the items in the collection should be identifiable and that they and their accompanying details can be located and retrieved quickly and without undue difficulty.

Herbarium indexes can use a mechanical system and its variations or a computerized system that comprises electronic data processing (EDP) methods. Each specimen must have a unique accession number against which relevant details are recorded. The ease of retrieval of information on host range, geographic range, taxonomic group, typification, etc. varies. In mechanical systems separate indexes or a punched card system have to be maintained according to the requirements of the user. With an EDP system suitable programming will allow a variety of combinations of recall and is more versatile and faster than mechanically based systems. A summary of EDP systems for taxonomic collections is given by Brenan (1974) and their value in the context of mapping etc. is provided by Cutbill (1971).

The specimens may be stored in drawers or cabinets. A common method is to store the specimens, microscope slides and dried cultures in folded paper packets and to fix these to standard-sized herbarium sheets of paper (26 × 41 cm). Specimens of one species can be collated together in a 'species folder', which is in turn included with other species folders in a 'genus folder'. For fuller details see Hawksworth (1974).

Depredations by insects are always a potential hazard to any herbarium. In large herbaria it is usual to fumigate all specimens with methyl bromide. An alternative method is to freeze the material at a temperature of −30°C for 36 h followed by gradual thawing to avoid condensation.

Drying down cultures

Tap-water agar (1.5%) is melted and poured on to the smooth side of 12 cm squares of ordinary commercial hardboard. In hot countries the water agar should contain 2.5% glycerol to prevent cracking. The culture to be dried down is first killed by placing formalin-soaked filter paper in the tube or the lid of the dish overnight, and is then carefully removed from the Petri dish or tube and placed on the melted agar. In the case of slope cultures agar from the thick part is sliced off to make them thinner and flatter. Old, dry tube cultures can sometimes be loosened by heating a little water in the bottom of the tube. The hardboard squares, with culture, are placed under raised pieces of cardboard as a protection and the tap-water agar allowed to dry. After 2–5 days the culture is ready to loosen with a razor blade, peel off and trim. If the cultures have been allowed to become too dry this may be remedied by placing them in a damp chamber for about an hour. Dried slope cultures are usually stuck down with gum in ordinary slide boxes – the edges are temporarily pinned down if there is any tendency to curl. Petri dish cultures are fixed to the underside of removable cardboard rings, which fit into flat, cardboard boxes of special design; the reverse as well as the obverse of these cultures can be examined easily at any time.

Dispatch of Specimens of Fungi

Adequate precautions must be taken to ensure the safety of all those persons involved in the packaging, transport and receipt of strains (Smith and Onions, 1994) and that material transported presents no hazard to plants, people, animals, or the environment. National and international regulations govern the shipment of biological materials (Anon., 1998; Smith, 1996) and must be followed by all those who are involved in this activity. Most governments place restrictions on the import of strains from abroad, especially of plant pathogens, and it is frequently necessary to obtain official permission before a culture can be imported and used.

Import and export restrictions for perishable non-infectious or infectious biological substances by national postal services can be found in the *Official Compendium of Information of General Interest Concerning the Implementation of the Convention and its Detailed Regulations* revised at Hamburg in 1984, International Bureau of the Universal Postal Union, Berne. This information has also been compiled by the DSMZ (Anon., 1998) and is subject to a World Federation for Culture Collections report (Smith 1996). Most national post offices are members of the Universal Postal Union (UPU) and will provide detailed regulations on the transport of cultures, which will include when permits are required and where they can be obtained.

The CABI *Bioscience* UK Centre provides a service for the identification of microfungi, plant bacteria, insects and nematodes. A charge is made for most material but under special circumstances these charges can be waived and for certain DFID priority countries diagnosis and advise on plant disease problems is currently free of charge. Information on the service is available from CABI Bioscience UK Centre, Bakeham Lane, Egham, Surrey TW20 9TY, UK.

To facilitate the identification of cultures and specimens one should adopt the recommendation on packing and dispatch given in Chapter 8. It is essential that cultures arrive unbroken and free from mites, and that specimens are not overgrown by contaminants.

Select Bibliography

Alexander, M., Daggett, P.M., Gherna, R., Jong, S.C., Simione, F. and Hatt, H. (1980) *American Type Culture Collection Methods I. Laboratory Manual on Preservation Freezing and Freeze Drying*. American Type Culture Collection, Rockville, Maryland, USA.

Anon. (1981–1994) *Index of Fungi*. International Mycological Institute, Egham, Surrrey, UK.

Anon. (1998) *Instructions for the Shipping of Infectious and Non-infectious Biological Substances*. DSMZ, Braunschweig, Germany.

von Arx, J.A. and Schipper, M.A.A. (1978) The CBS fungus collection. *Advances in Applied Microbiology* 24, 215–236.

Ashwood-Smith, M.J. and Grant, E. (1976) Mutation induction in bacteria by freeze drying. *Cryobiology* 13, 206–213.

Atkinson, R.G. (1954) Quantitative studies on the survival of fungi in 5 year old dried soil cultures. *Canadian Journal of Botany* 32, 673–678.

Badcock, E.C. (1943) Methods for obtaining fructifications of wood-rotting fungi in culture. *Transactions of the British Mycological Society* 26, 127–132.

Boeswinkel, H.J. (1976) Storage of fungal cultures in water. *Transactions of the British Mycological Society* 66, 183–185.

Booth, C. (1971) *The Genus Fusarium*. Commonwealth Mycological Institute, Kew, UK.

Brenan, J.P.M. (1974) International conference on the use of electronic data processing in major European plant taxonomic collections. *Taxon* 23, 101–107.

Brezhneva, L.I. and Khokhoyakov, M.K. (1971) Storing pure cultures of fungi. *Mikologi i Fitopatologi* 5, 297–298.

Butterfield, W., Jong, S.C. and Alexander, M.J. (1974) Preservation of living fungi pathogenic for man and animals. *Canadian Journal of Microbiology* 20, 1665–1673.

Carmichael, J.W. (1956) Frozen storage for stock cultures of fungi. *Mycologia* 48, 378–381.

Chu, D. (1970) Forest pathology, storing of agar slants and cultures. *Bi-monthly Research Notes* 26, 48.

Clark, C. and Dick, M.W. (1974) Long term storage and viability of aquatic oomycetes. *Transactions of the British Mycological Society* 63, 611–612.

Clifford, B.C. (1973) Preservation of *Puccinia hordei* uredospores by lyophilisation of refrigeration and their subsequent germination and infectivity. *Cereal Rusts Bulletin* 1, 30–34.

Coulson, G.E., Morris, G.J. and Smith, D. (1986) A cryomicroscopic study of *Penicillium expansum* hyphae during freezing and thawing. *Journal of General Microbiology* 132, 183–190.

Cutbill, J.L. (ed.) (1971) *Data Processing in Biology and Geology*. Academic Press, London and New York, 346 pp.

Dahmen, H., Staub, T. and Schwinn, F.T. (1983) Technique for long-term preservation of phytopathogenic fungi in liquid nitrogen. *Phytopathology* 73, 241–246.

Davison, A., Brebandere, J. de and Smith, D. (1998) Microbes, collections and the MOSAICC approach. *Microbiology Australia* 19(1), 36–37.

Dick, M.W. (1965) The maintenance of stock cultures of Saprolegniaceae. *Mycologia* 57, 828–831.

El-Badry, H. (1963) *Micromanipulators and Micromanipulation*. Springer-Verlag, Vienna, 333 pp.

Ellis, J.J. (1979) Preserving fungus strains in sterile water. *Mycologia* 71, 1072–1075.

Ellis, J.J. and Roberson, J.A. (1968) Viability of fungus cultures preserved by lyophilization. *Mycologia* 60, 399–405.

Figueiredo, M.B. and Pimentel, C.P.V. (1975) Metodos utilizados para conservacao de fungos na micoteca de Secao de Micologia Fitopatologica de Instituto Biologico. *Summa Phytopathologica* 1, 299–302.

Gale, A.W., Schmitt, C.G. and Bromfield, K.R. (1975) Cryogenic storage of *Sclerospora sorghi*. *Phytopathology* 65, 828–829.

Gams, W., Hennebert, G.L., Stalpers, J.A., Jansens, D., Schipper, M.A.A., Smith, J., Yarrow, D. and Hawksworth, D.L. (1988) Structuring strain data for the storage and retrieval of information on fungi and yeasts in MINE, Microbial Information Network Europe. *Journal of General Microbiology* 134, 1667–1689.

Gentles, J.C. and Scott, E. (1979) The preservation of medically important fungi. *Sabouraudia* 17, 415–418.

Gould, G.W. and Measures, J.C. (1977) Water relations in single cells. *Philosophical Transactions of the Royal Society* B 278, 151–166.

Grivell, A.R. and Jackson, J.F. (1969) Microbial culture preservation with silica gel. *Journal of General Microbiology* 58, 423–425.

Gutter, Y. and Bakai-Golan, R. (1967) Observations on some fungi and other microorganisms preserved under mineral oil for 15 years. *Israel Journal of Botany* 16, 105–107.

Hawksworth, D.L. (1974) *Mycologist's Handbook*. Commonwealth Mycological Institute, Kew, UK, 231 pp.

Hawksworth, D.L. (1985) Fungus culture collections as a biotechnological resource. *Biotechnological and Genetic Engineering Reviews* 3, 417–453.

Hawksworth, D.L. and Kirsop, B. (1988) *Living Resources for Biotechnology*. Cambridge University Press, Cambridge, UK.

Hawksworth, D.L., Sastramihardja, I., Kokke, R. and Stevenson, R. (1990) *Guidelines for the Establishment and Operation of Collections of Cultures of Microorganisms*. World Federation for Culture Collections, Campinas, Brazil.

Heckly, R.J. (1978) Preservation of Microorganisms. *Advances in Applied Microbiology* 24, 1–53.

Hwang, S.-W. (1966) Long term preservation of fungus cultures with liquid nitrogen refrigeration. *Applied Microbiology* 14, 784–788.

Hwang, S.-W. (1968) Investigation of ultra-low temperature for fungal cultures. I. An evaluation of liquid nitrogen storage for preservation of selected fungal cultures. *Mycologia* 60, 613–621.

Hwang, S.-W., Kwolek, W.F. and Haynes, W.C. (1976) Investigation of ultra-low temperature for fungal cultures. III. Viability and growth rate of mycelial cultures following cryogenic storage. *Mycologia* 68, 377–387.

Ito, T. (1991) Frozen storage of fungal cultures desposited in the IFO culture collection. *Institute for Fermentation Osaka Research Communications* 15, 119–128.

Jong, S.C. (1978) Conservation of reference strains of *Fusarium* in pure culture. *Mycopathologia* 66, 153–159.

Kilpatrick, R.A. (1976) Fungal flora of crambe seeds and virulence of *Alternaria brassicola*. *Phytopathology* 66, 945–948.

Kilpatrick, R.A., Harmon, D.L., Loegering, W.Q. and Clark, W.A. (1971) Viability of uredospores of *Puccinia graminis* f. sp. *tritici* stored in liquid nitrogen. *Plant Disease Reporter* 55, 871–873.

Kobayashi, T. (1984) Maintaining cultures of Basidiomycetes by mineral oil method I. *Bulletin of the Forest and Forest Product Research Institute* 325, 141–147.

Kokke, E.G. and Elliot, M.E. (1977) Preservation of *Moellerodiscus lentus* (Sclerotiniaceae) by freeze-drying whole apothecia. *Mycologia* 69, 1206–1209.

Last, F.T., Price, D., Dye, D.W. and Hay, E.M. (1969) Lyophilization of sterile fungi. *Transactions of the British Mycological Society* 53, 328–330.

Little, G.N. and Gordon, M.A. (1967) Survival of fungus cultures maintained under mineral oil for twelve years. *Mycologia* 59, 733–736.

Loegering, W.Q. (1965) A type culture collection of plant rust fungi. *Phytopathology* 55, 247.

Long, R.A., Wood, J.M. and Schmitt, G.C. (1978) Recovery of viable conidia of *Sclerospora philippinensis*, *S. sacchari* and *S. sorghi* after cryogenic storage. *Plant Disease Reporter* 62, 479–481.

Marx, D.H. and Daniel, W.J. (1976) Maintaining cultures of ectomycorrhizal and plant pathogenic fungi in sterile water cold storage. *Canadian Journal of Microbiology* 22, 338–341.

Morris, G.J. (1981) *Cryopreservation: an Introduction to Cryopreservation in Culture Collections*. Culture Centre of Algae and Protozoa, Cambridge, UK.

Morris, G.J., Smith, D. and Coulson, G.E. (1988) A comparative study of the changes in the morphology of hyphae during freezing and viability upon thawing for twenty species of fungi. *Journal of General Microbiology* 134, 2897–2906.

Nei, T. (1964) Freezing and freeze-drying of microorganisms. *Cryobiology* 1, 87–93.

Onions, A.H.S. (1971) Preservation of fungi. In: Booth, C. (ed.) *Methods in Microbiology* 4. Academic Press, London and New York, pp. 113–151.

Onions, A.H.S. (1977) Storage of fungi by mineral oil and silica gel for use in the collection with limited resources. In: *Proceedings of the Second International Conference on Culture Collections*. World Federation for Culture Collections, Brisbane, Australia.

Onions, A.H.S. and Smith, D. (1984) Current status of culture preservation and technology. In: Batra, L.R. and Ligima, T. (eds) *Critical Problems of Culture Collections*. Institute of Fermentation, Osaka.

Onions, A.H.S., Allsopp, D. and Eggins, H.O.W. (1981) *Smith's Introduction to Industrial Mycology*, 7th edn. Edward Arnold, London, UK, 398 pp.

Perrin, P.W. (1979) Long term storage of cultures of wood inhabiting fungi under mineral oil. *Mycologia* 71, 867–869.

Prescott, J.M. and Kernkamp, M.F. (1971) Genetic stability of *Puccinia graminis* f. sp. *tritici* in cryogenic storage. *Plant Disease Reporter* 55, 695–696.

Reinecke, P. and Fokkema, N.J. (1979) *Pseudocercosporella herpotrichoides*: storage and mass production of conidia. *Transactions of the British Mycological Society* 72, 329–331.

Rhoades, M.S. (1970) Effects of 20 years storage on lyophilized cultures of bacteria, molds, viruses and yeasts. *American Journal of Veterinary Research* 31, 1867–1870.

Rowe, T.W.G. and Snowman, J.W. (1978) *Edwards Freeze-Drying Handbook*. Edwards High Vacuum, Crawley, UK.

Schipper, M.A.A. (1984) Collecting mycological cultures and improving cooperation: problems, impediments, opportunities. In: Batra, L.R. and Ligima, T. (eds) *Critical Problems of Culture*. Institute of Fermentation, Osaka, Japan.

Shain, L. (1979) Long term storage of fungi. *Transactions of the British Mycological Society* 63, 368–369.

Sharp, E.L. and Smith, F.G. (1952) Preservation of *Puccinia* uredospores by lyophilization. *Phytopathology* 42, 263–264.

Shearer, B.L., Zeyen, R.J. and Ooka, J.J. (1974) Storage and behaviour in soil of *Septoria* species isolated from cereals. *Phytopathology* 64, 163–167.

Sherf, A.F. (1943) A method for maintaining *Phytomonas sepedonica* in culture for long periods without transfer. *Phytopathology* 33, 330–332.

Smith, D. (1982) Liquid nitrogen storage of fungi. *Transactions of the British Mycological Society* 79, 415–421.

Smith, D. (1983) A two stage centrifugal freeze-drying method for the preservation of fungi. *Transactions of the British Mycological Society* 80, 333–337.

Smith, D. (1986) The evaluation and development of techniques for the preservation of living filamentous fungi. PhD thesis, London University, UK.

Smith, D. (1993) Tolerance to freezing and thawing. In: Jennings, D.H. (ed.) *Stress Tolerance in Fungi*. Marcel Dekker, New York, pp. 145–171.

Smith, D. (1996) Quality systems for management of microbial collections. In: Samson, R.A., Stalpers, J.A., van der Mei, D. and Stouthamer, A.H. (eds) *Culture Collections to Improve the Quality of Life*. Centraalbureau voor Schimmelcultures, Baarn, The Netherlands, pp. 137–142.

Smith, D. (1998) The use of cryopreservation in the ex-situ conservation of fungi. *Cryo-Letters* 19, 79–90.

Smith, D. and Kolkowski, J.A. (1996) Fungi. In: Belt, A. and Hunter-Cervera, J. (eds) *Maintaining Cultures for Biotechnology and Industry*. Academic Press, New York, pp. 101–132.

Smith, D. and Onions, A.H.S. (1983) A comparison of some preservation techniques for fungi. *Transactions of the British Mycological Society* 81, 535–540.

Smith, D. and Onions, A.H.S. (1994) *The Preservation and Maintenance of Living Fungi*, 2nd edn. IMI Technical Handbooks 2. CAB International, Wallingford, UK.

Smith, D. and Thomas, V.E. (1998) Cryogenic light microscopy and the development of cooling protocols for the cryopreservation of filamentous fungi. *World Journal of Microbiology and Biotechnology* 14, 49–57.

Smith, D., Coulson, G.E. and Morris, G.J. (1986) A comparative study of the morphology and viability of hyphae of *Penicillium expansum* and *Phytophthora nicotianae* during freezing and thawing. *Journal of General Microbiology* 132, 2013–2021.

Smith, D., Ryan, M.J. and Daly, J.G. (eds) (2001) *The UK National Culture Collection Biological Resource: Properties, Maintenance and Management*. UK National Culture Collection, Egham, 382pp.

Smith, D.H., Lewis, F.H. and Fergus, C.L. (1970) Long term preservation of *Botryosphaeria ribis* and *Dibotryon morbosum*. *Plant Disease Reporter* 54, 217–218.

Stack, R.W., Sinclair, A.W. and Larsen, A.O. (1975) Preservation of basidiospores of *Lacaria laccata* for use as mycorrhizal inoculum. *Mycologia* 67, 167–170.

Stevenson, R.E. and Jong, S.C. (1992) Application of good laboratory practice (GLP) to culture collections of microbial and cell cultures. *World Journal of Microbiology and Biotechnology* 8, 229–235.

Sugawara, H., Ma, J., Miyazaki, S., Shimura, J. and Takishima, Y. (1993) *World Directory of Collections of Cultures of Microorganisms*, 4th edn. WFCC World Data Center on Microorganisms, Japan.

Tamblyn, N. and Da Costa, E.W.B. (1958) A simple technique for producing fruit bodies of wood-destroying Basidiomycetes. *Nature* 181, 578–579.

Tan, C.S., Stalpers, J.A. and van Ingen, C.W. (1991) Freeze-drying of fungal hyphae. *Mycologia* 83, 654–657.

Thomas, V. and Smith, D. (1994) Cryogenic light microscopy and the development of long-term cryopreservation techniques for fungi. *Outlook on Agriculture* 23, 163–167.

Tommerup, I.C. and Kidby, D.K. (1979) Preservation of spores of vesicular–arbuscular endophytes by L-drying. *Applied and Environmental Microbiology* 37, 831–835.

Watling, R. (1963) Germination of basidiospores and production of fructifications of members of the agaric family Bolbitiaceae using herbarium material. *Nature* 197, 717–718.

Webster, J. and Davey, R.A. (1976) Simple method for maintaining cultures of *Blastocladiella emersonii*. *Transactions of the British Mycological Society* 67, 543–544.

Yukio, T. (1987) Preservation of plant pathogenic Mastigomycotina under super-low or ultra-low temperature conditions. *Bulletin of the Japan Federation for Culture Collections* 3, 1–7.

Mycological Media and Methods

B.J. Ritchie

CABI Bioscience UK Centre, Bakeham Lane, Egham, Surrey TW20 9TY, UK

General

The usual practice for fungi grown in the laboratory is for them to be maintained as single organism units. It is rare for fungi to grow in pure culture as in their natural environment they frequently interact in digestion of the substratum. As a result it is impossible to maintain fungi on one standard medium. If growth is poor on one medium others should be tried. Purely synthetic media are desirable when making biochemical and enzyme studies or in assay work, but most fungi do best on media from natural ingredients. The commonest method of cultivation is on a sterile jelly (usually a water agar gel with added nutrients) contained in a test tube, bottle or Petri dish. Fungi may be grown in liquid media, either on the surface of the liquid or, when constantly shaken, throughout the medium. They can also be grown on sterilized solid materials such as twigs, straw, grains, dung, leaves and soil. Fungi usually prefer a slightly acid reaction, pH 6–6.5, whereas bacteria prefer a more neutral pH of about 7. Carbohydrates and proteins in acid and alkaline solutions are decomposed by heat so avoid oversterilizing or alternatively, add at a later stage. Agar does not solidify satisfactorily in very acid or alkaline solutions. A perfectly clear medium is rarely required. Peptone is generally omitted from fungus cultivation media. Tap water is often preferable to distilled water as it contains useful trace elements although in some areas it may be slightly toxic. *Phytophthora* species are particularly sensitive and glass-distilled water is better for these organisms.

©CAB *International 2002. Plant Pathologist's Pocketbook*
(eds J.M. Waller, J.M. Lenné and S.J. Waller)

Preparation of Media

Correct preparation of agar media is a basic but necessary skill. Incorrectly prepared media leads to inaccurate results and a waste of time and expensive chemicals. A selection of glassware should be kept solely for making media; Pyrex® glassware is better than soda-glass as it is not subject to corrosion. Glassware can be cleaned by soaking overnight in 'cleaning' (chromic) acid and then repeatedly rinsed in running water and finished with a distilled water rinse. The acid can be reused many times. Clean distilled water-washed beakers or flasks should be used when mixing and dissolving chemicals; critical chemically defined media should always be made in acid/distilled water-washed glassware. Screw caps of bottles should either be metal without rubber liners or plastic so that they too can be thoroughly washed and distilled water-rinsed before use. Always use a glass rod to stir media; do not use metal or wooden spoons. Agar is slow to dissolve thoroughly and may take up to 1 h. Always dissolve agar and other constituents using the double saucepan method or in a water bath as this prevents the ingredients sticking to the bottom of the vessel and subsequently overheating and burning. To avoid caramelization of sugars or other heat-labile constituents such as peptone, add these constituents after the agar has dissolved prior to autoclaving. Some components will require filter sterilization. These are aseptically added to 'hand hot' autoclaved media immediately before pouring. Mycological growth media commonly use agar as the setting agent; always use a reliably pure proprietary brand to avoid problems with contaminants or poor setting. Dispense molten media into suitable screw cap bottles. Do not fill bottles more than two-thirds full. This is especially important with thick vegetable-based decoctions as the agar bubbles up during sterilizing and may overflow.

Proprietary media

Powdered media are available in dehydrated form to which it is only necessary to add water. Many of the commonly required media can be obtained in this form. Laboratory supply companies, e.g. Oxoid, Difco, Sigma, etc., have a wide choice of ready-prepared media.

Sterilization

Most media can be sterilized by heating at 115–121°C (pressure of 0.7–1 bar) for 15 min in an autoclave or domestic pressure cooker. Units used on the dials of these vessels vary according to type, source, age, etc. The relative pressures (at sea level), in commonly used units, needed to reach particular temperatures in pressure vessels are as shown in Table 38.1.

Table 38.1.   Pressure equivalents.

Temperature (°C)	bar	kg cm^{-2}	lb in^{-2} (p.s.i.)
107	0.34	0.35	5
110	0.48	0.49	7
115	0.68	0.7	10
121	1.02	1.05	15
126	1.36	1.4	20

Dry heat sterilization

Dry heat may be used for glass and other materials. Death of even the most resistant bacterial spores takes place within an hour in a hot air oven at 160°C. Glassware can be sterilized in a hot air oven or a domestic one can be used and will undoubtedly be cheaper to purchase than one advertised as solely for laboratory use. Glass dishes must be thoroughly washed and individually wrapped in aluminium foil and placed in a container suitable for use in an oven. Avoid wrapping dishes in paper as it either becomes very brittle or burns during sterilization. Do not use dry heat to sterilize bottles with plastic lids as they will melt – remember, metal caps only! The apparatus must not be too closely packed to facilitate circulation of hot air, and the heat must penetrate all contaminated parts to be effective. Large metal biscuit tins are suitable, as they are easy to obtain, reusable and are suitable for long-term storage to retain sterility after treatment. Sterilization should be complete after the times and temperatures shown below.

120°C	8 h
140°C	3 h
160°C	1 h
180°C	20 min

Media

Corn (maize) meal agar (CMA)

Maize meal	30 g
Oxoid agar no. 3	20 g
Tap water	1000 ml

Place the maize and water in a saucepan, heat using double saucepan until boiling; continue heating for 1 h, stirring occasionally. Filter the decoction through muslin, add agar and dissolve. Autoclave at 121°C for 20 min.

Czapek (Dox) agar (CZ)

Suitable for cultivation and identification of *Penicillium* and *Aspergillus* species.

Sucrose (Analar)	30 g
Oxoid agar no. 3	20 g
Czapek stock solution A	50 ml
Czapek stock solution B	50 ml
Distilled water	900 ml

Czapek stock solution A

Sodium nitrate ($NaNO_3$)	40 g
Potassium chloride (KCl)	10 g
Magnesium sulphate ($MgSO_4.7H_2O$)	10 g
Ferrous sulphate ($FeSO_4.7H_2O$)	0.2 g
Distilled water	1000 ml
Store in refrigerator	

Czapek stock solution B

Dipotassium hydrogen phosphate (K_2HPO_4)	20 g
Distilled water	1000 ml
Store in refrigerator	

Dissolve agar in distilled water using a double saucepan; add sucrose and stock solutions prior to autoclaving at 121°C for 20 min.

Emersons' yeast phosphate soluble starch (YPSS)

Soluble starch	15 g
Difco yeast extract	4 g
Potassium phosphate (K_2HPO_4)	1 g
Magnesium sulphate ($MgSO_4.7H_2O$)	0.5 g
Oxoid agar no. 3	20 g
Water	1000 ml

Dissolve agar using a double saucepan, add other constituents and dissolve. Autoclave at 121°C for 15 min.

Lima bean agar (LBA)

Lima beans (butter bean)	100 g
Oxoid agar no. 3	10 g
Distilled water	1000 ml

Soak beans overnight in 500 ml distilled water. Steam for 30 min, strain beans and retain liquid. Add agar to 500 ml distilled water and dissolve using a double saucepan. Add bean liquid and make up to 1000 ml. Autoclave at 121°C for 20 min.

Ready prepared lima bean agar powder is available commercially.

Malt-Czapek agar (MCZ)

Suitable for cultivation and identification of *Penicillium* and *Aspergillus* species.

Czapek stock solution A	50 ml
Czapek stock solution B	50 ml
Sucrose	30 g
Malt extract (sticky or toffee)	40 g
Oxoid agar no. 3	20 g
Distilled water	900 ml

Dissolve malt and agar using a double saucepan, add sucrose and dissolve. Add stock solutions and adjust pH to 5.0. Autoclave at 121°C for 20 min.

Malt extract agar (MA)

Malt extract	20 g
Oxoid agar no. 3	20 g
Tap water	1000 ml

Add agar to water and dissolve using a double saucepan, add malt and dissolve. pH should be 6.5; adjust with NaOH if required. Autoclave at 121°C for 20 min.

Malt extract agar plus sucrose

Modifications suitable for inducing sporulation in osmo- and xerotolerant organisms. Add the following amounts of sucrose to the basic MA recipe prior to autoclaving to reduce risk of caramelization.

M20:   add 200 g sucrose.
M40:   add 400 g sucrose.
M60:   add 600 g sucrose.

Malt extract agar plus antibiotics (penicillin/streptomycin)

Used as a general-purpose media for isolations from soil. Many other recipes and methods have been published; refer to Hall (1996) or Pankhurst *et al.* (1997) for further information.

Oatmeal agar (OA)

Oatmeal	30 g
Oxoid agar no. 3	20 g
Tap water	1000 ml

Add oatmeal to 500 ml of water and heat for 1 h. Add agar to remaining 500 ml of water and dissolve using a double saucepan. Pass cooked oatmeal through a fine strainer and add to agar mixture and stir thoroughly. Autoclave at 121°C for 20 min.

Potato carrot agar (PCA)

Potato	20 g
Carrot	20 g
Oxoid agar no. 3	20 g
Tap water	1000 ml

Wash, peel and grate vegetables, and boil for 1 h in 500 ml tap water. Strain through fine sieve retaining the liquid. Dissolve agar in 500 ml of tap water using a double saucepan, add strained liquid and mix. Autoclave at 121°C for 20 min.

A strip of sterile filter paper placed on the surface of the medium is very good for cellulose-destroying fungi, e.g. *Chaetomium* spp. Many mitosporic fungi produce good sporulation when grown on PCA plus filter paper or other sterilized cellulose source and subjected to near UV (Black Light) irradiation.

Potato dextrose agar (PDA)

Potatoes	200 g
Dextrose	15 g
Oxoid agar no. 3	20 g
Tap water	1000 ml

Scrub the potatoes clean, but do not peel. Cut into 12 mm cubes. Weigh out 200 g, rinse rapidly in running water, place in 1000 ml of tap water and boil until soft (approximately 1 h), then put through a blender. Add agar and dissolve using a double saucepan. Add dextrose and stir until dissolved. Make up to 1000 ml. Agitate stock while dispensing to ensure that each bottle has a proportion of solid matter. Autoclave at 121°C for 20 min.

Similar vegetable or cereal decoction agars can be made from locally available commodities using equivalent concentrations of ingredients to those used above. If a weak vegetable decoction medium such as PCA is not available, commercial potato dextrose agar can be modified so that the sugar content is reduced to a minimum and can therefore be used as a substitute. For example, to make 25% strength PDA use 25% of the quantity of PDA powder recommended and add an extra 75% of the quantity of the equivalent brand of plain agar used for making water agar, i.e.:

25% strength PDA	
Potato dextrose agar powder	9 g
Plain agar	11.25 g
Distilled water	1000 ml

Dissolve and sterilize as original instructions.

Potato sucrose agar (PSA)

PSA is suitable for expression of pigment production in *Fusarium*. This is an important diagnostic character within some species.

Sucrose	20 g
Agar	20 g
Distilled water	500 ml
Potato water	500 ml

Dissolve agar, add sucrose, dispense and autoclave 121°C for 20 min. If necessary adjust pH to 6.5 with calcium carbonate before sterilizing.
Potato water:
Peel and dice 1800 g of potatoes. Suspend in a double cheesecloth bag and boil in 4500 ml of water until potatoes are almost cooked. Discard potatoes.

Rabbit dung agar (RDA)

Suitable for cultivation and identification of coprophilous fungi. Rabbit dung must be from wild rabbits and dried before use.

| Oxoid agar no. 2 | 15 g |
| Tap water | 1000 ml |

Dissolve agar in water using a double saucepan. Dispense into bottles containing several rabbit dung pellets. Autoclave at 126°C for 20 min.

Sabouraud dextrose agar (SDA)

This is the classical medium for culturing dermatophytes (ringworm fungi).

Dextrose (or maltose)	40 g
Peptone	10 g
Oxoid agar no. 3	20 g
Distilled water	1000 ml

Dissolve agar using a double saucepan, add dextrose and dissolve. Add peptone and adjust pH to 5.6. Autoclave at 121°C for 10 min. Peptone caramelizes if overheated during sterilization.

Spezieller–Nährstoffarmer agar (SNA)

SNA is suitable for good spore production without encouraging excessive mycelial growth in the genus *Fusarium*.

Potassium dihydrogen phosphate (KH_2PO_4)	1 g
Potassium nitrate (KNO_3)	1 g
Magnesium sulphate ($MgSO_4.7H_2O$)	0.5 g
Potassium chloride (KCl)	0.5 g
Glucose (Analar)	0.2 g
Sucrose (Analar)	0.2 g
Oxoid agar no. 3	20 g
Distilled water	1000 ml

Dissolve all ingredients except agar in the distilled water and adjust pH to 6–6.5. Add agar and dissolve. Autoclave 121°C for 15 min.

Strips of filter paper (approximately 3 × 1 cm) sterilized by autoclaving or by dry heat (oven) can be aseptically placed on to the agar surface when set. The addition of filter paper improves sporulation with most *Fusarium* species.

SNA and PSA are used routinely at CABI *Bioscience* and predominate within the 'Fusarium working community' as the preferred media for identification and subsequent culturing.

Tap water agar (TWA)

Oxoid agar no. 3	15 g
Tap water	1000 ml

Dissolve agar in water. Sterilize at 121°C for 20 min.

Many fungi will sporulate on this medium if sterilized wheat straw, rice grains, or other suitable plant material has been added. Distilled water can be substituted if the tap water is known to contain heavy metals or other gross impurities that may prove toxic to sensitive strains.

V8 agar

'V8' vegetable juice	200 ml
Oxoid agar no. 3	20 g
Distilled water	800 ml

Dissolve agar in water using a double saucepan and add vegetable juice. Adjust to pH 6.0 with 10% sodium hydroxide. Autoclave at 121°C for 20 min. For cultivation of *Actinomycetes*, formula as above with addition of 4 g calcium carbonate. Adjust pH to 7.3 prior to autoclaving.

Antibacterial supplements

Antibiotics should always be kept in a refrigerator. Stock solutions are prepared for use in agar media and may be sterilized by passing through a bacterial filter (e.g. Millipore), although this is not usually necessary if aseptic techniques are employed during preparation of stock solutions from an original sterile commercial preparation. Do not prepare large stock solutions as potency can degrade quite quickly with some antibiotics.

Antibiotics should be added to 'hand-hot' media as excessive heat denatures most antibiotics.

Substances marked with an asterisk (*) are harmful, irritants, toxic or carcinogens. ALWAYS read the label for safety precautions before use.

Chloramphenicol* (Chloromycetin*)

Use at about 200 p.p.m. Stock solution prepared in 95% alcohol and can be added to media before autoclaving. Active against Gram-positive and Gram-negative bacteria, but also tends to inhibit *Pythium* and some *Phytophthora* species.

Penicillin G*

Use at 50–500 p.p.m. Stock solution: 1 g dissolved in 100 ml of sterile distilled water (1%). Addition of 1 ml of stock solution to 100 ml of medium gives 100 p.p.m. Add to media after autoclaving. Active against Gram-positive bacteria; should be used in conjunction with streptomycin sulphate.

Pimaricin*, Nystatin* (polyene antibiotics)

Antifungal antibiotics that do not affect *Pythium* and *Phytophthora* species. Used at 10–100 p.p.m. and added after autoclaving. Stock suspension: 4 ml of 2.5% trade suspension in 10 ml of sterile distilled water (1%). Addition of 0.1 ml of this solution to 100 ml of medium gives a final concentration of 10 p.p.m. Usually combined with a broad-spectrum antibacterial, e.g. vancomycin.

Polymixin B sulphate*

Use at 50–100 p.p.m. Stock solution: 0.5 g in 100 ml distilled water (0.5%). Addition of 1 ml of stock solution to 100 ml of medium gives a final concentration of 50 p.p.m. Add to media after autoclaving. Active against Gram-negative bacteria.

Rose Bengal*

A stain, which is effective against many bacteria and reduces the rate of spread of fast-growing fungi. Used at about 50 p.p.m. and can be added before autoclaving.

Streptomycin sulphate*

Use at about 200 p.p.m. Add to media after autoclaving. Stock solution: 1 g dissolved in 100 ml of sterile distilled water to give a 1% solution. Addition of 1 ml of stock solution to 100 ml of medium gives final concentration of 100 p.p.m. Broad-spectrum antibacterial but may also inhibit *Pythium* and some *Phytophthora* species. Should be used in conjunction with penicillin.

Tetracyclines*

Broad-spectrum antibacterial but inhibits some fungi. Used at about 200 p.p.m. Add after autoclaving.

Vancomycin*

Broad-spectrum antibacterial with little effect on fungi. Used at 100 p.p.m. and added after autoclaving. Stock solution: 2 g in 100 ml of distilled water and filter sterilise (2%). Addition of 1 ml of stock solution to 100 ml of medium gives a final concentration of 200 p.p.m. A more recent substitute for penicillin (active against Gram-positive bacteria) and polymixin B sulphate (active against Gram-negative bacteria) in selective media.

Selective fungicides

Cupric sulphate*

Can be used at up to 5% in selective media for dematiaceous fungi, added as a sterile solution after autoclaving as cupric sulphate denatures the setting power of the agar if autoclaved together.

Ethanol*

Used at 0.5–2% in selective media for *Verticillium* species and *Basidiomycota*, added after autoclaving.

Methyl benzimidazole carbamate* (MBC) (benomyl)

Can be used at 10–25 p.p.m. in selective media for *Oomycota* and *Basidiomycota*. Benomyl stock suspension: 1 g of commercial preparation (Benlate 50 wp) in 500 ml of distilled water (0.1%), autoclave at 115°C for 5 min. Use 2.5 ml of stock suspension to 100 ml of medium to give 25 p.p.m.

This and other fungicides can be used in media to isolate strains of fungi that are tolerant of (resistant to) the particular fungicide.

Pentachloronitrobenzene* (PCNB-quintozene)

A fungicide with little effect against *Pythium*, *Phytophthora* and *Fusarium* species at low dilutions. Used at about 100 p.p.m. and added before autoclaving. Will also reduce colony growth of bacteria.

Phytophthora isolation media

OA, CMA or LBA are suitable as base media.

BSPP medium

These quantities of stock solutions should be added to 100 ml of base medium:

  2.5 ml Benomyl* = 25 p.p.m.
  0.5 ml Streptomycin* = 50 p.p.m.
  1.0 ml Polymixin* = 50 p.p.m.
  0.6 ml Penicillin G* = 60 p.p.m.

3P medium

Quantities of stock solutions to be added to 100 ml of base medium:

> 1.0 ml Pimaricin* = 100 p.p.m.
> 1.0 ml Polymixin* = 50 p.p.m.
> 0.5 ml Penicillin G* = 50 p.p.m.

VP medium

Quantities of stock solution to be added to 100 ml of base medium:

> 1.0 ml Vancomycin* = 200 p.p.m.
> 0.1 ml Pimaricin* = 10 p.p.m.

Modifications to Culture Media to Induce Sporulation

Sporulation of fungi is essential for their identification and is also needed for the production of inoculum, e.g. for pathogenicity tests. Sporulation is greatly affected by the environment in which the fungus is growing. For the majority of fungi, spore production is the main method of reproduction and dispersal in their natural habitat; consequently 'unnatural' environments such as nutrient-rich agar media and warm dark incubators are not optimal for sporulation of plant pathogenic fungi. The following procedures, often in combination, usually induce sporulation of vegetative mycelium. Media such as PCA or OA, or similar vegetable decoctions with no added sugars should be used. Fungi grown on nutrient-rich media tend to produce excessive mycelium growth at the expense of sporulation. Some groups do require high nutrients to sporulate and it is sometimes a matter of 'trial and error' to find optimum growth requirements. Sterilized filter paper or a similar source of cellulose-containing material (i.e. wheat straw) should be added; small pieces of host material sterilized by steaming for 0.5 h or autoclaved at 107°C for 5 min can be utilized. These may work best when added to molten TWA. Physical wounding of the culture surface using a 'red-hot' inoculating needle followed by reincubation can induce sporulation as can incubation in daylight or with additional near UV irradiation and/or diurnal temperature fluctuations, particularly cool nights. Frustratingly, and despite all attempts, some fungi will not sporulate in artificial culture.

Safe Use of Chemicals

All chemicals must be treated with respect and all appropriate precautions taken to avoid contact with skin or clothing. This is essential not only for safety of the user and other colleagues, but also to ensure the purity of the chemical. Careful handling of chemicals is most important – always decant chemicals directly from their original containers into a suitable preweighed vessel, never directly on to the pan of the balance. Do not put unused or spilt chemicals back into the original container as this could easily introduce contamination rendering the chemical useless for further critical work. Avoid using a spatula to loosen hardened chemicals; if strictly necessary, use a distilled (deionized) water-washed glass rod, remembering to wash it thoroughly afterwards in distilled water. Correctly stored chemicals should not need to be 'dug out'! Extra care needs to be observed when making up solutions using concentrated acids and alkalis. Wear safety spectacles and proper laboratory clothing and take care not to contaminate door handles etc. when wearing gloves. Good ventilation is essential to avoid build-up of toxic fumes.

Under current UK and EU legislation, suppliers of laboratory chemicals must provide safety data sheets with all chemicals giving information on usage and disposal. Safety requirements should be determined *before* a new chemical is bought or process started.

Preparation of Stains and Other Reagents

A selection of glassware should be kept solely for making stains or reagents. Pyrex® glassware is better than soda-glass as it is not subject to corrosion. Glassware can be cleaned by soaking overnight in 'cleaning' (chromic) acid and then repeatedly rinsed in running water, finished with a distilled water rinse. The acid can be reused many times. Clean distilled water-washed beakers or flasks should be used when mixing and dissolving chemicals. Critical chemically defined reagents should always be made in acid/distilled water-washed glassware. Always use a double saucepan or water bath when dissolving as this prevents 'bumping' and avoids material sticking to the bottom of the vessel and becoming overheated. Ensure good ventilation and isolation from sources of ignition when working with volatile chemicals; a fume cupboard should be used if available. A clean distilled water-washed glass rod must be used for mixing chemicals when necessary; do not use wooden or metal spoons. Use non-absorbent 'fleas' ('flies') when using a magnetic hotplate/stirrer.

Never pipette by mouth, always use a proper pipette filler.

Add chemicals in the order listed if the recipe indicates the necessity to do so, otherwise unstable reactions may occur. Extra care needs to be observed when making up solutions using concentrated acids and alkalis. **Always slowly add the acid to the water.**

Stains should be stored in tightly stoppered dark bottles, correctly labelled, dated and with a hazard warning label if necessary.

Substances marked with an asterisk (*) are harmful, irritants, toxic or carcinogens. ALWAYS read the label for safety precautions before use.

Stains

Aniline blue stain for Bruzzese and Hasan (1983) technique

Ethanol* 95% (C$_2$H$_5$OH)	300 ml
Chloroform* (CHCl$_3$)	150 ml
Lactic acid* 90% (CH$_3$.CHOH.COOH)	125 ml
Phenol* (C$_6$H$_5$OH)	150 g

| Chloral hydrate* (CH$_3$.CH (OH)$_2$) | 450 g |
| Aniline blue (water soluble) | 0.6 g |

Add components, in order listed above, and mix thoroughly (with magnetic stirrer if available) until all solids are dissolved.

Bell's reagent (Bell, 1951)

Chloral hydrate*	350 g
Distilled water	140 ml
Cotton or trypan blue	0.5 g

Warm distilled water, remove from heat and add components. Allow to dissolve then transfer to a well-stoppered bottle to prevent crystallization.

Chloral hydrate solution (concentrated) for Bruzzese and Hasan (1983) technique

| Chloral hydrate* | 50 g |
| Distilled water | 20 ml |

Warm distilled water, remove from heat, and add chloral hydrate and dissolve. Avoid breathing harmful vapour.

Cotton blue (or trypan blue) in lactophenol

| Cotton blue or trypan blue | 0.1 g |
| Lactophenol | 100 ml |

This gives a 0.1% stain; the strength of the stain may be varied to suit individual requirements. The structures of many hyaline fungi can, over a period of time, absorb excessive amounts of stain if the initial concentration is too high. This can mask fungal structures when re-examined at a later date. Lactic acid may be used as a substitute for mounting and as a stain with cotton blue, but this will not clear plant tissues as efficiently as lactophenol.

Erythrosin

| Erythrosin* | 1 g |
| Ammonia* 10% | 100 ml |

This can be used for temporary mounts and gives a very clear outline of septation and structure of the sporogenous cell in hyaline fungi.

Lacto-fuchsin

| Acid-fuchsin* | 0.1 g |
| Lactic acid* | 100 ml |

Mix equal parts of this solution with 'Gurr's' water mounting medium.

Lactophenol

| Phenol* | 100 g |
| Lactic acid* | 100 ml |

Glycerol	200 ml
Distilled water	100 ml

Warm phenol in distilled water until dissolved, add lactic acid and glycerol. If a stain is required 0.01–0.1% cotton or trypan blue can be added.

Orange G solution

Saturated solution in absolute alcohol. Keep refrigerated.

Potassium metabisulphite solution

Potassium metabisulphite* 10% aqueous ($K_2S_2O_5$)	5 ml
1 M Hydrochloric acid* (HCl)	5 ml
Distilled water	90 ml

This solution is usable while it smells fairly strongly of sulphur dioxide. Keep refrigerated.

Schiff's reagent

Pour 100 ml boiling distilled water over 0.5 g basic fuchsin* or para-rosaniline* to dissolve it. Allow solution to cool to 50°C, filter, and add 10 ml 1 M hydrochloric acid* and 0.5 g anhydrous potassium metabisulphite* ($K_2S_2O_5$). Leave the reagent overnight, or longer, to clear, when it should be colourless or slightly straw-coloured. The solution remains usable for at least 6 months if kept in a well-filled, tightly stoppered bottle in a refrigerator. It must be stored away from light and air.

Thionin

Thionin	0.1 g
Phenol*	5 g
Distilled water	100 ml

Dissolve phenol in warm distilled water and add thionin.

0.05% Toluidine blue O in 0.1 M phosphate buffer at pH 6.8 (see p. 429)

Phosphate buffer Solution A	12.75 ml
Phosphate buffer Solution B	12.25 ml
Toluidine blue O	0.05 g
Distilled water to make up to	50 ml

Techniques

Bell's clearing and staining (Bell, 1951)

- Cut leaf into small pieces, approximately 0.5 cm square.
- Boil in distilled water for approximately 1 min.
- Wash while still warm in industrial alcohol until mostly decolourized.

- Immerse in a few millilitres of Bell's reagent for 12–24 h then mount in plain lactophenol or lactic acid. If kept warm (40°C) the staining procedure is accelerated.

Bruzzese and Hasan clearing and staining (Bruzzese and Hasan, 1983)

This is a clearing and staining technique that causes a minimum amount of disturbance of fungal structures on leaf surfaces. Leaf portions containing high concentrations of tannins need to be bleached before clearing and staining, e.g. soak in NaOCl or H_2O_2, but do not boil in water. The method allows fungal hyphae and spores to be readily distinguished from plant cells by their generally deeper blue staining.

- Immerse small squares of leaf tissues in clearing stain (2 ml of stain per cm^2 of tissue), in a stoppered glass tube for 48 h at approximately 20–25°C.
- Remove leaf pieces and place in chloral hydrate clearing solution for 12–24 h. This step may need to be repeated with fresh solution for another 12 h if the sample has taken up large amounts of stain.
- Rinse thoroughly and rapidly in distilled water. Mount in polyvinyl alcohol mounting medium.

Giemsa (for fungal nuclei) (Shaw, 1953; Hrushovetz, 1956; Ward and Ciurysek, 1961)

Different commercial brands of Giemsa stain are available. The proportion of dyes present varies but, in general, all give satisfactory results.

1. Preparation of staining solution: triturate 3.8 g Giemsa powder with 250 ml pure glycerine and 250 ml absolute methyl alcohol. Giemsa stain prepared in this manner or obtained in solution form has to be diluted by adding 15 ml of phosphate buffer (pH 6.9) to every 1 ml stain before use. Use new or acid-washed slides.
2. Staining procedure:

- Fix spores and mycelia in a mixture of 3 parts absolute ethyl alcohol and 1 part glacial acetic acid (EAA) or in a 3:1:1 mixture of absolute ethyl alcohol, glacial acetic acid and lactic acid (EAL) for 10–12 min.
- Rinse in 95% ethyl alcohol.
- Transfer to 70% ethyl alcohol where it can be stored for several days before staining. Rinse in distilled water before continuing.
- Immerse in 1 M HCl for 5 min at room temperature.
- Remove material and immerse in 1 M HCl at 60°C for 7 min. Depending on the nature of the material the temperature and duration of this treatment has to be varied for successful results.
- Rinse in distilled water and then in phosphate buffer (pH 6.9), five changes in each.
- Place in staining solution (stock solution diluted by adding 15 ml phosphate buffer to every 1 ml stain) for 2 h.
- After staining rinse in phosphate buffer followed by distilled water.
- Air dry.

- Mount in euparal for a permanent, or lactic acid for a semipermanent, preparation.

Nuclei stain rose to pink, cytoplasm generally remains clear or light blue.

Periodic acid–Schiff (for staining fungi in higher plants) (Dring, 1955)

- Immerse material in 1% w/v aqueous periodic acid solution for 2–5 min; if the plant tissue is to be counterstained do not exceed 3 min.
- Wash in running tap water for 10 min.
- Immerse in Schiff's reagent (q.v.) for 10 min. The reaction will be complete by this time, but the sections will remain colourless or slightly pink.
- Transfer the slides directly to potassium metabisulphite solution.
- Use at least two changes for a total period of 10 min. This solution, which should be kept tightly stoppered, is usable while it smells fairly strongly of sulphur dioxide.
- Wash in running tap water for 10 min.
- Dehydrate, clear and mount.

Fungal material stains magenta; other cellulosic plant material also stains magenta, but lignified material is little affected.

Toluidine blue O (Ghemawat, 1977)

- Remove air bubbles from leaf strips by immersing in industrial alcohol for 10 min.
- Rinse off industrial alcohol with a solution of 0.05% toluidine blue O.
- Stain for 15 min in fresh solution of 0.05% toluidine blue O.
- Mount in 0.005% toluidine blue O, examine immediately or keep in a humid chamber for storage of more than 24 h and seal coverslip.

Sealed slides store well for several years, the polychromatic staining being retained almost unchanged.

Fixatives

The simplest fixative to make is 2% formaldehyde* (= 5% formalin*) and will give perfectly satisfactory results. Formalin is commonly available and used as a fumigant. It is important to realize that 'formalin' is the name given to a 40% solution of formaldehyde, i.e. it is not pure formaldehyde, which is gaseous.

Ethyl alcohol–acetic acid (EAA)

3 parts absolute ethyl alcohol*
1 part glacial acetic acid*

Ethyl alcohol–acetic acid–lactic acid (EAL)

3 parts absolute ethyl alcohol*
1 part glacial acetic acid*
1 part lactic acid*

Formal–acetic–alcohol (FAA)

Formalin*	13 ml
Glacial acetic acid*	5 ml
Ethyl alcohol* 50%	200 ml

FA 4:1

40% Formaldehyde* (formalin)	10 ml
Glacial acetic acid*	1 ml
Distilled water	89 ml

TAF

40% Formaldehyde* (formalin)	7 ml
Triethanolamine*	2 ml
Distilled water	91 ml

Mounting Media

Lactic acid

Lactic acid* may be used by itself as a mounting medium; 0.05 g of cotton blue can be added.

Lactophenol

Phenol* (pure crystals)	20 g
Lactic acid* (SG I 21)	20 g
Glycerol	40 g
Water	20 ml

Heat water using a double saucepan or water bath, add phenol and dissolve, add lactic acid and glycerol.

'Necol'

Acetone*	4 parts
Diacetone alcohol*	1 part
(containing 1% benzyl abietate; 1% triacetin)	

Add cellulose acetate to the above solution until of a suitable consistency. Thin the solution with acetone until it flows like glycerine. Place a small drop of this fluid on the fungus colony. Allow to dry, usually 15–30 min, until a thin transparent film is formed in which the fungus is embedded. Remove the dry film carefully with forceps and place on a microscope slide. Add drops of acetone with a pipette. This causes the cellulose acetate film to dissolve and spread out in a ring leaving the fungus colony behind; wipe off the ring of residue with a cloth. Repeat until all the cellulose acetate is removed, mount and stain.

Polyvinyl alcohol mounting medium (Omar et al., 1978)

This is a permanent mounting medium with a high refractive index (1.39) for microscopic preparations of fungi; it is rapid setting and no sealing is required.

Polyvinyl alcohol	1.66 g
Distilled water	10 ml
Lactic acid*	10 ml
Glycerin	1 ml

Dissolve polyvinyl alcohol crystals in distilled water, add lactic acid and stir vigorously, then add glycerin. Filter if necessary. Leave solution to mature for 24 h before using.

Sealed and permanent slides

The edges of microscope coverslips on lactophenol or lactic acid mounts can be sealed with two or three coats of nail polish (varnish). Nail polish has the advantage of being cheap and easily obtainable around the world. This should keep the slide from drying out or otherwise deteriorating for several years. 'Glyceel', a cellulose-based sealant for permanent or semipermanent slides obtainable from laboratory supply houses is also an effective sealant. A solution of polyvinyl alcohol/lactophenol* can be used as a mountant; this gradually sets to make a rigid, permanent mount.

Permanent mounts can be produced using dehydration procedures. These are normally prepared by immersing the specimen in a series of progressively stronger ethanol solutions, e.g. 70%, 80%, 90% and absolute, each for a few minutes so that all water is removed. The specimen can then be cleared in xylene* and mounted in Canada balsam.

Miscellaneous solutions

Boric acid–borax buffer (Cruickshank, 1960)

It is essential that distilled or deionised water is used to prepare buffer solutions.

STOCK SOLUTION A: 0.2 M SOLUTION

Boric acid*	12.4 g
Distilled water	1000 ml

STOCK SOLUTION B: 0.05 M SOLUTION; 0.2 M IN TERMS OF SODIUM BORATE

Borax*	19.05 g
Distilled water	1000 ml

To 50 ml of solution A add x ml of solution B, and dilute to a total of 200 ml to give buffer with the following pH values:

B	pH
2.0	7.6
3.1	7.8
4.9	8.0
7.3	8.2
11.5	8.4
17.5	8.6
30.0	8.8
59.0	9.0
115.0	9.2

Citrate buffer (Cruickshank, 1960)

It is essential that distilled or deionized water is used to prepare buffer solutions.

STOCK SOLUTION A: 0.1 M SOLUTION

Citric acid*	19.21 g
Distilled water	1000 ml

STOCK SOLUTION B: 0.1 M SOLUTION

Sodium citrate ($C_6H_5O_7Na_3.2H_2O$)	29.41 g
Distilled water	1000 ml

To x ml of solution A add y ml of solution B, and dilute to a total of 100 ml to give buffer with the following pH values:

A	B	pH
46.5	3.5	3.0
43.7	6.3	3.2
40.0	10.0	3.4
37.0	13.0	3.6
35.0	15.0	3.8
33.0	17.0	4.0
31.5	18.5	4.2
28.0	22.0	4.4
25.5	24.5	4.6
23.0	27.0	4.8
20.5	29.5	5.0
18.0	32.0	5.2
16.0	34.0	5.4
13.7	36.3	5.6
11.8	38.2	5.8
9.5	41.5	6.0
7.2	42.8	6.2

Chromic or 'cleaning' acid

Potassium dichromate* ($K_2Cr_2O_7$)	75 g
Distilled water	500 ml
Concentrated sulphuric acid* (H_2SO_4)	500 ml

Dissolve potassium dichromate in warm water, allow to cool and very slowly add the sulphuric acid, stirring continuously with a glass rod.
EXTREME CARE must be exercised when working with concentrated acids. Wear safety glasses and proper laboratory clothing. Work only in a well-ventilated room.

Hydrochloric acid* (HCl)

Concentrated hydrochloric acid is 10 molar (M). To obtain weaker solutions dilute with distilled water.

1 ml conc. HCl + 9 ml distilled water = 1 M solution
2 ml conc. HCl + 8 ml distilled water = 2 M solution

ALWAYS add the acid to the water. EXTREME CARE must be exercised when working with concentrated acids. Wear safety glasses and proper laboratory clothing.

Knop's solution

Calcium nitrate ($Ca(NO_3)_2$)	0.5 g
Potassium nitrate (KNO_3)	0.125 g
Magnesium sulphate ($MgSO_4.7H_2O$)	0.125 g
Potassium phosphate (K_2HPO_4)	0.125 g
Ferrous chloride ($FeCl_2$)	0.005 g
Distilled water	1000 ml

Dissolve all constituents and store solution in a refrigerator.

Normal saline solution

Sodium chloride (NaCl)	8.5 g
Distilled water	1000 ml

Peroxide–ammonia bleaching solution (Johansen, 1940)

For bleaching of plant tissue prior to staining.

Hydrogen peroxide* 10% strength (H_2O_2)	10 ml
Distilled water	200 ml
Aqueous ammonia* (NH_4OH)	1 ml

Soaking time required will depend on thickness of material and amount of tannins, gums, etc. present. Wash specimen thoroughly in water after bleaching.

Phosphate buffer (Cruickshank, 1965)

It is essential that distilled or deionized water is used to prepare buffer solutions.

STOCK SOLUTION A: 0.2 M SOLUTION

Sodium dihydrogen orthophosphate (NaH$_2$PO$_4$.2H$_2$O)	31.2 g
Distilled water	1000 ml

STOCK SOLUTION B: 0.2 M SOLUTION

Disodium hydrogen orthophosphate (anhydrous) (Na$_2$HPO$_4$)	28.39 g

or

Disodium hydrogen orthophosphate dodecahydrate (Na$_2$HPO$_4$.12H$_2$O)	71.7 g
Distilled water	1000 ml

To x ml of solution A add y ml of solution B, diluted to a total of 200 ml (to give a 0.1 M solution) to give buffer of the following pH values:

A	B	pH
92.0	8.0	5.8
87.7	12.3	6.0
81.5	18.5	6.2
73.5	26.5	6.4
62.5	37.5	6.6
51.0	49.0	6.8
39.0	61.0	7.0
28.0	72.0	7.2
19.0	81.0	7.4
13.0	87.0	7.6
8.5	91.5	7.8
5.3	94.7	8.0

Sodium hypochlorite solution

Suitable for surface sterilization or as a bleaching solution for clearing plant tissues.

Sodium hypochlorite* (full strength commercial bleach) (NaOC1)	10 ml
Distilled water	90 ml
Detergent or Tween 80	few drops

Keep refrigerated to prevent rapid loss of active chlorine. Usable while it smells fairly strongly of chlorine.

References and Bibliography

Bell, F.H. (1951) Distribution of hyphae of several plant pathogenic fungi in leaf tissue. *Phytopathology* 41, 3.

Booth, C. (1971) *Methods in Microbiology*, Vol. 4. Academic Press, London and New York, 795 pp.

Bruzzese, E. and Hasan, S. (1983) A whole leaf clearing and staining technique for host specificity studies of rust fungi. *Plant Pathology* 32, 335–338.

Cruickshank, R. (1960) *Handbook of Bacteriology*, E. & S. Livingston Ltd, Edinburgh and London, pp. 281–285.

Cruickshank, R. (1965) *Medical Microbiology. A Guide to the Laboratory Diagnosis and Control of Infection*, 11th edn, E. & S. Livingston Ltd, Edinburgh and London, 852 pp.

Dring, D.M. (1955) A periodic acid–Schiff technique for staining fungi in higher plants. *New Phytologist* 54, 277–279.

Ghemawat, M.S. (1977) Polychromatic staining with toluidine blue O for studying the host–parasite relationships in wheat leaves of *Erysiphe graminis* f. sp. *tritici*. *Physiological Plant Pathology* 11, 251–253.

Hall, G.S. (1996) *Methods for the Examination of Organismal Diversity in Soils and Sediments.* CAB International, Wallingford, UK, 307 pp.

Hrushovetz, S.B. (1956) Cytological studies of *Helminthosporium sativum*. *Canadian Journal of Botany* 34, 321–327.

Johansen, D.A. (1940) *Plant Microtechnique.* McGraw-Hill, New York, USA.

Omar, M.B., Bolland, L. and Heather, W.A. (1978) A permanent mounting medium for fungi. *Stain Technology* 53, 293–294.

Onions, A.H.S., Allsopp, D. and Eggins, H.O.W. (1981) *Smith's Introduction to Industrial Mycology*, 7th edn. Edward Arnold, London, UK, 372 pp.

Pankhurst, C.E., Doube, B.M. and Gupta, V.V.S.R. (1997) *Biological Indicators of Soil Health.* CAB International, Wallingford, UK, 451 pp.

Shaw, D.E. (1953) Cytology of *Septoria* and *Selenophoma* spores. *Proceedings of the Linnean Society of New South Wales* 78, 122.

Stoughton, R.H. (1930) Thionin and orange G for the differential staining of bacteria and fungi in plant tissues. *Annals of Applied Biology* 17, 162–164.

Waller, J.M., Ritchie, B.J. and Holderness, M. (1997) *Plant Clinic Handbook.* IMI Technical Series No. 3. CAB International, Wallingford, UK.

Ward, E.W.B. and Ciurysek, K.W. (1961) Somatic mitosis in a basidiomycete. *Canadian Journal of Botany* 39, 1497–1503.

Photography

39

E. Boa

CABI Bioscience UK Centre, Bakeham Lane, Egham, Surrey TW20 9TY, UK

Photography is an important tool for use in plant pathology. It is used to record disease symptoms in the field, and in the laboratory to record microscopic features of pathogens, plant tissues and banding patterns in gels. These images provide a personal history of events and a means of sharing information with others. Time sequences show the symptom development or appearance of a pathogen. Photographs are an essential part of research, teaching and publicity campaigns aimed at plant quarantine or control of an important plant disease. They have a crucial role in facilitating diagnosis of ill-health problems, a process that often starts with a search for a photograph showing similar symptoms to those of the disease specimen.

Photographic hardware has improved dramatically since the last edition of this pocketbook, and has become available and affordable to a much larger group of scientists. Many of the mechanical steps in taking pictures have been automated: most cameras have automatic exposure and automatic focus. The quality of zoom lenses has increased and films that require less light for exposure have improved sharpness. It has never been easier or cheaper to take good photographs yet the evidence for increased standards in photography is hard to find. The quality of original photographs shown in lectures and at conferences is generally poor, and colour reproduction in printed material often fails to reach the high standards of which it is capable. The purpose of this chapter is not to examine the causes for these low standards but to provide some general advice on how to improve them. The days of the departmental or institute photographer, who could be reliably called upon to take, develop and print a photograph, have gone. Individual scientists have to take their own photographs and make the decisions previously taken by a professional photographer.

©CAB *International* 2002. *Plant Pathologist's Pocketbook*
(eds J.M. Waller, J.M. Lenné and S.J. Waller)

This chapter helps to explain the different types of photographic equipment and material available to basic photographers and which is best suited to typical needs of plant pathologists. An increase in choice of camera and films offers potential advantages yet confusion has also arisen concerning the need to use digital cameras rather than conventional film, or whether to buy a compact camera or one that takes interchangeable lenses. These are some of the points that will be discussed here. Some simple advice on how to improve the quality of photographs is also provided.

Any discussion of photography must now include various electronic means for recording and printing photographs. Light-sensitive film is still the most common means for capturing an image but digital cameras, scanners and computers are also regularly used. Digital cameras offer new possibilities for storing and manipulating images; scanners and computers allow digital copies of prints and slides to be manipulated and included directly in computer files. Digital cameras still lag behind other cameras in quality and are more costly to buy, though both these disadvantages will diminish as new and better models become available. Digital cameras also require suitable computer media for temporary and permanent storage of files and a colour printer to print the photographs. A fuller discussion of digital cameras, electronic images and the use of software to manipulate such images is included under a later section. The correct storage of slides and negatives should not be forgotten and some notes are given.

New features include a short section on aerial photography. This explores various options without discussing technical details. The section on photomicrography has been reduced from the previous edition, mainly because of changed priorities in plant pathology. Other sources of photographic information are given at the end of this chapter, including selected web sites on the Internet. A short list of reference books is appended.

Film Formats

The most common film format in use is 35 mm. This provides an image 36 mm × 24 mm. Many other formats also exist, some more or less consigned to history and others with specialist uses. A new format known as APS (Advanced Photography System) is being increasingly adopted but does not appear to have had any noticeable impact in scientific circles. The 35 mm format continues to offer many features and advantages and will satisfy most scientific needs. Larger format films such as 120 film, often referred to as roll film, offer higher-quality images. However, the cameras and equipment required to use this film are more expensive to buy and heavier to carry. The increased quality available from 120 film is rarely needed by plant pathologists, and slides require a special projector.

Film Types

When buying 35 mm film, which is widely available, avoid buying films within a few months of their expiry date (usually printed on the side of the packaging). There are three main types of film: black and white; colour negative; and colour transparency or slide film. Black and white films are still widely available and are simple to process. In practice, plant pathologists will most commonly use colour transparency or colour negative film, sometimes also called colour print film. Colour prints, slides and negatives can be scanned and printed in black and white on laser printers and inkjet printers if required.

Remember that the cost of buying films and processing is often a very small proportion of project or experimental costs. Always take spare films with you to the field; it is surprising how often a particular example of a symptom is seen once and never again and the worse excuse for not having a photograph is that there was no film.

Plant pathologists, like many other scientists, show a strong preference for transparency film, despite the fact that the images need to be projected for details to be seen clearly. Prints can be made from slides but these are expensive and time-consuming to organize. The make of colour transparency film can influence the appearance of different features. As a general rule, Fuji film is good for vegetation whereas Kodak film is often recommended for portraits. Personal tests are advised if colour rendition is crucial. Colour negative film has many unsung advantages: the images are easier to show to others, enlargements are cheaper and more convenient to produce, and prints are readily scanned and transferred for use on a computer. It is also possible to scan transparencies but suitable scanners are less widely available. A suitable compromise is to transfer important transparencies on to CD-ROM using the Photo CD facility developed by Kodak. This provides a portable source of images for use in reports and digital presentations and is a convenient means of sending images to colleagues via electronic mail without risking the loss of the original transparency. Each CD-ROM can hold up to 100 transparencies. When selecting transparencies for transfer to CD-ROM or other crucial work, examine them first with a hand lens or photographic loupe (usually ×4) to make sure that the image is sharp and in focus. If transparencies and prints are required for your work, consider using two cameras loaded with different types of film. This increases the weight of equipment to be carried and also the time required to take photographs, but the additional weight (and space) can be minimized by using a compact camera loaded with the colour negative film.

The speed rating of films is normally expressed in ISO units and the most common speeds available are 100 ISO and 200 ISO. Scanning equipment in airports should not affect films with a low speed rating. The faster films – those with a rating of 400 ISO or more – are more prone to damage by scanning equipment. The risk of damage or fogging in films can be reduced by using widely available special bags. Camera bags may be examined separately though some authorities do not allow this. Films with a lower speed rating produce a higher-quality image capable of greater enlargement without reduction in sharpness of the image. The disadvantage is that they require more light and slower shutter speeds. Camera-shake at lower shutter speeds can be important in blurring images. Manufacturers are constantly improving the quality of faster films and 200 ISO film will generally be capable of meeting most needs.

Care and Processing of Films

There is a general belief that the name of the film manufacturer is the most important guarantee of quality. In reality, quality of the recorded image will depend more on how the film was stored before it was bought and after it is processed. The colour balance of transparency films is most affected by high temperatures but colour negative film can also deteriorate if kept on open shelves for too long. Keep unused films in domestic refrigerators and allow sufficient time for them to reach room temperature before loading in the camera. Avoid buying film from shops where it might have been stored on open shelves for prolonged periods of time or kept at high temperatures. Similarly, avoid leaving camera bags with films in hot places, such as cars parked in unshaded places. If you have spare film from a field visit be careful not to leave it in the camera bag for long periods of time. Similarly, do not keep a film in the camera to 'finish off the last few exposures' since this may result in a delay of weeks or months before being sent for processing. Finally, ensure that films are sent for processing as soon as they have been exposed.

Colour transparency film is not as widely available as colour negative film and can be more difficult to process. Most colour transparency film is developed by a process known as E6. Processing facilities are widely available, even in remote places. Kodachrome requires a different development process and specialist laboratories are more difficult to find, particularly in less-developed countries. Photographic or photo laboratories exist everywhere for processing and printing colour negative film, many offering a rapid service at an increased price. Although photo laboratories use standard automatic equipment to process films and print negatives, in practice the quality of prints is variable. It is good to standardize on a particular make of film, whether colour negative or transparency, so that it is easier to spot changes in quality due to processing.

It is often helpful to have two sets of prints produced from colour negative film. This is often cheaper when the film is first developed. (Similarly, it is easier to take two slides of a scene than to copy slides later.) Keep a record of all films you send by post to help trace any material that goes missing. Many mail order companies provide tags, which help to identify your films.

Use of Photographs in Publications

Commercial printers prefer transparencies since these reproduce best in publications. Check the quality of the transparency first with a suitable hand lens or loupe. Blurred images due to camera-shake and poor focusing are sometimes difficult to identify, even when the transparency is projected on to a screen. Commercial printers will accept digital scans of transparencies produced either from a Photo CD or from a slide scanner but will usually prefer scans from professional (drum) scanners. The cost of such professional scans can often be high and it is worth enquiring about prices when calculating the budget for a publication.

The quality of scans needed to produce an acceptable image in reports and other self-published documents is less critical. The original image can be a scan from a print or transparency or from a Photo CD. The quality of information stored in the digital file is often referred to in dots per inch (dpi). Some simple tests will help to determine what resolution to scan the original file at, bearing in mind that higher resolutions mean larger files that take longer to print.

Dramatic improvements in print quality can be achieved by using coated or glossy paper, but check again that the selected paper is suitable for images and text. Some inkjet printers are specifically designed to print photographs and special 'photo' inkjet cartridges can be used in other models. Colour laser printers are becoming more affordable and threaten to replace the use of commercial printers for small-scale printing runs. Although inkjet printers are good for producing proofs of documents their running costs remain high.

The dominant word-processing program is Microsoft Word. This has various features for including photographs in publications but they have a limited use. Photographs are best combined with text using desktop publishing programs such as Adobe Pagemaker or InDesign or even Quark Express. These programs have the disadvantage of being more difficult to use for the infrequent publisher and many will rely on Microsoft Word. Where multiple photographs are to appear in a publication, it may be easier to print all the photographs on a separate sheet than try to incorporate photographs and text in a document. If you decide to do this, create a table and insert the digital images into separate cells. The captions can be written in cells below each image.

Equipment: Cameras, Lenses, Accessories

Cameras

Most plant pathologists will use a 35 mm camera to take photographs. The cost of digital cameras is decreasing all the time but they are still relatively expensive. The main choice, therefore, is the type of 35 mm camera and choice of lenses.

Too much emphasis is placed on buying the right make of camera. Many expensive cameras are used to take bad photographs – do not confuse price with quality of photographs. There are two main forms of 35 mm camera: the single lens reflex or SLR and the compact or rangefinder camera. The SLR camera shows the photographer the exact image through the lens (in fact usually slightly less than appears on the film). The compact camera has a separate viewfinder with guidelines to indicate what the lens sees – the so-called 'parallax error'. This is not a problem for general scenes but does impose restrictions on close-up photographs. If you do not want to carry a lot of camera equipment consider the use of a small compact camera, bearing in mind that your ability to take close-ups will be limited. Compact cameras are easy to use and light to carry, and are suitable for plant pathologists with modest photographic needs. All bar the simplest model will have autofocus, a feature commonly available on SLRs. Compact cameras come with fixed focal length lenses, usually providing a slightly wide-angle view, and with zoom lenses. The zoom feature allows you to change between telephoto and wide-angle views but does increase the price of the camera. Indeed, compact cameras have become burdened with so many features that the price of the more elaborate models is the same as some SLRs. Consider buying a second-hand compact camera, particularly some of the earlier models (Olympus, Canon, Yashica). These may not have automatic focusing or exposure but usually have a rangefinder system and a built-in exposure meter, some of which do not require batteries.

The features available on modern electronic SLRs offer a bewildering variety of possibilities, most of which are rarely used. The ability to change lenses and obtain accurate close-ups of scenes remain the main advantages of the SLR. There is a large second-hand market in 'manual' SLRs, and models such as the Canon AE-1 and early Nikons offer good value for money. It is important that cameras and lenses are regularly serviced and cleaned, particularly if they are used in tropical climates. Fungal growth on lenses is particularly difficult to get rid of if ignored for too long. Examine lenses by opening them to maximum aperture and looking through them at an oblique angle. Equipment can be stored in dry places but this is rarely effective for cameras in regular use.

Lenses

Protect all lenses with filters. UV and skylight filters are the most common types used and help to reduce unwanted hazes. These can be kept in place all the time. Polarizing filters reduce unwanted reflections. Many other types of filter are available but have little use for general photography. Always use a lens hood. They help to eliminate unwanted reflections that reduce the contrast of photographs. Some lenses have built in lens hoods.

Lenses have either a fixed or variable focal length. The latter are also called zoom lenses and offer a range of different views without having to change lenses. Focal length is expressed in millimetres and indicates the angle of view. Do not confuse focal length with magnification. A telephoto lens does not increase the size of an object on the negative but allows the photographer to take photos from further afar. A wide-angle lens decreases the apparent size of objects by increasing the angle of view. Aperture or 'f-stop' indicates the amount of light passing through a lens: the lower the f number the larger the aperture and the greater the amount of light passing through the lens. A 35 mm f.2 lens is one with a maximum aperture of f.2. The advantage of having a larger maximum aperture is that faster shutter speeds can be used. This helps to minimize camera shake, albeit at an extra cost for the lens.

Different lenses can be used on an SLR though each manufacturer has a different type of fitting. Many SLRs are sold with a 50 mm lens, often referred to as a standard lens. A moderate wide-angle lens, either 35 mm or 28 mm, offers many advantages when taking pictures of plants or other objects found in confined spaces. Many photographers use a wide-angle lens as an alternative standard lens. Wide-angle lenses less than 28 mm are prone to distortion when the lens is tilted downwards or upwards and are not suitable for normal use in plant pathology. A telephoto lens of up to 200 mm focal length will isolate parts of plants, particularly trees.

Good-quality zoom lenses are available that provide a range of focal lengths between 28 mm and 200 mm. The greater the range, the higher the price you will pay. The quality of zoom lenses has improved considerably from initial designs but the maximum aperture is less than that available on fixed focal length lenses. This is not an important consideration in most circumstances. The price of all lenses increases dramatically as the maximum aperture increases.

Focal lengths longer than 200 mm and shorter than about 24 mm are difficult to use and have only a limited application in normal plant pathology. The longer lenses need to be held carefully to avoid camera-shake whereas the shorter wide-angle lenses distort perspective when not held strictly horizontal.

Macro lenses increase the size of objects on the film. Zoom lenses may have a macro facility but the writer has a personal preference for a fixed length, moderately telephoto lens with macro features. A 90 mm or 100 mm macro lens allows you to remain some distance from the object and thus avoids reducing available light and interference with the subject. Macro lenses can of course also be used for normal photography. Extension tubes and bellows permit even greater magnification of objects but normally require tripods to reduce the risk of camera-shake and keep the object in focus. Supplementary light may also be necessary to achieve the small apertures required to give an adequate depth of focus. The whole process of macrophotography thus becomes more complicated. A good macro lens avoids this complication while allowing the full details of plant symptoms to be captured.

Flash

Many pictures are spoilt by poor illumination. Invest in a good flashgun. There is a huge variety to choose from but try to ensure that the flash is automatic or 'dedicated' for your particular make of camera. Such flashguns calculate exposure via the camera and allow the photographer to concentrate on taking the picture. The power of the flashgun determines the maximum distance for correct exposure of an object. Also note that the greater the power (usually expressed as a 'guide number'), the more expensive the flashgun. A word of caution concerning the built-in flash: these have a limited power and are only intended for use with small groups of people. If a flashgun is not available, consider the use of reflectors to increase the available light. Simple reflectors can be made from tin foil wrapped around boards or from sheets of white cloth. Reflectors help to fill in dark areas or shadows. Portable reflectors are made by companies such as Lastolite in the UK.

Ring flashes fit around the front circumference of a lens (check the maximum diameter of lens that can be accommodated), and provide an even illumination suitable for close-up photography. This even illumination avoids the shadows created by lateral light sources but also reduces contrast and gives the scene a slightly dull appearance. A suitable compromise, particularly if a ring flash is not available, is to hold a normal flashgun slightly to the side of the camera, thus providing some contrast with the minimum of shadowing. Shadowing can also be reduced by the use of reflectors.

Tripods

All photographs are improved by the use of a tripod. Many photographs are spoiled by camera-shake though this is not always apparent on colour transparencies viewed from a distance, either on a screen or at arm's length. The general rule is that the heavier the tripod the more it will reduce unwanted vibrations. However, tripods add to the weight of field equipment and effective compromises can be made through the use of mini-tripods or sturdy table-top versions. Support for cameras can also be improvised by using bags, clothing and so on. Use a cable release or the delayed shutter facility to eliminate unwanted vibrations when taking photographs from a tripod.

Taking Photographs

The number one rule is not to run out of film! Always have plenty of film available. This is a minor cost relative to other expenses and you may be unable to repeat an experiment or find a similar specimen again.

Taking photographs in the field requires time. Try to avoid casual and careless technique by remembering some simple points. Make sure lens surfaces are clean – a well-washed handkerchief is often sufficient to remove ambient dust and small marks. Switch the camera off between photos to conserve batteries (and carry spares at ALL times). Above all, remember that you will often only have one opportunity to take a good photograph of a particularly good specimen. Perhaps most simply of all, take a camera with you at all times.

Photographs taken at the beginning and end of the day are more likely to show better contrast, though colour balance will alter with the onset of twilight. This is particularly true in tropical countries, where photographs taken in the middle of the day often look washed out. Rain showers help to remove dust from the atmosphere and photos taken after the rain has stopped have an enhanced contrast and depth of colour. Always try to keep the sun behind you when taking photographs and be aware of deep shadows, particularly in tropical conditions. Most films have an exposure latitude of two (aperture) stops. In other words, if the sunny area requires f.8 at 1/60th of a second, and the shadows require f.4 at 1/60th, detail in both areas will be visible. If in doubt, expose for the shadows, or minimize the area in sunlight.

The simplest guide to taking good photographs is to look at the results obtained by professionals. There are many excellent illustrated books on natural history that demonstrate the general principles of good photography. Photographic societies exist around the world and these can be a very useful source of advice. The Royal Photographic Society in the UK, for example, has an extensive range of specialist groups.

Opinions vary on the use of small objects to indicate scale. Coins, knives and large rulers can be obtrusive and, if necessary, you might consider including a short scale showing a few centimetres only at the edge of the scene. This can subsequently be edited out or 'cropped' as required.

Look carefully at the background. Is this obtrusive and does it detract from the object you are trying to photograph? Remove leaves and other small plant parts and photograph separately, placed against a neutral background. A (quiet) road surface or path allows the photographer to look directly down on the specimens and reduces movement due to wind. Take several photographs of symptoms or other features you are trying to record and select the best ones when they are developed.

With some categories of symptoms, such as those associated with viruses or phytoplasmas, it is good practice to photograph a healthy and diseased specimen together. Again, either use a neutral background or get a colleague to hold both at arm's length, slightly apart.

Photography in the field shows symptoms as they occur in nature, but is subject to weather conditions and quality of the light. Whole plants can also be difficult to photograph as it may be impossible to isolate them from neighbouring plants or vegetation. Laboratory photography is easier to organize though field specimens will not be fresh. Artificial light may be required, particularly in dimly lit rooms, and careful attention needs to be paid to removing unwanted shadows. Copy stands have a flat base with a column at one edge with a mounted camera that can be moved up and down for correct focus. Lights or flashguns are mounted either side on extensible arms, to reduce shadowing. Special photographic lights (photofloods) must be used to reduce the colour shift produced with artificial light. In technical terms, the colour temperature of the film should match that of the lights. Fluorescent lights give a green cast whereas normal table lamps produce a brown cast on film designed for daylight conditions. When photographing pot plants, for example from artificial inoculations, use a neutral background to isolate the plant features and reduce shadowing. Do not use a white sheet since this will 'trick' the automatic exposure system in the camera and underexpose the objects. Shadows will be clearly visible. Black backgrounds will result in overexposure and are best used with artificial light: shadows are not visible. Light grey backgrounds are most suitable.

A common frustration is knowing exactly what has been photographed. Some cameras provide a date stamp whereas other more sophisticated models store other information. A simple trick is to photograph a general scene at the beginning of a film, to remind you of the general location and occasion, taking another general scene when you move to a new location. Or you may wish to write down particular details on a piece of paper and photograph this. The simplest method is to note details of photos in a notebook in the sequence they are recorded on film. This will usually be sufficient to label photographs once they are developed.

Exposure

Modern cameras have built-in exposure meters, which can often be pro-grammed to adapt to particular conditions. The most common mistakes are made when plants are photographed against the light or large amounts of sky appear, thus reducing the exposure and rendering foreground detail too dark. Many cameras automatically set the film speed (DX) when it is loaded but, if not, always check this is correctly set. If you do set the wrong film speed this can often be corrected at the processing stage. Discuss with your photographic laboratory.

Exposure is controlled by the amount of light coming through the lens, the aperture size, and the time the film is exposed or shutter speed. For a given amount of light, aperture and shutter speed can be adjusted according to the needs of the photographer. To give an example of adjusting aperture against shutter speed: f.2.8 at 1/125th of a second provides the same exposure as f.4 at 1/60th of a second.

The main limiting factor is usually the slowest shutter speed that can be hand-held by the photographer. This depends on the focal length of the lens and the photographer. As a general rule, it is difficult to avoid camera-shake at shutter speeds less than 1/60th of a second for a 50 mm lens and wide-angle lenses. A minimum speed of 1/125th of a second is required for short telephoto lenses (135 mm) and so on as the focal length increases. If you must use a slow shutter speed either use a tripod or brace yourself against an object with your elbows tucked in against your abdomen for further rigidity. You can also use the shoulder of a colleague on which to rest a telephoto lens at slow shutter speeds.

The largest or maximum aperture is confusingly that with the lowest numerical value. Thus f.2.8 allows twice as much light to pass through as f.4 and four times as much as f.5.6. The disadvantage of using larger apertures is that these have the smallest depth of focus (also called depth of field), or distance in which objects will be sharp. This can be important for macro-photography, where the greatest depth of focus is often required and apertures of f.16 or lower are often used (and long shutter speeds or flash). Lenses are designed to give the best-quality image at lower apertures, usually around f.8 or f.11.

Short focal length lenses have a greater depth of focus for a given aperture compared to longer lenses. A similar effect can be seen on a normal compound microscope: lower-power lenses allow you to see more of an object sharply compared to higher-power lenses.

The plusses and minuses of particular combinations of aperture and shutter speed can be difficult to understand and juggle with, hence the reason that modern cameras have automatic exposure. Some can also be programmed for maximum depth of focus and so on. The manufacturers have tried to take much of the decision making away from the photographer, even preventing shutter release when there is not enough light or automatically firing the flash. This is fine for general purposes but does not help when trying to take a close-up of plant symptoms. The plant pathologist will have to pay closer attention to detail and perhaps override the automatic features of the camera in order to get the best photograph.

Fast films allow the photographer to use faster shutter speeds but there is a trade-off with reduced quality (sharpness) of images. However, this may be perfectly acceptable and the use of 400 ISO films is one way to minimize camera-shake. Modern films have considerable latitude for over- and underexposure but slide films are the least forgiving of wrong exposure, particularly overexposure.

If you are in any doubt about the correct exposure, take a series of photographs at small exposure intervals either side of that indicated by the camera's exposure system. This is called 'bracketing' and is best achieved by keeping the same shutter speed and altering the aperture on the lens accordingly.

Digital Photography

The most common type of digital camera resembles a compact 35 mm model in appearance and basic operation but captures images and stores them on computer media as digital files. Most digital cameras have a zoom lens, albeit of a limited range, though only the more expensive models have a true optical zoom lens. They have automatic focus and integrated flash. Most allow you to review photographs and delete them immediately. There is a direct relation between the cost of a digital camera and the quality of the image it is capable of taking.

Early models accepted standard computer diskettes but these had a limited capacity and have largely been replaced by flash cards. These are the size of several postage stamps and only a few millimetres thick. They can be used repeatedly once images have been transferred to another form of computer storage or deleted. Digital cameras can be connected directly to a computer for image transfer but this can quickly drain expensive batteries that are required to operate the camera. The alternative is to use a flash card reader, a separate accessory, which connects to a computer and acts like a disk drive. Flash cards have a minimum capacity of 4 MB increasing up to 124 MB and more.

The number of images that can be stored depends on the 'resolution' selected via the camera controls and the quality of the camera. Cameras are advertised on the number of 'megapixels' they are capable of storing. The 'resolution' of an image determines the maximum size that an image can be enlarged to without unacceptable loss of detail. Details vary between cameras but an example from the Canon PowerShot S10 illustrates the capacity for storing pictures and quality that can be obtained. On this camera a 16 MB flash card can store 34 images of highest quality with moderate compression of files. These images will yield a satisfactory print on A4 size photo paper when printed at best quality on a colour (inkjet) printer.

The initial financial outlay for a digital camera is relatively high. They can be directly connected to specially adapted colour inkjet printers but most people will print files from a computer. This means that you either have to buy several flash cards (or other similar media) to store images in the field, or carry a laptop computer to regularly download images. Improvements are being made all the time in the technology of digital photography but this will continue to be an expensive option for many. The advantages of having immediate digital images are, however, considerable. Photos of plant symptoms can be sent to colleagues quickly. Some extension services in developed countries already use digital cameras to transmit photographs of symptoms taken in the field direct to a diagnostic lab. This can save valuable time when dealing with valuable crops and disease forecasting.

There is an alternative to buying a digital camera, which combines traditional photography with computers. Transparencies and prints are scanned and the digital images treated in the same way as those from a digital camera. This is a slower route to digital photography but one available to anyone with a camera and access to a scanner. Scanners are widely available and purchase costs have decreased dramatically in recent years. The scanner can also be used as a simple digital camera and you may wish to experiment with plant samples placed on the glass of a flatbed scanner, ensuring that extraneous light is excluded. This technique works best with thin samples, for example leaves and leaf symptoms, but it is worthwhile exploring with other plant material.

The final point to make about digital photography is that good-quality paper must be used to get the best results from colour printing. The more expensive colour inkjet printers are capable of producing the best results and A3 models are increasingly common at a reasonable price. As digital images increase in file size so the need for more powerful computers will become increasingly apparent. Buying a digital camera is only the start of the process required to convert digital images into hard copy.

Aerial Photography

Taking photographs of plant diseases from above is an intriguing option but needs to be carefully considered for individual purposes. The cost is usually high and special permission may be needed in some countries. There are alternative options to hiring light aircraft. These include the use of model aircraft, kites and balloons, all adapted to carry cameras. This is very much the realm of the dedicated enthusiast and such services have a limited availability. If there is a confirmed need for aerial photography, hire someone to do it for you. There are new methods available at affordable prices, which consist of attaching a special camera to the door of a light aircraft.

Storage of Negatives and Slides

The process of logging photographs and storing them safely in a way that allows easy access is a time-consuming process. Priorities will differ among plant pathologists but obviously the more photographs you accumulate, the greater the need to organize and label them. Negatives and transparencies are prone to fungal growth if humidity and temperature are consistently high. At the very minimum, it is important to store material in dry places where extremes of temperature are avoided. This presents very real problems in some tropical countries. Silica gel offers some relief but this is not a permanent solution to a difficult problem.

Colour negatives are commonly stored in transparent sleeves held in ring folders. Write the date and details of the negatives directly on to the storage sheet. The details of each negative are best written on the back of its print using sticky labels.

Slides are best kept in hanging trays stored in filing cabinets. This is the most convenient method for organizing and viewing slides. A less expensive method involves using clear sheets with pockets held in ring folders, though this is less secure. There are numerous methods for cataloguing slides and recording details but it is simplest to write directly on to the slide mount with an indelible pen. Software programs are available for keeping track of your slide collection and printing sticky labels. These require consistent effort and dedication to keep up to date and the writer prefers a simpler manual method, which involves giving a two-letter code to indicate the general topic, followed by a number to denote the particular subject. Slides are numbered sequentially. Thus PD 1.5 denotes Plant Diseases (PD), Trees (1), slide number 5.

A light box is an invaluable tool for examining and organizing transparencies. Holding slides up to the nearest source of light and examining by eye is a poor method for determining quality and content. Edit your photographs and be ruthless with those that are of poor quality. Use a magnifier to reject transparencies that are not sharp or are poorly exposed.

Process-paid slide films are returned with the slides already mounted and date-stamped. If you mount your own slides ensure that the date is written on the mount. Glass mounts may be required for projection at large conferences but plastic mounts are suitable for most general purposes.

Photomicrography

The larger manufacturers of microscopes sell complete systems that incorporate a camera and automatic exposure system. These are expensive but worth the outlay if photomicrography is an important part of your work. If you do not have access to such a system it may be possible to use facilities elsewhere. Cameras can be attached to most microscopes but will require some initial tests to determine the correct exposure. The automatic exposure meter in SLR cameras will help but do not rely on this absolutely, particularly where long shutter speeds are required. You will need to ensure that the internal mirror of the camera can be locked in place to avoid unwanted vibrations. Ensure that the microscope is mounted on a solid and rigid surface.

There is a frustrating gap between the maximum magnification that can be obtained from a macro lens and the magnifications that can be achieved with a camera attached to a stereo microscope. Manufacturers do provide complete systems for photomacrography but these are expensive. Again, ensure that the microscope is mounted on a solid surface. Lighting of the sample is important and initial tests using different types of light and exposures will allow you to use the equipment to maximum effect in the future.

Magnifications greater than life size can be obtained with bellows and extension tubes attached to a camera. This process requires careful attention and patience, which is beyond the scope of this article. A specialist text is given at the end of this chapter for further reference. Macro lenses represent the greatest magnification required by the majority of plant pathologists and these can be simply operated using the normal features of the camera.

Web Sites

www.photo.net is the most comprehensive and useful web site the author has found. www.rps.org.uk Royal Photographic Society. Based in the UK this has an international membership with special interest groups that include those with an interest in natural history and scientific topics. The emphasis, though, is mainly on general photography and the organization is best seen as a point of contact for preliminary enquiries.

Further Reading

There are many basic photographic texts that provide guidance on using cameras and taking photographs. Some contain information on close-up photography and other specialized aspects that could be of use to plant pathologists. The following texts are comprehensive guides to techniques, equipment and materials, ultimately aimed at the professional or serious non-professional photographer. However, they can be relied on to give a comprehensive guide and may be worth consulting for a particular type of photography. They are all published by the Focal Press (www.focalpress.com).

Constant, A. (2000) *Close-up Photography*. Focal Press, Woburn, Massachusetts, USA, 144 pp.

Jacobson, R.E., Ray, S.F., Axford, N.R. and Attridge, G.G. (2000) *Manual of Photography*. Focal Press, Woburn, Massachusetts, USA, 496 pp.

Langford, M. (2000) *Basic Photography*. Focal Press, Woburn, Massachusetts, USA, 368 pp.

Ray, S.F. (1999) *Scientific and Photography and Applied Imaging*. Focal Press, Woburn, Massachusetts, USA, 584 pp.

Zakia, R.D. (2000) *Basic Photographic Materials and Processes*. Focal Press, Woburn, Massachusetts, USA, 400 pp.

Standards and Measurements | 40

Compiled by J.M. Waller

CABI Bioscience UK Centre, Bakeham Lane, Egham, Surrey TW20 9TY, UK

Units of Measurement

The SI (Systéme International d'Unités) standards for measurement are now used. Some common units are as follows.

Basic SI (Système International) Units

Physical quantity	Name of unit	Symbol for unit
Length	metre	m
Mass	kilogram	kg
Time	second	s
Electric current	ampere	A
Temperature	kelvin	K
Luminous intensity	candela	cd

Derived SI units

The unit of force is the *newton* (N) (kg m s^{-2}), that of energy the *joule* (J) (N × m), and of power the *watt* (W) (J s^{-1}). The unit of electric potential is the *volt* (V) and that of electric resistance the *ohm* (Ω). Illumination is measured by the *lux* and customary temperature by the *degree Celsius* (°C).

Other derived units are:

area	square metre	m^2
volume	cubic metre	m^3
density	kilogram per cubic metre	$kg\ m^{-3}$
velocity	metre per second	$m\ s^{-1}$

Allowed units

volume	litre	l	$10^{-3}\ m^3 = dm^3$
mass	tonne	t	$10^3\ kg = Mg$
area	hectare	ha	$10^4\ m^2$

Also the common units of time, e.g. hour, year may be used.

Fractions and Multiples

Multiples of units are normally restricted to steps of a thousand and fractions to steps of a thousandth.

Fractions

Fraction	Prefix	Symbol
10^{-1}	deci	d
10^{-2}	centi	c
10^{-3}	milli	m
10^{-6}	micro	µ
10^{-9}	nano	n
10^{-12}	pico	p

Multiples

Multiple	Prefix	Symbol
10	deca	da
10^2	hecto	h
10^3	kilo	k
10^6	mega	M
10^9	giga	G

Control of Temperature and Humidity

Normal laboratory incubators give a relatively crude control of temperature; those with air circulation give better control but can lead to faster drying of agar in Petri dishes. Tolerances of 0.5–1.0°C are usually adequate. Incubators operating at sub-ambient temperatures have a cooling device added. Direct cooling can lead to ice formation on the cooling coils in the incubator and increased drying of material in the incubator. Indirect cooling, where the coolant operates via a secondary circulation system (e.g. water jacket), avoids this problem but is more complex and bulky.

Where humidity control is required temperature control is critical. Even a very small drop in temperature can lead to condensation of liquid water at humidities above 95%. Accurate control of humidity can be maintained in small closed containers by using different concentrations of sulphuric acid (Table 40.1) or saturated solutions of certain salts (Table 40.2).

Measurement of Concentration of Spores

This is usually done by using a haemocytometer. The central area between the two channels on the haemocytometer slide is 0.1 mm below the level of the sides so that when the specially machined coverslip is placed on the slide the space between the coverslip and central portion of the slide is exactly 0.1 mm deep. The grid rulings on the central areas are 1 mm apart giving a series of 1 mm squares. The central square is subdivided into 25 smaller squares and each of these subdivided again into 16 squares producing, respectively, $1/25$ mm^2 and $1/400$ mm^2. As the chamber beneath the coverslip is 0.1 mm deep, the volume above the 1 mm^2 is $1/10,000$ cm^3, that above the $1/25$ mm^2 is $1/250,000$ cm^3 and that above the $1/400$ mm^2 is $1/4,000,000$ cm^3. Alternatively the turbidity of a suspension can be accurately measured by means of a photoelectric colorimeter or spectrophotometer.

pH and its Measurement

The pH of a solution is the measure of its acidity or alkalinity. The classical definition of pH is the negative value of the logarithm of the concentration of hydrogen ions in gram ions per litre, which may be written $pH = -\log_{10}cH^+$. If an acid is dissolved in the water the concentration of hydrogen ions increases, that of the hydroxyl ions falls and the solution becomes acid. It will

Table 40.1.   Control of relative humidity (RH) using concentrations of sulphuric acid.

H$_2$SO$_4$ (%)	RH (% at 25°C)	H$_2$SO$_4$ (%)	RH (% at 25°C)
0	100.0	45	46.8
5	98.5	50	36.8
10	96.1	55	26.8
15	92.9	60	17.2
20	88.5	65	9.8
25	82.9	70	5.2
30	75.6	75	2.3
35	66.8	80	0.8
40	56.8		

Table 40.2. Control of humidity using saturated solutions of salts.

Salt, sat. soln at 20°C	Relative humidity (%)
$LiCl.H_2O$	15
$KC_2H_3O_2$	20
$CaCl_2.6H_2O$	32
CrO_3	35
$Zn(NO_3)_2.6H_2O$	42
KNO_2	45
KCNS	47
$Na_2Cr_2O_7.2H_2O$ or $NaHSO_4.H_2O$	52
$NaBr.2H_2O$	58
$Mg(C_2H_3O_2)_2.4H_2O$	65
$NaNO_2$	66
$NH_4Cl + KNO_3(1:1)$	73
$NaClO_3$	75
$H_2C_2O_4.2H_2O$ (oxalic acid) or $NaC_2H_3O_2.3H_2O$	78
$Na_2S_2O_3.5H_2O$	76
NH_4Cl	79
$(NH_4)_2SO_4$	81
KBr	84
$KHSO_4$	86
K_2CrO_4	88
$ZnSO_4.7H_2O$	90
K_2HPO_4 or $NaBrO_3$	92
$NH_4H_2PO_4$ or $Na_2SO_4.10H_2O$	93
$Na_2HPO_4.12H_2O$ or $Na_2SO_3.7H_2O$	95
$CaSO_4.5H_2O$ or $Pb(NO_3)_2$	98

be seen that an increase of concentration of hydrogen ions from 1×10^{-7} to 1×10^{-6} will result in a fall of pH from 7 to 6. The greater the acidity, the lower the pH. The two types of method commonly used by biologists to measure pH are indicators and electrical methods.

Indicators

Indicators are substances in which a change of structure or ionization at different pH values causes a change of colour in their solutions. They vary widely in the pH ranges at which the colour changes occur (Table 40.3), and for this reason some care is necessary in the choice of indicator for various tasks. The most accurate way of using indicators is by comparison with standard solutions of known pH, e.g. buffer solutions, containing the same concentration of indicator, or by comparison with standard coloured glasses in a colour comparator (e.g. the Lovibond Comparator). With careful use these methods will often give results accurate to within 0.1–0.2 pH units. The simplest and most rapid method of all is to use indicator papers. These may be dipped quickly into the solution under test, or a drop may be removed and placed on the paper using a wire loop or a Pasteur pipette. The colour obtained is compared with the colour chart supplied with the papers. 'Universal papers'

Table 40.3.   Some indicators and the pH of their colour changes.

Indicator	Range of pH	Colour change
Thymol blue (acid range)	1.2–2.8	Red to yellow
Methyl yellow	2.8–4.0	Red to yellow
Bromophenol blue	3.0–4.6	Yellow to violet
Methyl orange	3.0–4.5	Red to yellow–orange
Bromocresol green	3.8–5.4	Yellow to blue
Methyl red	4.4–6.2	Red to yellow
Chlorophenol red	4.5–7.0	Yellow to red
Litmus	5.0–8.0	Red to blue
Bromocresol purple	5.2–6.8	Yellow to violet
Bromothymol blue	6.0–7.6	Yellow to blue
Phenol red	6.6–8.4	Yellow to red
Neutral red	6.8–8.0	Red to yellow
Cresol purple	7.6–9.2	Yellow to violet
Thymol blue (alkaline range)	8.0–9.6	Yellow to blue
Phenolphthalein	8.2–10.0	Colourless to red
Thymolphthalein	9.3–10.5	Colourless to blue
Alizarine yellow GG	10.3–12.0	Colourless to yellow
BDH Universal	4.0–11.0	Red, yellow, green, violet, red–violet

may be used for rough measurement in the range pH 4.0–11.0. More accurate measurement may then be made using the most suitable narrow range paper, or one impregnated with an individual indicator.

A number of factors such as the presence of proteins, alkaloids and salt in significant concentrations may affect the accuracy of the reading. Temperature should normally be about 20–25°C.

Electrical measurement

The hydrogen gas electrode is the ultimate standard of reference for pH measurement, but its limitations and experimental diffculties make it unsuitable for biological work. Its application depends on the observation that when solutions of differing pH are separated by a very thin glass boundary, an electromotive force (e.m.f.) is set up. Most pH meters consist of a glass electrode and a calomel electrode. The e.m.f. developed between the two electrodes, which is dependent on the pH, is measured with a microammeter in a high impedance circuit, or a null-type bridge circuit. No current is allowed to flow, as the resulting electrolysis would cause incorrect readings. Electrodes are delicate and care is needed when handling them.

Buffers

Buffers control the concentration of hydrogen ions and therefore resist changes of pH when acid or alkali is added. They are usually mixtures of a weak

Table 40.4.   Buffers.

Common name	Constituents	pH range
Citrate buffer	Citric acid and sodium citrate	3.0–6.2
Acetate	Acetic acid and sodium acetate	3.6–5.6
Citrate–phosphate	Citric acid and dibasic sodium phosphate	2.6–8.0
Phosphate	Monobasic and dibasic sodium phosphate	5.8–8.0
Barbitone (Veronal)	Sodium barbitone and HCl	6.8–9.2
Tris	Tris(hydroxymethyl)aminomethane and HCl	7.2–9.0
Boric acid–borax	Boric acid and borax	7.6–9.2
Carbonate–bicarbonate	Sodium carbonate and bicarbonate	9.2–10.6

acid and its salt, or occasionally a weak base and its salt (Table 40.4). The salt formed from a weak acid and a weak base has some buffering action alone, but salts formed from a strong acid and a strong base have none. The desired pH of a buffer is changed, usually within a range of 1–2 pH units, by varying the proportions of the components but the buffering power is greatest at one particular proportion, usually equimolar, and falls off rapidly as the proportions are changed. Dilution of a buffer mixture does not appreciably affect the pH. (See also Chapter 38.)

Provided that no chemical action occurs other than acid–base equilibria, a mixture of two different buffers is effective over the pH ranges of each individually. Thus citric acid and disodium hydrogen phosphate is equivalent to a system of five acids (three stages of citric and the first two of phosphoric) and therefore gives a wide range of pH.

Some Common Elements

Element	Symbol	Atomic weight
Aluminium	Al	26.97
Antimony	Sb	121.76
Arsenic	As	74.91
Barium	Ba	137.36
Bismuth	Bi	209.00
Boron	B	10.82
Bromine	Br	79.916
Cadmium	Cd	112.41
Calcium	Ca	40.08
Carbon	C	12.01
Chlorine	Cl	35.47
Chromium	Cr	52.01
Cobalt	Co	58.94
Copper	Cu	63.57
Fluorine	F	19.00
Gold	Au	197.2
Hydrogen	H	1.0081
Iodine	I	126.92

Continued

Element	Symbol	Atomic weight
Iron	Fe	55.84
Lead	Pb	207.21
Lithium	Li	6.94
Magnesium	Mg	24.32
Manganese	Mn	54.93
Mercury	Hg	200.61
Molybdenum	Mo	96.0
Nickel	Ni	58.69
Nitrogen	N	14.008
Osmium	Os	190.20
Oxygen	O	16.00
Phosphorus	P	30.18
Platinum	Pt	195.23
Potassium	K	39.096
Selenium	Se	79.96
Silicon	Si	28.06
Silver	Ag	107.88
Sodium	Na	22.997
Strontium	Sr	87.63
Sulphur	S	32.06
Tin	Sn	118.7
Tungsten	W	183.92
Uranium	U	238.07
Zinc	Zn	65.38

Colour Standards and Colour Nomenclature

Many names applied to colours have over the centuries been used in varying senses and today in everyday use many colour names are only precise in a particular locality or for a particular purpose. The Munsell Color System (Munsell, 1963) is a comprehensive and satisfactory system in common usage. The *Methuen Handbook of Colour* by Kornerup and Wanscher (1967), which is an English translation from the Danish of *Forver i Farver* by the same authors, provides a most useful, comprehensive and inexpensive colour chart for which in the 2nd edition (1967 for both versions) the equivalent Munsell notation is given for each colour. Dade (1949) attempted to relate the traditional use of latinized colour names with the Ridgway standard and compiled a long list of colour names. The British Mycological Society recommended that the Munsell system of notation should be used by mycologists and plant pathologists and that for nomenclature the terms used by Dade (1949) should be followed. Rayner (1970) produced a colour chart showing the midpoint of the range of colours included under each term and a restandardization of the Ridgway colour samples against Munsell. Also useful is the *Colour Identification Chart* published by Henderson *et al.* (1969) in their *Flora of British Fungi*. Accurate colour measurements for chemical or physical purposes are done with a spectrophotometer.

References

Dade, H.A. (1949) Colour terminology in biology, 2nd edn. *Mycological Papers* No. 6, 22 pp.

Henderson, D.M., Orton, P.D. and Watling, R. (1969) *Flora of British Fungi: Colour Identification Chart*. HMSO, Edinburgh, UK.

Kornerup, A. and Wanscher, J.H. (1967) *Methuen Handbook of Colour*, 2nd edn. Methuen, London, UK, 243 pp.

Munsell, A.H. (1963) *Munsell Book of Color*. Munsell Color Co., Baltimore, Maryland, USA. (Several different forms of this atlas are available)

Rayner, R.W. (1970) A *Mycological Colour Chart*. Commonwealth Mycological Institute, Kew, UK, 34 pp.

Sources of Names of Economic Plants

Bogdan, A.V. (1977) *Tropical Pasture and Fodder Plants*. Longman, London, UK, 475 pp.

Herklots, G.A.C. (1972) *Vegetables in South-East Asia*. George Allen and Unwin, London, UK, 525 pp.

Howes, F.N. (1974) *A Dictionary of Useful and Everyday Plants and Their Common Names*. Cambridge University Press, Cambridge, UK, 209 pp.

Mabberley, D.J. (1997) *The Plant Book*, 2nd edn. Cambridge University Press, Cambridge, UK, 857 pp.

Purseglove, J.W. (1968) *Tropical Crops. Dicotyledons*, Vol. 1, xiv + 332 pp. Vol. 2, 387 pp. Longman, London, UK.

Purseglove, J.W. (1972) *Tropical Crops. Monocotyledons*. Vol. 1, x + 334 pp. Vol. 2, 273 pp. Longman, London, UK.

Rehder, A. (1954) *Manual of Cultivated Trees and Shrubs Hardy in North America Exclusive of the Subtropical and Warmer Temperate Regions*, 2nd edn. Macmillan, New York, USA, 996 pp.

Spedding, C.R.W. and Diekmahns, E.C. (eds) (1972) Grasses and legumes in British agriculture. *Bulletin, Commonwealth Bureau of Pastures and Field Crops* No. 49, 511 pp.

Uphof, J.C.T. (1968) *Dictionary of Economic Plants*, 2nd edn. Cramer, Lehre, 591 pp.

Usher, G. (1974) *A Dictionary of Plants Used by Man*. Constable, London, UK, 619 pp.

Wellman, F.L. (1977) *Dictionary of Tropical American Crops and Their Diseases*. Scarecrow Press, Metuchen, 495 pp.

Willis, J.C. (revised by Airy Shaw, H.K.) (1973) *A Dictionary of the Flowering Plants and Ferns*, 8th edn. Cambridge University Press, Cambridge, UK, 1245 pp.

Plant names can also be found at the IPGR and Kew web sites at www.ipgr.org and www.rbgkew.org.uk

Publication

41

J.M. Waller

CABI Bioscience UK Centre, Bakeham Lane, Egham, Surrey TW20 9TY, UK

The publication of the results of research is the traditional and appropriate conclusion of a scientific investigation. Much of the work published in the scientific literature is repetitious, particularly because the acceptance of published papers are evidence of output and because of the ease with which modern word-processing facilities can churn out literature. Publication usually takes the form of scientific papers but other forms of literary output are often appropriate for plant pathologists bearing in mind that the products of the work may ultimately be used by farmers. Preparation of project proposals is also a critical area nowadays requiring concise and logical writing.

This is not the place to offer detailed guidance on style. A good style is usually attained only by long practice and much rewriting. It may perhaps be noted that short sentences are easier to handle than long. Although most modern plant pathology literature is in English, this is not the mother tongue of many readers. Complicated constructions should be avoided. Short words should be used whenever possible. Avoid technical jargon. Coin new terms only when essential. Respect the traditional meaning of words.

There are many publications offering guidance for scientific writing and a selection is:

Lindsay, D. (1996) *A Guide to Scientific Writing*. Addison Wesley Longman, Australia.

Anon. (1994) *Scientifc Style and Format: the CBE Manual for Authors, Editors and Publishers*. Cambridge University Press, Cambridge, UK.

Friedland, A.J. and Folt, C.L. (2000) *Writing Successful Science Proposals*. Yale University Press, New Haven, USA.

Day, R.A. (1998) *How to Write and Publish a Scientific Paper*. Cambridge University Press, Cambridge, UK.

Alley, M. (1996) *The Craft of Scientific Writing*. Springer-Verlag, New York, USA.

Williams, K. (1996) *Scientific and Technical Writing*. Oxford Brookes Univeristy, Oxford, UK.

The many word-processing packages now available include dictionaries but the standard 'spell check' can be misleading. Most journals and publishers give indications of which type of English they prefer (e.g. US or UK usage).

Many journals and publishers offer guidance to authors on the preparation of manuscripts and as the conventions adopted vary between one journal or publisher and another an author writing a paper with a particular periodical in mind should consult a recent issue and conform to the style.

The following notes merely draw attention to some general points that if acted on result in the saving of much editorial time.

Title

Keep the title as short and as informative as possible so that whenever cited it gives the maximum information regarding the content of the paper.

It is frequently neater to relegate a general title of a series of papers to a footnote, e.g. Effects of temperature and humidity on fungi of the rhizosphere of soybeans.*

*Studies on the rhizospheres of leguminous plants, 16.

Summary or abstract

Most of those consulting the original journal will do no more than glance at the summary or abstract of your paper. It may also be used with little or no alteration by abstract journals. Ensure that it gives a balanced indication of the contents.

Arrangement

Although there can be no hard and fast rule for subdividing a paper, three categories are often sufficient. Different hierarchies of subject headings are usually needed. Main headings are usually in bold, lesser ones italicized. Journal styles will indicate which are in current use by the editors.

Tables

Every table should be typed double-spaced on a separate page together with its explanation. Check that every table is referred to in the text and indicate as a text insert [often put in square brackets] the approximate place for its insertion. Tables should be concise and where relevant have appropriate statistical data indicating variances.

Illustrations (figures)

Photographs should be on glossy paper and show good contrast. With digital cameras it is becoming easier to prepare computer files of pictures, which can be directly used in computerized printing. For line drawings, graphs, etc. there are many computer packages enabling these to be prepared as computer files for direct printing. If hard copies have to be prepared they should be done so in indian ink on white bristol board and sufficiently large to allow a reduction to approximately one-half.

Print-outs of figures or original drawings should have, written in pencil on the back, the author's name and short title of the paper, the number of the figure, the reduction required, and when not self-evident the magnification of the original drawing.

Check that every illustration is referred to in the text.

Scientific names

The Latin (scientific) names of fungi and bacteria are governed by the International Codes of Botanical and Bacteriological Nomenclature, respectively. Every effort should be made to ensure that the names used are in line with the code. As nomenclatural changes may frequently take place, it is often advisable to consult specialist taxonomists on the most recent usage. Taxonomists in one's own country may be consulted and may be contacted through the local BioNet loop or recourse can be made to the taxonomists at CABI *Bioscience* for microfungi and plant bacteria and at the Royal Botanic Gardens, Kew, for macrofungi and plants.

Binomial Latin names are italicized (underline in typescript) and the specific epithet is always decapitalized. The names of orders and families are printed in italics with initial capital letters. A generic name is always italicized but may be abbreviated to the initial capital letter, after being spelt out where first used, whenever this can be done without ambiguity. Common names of fungi derived from generic names are printed in roman and decapitalized, e.g. '*Fusarium* species' but 'some fusaria'.

Abbreviations

A list of some commonly used and generally acceptable abbreviations is given in Table 41.1. Many other abbreviations are at times permissible so long as intelligibility is not reduced and the style does not become telegraphic.

Authors' names

The orthography of authors' names and consistency in their use is fraught with difficulties. It may be necessary to know the nationality of the author. For

Table 41.1.   Acceptable abbreviations for some common terms.

Absolute	abs.	millilitre(s)	ml
Academy	Acad.	millimetre(s)	mm
Agriculture	agric.	minute(s)	min
Applied	Appl.	minimum	min.
approximately	approx.	Ministry	Minist.
Association	Assoc.	molar	M
atmosphere(s)	atm	mole	mol
average	av.	Museum	Mus.
Bacteriology etc.	Bact.	Mycology etc.	Mycol.
Biology etc.	Biol.	nanometre	nm
boiling point	b.p.	National	Natl
Botany etc.	Bot.	normal (of solutions)	N
chapter(s)	chap.	North	N.
coefficient	coeff.	number(s)	no.
coloured (as in col.pl.)	col.	optimum	opt.
Commonwealth	Commw.	page, pages	p., pp.
compare	cf.	parts per million	p.p.m.
concentrated etc.	conc.	Pathology etc.	Pathol.
constant	const.	Physiology etc.	Physiol.
cultivar	cv.	Phytopathology etc.	Phytopathol.
decomposition	decomp.	plate(s)	pl.
diagram(s)	diag.	Protection	Prot.
diameter, in diameter	diam.	quintal(s)	q
Division	Div.	which see	q.v.
East	E.	reference(s)	ref.
edition	edn.	relative humidity	RH
Entomology etc.	Entomol.	Research	Res.
Experiment etc.	Expt.	revolutions per minute	r.p.m.
Faculty	Fac.	School etc.	Sch.
family	fam.	Science etc.	Sci.
figure(s)	fig.	second(s)	s
Forestry	For.	Section	Sect.
forma	f.	Society	Soc.
forma(e) specialis(es)	f.sp. (ff.sp.)	South	S.
gallon(s)	gal.	solution	soln.
genus	gen.	species (singular and plural)	sp., spp.
genus, new	gen.nov.	species, new	sp. nov
gram(s)	g	square	sq.
hectare	ha	Station	Sta.
Horticulture etc.	Hort.	strain	str.
hour(s)	h	subspecies	ssp.
inch(es)	in.	Technical	Tech.
Institute	Inst.	Technology etc.	Technol.
Institution	Instn.	temperature	temp.
kilogram(s)	kg	ultra violet	UV
kilometre(s)	km	University	Univ.
maximum	max.	Variety	var.
metre(s)	m	volume	vol.
Microbiology etc.	Microbiol.	weight	wt.
microgram(s)	μg	West	W.
micrometre	μm	Year(s)	yr
milligram(s)	mg	Zoology etc.	Zool.

example, in Spanish the father's name precedes the mother's (e.g. A. Gonzalez Fragoso, which should be listed in bibliographies and indexes as Gonzalez Fragoso, A.) whereas in Portuguese the mother's name is given first (e.g. A Sousa da Camara, indexed as Da (or da) Camara, A.S.). Transliteration by different systems into different languages from cyrillic characters can also cause confusion – Sholokhov, Scholochow, Cholokhov and Solochov are variants of one name and a German name such as 'Herzen' after transliteration into cyrillic characters and back again may be transformed into 'Gertsen'.

In cases of difficulty a librarian should be consulted or what appears to be custom followed.

Measurements

Use the metric system whenever possible. Otherwise give the metric equivalents of measurements for other systems (see Chapter 40).

References

The basic citation for a paper varies in different journals, but a common arrangement is: author's surname, initials of forenames, date of publication, title of paper, title of periodical, volume number, and first and last pages. The system of using abbreviated titles of journals has now been largely abandoned and the full title should be used.

The citation for books typically includes: author's name, date of publication, title of the book (either in italics or in quotes), edition, publisher, place of publication and number of pages.

Before submitting a paper always check that references referred to in the text are included in the bibliography and vice versa.

Acknowledgements

These can be important if support from outside funding agencies has been used.

Copyright and intellectual property rights issues

Publication of information puts it in the public domain and can compromise subsequent rights to patent etc. Many funding agencies are sensitive about IPR issues relating to outputs of research that they have supported, so it is essential to be clear about these issues before publication of any sort. Authors generally hold rights to material they have produced including photos, art work, etc. unless otherwise designated, e.g. by employer or funding agency.

Certain restrictions on copying are imposed on material published in all countries that are signatories to the Berne copyright convention.

To ensure that a request to a library for copies of articles from periodicals is dealt with without delay it is essential that it be accompanied by a copyright declaration. The declarations required by different organizations vary in detail. The essential points are covered by such wording as:

1. I the undersigned hereby request you to make and supply to me a copy of the item(s) listed below which I require for the purposes of research or private study.

2. I have not previously been supplied with a copy of these items.

3. I undertake that if a copy is supplied to me in compliance with the request made above, I will not use it except for the purposes of research or private study.

Date .... Signature*

*This must be the personal signature of the person making the request. A stamped or typewritten signature or the signature of an agent is not sufficient. (A list of the items required follows.)

Only one copy of each article can be supplied and not more than one article from any one issue of a publication can be sent to the same individual. The supply of material does not imply any freedom to make further copies. To do so may be an infringement of copyright.

For extracts from publications, other than periodicals, which are still in copyright, the permission of the copyright holder must be obtained.

For a general discussion of copyright as regards learned journals see St Aubyn (1981) *Journal of Research Communication Studies* 3, 65–73.

Research Proposals

The writing of these can take up much time and it is necessary to be clear on what is required by the funding agency. Some agencies have extensive forms requiring detailed attention, whereas others may leave the construction of the proposal to the proposer. Competitive judgements can be swayed by the ease with which a proposal can be read and assimilated. In general terms most proposals require some initial (executive) summary of the proposal with cost, a brief background, who will be involved, where will it be done, and how it links with related research both past and present. The objectives need to be clearly explained together with the outputs that the research will produce to satisfy these objectives, the activities that will be undertaken to generate the outputs and over what time frame. There is usually a need to demonstrate how the outputs will be used to address some perceived need and some evidence for the need for the work, who will be the beneficiaries and how will the outputs of research reach and be used by them. Detailed budgets are invariably required.

Oral Presentations

Although at one time or another every plant pathologist has to give informal talks to growers, papers at scientific meetings or more formal lectures, very few receive any training in elocution or public speaking. The following notes merely offer guidance on how to avoid some of the principal hazards.

Only the most formal paper, such as a Presidential Address, should be read from manuscript and even then there is usually opportunity to interpolate asides or to summarize in other words passages that refer to slides or other illustrative material. Inexperienced, and many experienced, speakers do, however, find it helpful to write out their contributions in full and to have the text or a summary before them as they speak. Short notes or cards with headings on are often useful.

Delivery of a paper in public takes appreciably longer than to read it aloud alone in private. Allow approximately 400 words per 5 min and rate each slide at 50 words. It is a useful and salutary experience to tape your paper in private and to listen to the recording.

Papers as prepared for publication are usually unsuitable for public presentation. In view of the increasing specialization of workers, even in one field, many members of an audience usually welcome having the topic to be dealt with set in a wider context – but come to the point of the paper without undue delay.

Do not overload the talk with data or use laboratory jargon. Detailed descriptions of techniques should be avoided. Accounts of work in progress tend to be more stimulating than those of completed investigations. A contribution should never exceed the allotted time.

Slides for projection should be relevant and if of tabulated data must not be overcrowded. (Graphs and histograms are more readily comprehensible than extensive numerical data.) As a rough guide a 35 mm slide will be legible to those at the back of the lecture theatre if the slide can be read when held 35 cm (14 in) from the eye. Avoid overloading with data, and do not have more than six lines of text or data. Similar recommendations apply to powerpoint demonstrations. These allow construction of artwork and background effects, but care is needed to avoid distracting the audience with superfluous effects. Ensure that the text/data contrasts and stands out from any background effects. Make sure all visual aid equipment is operating properly beforehand, e.g slides the right way round for projection, computer and projector for powerpoint are functioning correctly.

Chairmen, unless specially commissioned to introduce a topic, should remember that the audience has come to hear the speakers and that a chairman's duty is to allow the speakers fair play and to regulate the discussion. Remain at the rostrum during any discussion of your contribution.

Finally, be audible. Address those at the back of the room rather than those in the front row. Do not direct important statements at the screen or the blackboard or wave scripts in front of you.

Additional suggestions for speakers are given in the following papers:

Norris, J.R. (1978) How to give a research talk: notes for inexperienced lecturers. *Biologist* 25, 68–74.
Wheatley, G.A. (1981) Notes on the preparation, illustration and presentation of papers for scientific meetings. *Annals of Applied Biology* 99, 191–194.

Electronic Databases and Information Technology in Plant Pathology

H.L. Crowson and L.A. McGillivray

CAB International, Wallingford, Oxfordshire OX10 8DE, UK

Introduction

The application of information technology (IT) can now provide a variety of tools for plant pathologists, whether they are working in the laboratory or field, or involved in making decisions or guiding policy. Databases can be used routinely for storing experimental data, recording field conditions, mapping the distribution of pests, searching the literature, and in making decisions for quarantine and pest-risk analysis. A database can be defined as a collection of structured data independent of any particular application and can comprise a simple 'electronic card index' or a sophisticated multimedia tool for managing knowledge and exchanging information. A list of electronic databases that is still largely relevant to crop protection was prepared by Gooch and Cowie (1989).

Wider accessibility to computers and related software, both in terms of cost and user-friendliness, combined with continuous improvements in speed, power and memory have increased the importance of electronic data as an information resource for plant pathologists. Such databases enable users to organize stored information in a manner that can be easily searched, retrieved, analysed and updated. These attributes have become invaluable in monitoring constantly changing variables, such as pathogen nomenclature, host range, distribution and molecular biology. Rapid developments in scientific techniques, ideas and information, such as those in the field of molecular biology, have made electronic systems of storage essential to the laboratory scientist. The shift towards the publication of databases and information on the World Wide Web will, by facilitating communication between colleagues in different disciplines and countries, inevitably further accelerate the pace of both research and the development of more sophisticated software tools to order information and enable its exchange.

The practical implementation of IT helps not only those involved directly in research. The transformation of methods of recording and interpreting research data into knowledge-based systems has led to improvements in the quality and ease of access of information available to scientists working at the decision-making level, so that policies can be based on real events and the prediction of potential outcomes.

Plant pathologists now have a vast array of tools to manage information and access to a wide range of sources of information. The scale and diversity of this information can seem quite daunting. This chapter attempts to provide a brief overview of the main uses of IT in relation to plant pathology and some of the future prospects for developments in this field. For further information, see Bridge *et al.* (1998). Any printed record of the resources available over the Internet will inevitably become out of date: instead, interested readers should refer to Kraska (2001) for an up-to-date guide to sites and resources of interest to plant pathologists.

Handling Facts to Produce Information

Data management

All scientists need to be able to store, retrieve, analyse and manipulate their data. The routine use of electronic databases and graphic and statistical software packages has, for many, become routine, replacing card indexes, graph paper and calculators. Data can be kept on a stand-alone computer for individual access or networked so that it is available to colleagues. Although the means of manipulating data has become ever more sophisticated, it is important to remember that the quality of the raw data will determine the accuracy of subsequent analysis and there is still no substitute for good experimental design and technique. The application of IT to general data management can be summarized under the following headings:

- data capture
- data monitoring
- data storage (e.g. image galleries)
- data manipulation
- data analysis (e.g. statistics programs)
- stock control
- record keeping

The following are some examples of large databases that may be of use to those working in plant pathology research.

Nomenclature systems

Databases of pest names and synonymy are very important to pathologists searching the literature and embarking on research projects. Pest nomenclature is a constantly changing field. Species 2000 (www.sp2000.org/) has the objective of enumerating all known species of plants, animals, fungi and microbes on earth as the baseline dataset for studies of global biodiversity. It will also provide a simple access point enabling users to link to other data systems for all groups of organisms, using direct species-links.

Culture collections

Sources of culture collections are available electronically and are valuable to pathologists looking for sources of cultures for their research. For example, databases developed at the US National Fungus Collections (http://nt.ars-grin.gov/) provide access to information about fungi, primarily those associated with plants or otherwise of agricultural importance. The United Kingdom National Culture Collection (www.ukncc.co.uk/) coordinates the activities of the UK national service collections. Index Herbariorum (www.nybg.org/bsci/ih/), a joint project of the International Association for Plant Taxonomy and The New York Botanical Garden, is a detailed directory of over 3000 public herbaria of the world and the 8000 staff members associated with them.

Molecular data

Access to molecular biological data, e.g. protein and DNA sequences and structural models, has proved extremely valuable in allowing plant pathologists to keep abreast of the rapid developments in molecular biology. See Kraska (2001) for an up-to-date guide to further sources of information.

Quarantine databases

Several computerized relational databases, developed in the 1980s and 1990s to compile data on pest distributions and make these searchable by pest, host or country basis, are now available to plant pathologists. The European and Mediterranean Plant Protection Organization (EPPO) database PQR is one example of this type of database. It gives access to data on all the quarantine pests of the EPPO A1 and A2 lists, contains data on many other quarantine pests of interest to other regional plant protection organizations, and is updated regularly. For further information, see EPPO (2000). Information relating to international aspects of plant quarantine can be found at the International Plant Protection Convention (IPPC) area of the Food and Agriculture Organization (FAO) web site (www.fao.org).

Bibliographic databases

Bibliographic databases usually include references and abstracts, indexed by subject-specific keywords so users can retrieve searches for a particular topic or species of interest. For example, the CAB ABSTRACTS database, published by CAB *International*, provides abstracts of internationally published scientific research literature in agriculture and the biosciences. The Review of Plant Pathology is compiled from this database and is available as a CD-ROM, online and as a printed journal (see www.cabi.org or http://pest.cabweb.org/ for further information). AGRICOLA (AGRICultural OnLine Access) is a machine-readable database of bibliographic records created by the National Agricultural Library of the US Department of Agriculture and its cooperators. See www.nalusda.gov/ for further information. Some bibliographic databases include the full text of the original paper, allowing users greater access to the world's literature. Software packages are also available to help scientists manage databases of references they need for writing reports.

Interpreting Information to Produce Knowledge

The interpretation of stored scientific data leads to knowledge-based systems. There is wide scope for the development of these systems in the future and for their ultimate transfer to the web. Examples of such systems include the following.

Taxonomic information systems

One example of a taxonomic information system is the Universal Virus Database which uses DELTA (the DEscription Language for TAxonomy), a method of coding the descriptions of taxa, their characters and states in a data matrix using ASCII text (Dallwitz *et al.*, 1993). The resulting database is authorized by the International Committee on Taxonomy of Viruses and is available on the Internet (www.ncbi.nlm.nih.gov/ICTVdb/).

Molecular information systems

These systems record, interpret and model molecular information. Molecular structures, produced from primary data, are available. Examples of this type of system are the Protein Data Bank (maintained by the Research Collaboratory for

Structural Bioinformatics; see www.rcsb.org/pdb/) and the European Molecular Biology Laboratory's EMBL Nucleotide Sequence Database (Stoesser *et al.*, 1999; see also www.ebi.ac.uk/embl/).

Geographical information systems

Geographical Information Systems (GIS) present digital information on a geographical map of the area to which the information pertains. CLIMEX, a program developed by CSIRO Division of Entomology (Australia) in 1985 and updated since, is an example of a computer-based system for estimating or predicting the potential distribution and relative abundance of species in relation to climate. See www.ento.csiro.au/research/climex/climex.htm for further information.

Multimedia systems

Multimedia systems incorporate various forms of media in a single program; such systems may resemble 'encyclopaedias', 'picture galleries' or 'film archives'. Within the discipline of crop protection there is considerable scope for the further development of multimedia systems to help diagnose and manage pest and disease problems. Many such systems are already available; some focus on specific crop hosts, whereas others aim to present a broader coverage.

The *Crop Protection Compendium* is an example of one such multimedia resource that will also be regularly updated. Currently available on CD-ROM (with options to link to Internet information) and via the Internet, it incorporates a hierarchical taxonomic structure, text, images, GIS, identification keys, diagnostic aids, glossaries, details on pesticides and biopesticides and bibliographic information. The Compendium was developed by CAB *International*, in partnership with many organizations from both the public and private sectors and hundreds of international pest specialists (Sweetmore *et al.*, 1998). Users can search the relational database for data sheets on specific pests using a list of names, including scientific names, synonyms and common names in a number of languages. Data sheets on pests, diseases, weeds, nematodes and natural enemies each consist of comprehensive texts, taxonomic information, pictures and a distribution map. Crop and country data sheets link to information on crop production from the FAO. Country data sheets contain information on climate, land area, population, etc. and selected statistics from the World Bank including gross national product (GNP) per capita and gross domestic product (GDP) from agriculture. As an encyclopaedic source of worldwide information on pests and crops, with links to many other information sources, the Compendium is an extensive, authoritative knowledge base with potential as an aid for research, training and decision making. See www.pest.cabicompendium.org/cpc/ for further details.

Using Knowledge to Support Decision Making

The knowledge acquired from using electronic media and database sources can be used to improve training and support decision making.

Diagnostics

The identification of unknown species by interactive computer software offers the advantage of allowing the user to enter attributes in order to identify a specimen; the program eliminates the taxa that do not match the specimen. There are many keys to insect groups but there is also great scope for development in this field and in symptom-based diagnostics. The *Crop Protection Compendium* (described above) includes a diagnostic tool that uses information on crop attacked, country or symptoms caused to list potential pests (including plant pathogens). DIAGNOSIS (see below) also includes symptom-based diagnostic training aids.

Over-reliance on computer-based diagnosis without adequate knowledge of field situations could present pitfalls; a computer package should be viewed as an additional tool and not a substitute for trained personnel.

Models of Epidemics

The impact of IT on modelling epidemics of plant pathogens has yet to be fully exploited. Electronic media offer great potential in this field.

The Centre for Pest Information Technology and Transfer (based at the University of Queensland, Australia; see also www.cpitt.uq.edu.au/) has developed several models of epidemics. 'RiceIPM' is aimed at providing training and decision support for integrated pest management of rice; it includes a model that relates pest damage and yield loss. 'Mouser' is an information transfer and decision support system for the management of mouse plagues. It contains information for farmers and extension officers and includes a population model.

Using Knowledge to Make Predictions

Forecasting

Computerized decision support has been applied to risk analysis in software such as CLIMEX, which can be used in Pest Risk Analysis. To date, CLIMEX has been used to estimate the distribution of insects, but could equally be applied to pathogens.

Some extension services now provide forecast services via the Internet. For example, the North American Plant Disease Forecast Center at North Carolina State University maintains the Blue Mold Forecast service for *Peronospora tabacina* on tobacco (www.ces.ncsu.edu/depts/pp/bluemold/index.html). Forecasts are generated several times each week from March through September. Additional forecasts are provided during peak epidemic periods.

Conveying Knowledge: Education and Training

One of the main areas where electronic media can offer a significant advantage is in the development of interactive training media. IT can offer the major advantage of providing access to high-quality images such as symptoms on diseased plants and electron micrographs of pathogens. Software packages and Internet links and discussion groups can also facilitate distance learning. Interactive electronic media allow flexibility in training, the speed and direction of training being governed by the individual.

DIAGNOSIS (developed as a joint project between Massey University, New Zealand, and the Co-operative Research Centre for Tropical Pest Management, Queensland, Australia) is a multimedia training resource that aids in teaching the process of diagnosing crop problems. The user is presented with a problem scenario and must interrogate the program to diagnose the situation. See Stewart *et al.* (1995) or www.diagnosis.co.nz/ for further information.

In addition to teaching plant pathology students, electronic teaching aids or tutorials can be used to transfer information from research to extension staff and growers. Several tutorial programs have been developed that involve simulations of plant diseases for training purposes.

Storing and Disseminating Information

Electronic publishing

Nowhere has the move towards electronic media become more apparent than in the published literature. Increased usage of electronic publishing has improved access to original research articles and bibliographic databases, which are now available in a variety of electronic formats such as CD-ROMs, and on the Internet. The electronic dissemination of this information increases the timeliness of the literature and reduces the reliance on library services. Electronic information can also be easily downloaded and transferred between electronic systems, and this flexibility is one of the major benefits of the electronic medium. Publishing information on the Internet also has the major advantage of frequent updating; access to archival information should, however, be maintained.

The World Wide Web

The Internet is becoming increasingly important as a source of plant pathology information. The range of information available on the web is increasing, with more organizations moving their databases and publications to the web. Member societies such as the British Society for Plant Pathology (www.bspp.org.uk/) and the American Phytopathological Society (www.apsnet.org/) aim to create 'virtual communities' that complement their publishing and conference activities. The International Society for Plant Pathology (www.isppweb.org/) maintains a searchable database of plant pathologists around the world and links to national plant pathology societies.

There are also sites available on the web advising plant pathologists on useful links to help navigate through the array of information available (e.g. Kraska, 2001).

The University of Florida, the Smithsonian National Museum of Natural History, USA and the FAO are developing an Internet-based access portal to Ecological Knowledge (www.ecoport.org/). Among other features, it provides information on host plants (including uses, ethnobotany, distribution and recorded pests) and on pests, detailing each pest's symptoms, ecology, dispersal/vectors, control and distribution.

IT in Developing Countries

There are obvious limitations to the use of IT in developing countries such as the phone network, electricity generation, cost, access and problems as the speed of computer development gives rise to redundant systems that need to be replaced. Some countries, however, have side-stepped stages of development and gone straight to cable systems and mobile phones without reliance on the traditional phone systems. There are arguable cost benefits to using electronic sources of information when setting up new libraries. Electronic forms of data are easier to search and less costly to maintain and update than hard copies of books and journals. The fact that the web is becoming a major resource in scientific information is also a serious consideration when decisions are being made with regard to future information management.

References

Bridge, P., Jeffries, P., Morse, D.R. and Scott, P.R. (1998) *Information Technology, Plant Pathology and Diversity*. CAB International, Wallingford, UK.

Dallwitz, M.J., Paine, T.A. and Zurcher, E.J. (1993) *DELTA User's Guide: a General System for Processing Taxonomic Descriptions*, 4th edn. CSIRO, Canberra, Australia, 136 pp.

EPPO (2000) *PQR – the EPPO Plant Quarantine Data Retrieval System*. European and Mediterranean Plant Protection Organization, Paris, France. www.eppo.org/

Gooch, P. and Cowie, A. (1989) Survey of electronic databases in crop protection. In: Harris, K.M. and Scott, P.R. (eds) *Crop Protection Information*. CAB International, Wallingford, UK, pp. 237–291.

Kraska, T. (2001) *The Plant Pathology Internet Guide Book*. Institute for Plant Diseases, University of Bonn, Bonn, Germany. www.ifgb.uni-hannover.de/extern/ppigb/ppigb.htm

Stewart, T.M., Blackshaw, B.P., Duncan, S., Dale, M.L., Zalucki, M.P. and Norton, G.A. (1995) Diagnosis: a novel, multimedia, computer-based approach to training crop protection practitioners. *Crop Protection* 14(3), 241–246.

Stoesser, G., Tuli, M.A., Lopez, R. and Sterk, P. (1999) The EMBL Nucleotide Sequence Database. *Nucleic Acids Research* 27 (1), 18–24.

Sweetmore, A., Schotman, C.Y.L., Zhang, B.C., Rudgard, S.A. and Scott, P.R. (1998) Integrated information management: a multimedia system for crop protection. In: Bridge, P., Jeffries, P., Morse, D.R. and Scott, P.R. (eds) *Information Technology, Plant Pathology and Biodiversity*. CAB International, Wallingford, UK, pp. 117–128.

Glossary of Plant Pathological Terms

43

Compiled by J.M. Lenné

ICRISAT, Patancheru, Andhra Pradesh 502 324, India

Periodically plant pathologists develop a cooperative concern on the misuse of plant pathological terms. In an attempt to increase precision in terminological usage the American Phytopathological Society published definitions of some common phytopathological terms (*Phytopathology* 30, 361–368, 1940) and of terms applied to fungicides (*Phytopathology* 33, 624–626, 1943). Subsequently the Plant Pathology Committee of the British Mycological Society published its own definitions of terms used in plant pathology (*Transactions of the British Mycological Society* 33, 154–160, 1950). In addition, many authors have added short glossaries to their texts and a number of phytopathological terms are included in Kirk, P.M., Cannon, P.E., David, J.C. and Stalpers, J.A. (eds) (2001) *Ainsworth and Bisby's Dictionary of the Fungi* (9th edn, CAB International, Wallingford, UK) and in Snell, W.H. and Dick, E.A. (1971) *A Glossary of Mycology* (Harvard University Press, Cambridge, Massachusetts, USA). In 1973 the Terminology Subcommittee of the Federation of British Plant Pathologists prepared *A Guide to the Use of Terms in Plant Pathology* (*Phytopathological papers* no. 17, 55 pp.). *A Dictionary of Plant Pathology*, 2nd edn (P. Holliday, 1989, Cambridge, University Press, Cambridge, UK) provides the most comprehensive treatment and reference to this is recommended. The above-mentioned publications were all consulted during the preparation of the following select glossary.

Multilingual Glossaries

Bos, L. (1978) *Symptoms of Virus Diseases in Plants*, 3rd edn. Centre for Agricultural Publishing and Documentation, Wageningen, The Netherlands, 225 pp.

Merino-Rodriguez, M. (1966) *Elsevier's Lexicon of Plant Pests and Diseases*. Elsevier, Amsterdam, 351 pp.

Miller, P.R. and Pollard, H.L. (1976) *Multilingual Compendium of Plant Diseases*. American Phytopathological Society, St Paul, Minnesota, USA, 457 pp.

Miller, P.R. and Pollard, H.L. (1977) *Multilingual Compendium of Plant Diseases. Viruses and Nematodes*. American Phytopathological Society, St Paul, Minnesota, USA, 434 pp.

Glossary

acervulus, an erumpent, cushion-like mass of hyphae having conidiophores and conidia, and sometimes setae, characteristic of the *Melanconiales*.

acid-fast, (of bacteria) having cells that are relatively impermeable to simple stains, but when stained with a strong reagent (e.g. basic fuchsin in aqueous 5% phenol applied with heat), subsequently resist decolorization by strong acids (e.g. 20% sulphuric acid).

acquisition feeding, the feeding of a vector on a virus source in transmission tests.

active ingredient, the active component in a formulated product.

adherence, the property of a fungicide to adhere or stick to a surface.

adjuvant, material added to a fungicide to improve some chemical or physical property.

aerobe, an organism needing free oxygen for growth.

aetiology (etiology), the science of the causes of disease; the study of the causal factor, its nature, and relations with the host.

agglutinin, an antibody that causes a particulate antigen to clump and settle out of suspension.

air spore, *see* spore.

alternate host, either of the two hosts of a heteroecious rust.

alternative host, one of several plant species hosts of a given pathogen.

amphitrichous, having one flagellum at each pole.

anaerobe, an organism able to grow without free oxygen.

anamorph, asexual or imperfect state of a fungus.

antagonism, a general term for counteraction between organisms or groups of organisms.

anthracnose, a plant disease having characteristic limited black lesions, usually sunken, generally caused by one of the *Melanconiales*; **spot anthracnose,** a disease caused by *Elsinoë* or *Sphaceloma*.

antibiotic, a substance produced by a microorganism and able to inhibit the growth of other microorganisms, or to destroy them.

antibody, a substance that is produced in response to injection of a foreign substance (antigen) into an animal body, and that reacts specifically with the foreign substance. Antibodies are modified serum globulins.

antigen, a substance that, when injected into an animal body, stimulates the production of a substance (antibody) antagonistic to the substance injected.

antiserum, serum that contains antibodies.

antisporulant, substance preventing spore production without killing vegetative growth of a fungus.

apothecium, the generally cup- or saucer-like ascocarp of *Discomycetes*.

appressorium, a thick-walled fungal cell formed prior to penetration.

asexual, having no sex organs or sex spores; vegetative.

atomize, to reduce a liquid to fine droplets by passing it under pressure through a suitable nozzle, or by applying drops to a spinning disc.

attenuation, lessening of the capacity of a parasitic organism or virus to cause disease; reduction of its virulence.

autoecious, completing the life cycle on one host (especially of rusts), cf. heteroecious.

autolysis, the dissolution of tissues by enzymes within them; self digestion; cf. lysis.

autotrophic, living on inorganic materials as nutrients, cf. heterotrophic.

avirulent, lacking virulence.

axenic, culture of a single species in the absence of others, pure culture.

axeny, inhospitality, 'passive' as opposed to 'active' resistance of a plant to a pathogen.

bactericide, a substance causing death of bacteria.

bacteriophage, a virus causing lysis of bacterial cells.

bioassay, quantitative estimation of biologically active substances by the extent of their actions under standardized conditions on living organisms.

biological control, total or partial destruction of pathogen populations by other organisms.

biological specialization, manifestation by members of a species of physiological specialization of infective ability on a host or group of hosts. See physiologic race.

biologic form or race, = physiologic race (q.v.)

biotechnology, the use of genetically modified organisms and/or modern techniques and processes with biological systems for industrial production.

biotroph, an organism that can live and multiply only on another living organism.

biotype, a subgroup within a species or race usually characterized by the common possession of a single or a few new characters.

blight, a disease characterized by general and rapid killing of leaves, flowers and stems.

blotch, a disease characterized by large, irregularly shaped, spots or blots on leaves, shoots and stems.

breeding, the use of controlled reproduction to improve certain characteristics in plants (and animals).

budding, a method of vegetative propagation of plants by implantation of buds from the mother plant on to a rootstock.

bunt, a disease of wheat caused by the fungus *Tilletia* in which the contents of the wheat grains are replaced by odorous smut spores.

callus, a mass of thin-walled undifferentiated cells, developed as the result of wounding or culture on nutrient media.

canker, a plant disease in which there is a sharply limited necrosis of the cortical tissue and malformation of the bark.

carrier, (1) an organism harbouring a parasite without itself showing disease, **(2)** the material used to convey a fungicide to its target.

chemotherapy, treatment of disease by chemicals.

chlorosis, absence, partial or complete, of normal green colour.

chronic symptoms, symptoms that appear over a long period of time.

circulative viruses, viruses that are acquired by their vectors through their mouthparts, accumulate internally, then are passed through their tissues and introduced into plants again via the mouthparts of the vectors.

clone, the group of genetically identical individuals produced asexually from one individual.

cloning, the multiplication of a group of DNA molecules derived from one original length of DNA sequences and produced by a bacterium or virus into which it was introduced using genetic engineering techniques, often involving plasmids.

collar rot, rotting of the stem at or about soil level.

colony, (of bacteria and yeasts) a mass of individuals, generally of one species, living together; (of mycelial fungi) a group of hyphae (frequently with spores), which, if from one spore or cell, may be one individual.

compatible, (of pesticides) able to be mixed without deleterious effect.

complement, a thermolabile component of animal serum, which reacts non-specifically with antigen–antibody complexes.

complementary DNA (cDNA), DNA synthesized by reverse transcriptase from an RNA template.

complement fixation, the combination of complement with an antigen–antibody complex.

conidioma, any hyphal structure bearing conidia.

conidium (conidiospore), any asexual spore (other than a sporangiospore or intercalary chlamydospore), especially a *conidium verum*, an asexual spore (or pycnidiospore), which comes away from its conidiophore when mature.

conjugation, a process of sexual reproduction involving the fusion of two gametes; (in bacteria) the transfer of genetic material from a donor cell to a recipient cell through direct cell-to-cell contact.

constitutive, a substance, usually an enzyme, whose presence and concentration in a cell remains constant, unaffected by the presence of its substrate.

control, (1) to prevent or retard the development of disease; **(2)** untreated subject for comparison with experimental treatment.

coverage, distribution of a fungicide over a discontinuous area such as leaves of a tree.

cross protection, the phenomenon in which plant tissues infected with one strain of a virus are protected from infection by other more severe strains of the same virus.

cryptogram, a descriptive code summarizing some of the main properties of a virus.

cultivar, a variety of a cultivated plant.

culture, to grow microorganisms or plant tissue in a pure form, generally on a prepared food material; a colony of microorganisms or plant cells artificially maintained on such food material (or for biotrophic pathogens on a host plant).

cuticle, a thin, waxy layer on the outer wall of epidermal cells consisting primarily of wax and cutin.

cyst, a sac, especially a resting spore or sporangium-like structure.

damping-off, a disease that rots seedlings at soil level, or prevents their emergence (pre-emergence damping-off).

dark mildew, *see* mildew.

-deme, a suffix denoting any group of individuals of a specific taxon.

density-gradient centrifugation, a method of centrifugation in which particles are separated in layers according to their density.

deposit, quantity of dry fungicide deposited on a unit area of material treated.

dieback, progressive death of shoots, branches and roots generally starting at the tip.

dikaryotic, mycelium or spores containing two sexually compatible nuclei per cell. Common in the *Basidiomycetes*.

diluent, an inert material added to a fungicide to reduce its concentration.

dilution end-point, the extent to which sap from virus-infected plants can be diluted with water before its infectivity is lost.

dimorphic, having two states, particularly of pathogenic fungi that are mycelial in culture and yeast-like in the host.

disease, (1) any malfunctioning of host cells and tissues that results from continuous irritation by a pathogenic agent or environmental factor and leads to development of symptoms; **(2)** harmful deviation from normal functioning of physiological processes.

disease cycle, the chain of events involved in disease development, including the stages of development of the pathogen and the effect of the disease on the host.

disinfectant, a physical or chemical agent that frees a plant, organ, or tissue from infection.

disinfestant, an agent that kills or inactivates pathogens in the environment or on the surface of a plant or plant organ before infection takes place.

disorder, a harmful non-pathogenic deviation from normal growth.

dose, dosage, quantity of toxicant applied per unit of material treated.

downy mildew, *see* mildew.

dressing (seed), (1) the process of covering seeds with a fine coating of fungicide; **(2)** the fungicide applied to seed.

ecoclimate, climate within a plant community, e.g. a crop.

ectotrophic mycorrhiza, *see* mycorrhiza.

egg, a female gamete. In nematodes, the first stage of the life cycle containing a zygote or a juvenile.

ELISA, a serological test in which one antibody carries with it an enzyme that releases a coloured compound.

elicitors, molecules produced by the pathogen that induce a response by the host.

endemic, of a disease, permanently established in a defined area, e.g. a country.

endoparasite, a parasite that enters a host and feeds from within.

endotrophic mycorrhiza, *see* mycorrhiza.

entomogenous, produced in or upon insects.

enzyme, a protein produced by living cells that can catalyse a specific organic reaction.

epidemic, a disease increase in a population. Usually a widespread and severe outbreak of a disease.

epidemiology, study of the factors affecting outbreaks of disease and spread of infectious diseases.

epidermis, the superficial layer of cells occurring on all plant parts.

epiphytically, existing on the surface of a plant or plant organ without causing infection.

epiphytotic, a widespread and destructive outbreak of a disease of plants; epidemic.

eradicant, a chemical substance that destroys a pathogen at its source.

eradicant fungicide, *see* fungicide.

eradication, control of plant disease by eliminating the pathogen after it is established or by eliminating the plants that carry the pathogen.

etiology of disease, *see* aetiology.

facultative parasite, one able to live as a saprophyte and to be cultured on laboratory media, not obligate (q.v.).

field immune, (of plants) not becoming infected by a pathogen in the field, although susceptible under experimental conditions.

field resistance, resistance observed under natural field conditions but not always detected under experimental conditions.

filler, a diluent in powder form.

flagellum (pl. flagella), a flexible, whip-like appendage used as an organ of locomotion.

foot rot, rot involving the lower part of the stem–root axis.

form (Latin *forma*), a subdivision of a species below the rank of variety.

form genus (or species), a genus or species for imperfect states, e.g. in *Fungi Imperfecti* and *Uredinales*.

forma specialis (f.sp.), a group of races and biotypes of a pathogen species that can infect only plants within a certain host genus or species.

freeze-drying, preservation of living microorganisms etc. by removing water under vacuum while tissue remains in frozen state (lyophilization).

fruiting body, a complex fungal structure containing spores.

fumigant, (1) a toxic gas or a volatile substance that is used to disinfest certain areas from various pests; (2) a chemical toxicant used in volatile form.

fumigation, the application of a fumigant for disinfestations of an area or contained volume.

fungicidal, able to kill fungus spores or mycelium.

fungicide, a substance that kills fungus spores or mycelium. **Eradicant fungicide,** (1) one applied to a substratum in which the fungus is already present; (2) a fungicide used in disease control after infection has been established. **Protective fungicide,** one used as a protectant (q.v.).

fungistasis, the prevention of fungal growth.

fungistatic, able to prevent the growth or development of fungus spores without killing them; preventing fungus growth without being fungicidal.

gall, a localized proliferation of plant tissue producing a swelling or outgrowth.

gene, a linear portion of the chromosome that determines or conditions one or more hereditary characters. The smallest functioning unit of the genetic material.

gene cloning, the isolation and multiplication of an individual gene sequence by its insertion into a bacterium where it can multiply.

gene-for-gene concept, the concept that corresponding genes for resistance and virulence exist in host and pathogen, respectively.

genetic engineering, the alteration of the genetic composition of a cell by various procedures (transformation, protoplast fusion, etc.) in tissue culture.

genotype, the genetic constitution of an organism.

gibberellins, a group of plant growth-regulating substances with a variety of functions.

gnotobiotic growth, growth of an organism in the absence of other organisms; 'pure culture'.

grafting, a method of plant propagation by transplantation of a bud or a scion of a plant on another plant. Also, the joining of cut surfaces of two plants so as to form a living union.

Gram-positive or -negative, (of bacteria) staining or not staining by Gram's staining method.

group, any taxonomic group; taxon.

growth regulator, a natural substance that regulates the enlargement, division, or activation of plant cells.

gummosis, a plant disease having secretion of gum as a well-marked symptom.

habitat, natural place of occurrence of an organism.

haustorium, a special hyphal branch, especially one within a living cell of the host, for absorption of food.

heteroecious, undergoing different parasitic stages in two unlike hosts, as in the *Uredinales*; cf. autoecious.

heterothallism, condition of sexual reproduction in which conjugation is possible only through the interaction of different thalli; cf. homothallism.

heterotrophic, living on food substances that are made by other organisms; cf. autotrophic.

holomorph, a whole fungus in all its states.

homothallism, the condition in which sexual reproduction can occur without the interaction of two different thalli; cf. heterothallism.

horizontal resistance, resistance evenly spread against all races of a pathogen.

hormone, a growth regulator. Frequently referring particularly to auxins.

host, an organism harbouring a parasite. Potential host, an organism capable of harbouring a particular parasite.

host range, the various kinds of host plants that may be attacked by a parasite.

hybrid, the offspring of two individuals differing in one or more heritable characteristics.

hybridization, the crossing of two individuals differing in one or more heritable characteristics.

hyperparasite, a parasite parasitic on another parasite.

hyperplasia (hyperplasic), enlargement of a tissue as a result of excessive production of cells; cf. hypoplasia.

hypersensitive, giving violent local reaction to attack by a pathogen, the prompt death of tissue round the points of entry preventing further spread of infection.

hypersensitivity, excessive sensitivity of plant tissues to certain pathogens. Affected cells are killed quickly, blocking the advance of obligate parasites.

hypertrophy, excessive growth due to enlargement of individual cells.

hypha (pl. hyphae), one of the threads of a mycelium.

hypoplasia, a pathologically subnormal cell multiplication, as in dwarfing.

hypovirulence, reduced virulence of a pathogen strain as a result of the presence of transmissible double-stranded RNA.

immune, exempt from infection.

immunity, the state of being immune.

imperfect state, the state in which asexual spores (such as conidia) or no spores are produced; anamorph.

inclusion bodies, crystalline or amorphous structures in virus-infected plant cells that are produced by and consist largely of viruses and are visible under the compound microscope.

incompatible, not cross-fertile.

incubation period, the period of time between penetration of a host by a pathogen and the first appearance of symptoms on the host.

indexing, any procedure for demonstrating the presence of known viruses in suscept plants.

indicator plant, one which reacts to certain viruses or environmental factors with specific symptoms, used for identification of the viruses or the environmental factors.

inducible or induced, a substance, usually an enzyme, whose production has been or may be stimulated by another compound, often a substrate or a structurally related compound called an inducer.

infect, to enter and establish a pathogenic relationship with an organism; to enter and persist in a carrier.

infection court, the place of invasion of a host by a pathogen.

infective, able to infect.

infested, (1) attacked by animals, especially insects; **(2)** (of soil or other substrata) contaminated by noxious organisms.

injury, damage of a plant by an animal, physical or chemical agent.

inoculate, to introduce a microorganism or virus into an organism or into a culture medium.

inoculum, the substance used for inoculating.

inoculum potential, the energy of growth of a fungus (or other microorganism) available for colonization of a substratum at the surface of the substratum to be colonized.

integrated control, the complementary use of biological, chemical and cultural methods to control pathogens.

integrated pest management, the attempt to prevent pathogens, insects, and weeds from causing economic crop losses by using a variety of management methods that are cost-effective and cause the least damage to the environment.

intercellular, between cells.

intracellular, within a cell.

invasion, the spread of a pathogen into the host.

in vitro, in culture. Outside the host.

in vivo, in the host.

isolate, (1) (verb) to separate a microorganism from host or substrate and establish it in pure culture; **(2)** (noun) a single spore or pure culture and the subcultures derived from it. The term is also applicable to viruses.

isolation, the separation of a pathogen from its host and its culture on a nutrient medium.

isozymes, the different forms of an enzyme that carry out the same enzymatic reaction but require different conditions (pH, temperature, etc.) for optimum activity.

kilobase, one thousand continuous bases (nucleotides) of single-stranded RNA or DNA.

klendusity, ability of a susceptible variety to escape infection because of possession of some quality preventing or hindering successful inoculation under conditions conducive to infection on other varieties.

Koch's postulates, three criteria proposed by R. Koch for proving pathogenicity.

latent infection, the state in which a host is infected with a pathogen but does not show any symptoms.

latent period, (1) time between infection and appearance of disease symptoms; **(2)** period after acquisition of virus by vector before it becomes infective.

latent virus, a virus that does not induce symptom development in its host.

leaf spot, a self-limiting lesion on a leaf.

lesion, a localized area of diseased or disordered tissue.

life cycle, life history, (of fungi) the stage or stages (states) between one spore form and its recurrence. There are commonly two stages, the imperfect, with one or more spore forms, and the perfect, but one or the other may not be known.

line, (1) an inbred homozygous strain; **(2)** an isolate; **(3)** a subdivision of a physiologic race.

lophotrichous, having a tuft of flagella at one or both poles.

lyophilization, *see* freeze-drying.

lysis, a breaking down or dissolution of cells by enzymes or viruses; cf. autolysis.

macerate, to soften by soaking.

macroscopic, visible to the naked eye without the aid of a magnifying lens or a microscope.

marbled, stained with irregular streaks of colour.

masked symptoms, virus-infected plant symptoms that are absent under certain environmental conditions but appear when the host is exposed to certain conditions of light and temperature.

masked virus, one carried by a plant that does not show symptoms of its presence.

mass median diameter (MMD), the figure dividing a total volume of spray into two equal parts; half the mass of spray is of droplets of smaller diameter than the MMD, half of droplets of larger diameter.

matrix, the material in which an organism or organ is embedded.

mechanical inoculation, inoculation of a plant with a virus through transfer of sap from a virus-infected plant to a healthy plant.

medium, culture medium, a substance or solution for the culture of microorganisms.

messenger RNA (mRNA), a chain of ribonucleotides that codes for a specific protein.

metabolism, the process by which cells or organisms utilize nutritive material to build living matter and structural components or break down cellular material into simple substances to perform special functions.

μm (micrometre), a unit of length equal to 0.001 of a millimetre.

micron, 0.001 mm (1 μm).

microscopic, very small; can be seen only with the aid of a microscope.

mildew, a fungal disease of plants in which the mycelium and spores of the fungus are seen as a growth on the host surface (usually of a whitish colour). A **powdery mildew** is caused by one of the *Erysiphaceae*, a **downy mildew** by one of the *Peronosporaceae*; a **dark (sooty) mildew** by one of the *Meliolaceae* or *Capnodiaceae*.

mm (millimetre), a unit of length equal to 0.1 of a centimetre (cm) or 0.03937 inch.

mist spraying, method in which concentrated spray is atomized into a high-velocity air stream, the air then acting as diluent and carrier.

monoclonal antibodies, identical antibodies produced by a single clone of lymphocytes and reacting only with one of the antigenic determinants of a pathogen or protein.

monotrichous, having one polar flagellum.

mosaic, symptom of certain viral diseases of plants characterized by intermingled patches of normal and light green or yellowish colour.

mottle, arrangement of spots or confluent blotches of colour, often symptomatic of virus diseases.

mould, a mycelial microfungus or a visible growth of such a fungus.

mummy, a dried, shrivelled fruit.

mutant, an individual possessing a new, heritable characteristic as a result of a mutation.

mutation, an abrupt appearance of a new characteristic in an individual as the result of an accidental change in a gene or chromosome.

mycelium, the hypha or mass of hyphae that make up the body of a fungus.

mycoplasmas, pleomorphic prokaryotic microorganisms that lack a cell wall (*see* phytoplasma).

mycorrhiza, a symbiotic association of a fungus with the roots of a plant: **ectotrophic mycorrhiza**, on the surface of the roots; **endotrophic mycorrhiza**, within the roots; **pseudotrophic mycorrhiza**, in which the fungus is parasitic; **tolypophagous mycorrhiza**, in which the fungus is killed and digested by the host.

mycosis, an infection by a parasitic fungus, or a disease so caused.

mycostatic, fungistatic.

nanism, dwarfism.

nm (nanometre), a unit of length equal to 0.001 of a micrometre.

necrophyte, an organism living on dead material; cf. perthophyte, saprophyte.

necrosis, death of plant cells, especially when resulting in darkening of the tissues; a common symptom of fungus infection.

necrotic, dead and discoloured.

needle cast, (of conifers) loss of leaves, caused generally by species of *Phacidiales*.

nematicide, a chemical compound or physical agent that kills or inhibits nematodes.

nematode, generally microscopic, worm-like animals that live saprophytically in water or soil, or as parasites of plants and animals.

non-host resistance, inability of a pathogen to infect a plant because the plant is not a host of the pathogen due to lack of something in the plant that the pathogen needs or due to the presence of substances incompatible with the pathogen.

non-infectious disease, a disease that is caused by an abiotic agent, i.e. by an environmental factor, not by a pathogen.

non-persistent virus, one which remains infective within its insect vector for only a short period.

nucleic acid, an acidic substance containing pentose, phosphorus, and pyrimidine and purine bases. Nucleic acids determine the genetic properties of organisms.

nucleoprotein, referring to viruses: consisting of nucleic acid and protein.

nucleoside, the combination of a sugar and a base molecule in a nucleic acid.

nucleotide, the phosphoric ester of a nucleoside. Nucleotides are the building blocks of DNA and RNA.

obligate parasite, a parasite that in nature can grow and multiply only on or in living organisms. cf. facultative parasite.

oidium (pl. oidia), (1) spermatium formed on a hyphal branch, especially in heterothallic *Hymenomycetes*; **(2)** flat-ended asexual spore formed by the breaking up of a hypha into cells; arthrospore; **(3)** a mildew.

omnivorous, (of parasites) attacking a number of different hosts.

osmosis, the diffusion of a solvent through a differentially permeable membrane from its higher concentration to its lower concentration.

oxidative phosphorylation, the utilization of energy released by the oxidative reactions of respiration to form high-energy ATP bonds.

ozone (O_3), a highly reactive form of oxygen that in relatively high concentrations may injure plants.

parasexual cycle, a mechanism whereby recombination of hereditary properties is based on mitosis.

parasite, an organism or virus living on or in, and getting its food from, its host, another living organism.

particle size, average, (1) the arithmetical mean diam.; **(2)** diam. of particle of average surface or **(3)** volume; **(4)** diam. of particles of same specific surface as the material. (It should always be stated which of these diam. is measured.)

pathogen, a parasite able to cause disease in a particular host or range of hosts.

pathogenesis, the sequence of processes in disease development from initial contact between pathogen and host to completion of syndrome.

pathogenicity, the characteristic of being able to cause disease.

pathotype, pathovar, a subdivision of a species distinguished by common characters of pathogenicity, particularly in relation to host range. In bacteriology, where pathovar is the preferred term, pathotype is used to describe the type (or reference) culture of a pathovar.

pelleting, coating of seed with inert material, often incorporating pesticides, to ensure uniform size and shape.

perfect state (stage or phase), the state of the life cycle in which spores (such as ascospores and basidiospores) are formed after nuclear fusion or by parthenogenesis; teleomorph.

perithecium, the subglobose or flask-like ascocarp of the *Pyrenomycetes*, limited by some to the thin-walled envelope and contents developed from an archicarp.

peritrichous, having flagella distributed over the whole surface.

persistent virus, one which remains infective within its insect vector for a long period.

perthophyte, an organism feeding on dead tissues of living hosts; cf. saprophyte.

phage, *see* bacteriophage.

phyllody, the replacement of floral parts by leaf-like structures.

physiologic race, a subspecific group of parasites characterized by specialization to different cultivars of one host species.

phytiatry, the treatment of plant diseases, especially by chemical methods.

phytoalexin, a substance that inhibits the development of a microorganism, produced in higher plants in response to certain stimuli.

phytoncide, a chemical substance produced by higher green plants, which can inhibit the growth of microorganisms.

phytoplasma, a mycoplasma-like organism (MLO) parasitic in plants (*see* mycoplasma).

phytosanitary certificate, a certificate of health of plants or plant products to be exported.

phytosanitation, any measures involving the removal or destruction of infected plant material likely to form a source of reinfection by a pathogen.

phytotoxic (phytotoxicity), toxic to plants.

phytotoxin, a toxin toxic to a plant.

pleomorphic, having more than one independent form or spore stage in the life cycle.

powdery mildew, *see* mildew.

precipitin, an antibody that causes precipitation of soluble antigens.

primary symptom, in a virus disease, the first to appear in a case in which more than one type of symptom may be produced.

promoter, a region on a DNA or RNA sequence that is recognized by RNA polymerase in order to initiate transcription.

promycelium, the short hypha produced by the teliospore; the basidium.

propagative virus, a virus that multiplies in its insect vector.

propagule, that part of an organism by which it may be dispersed or reproduced.

protectant, a substance that protects an organism against infection by a pathogen.

protective fungicide, *see* fungicide.

pseudostroma, a mass of fungal cells combined with host cells to produce a distinct tissue.

pseudotrophic mycorrhiza, *see* mycorrhiza.

pustule, a blister-like spot, on a leaf etc., from which erupts a fruiting structure of a fungus.

pycnidiospore, a conidium in or from a pycnidium.

pycnidium, the fruit body of the *Sphaeropsidales*, frequently globose or flask-like.

pycnium, (in *Uredinales*) the pycnidium-like haploid fruit-body, or spermogonium.

quarantine (plant), (1) control of import and export of plants to prevent spread of diseases and pests; **(2)** holding of imported plants in isolation for a period to ensure their freedom from diseases and pests.

race, (1) a genetically, and as a rule geographically, distinct mating group within a species; **(2)** physiologic race.

recognition factors, specific receptor molecules or structures on the host (or pathogen) that can be recognized by the pathogen (or host).

resistance, the power of an organism to overcome, completely or in some degree, the effect of a pathogen or other damaging factor.

resistant, possessing qualities that hinder the development of a given pathogen.

resting spore, a thick-walled spore, usually formed as the result of a sexual process, which germinates after a resting period (frequently over winter).

restriction enzymes, a group of enzymes from bacteria that break internal bonds of DNA at highly specific points.

reverse transcription, copying of an RNA into DNA.

rhizobia, the nodule bacteria of leguminous plants.

rhizomorph, a thread-like or cord-like structure made up of aggregated hyphae.

rhizoplane, the surface of a root.

rhizosphere, the soil near a living root.

rickettsiae, microorganisms similar to bacteria in most respects but generally capable of multiplying only inside living host cells; parasitic or symbiotic.

ringspot, a circular area of chlorosis, the centre remaining green, symptomatic of many virus infections.

RNA (ribonucleic acid), a nucleic acid involved in protein synthesis; also, the most common nucleic acid (genetic material) of plant viruses.

RNase (ribonuclease), an enzyme that breaks down RNA.

roguing, critical examination of a crop and removal of unhealthy (e.g. virus-infected) or otherwise unwanted plants.

rosette, short, bunchy habit of plant growth.

rot, the softening, discoloration and often disintegration of a succulent plant tissue as a result of fungal or bacterial infection.

run-off, quantity of spray that runs off unit area of plant surface.

rust, (1) a disease caused by one of the *Uredinales*; **(2)** one of the *Uredinales*; **(3)** a disease with 'rusty' symptoms, e.g. red rust of tea (the alga *Cephaleuros*).

saltant, a discontinuous variation of unknown origin.

saltation, a mutation within an isolate known to be a pure genotype; dissociation.

sanitation, the removal and burning of infected plant parts, decontamination of tools, equipment, hands, etc.

saprobe, saprophyte when the organism is not a plant.

saprophyte, (1) a plant (organism) using dead organic material as food, and commonly causing its decay; **(2)** a necrophyte on dead material that is not part of a living host; cf. perthophyte, saprobe.

scab, a roughened, crust-like diseased area on the surface of a plant organ. A disease in which such areas form.

sclerotium, a compact mass of hyphae with or without host tissue, usually with a darkened rind, and capable of surviving under unfavourable environmental conditions.

scorch, 'burning' of leaf margins as a result of infection or unfavourable environmental conditions.

secondary infection, any infection caused by inoculum produced as a result of a primary or a subsequent infection; an infection caused by secondary inoculum.

secondary inoculum, inoculum produced by infections that took place during the same growing season.

secondary symptom, in a virus disease, one following the primary symptom (q.v.) in cases where more than one type of symptom is produced.

sectoring, 'mutation' in plate cultures resulting in one or more sectors of the culture having a changed form of growth.

senescent, growing old.

sensitive, reacting with severe symptoms to the attack of a given pathogen.

sensitivity, the tendency of an organism attacked by a disease to give more or less strong symptoms.

serology, a method using the specificity of the antigen–antibody reaction for the detection and identification of antigenic substances and the organisms that carry them.

serotype, (1) (of bacteria) an infrasubspecific subdivision of a species; **(2)** (of plant viruses) a group of viruses sharing only a few of its antigens in common with another group (serotype).

seta (pl. setae), a bristle or stiff hair, generally thick-walled and dark coloured.

sexual, participating in or produced as a result of a union of nuclei in which meiosis takes place.

shock symptoms, the severe, often necrotic symptoms produced on the first new growth following infection with some viruses; also called acute symptoms.

shot-hole, a symptom in which small diseased fragments of leaves fall off and leave small holes in their place.

signal molecules, host molecules that react to infection by a pathogen and transmit the signal to and activate proteins and genes in other parts of the cell and of the plant so they will produce the defence reaction.

slime molds, pseudofungi of the class *Myxomycetes*; also, superficial diseases caused by these pseudofungi on low-lying plants.

slurry, a thin, watery mixture.

smut, a disease caused by one of the *Ustilaginales* or the fungus itself; **covered smut,** one in which the mature spore mass keeps within the sorus, often till this is free of the host; **loose smut,** one in which the spores, as an uncovered mass of powder, are freed from the host by wind and rain; **stinking smut,** covered smut caused by *Tilletia* spp.

soft rot, a rot of a fleshy fruit, vegetable or ornamental in which the tissue becomes macerated by the enzymes of the pathogen.

soil solarization, attempt to reduce or eliminate pathogen populations in the soil by covering the soil with clear plastic so that the sun's rays will raise the soil temperature to levels that kill the pathogen.

somaclonal variation, variability in clones generated from a single mother plant, leaf, etc., by tissue culture.

somatic hybridization, production of hybrid cells by fusion of two protoplasts with different genetic make-up.

sooty mould, sooty coating on foliage and fruit formed by the dark hyphae of fungi (especially *Capnodiaceae* and *Meliolaceae*) that often live in the honeydew secreted by insects such as aphids, mealybugs, scales and whiteflies.

spawn, mycelium, especially that used for starting mushroom cultures; to put inoculum (spawn) into a mushroom bed or other substratum.

spore, a general name for a reproductive structure in cryptograms; **air spore,** the population ('spore flora') of airborne particles of plant or animal origin.

spore ball, a compound spore or ball of spores in certain genera of *Ustilaginales*, of varying structure.

sporodochium, a mass of conidiophores tightly placed together upon a stroma or mass of hyphae, as in the *Tuberculariaceae*.

sporophore, a spore-producing or spore-supporting structure, especially a conidiophore; a fruit body of larger fungi.

sporulate, to produce spores.

spot anthracnose, *see* anthracnose.

spread, uniformity and completeness with which a fungicide deposit covers a continuous surface.

spreader, a substance added to a spray to assist in its even distribution over the target.

stage, frequently = state but for fungi better reserved for circumstances where the succession of two or more states is regular.

staling, slowing of growth of fungus in pure culture as a result of accumulation of self-inhibiting metabolites, adverse pH, etc.

state, (1) (of fungi) one phase of a pleomorphic fungus, e.g. the anamorph characterized by asexual spores, the teleomorph characterized by sexual spores; **(2)** (of bacteria) 'the name given to the rough, smooth, mucoid and similar variants which arise in culture' (Bact. Code).

sterile fungi, a group of fungi that are not known to produce any kind of spores.

sterilization, the elimination of pathogens and other living organisms from soil, containers, etc., by means of heat or chemicals.

sticker, material added to a fungicide to increase tenacity; a substance added to a spray to make it adhere to the target.

strain, the many meanings include: **(1)** a group of similar isolates, race; form; **(2)** the descendants of a single isolation in pure culture, isolate; **(3)** (of bacteria) a cultivar; **(4)** (of plant viruses) a group of viruses having most of its antigens in common with another group (strain).

stroma (pl. stromata), a mass or matrix of vegetative hyphae, with or without tissue of the host or substratum, sometimes sclerotium-like in form, in or on which spores are produced.

stylet, a long, slender, hollow feeding structure of nematodes and some insects.

stylet-borne, a virus borne on the stylet of its vector; non-circulative virus.

subculture, a culture derived from another one.

substrate, (1) the material on which an enzyme acts; **(2)** the material or substance on which a microorganism feeds and develops.

substratum, the material on or in which a microorganism is living.

summer spore, a spore germinating without resting, frequently living only a short time; cf. resting spore.

sun scald, superficial damage to fruits resulting from the action of intense sunlight.

suppressive soils, soils in which certain diseases are suppressed because of the presence in the soil of microorganisms antagonistic to the pathogen.

surfactant, a surface active material, especially a wetter or spreader used with a spray.

suscept, organism affected or capable of being affected by a given disease.

susceptible, lacking the inherent ability to resist disease or attack by a given pathogen; non-immune.

symbiosis, a mutually beneficial association of two or more different kinds of organisms.

symptom, the external and internal reactions or alterations of a plant as a result of a disease.

symptomless carrier, a plant that, although infected with a pathogen (usually a virus), produces no obvious symptoms.

syndrome, the totality of effects produced in a plant by a disease.

synergism, (1) the association of two or more organisms acting at one time and effecting a change that one alone is not able to make; **(2)** increased fungicidal value of certain mixtures of fungicides or of fungicides and non-toxic materials.

systemic, (1) (of a plant pathogen) occurring throughout the plant; **(2)** (of a chemical) absorbed into the plant through root or foliage, and translocated elsewhere in the plant.

taxon (pl. taxa), any taxonomic group.

teleomorph, sexual or perfect state of a fungus.

tenacity, property of a fungicide deposit to resist removal by weathering etc.

teratology, the study of gross structural abnormalities.

thermal death-point, the lowest temperature at which heating for a limited period (usually 10 min) is sufficient to kill a microorganism.

thermal inactivation point, the lowest temperature at which heating for a limited period (usually 10 min) is sufficient to cause a virus to lose its infectivity or an enzyme its activity.

tissue, a group of cells of similar structure that performs a special function.

tolerant, able to endure infection by a particular pathogen, without showing severe disease, or giving little reaction to the effect of other factors (e.g. a virus tolerant of heat).

tolypophagous mycorrhiza, *see* mycorrhiza.

toxic, of, caused by, or acting as, poison.

toxicant, a toxic substance or preparation.

toxicity, the power of acting as a poison.

toxin, a toxic compound produced by a microorganism (*see* phytotoxin, vivotoxin).

tracheomycosis, a fungal disease in which the pathogen is mainly confined to the xylem.

transcription, copying of a gene into RNA. Also, copying of a viral RNA into a complementary RNA.

transduction, the transfer of genetic material from one bacterium to another by means of a bacteriophage.

transfer RNA (tRNA), the RNA that moves amino acids to the ribosome to be placed in the order prescribed by the messenger RNA.

transformation, the change of a cell through uptake and expression of additional genetic material.

transgenic (or transformed) plants, plants into which genes from other plants or other organisms have been introduced through genetic engineering techniques and are expressed, i.e. produce the expected compound or function.

translation, copying of mRNA into protein.

translocation, transfer of nutrients or virus through the plant.

transmission, the transfer or spread of a virus or other pathogen from one plant to another.

transpiration, the loss of water vapour from the surface of leaves and other above-ground parts of plants.

transposable element, a segment of chromosomal DNA that can move around (transpose) in the genome and integrate at different sites on the chromosomes.

tumour, an uncontrolled overgrowth of tissue or tissues.

tylose, a balloon-like intrusion into the lumen of a vessel.

tylosis, the process of tylose formation.

type, in taxonomy, the element to which the scientific name of a taxon is permanently attached.

variability, the property or ability of an organism to change its characteristics from one generation to the other.

variety, (1) a subdivision of a species below the level of subspecies; **(2)** cultivar.

vascular, term applied to a plant tissue or region consisting of conductive tissue; also, to a pathogen that grows primarily in the conductive tissues of a plant.

vector, an organism able to transmit a pathogen, especially an insect, nematode, etc. transmitting a virus.

vegetative, asexual; somatic.

vein banding, development in a virus disease of dark green bands along the veins.

vein clearing, development in a virus disease of pale bands adjacent to the veins of young leaves.

vertical resistance, resistance to some races of a pathogen but not to others.

viroid, a pathogenic RNA of low molecular weight.

virosis, a virus disease.

virulence, the degree or measure of pathogenicity.

virulent, strongly pathogenic.

viruliferous, (of a vector) carrying or containing a virus.

vivotoxin, a substance produced internally in an infected host or by the pathogen within it, responsible for some or all of the harmful changes produced in the course of the disease.

wetting agent, wetter, material that reduces the contact angle of a liquid on a surface.

white blister, characteristic pustular fructification of *Albugo* spp.

wilt, loss of rigidity and drooping of plant parts generally caused by insufficient water in the plant.

winter spore, *see* resting spore.

witches' broom, broom-like growth or dense clustering of branches of woody plants caused by abnormal proliferation of shoots.

yellows, applied to plant diseases of which yellowing is a conspicuous symptom, e.g. *peach yellows virus*, cabbage yellows (*Fusarium oxysporum* f.sp. *conglutinans*), etc.

zone lines, narrow, dark-brown or black lines in decayed wood (especially hard-woods) generally caused by fungi.

Useful Addresses

44

Addresses of major institutions or organizations involved with international or regional aspects of plant pathology are given here. Information on most is available through the Internet. No attempt is made here to list the many national institutions, but national societies involved with plant pathology are listed some of which can be accessed through the Internet. Addresses of these can be found through the International Society of Plant Pathology (ISPP) web site (www.isppweb.org), which also contains a world directory of plant pathologists.

Details of societies, national, international and commercial organizations can also be obtained from the online version of the Plant Pathology Internet Guide accessible through the British Society of Plant Pathology web site (www.bspp.org.uk/ppigb).

CAB *International*

CAB International, Wallingford, Oxon OX10 8DE, UK.
Email: cabi@cabi.org and bioscience-wallingford@cabi.org

CAB International, Africa Regional Centre (CABI-ARC), ICRAF Complex, PO Box 633, Village Market, Nairobi, Kenya.

CAB International South-East Asia Regional Centre (CABI-SEARC), Malaysian Agricultural Research and Development Institute (MARDI), PO Box 210, 43409 UPM Serdang, Malaysia.

CAB International, Caribbean and Latin America Regional Centre (CABI-CLARC), Gordon Street, Curepe, Trinidad and Tobago, West Indies.

CABI Bioscience Centre, Pakistan, CAB International, PO Box 8, Rawalpindi, Pakistan.

CABI Bioscience Centre, Switzerland, CAB International, 1 Chemin des Grillons, CH-2800 Delémont, Switzerland.

CABI Bioscience UK Centre, Egham, Bakeham Lane, Egham, Surrey TW20 9TY, UK.

BioNet-Interntaional, Technical Secretariat, Bakeham Lane, Egham, Surrey TW20 9TY, UK (www.bionet-intl.org).

International Institutes Involved with Crop Protection

Most of these are part of the system coordinated by the Consultative Group for International Agricultural Research (CGIAR), which has its headquarters at the World Bank.

CGIAR, World Bank Building, Washington, DC, USA.

CIAT, International Center of Tropical Agriculture, Apartado Aereo 6713, Cali, Colombia.
Email: ciat@cgnet.com

CIMMYT, International Maize and Wheat Improvement Center, APDO poatal 6-641 Mexico DF, Mexico.
Email: cimmyt@cgiar.org

CIP, International Potato Center, PO Box 1558, Lima 12, Peru.
Email: cip-web@cgiar.org

CIFOR, Centre for International Forestry Research, PO Box 6596 JKPWB, Jakarta 10065, Indonesia.
Email: Cifor@cgiar.org

ICARDA, International Centre of Agricultural Research in Dry Areas, PO Box 5466 Aleppo, Syrian Arab Republic.
Email: icarda@cgiar.org

ICRAF, International Centre for Research in Agroforestry, PO Box 30677, Nairobi, Kenya.
Email: icraf@cgiar.org

ICRISAT, International Crops Research Institute for the Semi-Arid Tropics, Patancheru 502 324, Andrah Pradesh, India.
Email: icrisat@cgiar.org

IPGRI, International Plant Genetic Resources Institute, Via dei Tre Denar: 472/a, 00057 Maccarese (Finmicino), Rome, Italy.
Email: Ipgr@cgiar.org

INIBAP, International Institute for Bananas and Plantains, Parc Scientifique, Agropolis 2, 34397 Montpellier, Cedex 5, France.

IITA, International Institute of Tropical Agriculture, PMB 5320, Ibadan, Nigeria.
Email: iita@cgiar.org

IRRI, International Rice Research Institute, MCPO Box 3127, Makati City 1271, Philippines.
Email: irri@cgiar.org

WARDA, West African Rice Development Association – WARDA/ADRAO, 01BP2551, Bouaké 01, Côte d'Ivoire, Africa.
Email: warda@cgiar.org

ISNAR, International Support to National Agricultural Research, PO Box 93375, 2509 AJ, The Hague, The Netherlands.
Email: isnar@cgiar.org

Other International and Regional Organizations

FAO, Plant Protection Service, Plant Production and Protection Division, FAO, Via delle Terme di Caracalla, 00100 Rome, Italy.

APPPC, Asia and Pacific Plant Protection Commission, FAO Regional Office for Asia and the Far East, Maliwan Mansion, Phra Atit Road, Bangkok-2, Thailand.

CPPC, Caribbean Plant Protection Commission, c/o FAO Subregional Office for the Caribbean, PO Box 631-C, Bridgetown, Barbados.

EPPO, European and Mediterranean Plant Protection Organization, 1 rue Le Notre, 75016 Paris, France.

IAPSC, Interafrican Phytosanitary Council, BP 4170 Nlongkak, Yaounde, Cameroon.

ISTA, International Seed Testing Association, Reckenholz, PO Box 412, CH-8046 Zurich, Switzerland.

PPPO, Pacific Plant Protection Organization, Plant Protection Service, Secretariat of the Pacific Community Private Mail Bag, Suva Fiji.

OIRSA, Organismo Internacional Regional de Sanidad Agropecuaria, Clle Ramon, Belloso, Col. Esacalon, San Salvador, El Salvador.

NAPPO, North American Plant Protection Organization, Observatory Crescent, Bldg no. 3, Central Experiment Station, Ottowa, Ontario, Canada.

SPC, South Pacific Commission, Plant Protection Officer, South Pacific Commission, Private Mail Bag, Suva, Fiji.

ICO, International Coffee Organization, 22 Berners Street, London W1P 4DD, UK.

ICCO, International Office for Chocolate and Cocoa, 22 Berners Street, London W1P 4DD, UK.

Burotrop, Bureau for the Development of Research on Tropical Perennial Oil Crops, 17 Rue de la Tour, Montpellier, 75116, Paris, France.

General Index

Diseases and Pathogens Index